GAMES AND DECISIONS

GAMES

A study of the Behavioral Models Project,
Bureau of Applied Social Research,
Columbia University

AND DECISIONS

introduction and critical survey

R. DUNCAN LUCE and HOWARD RAIFFA
Harvard University

John Wiley & Sons, Inc.
New York · London · Sydney

Dedicated to

the memory of

Professor John von Neumann

PREFACE

This book attempts to communicate the central ideas and results of
game theory and related decision-making models unencumbered by their
technical mathematical details: thus, for example, almost no proofs are
included.   It is a book about game theory, not a presentation of the
theory itself.   By laying bare the main structure of the theory—its assumptions and conclusions, its deficiencies and aspirations—we hope that the
book will serve as a useful critical introduction to the theory and a guide
to the literature.   We have tried, on the one hand, to make sufficiently
precise statements so that misunderstandings and misstatements will not
result from a reading of the book, but, on the other hand, we have striven
to keep the language and notation sufficiently familiar and simple that
there will be scientists who will benefit from it who would have found a
treatise on game theory unintelligible.   There are many mathematicians,
even among those sympathetic to social science applications, who feel that
these goals are incompatible, and we cannot deny having reached times
of despondency when we were ready to agree.

In many ways the overall outline of this book parallels the original
structuring given to the theory by von Neumann and Morgenstern in
Theory of Games and Economic Behavior [1944, 1947], but in detail it is
different: First, in the decade since the second edition of their book there

have been many additions to the theory, and we have tried to include most of these. Second, our emphasis is almost totally on the concepts, and so relatively little attention is given to the detailed "solutions" of specific games. Third, our critical discussion and our examples are strongly colored—at least as far as a mathematician is concerned—by a social science point of view. We have tried, insofar as we could, to indicate the major intuitive and empirical objections that social scientists could or have raised against the theory—objections not to the mathematics, but to the applicability of this mathematics to empirical problems.

This book may be correctly accused of having a critical tone. We do not, however, intend this to be carping, and we would hope that our readers do not use it for that purpose. Our aim is to warn and to challenge the reader at just those points where the theory is conceptually weak. We believe that this can be done with relatively little mathematics, and so we have gone to considerable pains to reduce the mathematical demands made upon the reader. If we have not failed completely, then there should be something of interest here for a wide group of scholars: economists concerned with economic theory, political scientists and sociologists having a methodological bent or a theoretical concern with conflict of interest, experimental psychologists studying decision making, management scientists interested in theories of "rational" choice and organization, philosophers intrigued with the axiomatization of portions of human behavior, statisticians and other professionally practicing decision makers, and finally mathematicians—those whose work, for the most part, we are reporting.

Still one may ask: what exactly are the prerequisites? It is not easy to say. Certainly neither the calculus nor matrix algebra as such are required, but neither will hinder, for probably the most important prerequisite is that ill-defined quality: mathematical sophistication. We hope that this is an ingredient not required in large measure, but that it is needed to some degree there can be no doubt. The reader must be able to accept conditional statements, even though he feels the suppositions to be false; he must be willing to make concessions to mathematical simplicity; he must be patient enough to follow along with the peculiar kind of construction that mathematics is; and, above all, he must have sympathy with the method—a sympathy based upon his knowledge of its past successes in various of the empirical sciences and upon his realization of the necessity for rigorous deduction in science as we know it.

Our primary topic can be viewed as the problem of individuals reaching decisions when they are in conflict with other individuals and when there is risk involved in the outcomes of their choices. In very general and intuitive terms this problem is described in Chapter 1. As a background to the theory of games itself, and to the other topics we shall discuss, we must examine the modern theory of individual decision making in risky situations—utility theory. This is done in Chapter 2. From Chapter 3 through 12 the theory of games is examined: Chapter 3 gives the general model; Chapters 4, 5, and 6 present theories of two-person games, and Chapters 7 through 12 the theories of games with more than two players. Chapter 13 turns to the problem of individual decision making when the outcomes are not simply risky but, rather, uncertain. This material, like that in Chapter 14, is included partly because of its inherent interest as part of the problem of decision making, but also partly because these models are related in various ways to game theory. The final chapter, 14, may be described as dealing with problems in group decision making, in contrast to all the preceding work, which is devoted to the individual in different "environmental" contexts. The eight appendices are concerned with more technical topics which arise naturally in various parts of the book, but which we chose not to present in the body of the book.

Depending upon his interests and background, the reader may elect not to read the chapters in order or not to read all of them. Certain plausible groupings come to mind, and these may be worth mentioning.

i. Chapters 1, 3, 4, 7, and 8 give the general coverage of game theory without going into some of the more special and controversial topics. Although this does not include utility theory (Chapter 2), it is probably an adequate program of reading for a novice who wants some background in the subject, but who does not care to go into it deeply or to explore the various related topics.

ii. Chapters 5 through 12 delve into the conceptually difficult and not fully satisfactory theory of general games—those which either have more than two players or are not zero-sum or both. The reader already quite familiar with two-person zero-sum theory may want to begin with Chapter 5, although we would also recommend that he read Chapter 3 where the basic postulates about the players are introduced and criticized.

iii. The bulk of the research activity has been on games with only two players, provided we include non-zero-sum, infinite, and recursive games as well as the more familiar zero-sum two-person games, and many readers can be expected to confine their attention to these topics. For them, Chapters 4, 5, and 6 and Appendices 2, 3, 4, 6, 7, and 8 are relevant.

iv. Chapter 4 is really a sufficient background for reading Chapter 13, provided the reader has already had some exposure to the axiomatic method, so if he is interested solely in the problem of decision making under uncertainty—including statistical decision making as a special case—then he need only read these two chapters.

v. What we have to say about linear programing and its relationship to games is largely conceptual, not computational, and it is covered in Chapters 1 and 4 and Appendices 5 and 6.   No attempt is made to instruct the reader in the delicate task of actually programing a computer to solve a linear-programing problem.

vi. Readers interested in such topics as arbitration, group decision making, social welfare planning, and processes of "fair" division may wish to concentrate their reading in Chapters 2, 4, 6, and 14.   Perhaps Chapter 11 can be added to this list, but then one should first read Chapters 7 and 8.

Other combinations are possible, and by consulting the table of contents it should not be difficult to work up one suitable for any particular set of needs.

ACKNOWLEDGMENTS

In 1952, through the initiative of Professors P. F. Lazersfeld and Herbert Solomon, the Behavioral Models Project was established as a part of the Bureau of Applied Social Research, Columbia University, and placed under the direction of a faculty committee composed of Professors T. W. Anderson, C. H. Graham, P. F. Lazersfeld, E. Nagel, H. Raiffa (chairman), H. Solomon, and W. Vickrey.   This committee set the preparation of systematic expositions and critiques of various applications of mathematics in the behavioral sciences as one of the primary goals of the project. A year later R. D. Luce joined the project, with the responsibility for the execution of that goal.   A number of studies have been completed, most of which have been distributed as technical reports to a limited audience. This volume is the first of these studies to receive wider distribution, and it is anticipated that the others, which are more limited in scope, will be published in forthcoming volumes.

Several organizations have contributed vitally to the preparation of the book.   The Office of Naval Research has generously provided the principal financial backing both through its contract supporting the Behavioral Models Project and through its support of basic research in the Department of Mathematical Statistics, Columbia University.   The Bureau of Applied Social Research has had administrative responsibility for the former contract, and its officers and staff have been unfailingly cooperative in supplying much of the clerical, administrative, and physical facilities essential to the work.   Finally, both of us owe a great deal to the Center for Advanced Study in the Behavioral Sciences, Stanford, California, where we each spent what can only be characterized as a delightful year unencumbered by the usual academic duties, yet surrounded by equally free colleagues whose many stimulating and sympathetically critical comments are more or less accurately incorporated into the book. In addition, the Center was generous in providing us with clerical assistance.

Most difficult to acknowledge accurately are the numerous individual contributions—some unwitting, we suspect—to the manuscript. Many have read it and offered comments and encouragement, but we shall have to content ourselves with mentioning only those who gave so freely of their knowledge and time that not to acknowledge them explicitly would be a gross lack of appreciation. Above all, we wish to thank Professor Harold W. Kuhn, who read carefully the next-to-final version of the manuscript and commented on it in detail. His care has helped to eliminate ambigious and misleading statements and some errors of substance. He is hardly to be held responsible for our opinions, some with which we know he disagrees, or for the errors that remain; however, to the extent that the book is free from error and ambiguity he deserves appreciable credit. In addition, our thanks go to Professors Ross Ashby, Robert Dahl, and Martin Shubik for their many useful comments, and to Professor A. W. Tucker for our title. As we said, this far from exhausts the list of colleagues who have contributed in varying degrees to the final version. Finally, like most books, this one would have been more difficult for the reader and extremely difficult for us to complete without extensive editorial help. Our especial thanks go to Miss Dorothy Wynne, who edited, criticized, and supervised the typing of the manuscript during its final year of preparation, and to Mrs. Dvora Frumhartz, who, at an earlier stage of writing, played a similar role with respect to the chapters on n-person games.

R. Duncan Luce
Howard Raiffa

New York, N.Y.
May, 1957

CONTENTS

xiii

6 Two-Person Cooperative Games   114

7 Theories of *n*-Person Games in Normal Form   155

8   Characteristic Functions   180

9   Solutions   199

10 ψ-Stability   220

11   Reasonable Outcomes and Value   237

chapter 1

GENERAL INTRODUCTION
TO THE THEORY OF GAMES

1.1 CONFLICT OF INTEREST

In all of man's written record there has been a preoccupation with conflict of interest; possibly only the topics of God, love, and inner struggle have received comparable attention. The scientific study of interest conflict, in contrast to its description or its use as a dramatic vehicle, comprises a small, but growing, portion of this literature. As a reflection of this trend we find today that conflict of interest, both among individuals and among institutions, is one of the more dominant concerns of at least several of our academic departments: economics, sociology, political science, and other areas to a lesser degree.

It is not difficult to characterize imprecisely the major aspects of the problem of interest conflict: An individual is in a situation from which one of several possible outcomes will result and with respect to which he has certain personal preferences. However, though he may have some control over the variables which determine the outcome, he does not have full control. Sometimes this is in the hands of several individuals who, like him, have preferences among the possible outcomes, but who in general do not agree in their preferences. In other cases, chance events (which are sometimes known in law as "acts of God") as well as other

1

individuals (who may or may not be affected by the outcome of the situation) may influence the final outcome. The types of behavior which result from such situations have long been observed and recorded, and it is a challenge to devise theories to explain the observations and to formulate principles to guide intelligent action.

The literature on such problems is so vast, so specialized, and so rich in detail that it is utterly hopeless to attempt even a sketch of it; however, the attempt to abstract a certain large class of these problems into a mathematical system forms only a small portion of the total literature. In fact, aside from sporadic forays in economics, where for the most part attempts have been made to reduce it to a simple optimization problem which can be dealt with by the calculus, or in more sophisticated formulations by the calculus of variations, the only mathematical theory so far put forth is the theory of games, our topic. In some ways the name "game theory" is unfortunate, for it suggests that the theory deals with only the socially unimportant conflicts found in parlor games, whereas it is far more general than that. Indeed, von Neumann and Morgenstern entitled their now classical book *Theory of Games and Economic Behavior*, presumably to forestall that interpretation, although this does not emphasize the even wider applicability of the theory.

1.2 HISTORICAL BACKGROUNDS

The modern mathematical approach to interest conflict—game theory— is generally attributed to von Neumann in his papers of 1928 and 1937 [1928; 1937]; although recently Frechet has raised a question of priority by suggesting that several papers by Borel [1953] in the early '20's really laid the foundations of game theory. These papers have been translated into English and republished with comments by Frechet and von Neumann [1953]. Although Borel gave a clear statement of an important class of game theoretic problems and introduced the concepts of pure and mixed strategies, von Neumann points out that he did not obtain one crucial result—the minimax theorem—without which no theory of games can be said to exist. In fact, Borel conjectured that the minimax theorem is false in general, although he did prove it true in certain special cases. Von Neumann proved it true under general conditions, and in addition he created the conceptually rich theory of games with more than two players.

Of more interest than a debate on priority is the fact that neither group of papers—the one in France and the other in Germany—attracted much attention on publication. There are almost no other papers than those mentioned before the publication in 1944 of von Neumann and Morgenstern's book,[1] and those were confined to the mathematical journals. Apparently little interest was stimulated in the empirical sciences most concerned with conflict of interest, but this is not surprising since the original papers were written for mathematicians, not for social scientists. Fortunately, von Neumann and Morgenstern attempted to write their book so that a patient scientist with limited mathematical training could absorb the motivation, the reasoning, and the conclusions of the theory; judging by the attention given it in non-mathematical journals, as well as in the mathematical ones, they were not without success in this aim.   Only a very few scientific volumes as mathematical as this one have aroused as much interest and general admiration.   Yet we know that much of the material had lain dormant in the literature for two decades.   Presumably the recent war was an important contributing factor to the later rapid development of the theory.   During that period considerable activity developed in scientific, or at least systematic, approaches to problems that had been previously considered the exclusive province of men of "experience."   These include such topics as logistics, submarine search, air defense, etc.   Game theory certainly fits into this trend, and it is one of the more sophisticated theoretical structures so far resulting from it.

Though it is not directly relevant to the theory itself, it is worth emphasizing again that game theory is primarily a product of mathematicians and not of scientists from the empirical fields.   In large part this results from the fact that the theory was originated by a mathematician and was, to all intents and purposes, first presented in book form as a highly formal (though, for the most part, elementary) structure, thus tending to make it accessible as a research vehicle only to mathematicians.   Indeed, we believe that so far the impact of game theory has been greater in applied mathematics, especially in mathematical statistics, than in the empirical sciences.

1.3 AN INFORMAL CHARACTERIZATION OF A GAME

Game theory does not, and probably no mathematical theory could, encompass all the diverse problems which are included in our brief characterization of conflict of interest.   In this introduction we shall try to cite the main features of the theory and to present some substantive problems included in its framework.   The reader will easily fill in examples not now in the domain of the theory, and as we discuss our examples we shall point out some other important cases which are not covered.

[1] The original edition of *Theory of Games and Economic Behavior* appeared in 1944, but the revised edition of 1947 is the more standard reference and it includes the first statement of the theory of utility, which we shall discuss in Chapter 2.   All of our references will be to the 1947 edition.

First, with respect to the possible outcomes of a given situation, it is assumed that they are well specified and that each individual has a consistent pattern of preferences among them.   Thus, if we ignore the fact that the player is in conflict with others and concentrate only on the outcomes, it is supposed that one way or another it can be ascertained what choice he would make if he were offered any particular collection of alternatives from which to choose.   This problem of individual decision making is crucial to the whole superstructure we shall discuss, and it is important that there be no confusion about the assumptions that are made. For this reason we have devoted the whole of Chapter 2 to the topic of modern utility theory.   In order that those familiar with the classical, and somewhat discredited, uses of the word "utility" not be misled, we shall expend some effort in establishing how the modern work on utility differs from the earlier ideas.   In brief, the current theory shows that if one admits the possibility of risky outcomes, i.e., lotteries involving the basic alternatives, and if a person's preferences are consistent in a manner to be prescribed, then his preferences can be represented numerically by what is called a utility function.   This utility has the very important property that a person will prefer one lottery to another if and only if the expected utility of the former is larger than the expected utility of the latter.   Thus, the assumed individual desire for the preferred outcomes becomes, in game theory, a problem of maximizing expected utility.

Second, the variables which control the possible outcomes are also assumed to be well specified, that is, one can precisely characterize all the variables and all the values which they may assume.   Actually, one may best think of them as partitioned into $n + 1$ classes if there are n individuals in the situation or, in the terminology of the theory, if it is an n-person game.   To each person is associated one of the classes, which represents his domain of choice, and the one left over is within the province of chance.

As we said earlier, in this type of conflict situation we are interested in only some of the resulting behavior.   Actually, our curiosity may encompass all of it—the tensions resulting, suicide rates or frequency of nervous disorder, aggressive behavior, withdrawal, changes in personal or business strategy, etc.—but of these, any one theory will, presumably, deal with only a small subset.   At present, game theory deals with the choices people may make, or, better, the choices they should make (in a sense to be specified), in the resulting equilibrium outcomes, and in some aspects of the communication and collusion which may occur among players in their attempts to improve their outcomes.   Although much of what is socially, individually, and scientifically interesting is not a part of the theory, certain important aspects of our social behavior are included.

A theory such as we are discussing cannot come into existence without assumptions about the individuals with which it purports to be concerned. We have already stated one: each individual strives to maximize his utility. Care must be taken in interpreting this assumption, for a person's utility function may not be identical with some numerical measure given in the game. For example, poker, when it is played for money, is a game with numerical payoffs assigned to each of the outcomes, and one way to play the game is to maximize one's expected money outcome. But there are players who enjoy the thrill of bluffing for its own sake, and they bluff with little or no regard to the expected payoff. Their utility functions cannot be identified with the game money payments. Indeed, there are many who feel that the maximization assumption itself is tautological, and that the empirical question is simply whether or not a numerical utility exists in a given case. Assuming that behavior is correctly described as the maximization of utility, it is quite another question how well a person knows the functions, i.e., the numerical utilities, the others are trying to maximize. Game theory assumes he knows them in full. Put another way, each player is assumed to know the preference patterns of the other players.

This, and the kindred assumptions about his ability to perceive the game situation, are often subsumed under the phrase "the theory assumes rational players." Though it is not apparent from some writings, the term "rational" is far from precise, and it certainly means different things in the different theories that have been developed. Loosely, it seems to include any assumption one makes about the players maximizing something, and any about complete knowledge on the part of the player in a very complex situation, where experience indicates that a human being would be far more restricted in his perceptions. The immediate reaction of the empiricist tends to be that, since such assumptions are so at variance with known fact, there is little point to the theory, except possibly as a mathematical exercise. We shall not attempt a refutation so early, though we feel we have given some defense in later chapters. Usually added to this criticism is the patient query: why does the mathematician not use the culled knowledge of human behavior found in psychology and sociology when formulating his assumptions? The answer is simply that, for the most part, this knowledge is not in a sufficiently precise form to be incorporated as assumptions in a mathematical model. Indeed, one hopes that the unrealistic assumptions and the resulting theory will lead to experiments designed in part to improve the descriptive character of the theory.

In summary, then, one formulation of a class of conflicts of interest is this: There are n players each of whom is required to make one choice from a well-defined set of possible choices, and these choices are made without any knowledge as to the choices of the other players. The domain of possible choices for a player may include as elements such things as "playing an ace of spades" or "producing tanks instead of automobiles," or, more important, a strategy covering the actions to be taken in all possible eventualities (see below). Given the choices of each of the players, there is a certain resulting outcome which is appraised by each of the players according to his own peculiar tastes and preferences. The problem for each player is: what choice should he make in order that his partial influence over the outcome benefits him most? He is to assume that each of the other players is similarly motivated. This characterization we shall come to know as the normalized form of an n-person game. Two other forms—the extensive and the characteristic function form—will play important roles in our subsequent discussion; but there is no need to go into them now.

1.4 EXAMPLES OF CONFLICT OF INTEREST

Next, we should consider what significant problems of conflict of interest are included in this formulation. Our brief examination will cite examples in each of three areas: economics, parlor games, and military situations. From these it is easy to generate analogous examples for other substantive disciplines.

One basic economic situation involves several producers, each attempting to maximize his profit, but each having only limited control over the variables that determine it. One producer will not have control over the variables controlled by another producer, and yet these variables may very well influence the outcome for the first producer. One may object to treating this as a game on the grounds that the game model supposes that each producer makes one choice from a domain of possible choices, and that from these single choices the profits are determined. It seems obvious that this cannot be the case, else industry would have little need for boards of directors and elaborate executive apparatus. Rather, there is a series of decisions and modifying decisions which depend upon the choices made by other members of the economy. However, in principle, it is possible to imagine that an executive forsees all possible contingencies and that he describes in detail the action to be taken in each case instead of meeting each problem as it arises. By "describe in detail" we mean that the further operation of the plant can be left in the hands of a clerk or a machine and that no further interference or clarification will be needed from the executive.

For example, in the game ticktacktoe, it is perfectly easy to write down all the different possible situations which may arise and to specify what shall be done in each case (and for this reason adults consider it a dull game). Such a detail specification of actions is called a (pure) strategy. There is, of course, no reason why the domains of action need be minor decisions; they may have as elements the various pure strategies of the players. Looked at this way, a player chooses a strategy that covers all possible specific circumstances which may arise. For practical reasons, it is generally not possible to specify economic strategies in full, and as a result a business strategy is usually only a guide to action with respect to pricing, production, advertising, hiring, etc., which neither states in detail the conditions to be considered nor the actions to be taken. The game theory notion of strategy is an abstraction of this ordinary concept in which it is supposed that no ambiguity remains with respect to either the conditions or the actions. With this concept one apparent difficulty in applying the game theoretic model to economic problems evaporates. The notion of a pure strategy, and some related concepts, will receive considerably more discussion in Chapters 3 and 4.

At least two problems render it difficult in practice to put many economic problems in game form. In general, it is hard to specify precisely the strategy sets available to the players. This may stem from a variety of causes, but one of the most striking is the possible modification of the strategy sets during the execution of the game. For example, this may result from a new invention or scientific discovery which opens a whole new range of activities to a producer. It is true that such complications can be encompassed formally by using the theories of decision making under uncertainty, which are discussed in Chapter 13, but this takes one outside the realm of games as we have been discussing them. Moreover, whether the current resolutions of, say, the invention problem are really useful is, at this time, debatable. If we restrict ourselves to just the formalism of risk, omitting uncertainty, then such a possibility causes trouble and so we may only hope at best to obtain limited predictions. This type of limitation seems to be regarded by many social scientists as a terrible inadequacy, and yet it is a common difficulty in all of physical science. It is analogous to a physical prediction based on boundary conditions which may be subject to change during the process, either by external causes or through the very process itself. The prediction will only be valid to the extent that the conditions are not changed, yet such predictions are useful and are used even when it is uncertain whether the assumed invariance actually holds. In many ways, social scientists seem to want from a mathematical model more comprehensive predictions of complex social phenomena than have ever been possible in applied physics and engineering; it is almost certain that their desires will never be fulfilled, and so either their aspirations will be changed or formal deductive systems will be discredited for them.

A second complication in describing the strategy sets of many economic situations is the fact that most decisions are not just described in terms of the obvious alternatives but require a specification of time. The importance of the timing of decisions is too obvious to need illustration. There is no real conceptual difficulty in enlarging the set of alternatives so as to include time as part of the choice, but any enlarging of these sets tends to cause severe practical difficulties. The strategy sets become gigantic very quickly, and so the determination of the quantities needed to describe the economic situation as a game becomes a practical impossibility. These are important difficulties, but, as we shall see, more can be said which will render them less crucial than they might now seem.

To turn from economics, it is well known that for parlor games there is always a clear-cut scoring procedure. In some games which are played for money, such as poker, there is a finely graded ordering of the possible outcomes. In others, such as chess, the outcome is simply winning or losing, to which one can assign a more or less arbitrary numerical scale, such as 1 and 0. Very often a player aims to maximize his expected gain as described by the numerical score of the game; but, as we pointed out earlier, there are cases when this score function cannot be identified with the person's utility, such as when an adult purposely loses to a child.

In a parlor game, as in our economic example, each player makes not one choice but a whole series whose order and nature depend upon the previous choices both he and the other players have made, that is, on the previous play of the game. In exactly the same way as in the economic situation one is able to show that the strategy notion allows this extensive form to be reduced to the above-mentioned normal form. In Chapter 3 we shall do this in some detail.

One common difference between a parlor game and an economic game is of the utmost importance in the theories developed. The rules, or at least the social mores, almost always specify that there shall be no collusion among the players of a parlor game. In economics, the concept of a coalition, i.e., of collusion among some of the producers so that each coalition member betters his position at the expense of the other producers not in the coalition or at the expense of the consumer, is widely recognized in theory, in the law, and in everyday discourse. It thus behooves a theory which purports to have application beyond parlor games to be concerned with this common phenomenon of conflict situations, and such is the case in game theory.

A military conflict is, by definition, a conflict of interest in which neither side has complete control over the variables determining the outcome, and in which the outcome is determined through a series of "battles." We may naively take the outcome to be winning or losing, to which we might assign the numerical values 1 and 0. More subtle interpretations of the outcomes are obviously possible, based on, say, the degree of destruction, etc. Again we have the same two practical difficulties as in the economic problem: there is actually a series of decisions on each side, the timing of which is of vital importance, and the domain of choices for these decisions is not usually well specified. The first problem can be surmounted as before by the notion of a strategy, and, indeed, the concept of a military strategy is common, even if it is not always clearly formulated. The second problem is again more profound, and it appears to render difficult a game theoretic analysis of many important military situations; but certainly some significant ones are subject to the theory. One of the simplest is the "duel," which in an elementary form consists of two players 1 and 2 having p and q "shots" respectively. For each player i there is a function which gives the probability that a shot fired by i at any time t will result in a "hit," let us suppose a fatal hit. We may suppose that the domain of t is limited, as it would be by the fuel supply in an air engagement. The problem is then to determine when each player best take each of his shots, assuming that he knows how many shots his opponent has already taken, so as to maximize the probability that he will hit his opponent before being hit. For most duel situations of interest, the probability of a hit increases with time, as, for example, in the classical duel of two men walking towards each other with guns leveled.

Political controversies are still another fertile source of situations involving conflicts of interest. In addition to the difficulties of the economic and military problems with respect to ill-defined domains of action, we know that here there is considerable ambiguity as to the outcome, or payoff, function even over a known domain of possible actions. This is to some extent true in the other situations we have described, but it is overwhelmingly obvious in the political realm, where, for example, the defeat of a candidate has sometimes been attributed (after the fact) to a single sentence out of the hundreds he spoke in a campaign.

A feature suggested by political and economic conflicts is the "social arbiter." Often it is felt that conflicts of interest should not be allowed to resolve themselves in, shall we say, the open market of threats and counter threats, but that there should exist social devices to take into account the preferences and strategic potentialities of each of the players and to arrive at a "fair" resolution of the conflict. Such a conciliation device—be it a voting scheme or an individual classed as an arbiter—must have the property that it is brought forth not to resolve a particular conflict but a wide class which potentially may arise; and its fairness is evaluated in the abstract with respect to this domain of possible conflicts.

Thus you have the compact summary that game theory is a model for situations of conflict among several people, in which two principal modes of resolution are collusion and conciliation. (Several of our colleagues convinced us that we should not use the tempting, but flip, title *Conflict, Collusion, and Conciliation* for this book.)

1.5 GAME THEORY AND THE SOCIAL SCIENTIST

From the above comments we see that there is some hope that the normalized form of a game includes some socially important phenomena, but it is clear that with respect to many situations there are serious practical difficulties. This, however, is not the entire picture. In developing the theory of coalition formation in n-person games, von Neumann and Morgenstern transformed the normal form into a mathematically simpler structure—simpler in that much of the detail of the normal form is condensed—which, it appears, will allow a broader application of the theory than the above discussion suggests. This is more appropriately discussed in Chapter 8 than here, and we shall confine ourselves to remarking that to attain such applications approximate estimates of the "characteristic function" will have to be obtained, presumably by empirical techniques. This does not appear to be beyond the scope of some of the techniques under development in social psychology and sociology, and it is to be hoped that some empiricists will be attracted to this problem. However, this is conjectural, and we have the historical fact that many social scientists have become disillusioned with game theory. Initially there was a naive band-wagon feeling that game theory solved innumerable problems of sociology and economics, or that, at the least, it made their solution a practical matter of a few years' work. This has not turned out to be the case.

What then is the significance of game theory to the social scientist? First, because there has not been a plethora of applications in a dozen years,[2] it does not follow that the theory will not ultimately be vital in applied problems. Judging by physics, the time scale for the impact of theoretical developments is often measured in decades. Second, although the present form of the theory may not be totally satisfactory—in part, presumably, because of its so-called normative character—this does not necessarily mean that abandoning it is the only possible course for a social scientist. Much of the theory is of very general importance, but some

[2] An important addition to the literature of applied game theory will soon be forthcoming: Martin Shubik's *Competition, Oligopoly and the Theory of Games* [1957].

revision may be required for fruitful applications. Attention to the theory is needed, and not attention from the mathematician alone, as is now the case. Third, game theory is one of the first examples of an elaborate mathematical development centered solely in the social sciences. The conception derived from non-physical problems, and the mathematics—for the most part elementary in the mathematical sense—was developed to deal with that conception. The theory draws on known mathematics according to need—on set theory, on the theory of convex bodies, etc.; furthermore, when known tools were not applicable new mathematics was created. Most other attempts at mathematization (with the exception of statistics which plays a special role) have tended to take over small fragments of the mathematics created to deal with physical problems. If we can judge from physics, the main developments in the mathematization of the social sciences will come—as in game theory— with the development of new mathematics, or significantly new uses of old mathematics, suited to the problem. No one of these theories should be expected to be a panacea, but their cumulative effect promises to be significant.

The achievement of von Neumann and Morgenstern is remarkable: In the first major publication on the subject, they formulated a clear abstraction, drawn from the relatively vague social sciences, having both considerable breadth and mathematical depth, and they developed an elaborate and subtle superstructure with masterful scope. The depth of their contribution can be partially appreciated from the fact that today the material still must be presented according to their outline; there have been additions, true, but the main concepts are unchanged.

chapter 2

UTILITY THEORY

2.1. A CLASSIFICATION OF DECISION MAKING

The modern theory of utility is an indispensable tool for the remainder of this book, and so it is imperative to have a sound orientation toward it. Apparently this is not easy to achieve, judging by the many current misconceptions about the nature of "utility." It is, perhaps, unfortunate that von Neumann and Morgenstern employed this particular word for the concept they created—unfortunate because there have been so many past uses and misuses of various concepts called utility that many people view anything involving that word with a jaundiced eye, and because others insist on reading into the modern concept meanings from the past. We certainly are not going to assert that there are no serious limitations to the von Neumann-Morgenstern theory, but it can be frustrating to hear devastating denunciations which, although relevant to theories of the past, are totally irrelevant to—or incorrect for—the modern theory.

Pedagogically, it might be wise to defer this discussion until it is forced upon us in the context of game theory. Certainly, the needs of game theory would provide excellent reason to study the concept; however, it would also necessitate a sizeable digression in what will prove to be an already long argument. Furthermore, utility theory is not a part of game theory. It is true that it was created as a pillar for game theory, but it can stand apart and it has applicability in other contexts. So we have elected to present it first. As background, we shall describe in this section how the problems of decision making have been classified, in this way showing where utility theory fits into the overall picture. In the next section we shall discuss the classical notion of utility and indicate how, through its defects, it led up to the modern concept. No attempt is made to trace the history of the concept in detail; an excellent history can be found in Savage [1954]. In sections 2.4 and 2.5 we shall present a version of the theory itself, and in 2.6 and 2.7 some of the more common fallacies surrounding it. The chapter closes with a brief discussion of the experimental problems. Appendix 1 describes a modification of utility theory in which preference is assumed to be probabilistic.

The field of decision making is commonly partitioned according to whether a decision is made by (i) an individual or (ii) a group, and according to whether it is effected under conditions of (*a*) certainty, (*b*) risk, or (*c*) uncertainty. To this last classification we really must add (*d*) a combination of uncertainty and risk in the light of experimental evidence. This is the province of statistical inference.

The distinction between an individual and a group is not a biological-social one but simply a functional one. Any decision maker—a single human being or an organization—which can be thought of as having a unitary interest motivating its decisions can be treated as an individual in the theory. Any collection of such individuals having conflicting interests which must be resolved, either in open conflict or by compromise, will be considered to be a group. These are not clearly defined formal words in the theory; rather they are vague classificatory concepts suggesting the identifications one might make in applications. Depending upon one's viewpoint, an industrial organization may be considered as an individual in conflict with other similar organizations or as a group composed of competing departments.

As to the certainty-risk-uncertainty classification, let us suppose that a choice must be made between two actions. We shall say that we are in the realm of decision making under:

(*a*) *Certainty* if each action is known to lead invariably to a specific outcome (the words prospect, stimulus, alternative, etc., are also used).

(*b*) *Risk* if each action leads to one of a set of possible specific outcomes, each outcome occurring with a known probability. The probabilities are assumed to be known to the decision maker. For example, an action might lead to this risky outcome: a reward of \$10 if a "fair" coin comes up heads, and a loss of \$5 if it comes up tails. Of course, certainty is a degenerate case of risk where the probabilities are 0 and 1.

(*c*) *Uncertainty* if either action or both has as its consequence a set of possible specific outcomes, but where the probabilities of these outcomes are completely unknown or are not even meaningful.

In terms of these notions, even though they are not precisely specified at present, the general structure and coverage of the book can be given. With the exception of Chapter 14, the book is concerned with individual decision making, admittedly much of it in a social or group context.   In the next two sections individual decision making under certainty is briefly characterized, and the scope of theories of this type is indicated. In particular, the problem of linear programing is described and its relationship to game theory sketched (for a more complete discussion see Appendix 5).   This we include to show that a topic in one part of the classificatory scheme may have strong formal relations to one in another part.   Then we turn to individual decision making under risk—utility theory.   This is followed by the main body of the book—Chapters 3 through 12—on game theory.   Intuitively, the problem of conflict of interest is, for each participant, a problem of individual decision making under a mixture of risk and uncertainty, the uncertainty arising from his ignorance as to what the others will do.   In game theory one attempts to idealize this problem in such a way as to transform it into interacting problems of individual decision making under risk.   In point of fact, assumptions about the motivations of the players are not sufficient to eliminate completely the uncertainty aspects of the problem, as we shall become acutely aware.

Chapter 13 offers a brief survey of the area traditionally known as decision making under uncertainty, which is typified by the problem of the statistician attempting to reach a decision when the "state of the world" is unknown, and the problem of mixed uncertainty and risk when the uncertainty can be reduced at the cost of experimentation.   One of the tasks will be to explicate the meaning of "unknown," and, unfortunately, this will force us into some remarks about the foundations of probability theory.   Although we have to touch on it, we cannot try to do this topic justice.

In Chapter 14 we turn from individual decision making to that of groups.   We shall review some of the issues raised and clarified by Arrow in *Social Choice and Individual Values* [1951 *a*], where he considered the problem of how best to amalgamate the discordant preference patterns of the members of a society to arrive at a compromise preference pattern for society as a whole.   Such material is appropriately included in this book because of its similarity to the problem of arbitration in bargaining models and two-person cooperative games (see Chapter 6) and to games against nature (see Chapter 13).

With this overall structure in mind, we may now say a few words about decision making under certainty and about the backgrounds of modern utility theory. For more comprehensive surveys of utility theory see Adams [1954], Edwards [1954 c], and Savage [1954].

2.2. INDIVIDUAL DECISION MAKING UNDER CERTAINTY

Decision making under certainty is a vast area! The bulk of formal theory in economics, psychology, and management sciences can be classed under this heading. Until quite recently, the mathematical tools used were largely the calculus to find maxima and minima of functions and the calculus of variations to find functions, production schedules, inventory schedules, and so on which optimize performance over time—dynamic programing, so to speak. We will not discuss these topics, for their connection with game theory is remote, but we shall sketch some of the ideas which led to utility theory.

Typically, decision making under certainty boils down to this: Given a set of possible acts, to choose one (or all) of those which maximize (or minimize) some given index. Symbolically, let $\mathbf{x}$ be a generic act in a given set F of feasible acts and let $f(\mathbf{x})$ be an index associated to (or appraising) $\mathbf{x}$; then find those $\mathbf{x}^{(0)}$ in F which yield the maximum (or minimum) index, i.e., $f(\mathbf{x}^{(0)}) \geqslant f(\mathbf{x})$ for all $\mathbf{x}$ in F.

Very often the heart of the problem is the appropriate choice of the associated index. In many economic contexts profit and loss are suitable indices, but in other contexts no such quantities are readily available. Consider, for example, a person who wishes to purchase one of several paintings. In a sense, we can assert that the essence of the problem is: how should the subject select an index function so that his choice reduces to finding the alternative with the maximum index?

Operationally, of course, we can suppress this problem, for all we need to do is observe his purchase. Alternatively, we can observe his behavior in a host of more restricted situations and from this predict his purchase. For example, in an experimental study one might instruct him as follows. "Here are ten valuable reproductions. We will present these to you in pairs and you will tell us which one of each pair you would prefer to own. After you have given your answers to all paired comparisons, we will actually choose a pair at random and present you with the choice you have previously made. Hence, it is to your advantage to record, as best as you are able, your own true tastes." Now, it may be possible to account for all his choices by assuming that he has a simple ranking of the paintings, from the least liked to the most liked, such that the subject always chooses, in any paired comparison, the one with the higher ranking. If so, his choices can be pithily summarized by assigning numbers to the paintings in such a way that their magnitudes reflect this preference ranking. For example, 1 to the least liked, 2 to the next, $\cdots$ and 10 to the best liked. This can be done provided his preferences satisfy the one condition of *transitivity:* if A is preferred in the paired comparison (A, B) and B is preferred to the paired comparison (B, C), then A is preferred in the paired comparison (A, C), and this holds for all possible triples of alternatives A, B, and C. This concept of transitivity is extremely important and should be well understood. It is a completely natural notion in much of our language: if A is larger than B and B is larger than C, then A is larger than C (or substitute "heavier," "better," etc.).

If we are able to rank the alternatives and if we assign a numerical index, then in a totally *tautological* sense we can assert that the subject has behaved as if he always chooses the painting with the higher index. From this it is easy to slip into saying that one was preferred to another because it has a larger latent index of "satisfaction" or "utility." This is an unrewarding slip—indeed, it is a trap one must be careful to avoid. This usage was once a burning issue in the economic literature, but it has been totally discredited. One of the reasons is the striking non-uniqueness of the index. For example, suppose we have only the three alternatives A, B, and C, where A is the most preferred, B the next, and C the least. Then, one may summarize this by saying that they are worth 3, 2, and 1 "utiles" respectively. Of course, had the associated "utiles" been 30, 20.24, and 3.14, the same manifest response pattern of preference would be observed. Indeed, any numbers a, b, and c such that $a > b > c$ would lead to the same manifest data. When it was conclusively shown that large segments of economic thought could be maintained by postulating merely an ordinal preference pattern—an ordering—for alternatives without including an underpinning of latent "utiles," the utility notion was not worth philosophizing about. Still, one may contend that introducing the numbers does no harm, that they summarize the ordinal data in a compact way, and that they are mathematically convenient to manipulate. But, in part, their very manipulative convenience is a source of trouble, for one must develop an almost inhuman self-control not to read into these numbers those properties which numbers usually enjoy. For example, one must keep in mind that it is meaningless to add two together or to compare magnitudes of differences between them. If they are used as indices in the way we have described, then the only meaningful numerical property is order. We may compare two indices and ask which is the larger, but we may not add or multiply them.

Before we turn to the classical and modern approaches to utility in the theory of choice under risk, it would be well to cite an example of a nontrivial problem of decision making under certainty which requires the use of non-classical mathematics. We choose the linear programing problem, which among other things is extremely closely related to the theory of (two-person zero-sum) games.

*2.3 AN EXAMPLE OF DECISION MAKING UNDER CERTAINTY:
LINEAR PROGRAMING [1]

This section is not needed in the following development, and one part of it will be completely meaningful only after Chapter 4 is read.

The following diet problem is an example of a linear programing problem. Suppose that there are given n specific foods $F_1, F_2, \cdots, F_n$. A "diet" simply prescribes the daily amount of each food to be consumed: x_1 units of F_1, x_2 units of F_2, and so on. To each diet $(x_1, x_2, \cdots, x_n)$ we can associate, i.e., determine, the nutrient yield of any given list of nutrients we care to be concerned about, say, iron, calcium, riboflavin, vitamin C, etc. We observe that the nutrient yield of the diet $(x_1, x_2, \cdots, x_n)$ for iron, say, is a linear expression of the form

$$a_1x_1 + a_2x_2 + \cdots + a_nx_n,$$

where a_1 represents the amount of iron in a unit amount of the food F_1, a_2 the amount in a unit of F_2, and so on. Of course, some of the a_i may be 0, but none may be negative. The same holds for the x_i. Now, medical research has established that there are certain minimal requirements of the several nutrients, say a units per day of iron. Thus, one wants to choose only from among those diets which provide these minimum requirements. So we demand that $(x_1, x_2, \cdots, x_n)$ satisfy linear inequalities of the form

$$a_1x_1 + a_2x_2 + \cdots + a_nx_n \geqslant a.$$

This is only the one for iron; there is a similar expression for each of the other nutrients. Obviously, these constraints can be met by choosing sufficiently large quantities of food without regard to cost; however, we often wish to choose "diets" so as to minimize cost—such might be the case in a hospital—and that creates a problem of some complexity. If $p_1, p_2, \cdots, p_n$ are the unit prices of the foods $F_1, F_2, \cdots, F_n$ respectively, then the cost of the diet $(x_1, x_2, \cdots, x_n)$ is

$$p_1x_1 + p_2x_2 + \cdots + p_nx_n.$$

[1] Throughout the book starred sections will be found. It is unnecessary to read these to be able to comprehend the rest of the book, and, in some cases, they require more mathematical sophistication to be understood. Within unstarred sections some paragraphs are in small print, and to these the same comment applies.

So, finally, the problem is to choose a diet so as to satisfy the nutrient requirements (the linear inequalities) at the minimum cost (the linear cost expression).   This is a typical linear-programing problem.

In the terms of the last section, the most general linear-programing problem consists of:

i. Acts, each of which consists of the specific choice of n real numbers (e.g., diets).

ii. The feasibility conditions, which are linear equalities and inequalities which constrain the possible acts (e.g., minimal nutrient requirements).

iii. An index associated to each act which is a weighted average of the n numbers constituting the act (e.g., the cost function).

The problem is to find an act satisfying (ii) and minimizing (iii).   It is clearly a decision-making problem under certainty; however, it cannot be handled by the traditional methods of the calculus.   What is known as the theory of convex bodies has proved crucial.

To the readers who know nothing of game theory, the following statements on the approximate relationship between linear programing and game theory will be little more than words; still, some of the ties may come across.   First, to each linear-programing problem there is associated a two-person zero-sum game, and conversely, such that whenever the linear-programing problem is soluble the solution to one problem can always be interpreted as providing a solution to the other; second, the proofs of the principal results of both theories use the same formal mathematical tools, e.g., the separation theorem of convex bodies; and, third, assuming the truth of the principal theorem of one theory, the truth of the principal theorem of the other follows readily.   Thus, when results of a game-theoretic type appear unexpectedly in a problem, one can often discover a natural linear-programing problem lurking in the background. For a more complete discussion of this relationship, see Appendix 5.

Two other famous problems of decision making under certainty, both of which are closely related to the linear-programing problem, should be mentioned.   The first, known as the personnel assignment problem, assumes there are n people and n jobs to be filled.   The "worth" or "yield" of man i in job j is assumed to be known and to be given by a number which we may denote a_{ij}.   The task is to find the assignment of individuals to jobs which maximizes the gross yield.   In this problem the set of feasible acts, F, consists of all those one-to-one assignments of people to jobs, and therefore there are $n! = 1 \cdot 2 \cdot 3 \cdot \cdot \cdot n$ such acts.   The second problem, the traveling salesman problem, is conceptually quite similar to, but more difficult than, the personnel problem.   A salesman starts from a given state capital, and he must visit each of the other state capitals.   What is the shortest route he can follow?   In this problem there are 47! different feasible acts (not 48! because the initial capital is fixed), where an act is a directed path which touches each state capital, and to each act the associated index is the total distance traveled.

Both of these problems have one aspect which is inherently different from the linear-programing problem, namely, the set of feasible acts is finite.   Thus, in a sense, these problems are conceptually trivial, for in a finite number of steps the indices of all the acts can be checked and the optimal one chosen.   In practice this will not do, not even with modern high speed computers, for $n!$ is a fantastically large number even for n of moderate size, e.g., $20! = 2,432,902,008,176,640,000$.

One way to solve the personnel assignment problem is to embed it into a linear-programing problem.   It can be shown that the finite feasible set F can be enlarged to an infinite set F^*, and that the association of indices to each act can be extended to the new set in such a manner that the enlarged problem is a linear-programing problem and the solution to the enlarged problem is actually an act in F.   Thus, the solution to the enlarged problem is also a solution to the original problem.   (The linear-programing formulation for 5 cities is given by Kuhn [1955] and for 7 by Norman [1955].)   Paradoxically, but very common in mathematics, by complicating the problem tremendously we have rendered it more amenable to analysis.   Since the linear-programing problem is in turn related to the zero-sum game, we can see in how devious a way game theory can enter into the picture.   Surprisingly, in this case and others, the induced game-theoretic problem has a neat substantive interpretation which can aid in making quick intelligent guesses as to approximate solutions of the original problem.   We shall meet this sort of thing again in Chapter 13 when we turn to the connections between statistical inference and game theory.

2.4 INDIVIDUAL DECISION MAKING UNDER RISK

The problems of making decisions under risk first appeared in the analysis of a fair gamble, and here again the desire for a utility concept arose.   Consider a gamble in which one of n outcomes will occur, and let the possible outcomes be worth a_1, a_2, $\cdots$, a_n dollars, respectively. Suppose that it is known that the respective probabilities of these outcomes are $p_1, p_2, \cdots , p_n$, where each p_i lies between 0 and 1 (inclusive) and their sum is 1.   How much is it worth to participate in this gamble?

The monetary expected value is

$$b = a_1 p_1 + a_2 p_2 + \cdots + a_n p_n,$$

and, so one argument goes, the "fair price" for the gamble is its expected value b. However, the famous St. Petersburg paradox, due to D. Bernoulli, casts serious doubt that for most people the money expectation formulates what they consider to be the "fair price." The paradox is this: A "fair" coin, which is defined by the property that the probability of heads is $\frac{1}{2}$, is tossed until a head appears. The gambler receives 2^n dollars if the first head occurs on trial n. The probability of this occurrence is simply the probability of a sequence of tails on the first $n - 1$ trials and a head on the nth, which is $\frac{1}{2}$ multiplied n times, i.e., $(\frac{1}{2})^n$. Thus, one receives 2 dollars with probability $\frac{1}{2}$, 4 dollars with probability $\frac{1}{4}$, 8 dollars with probability $\frac{1}{8}$, and so on. Therefore, the expected value is

$$2(\tfrac{1}{2}) + 4(\tfrac{1}{4}) + 8(\tfrac{1}{8}) + 16(\tfrac{1}{16}) + \cdots = 1 + 1 + 1 + 1 + \cdots,$$

which does not sum to any finite number. It follows, then, that one should be willing to pay any sum, however large, for the privilege of participating in such a gamble. As a description of behavior, this is silly! As Bernoulli emphasized, people do not, and will not, behave in accord with the monetary expected value of this gamble.

Bernoulli suggested the following modification of the analysis in order to rescue the principle that people behave according to an expected value. The pertinent variable to be averaged, he argued, is not the actual monetary worth of the outcomes, but rather the intrinsic worths of their monetary values. It is plausible to suppose that the intrinsic worth of money increases with money, but at a diminishing rate. A function having this property is the logarithm. Thus, if the "utility" of m dollars is $\log_{10} m$, then the fair price would not be the monetary expected value but the monetary equivalent of the utility expected value

$$b = (\tfrac{1}{2}) \log_{10} 2 + (\tfrac{1}{4}) \log_{10} 4 + (\tfrac{1}{8}) \log_{10} 8 + \cdots.$$

It can be shown that this sum does, in the limit, approach a finite value, which we have called b. Then the "monetary fair price" of the gamble is a dollars, where $\log_{10} a = b$.

There are certain obvious criticisms of Bernoulli's tack, and these suggest some of the ideas involved in von Neumann and Morgenstern's approach to risk. First, the utility association to money is completely *ad hoc*. There are an infinity of functions which increase at a decreasing rate, and, certainly, the association may vary from person to person—but how? Second, why should a decision be based upon the expected value of these utilities? The rationale generally given for using expected value involves an argument as to what will happen in the long run when the gamble is repeated many times. Although it is easy to see the merit of such a frequency interpretation for a gambling house, it is by no means clear that it should apply to an individual who participates in the gamble only once.

Thus, what we want is a construction of a utility function for each individual which, in some sense, represents his choices among gambles and which has as a consequence the fact that the expected value of utility represents the utility of the corresponding gamble. We know from our previous examination of utility that this would be a hopeless aim if we considered only a finite number of certain alternatives, but, once we admit all possible gambles among a set of alternatives, we are dealing with an infinite set and by its very size there will be many more constraints on the utility function. *Very roughly*, von Neumann and Morgenstern have shown the following: If a person is able to express preferences between every possible pair of gambles, where the gambles are taken over some basic set of alternatives, then one *can* introduce utility associations to the basic alternatives in such a manner that, if the person is guided solely by the utility expected value, *he is acting in accord with his true tastes*—provided only that there is an element of consistency in his tastes. For the moment, let us ignore the exact statement of this proviso, which is important; we will return to it in the next section where we present a formal statement of one form of their result.

There are two points to be emphasized about this result. First, the utility function so constructed reflects preferences about the alternatives in a certain given situation, and so it will reflect not only how the subject feels about the alternatives (prizes, outcomes, or stimuli) in the abstract, but how he feels about them in the particular situation. For example, the resulting function will incorporate his attitude towards the whole gambling situation. Second, the utility associations are introduced in such a manner as to justify the central role of expected value without any further argument, specifically, without any discussion of long run effects.

The essence of their idea can be illustrated simply. Suppose that our subject prefers alternative A to B, B to C, and A to C. Any three numbers a, b, and c which decrease in magnitude are suitable indices to reflect this ordinal preference. But, remember, we are admitting gambles. Suppose we ask his preference between: (i) obtaining B for certain, and (ii) a gamble with A or C as the outcome, where the probability that it is A is p and the probability that it is C is $1 - p$. We refer to these as the "certain option" and the "lottery option." It seems plausible that if p is sufficiently near to 1, so that the outcome of the lottery option is very likely to be A, the lottery will be preferred. But, if p is near 0, then the certain option will be preferred. As p changes continuously from 1 to 0 the preference for the lottery option must change into preference for the certain option. We idealize this preference pattern by supposing that there is one and only one point of change and that at this point the two options are indifferent. Let us suppose that it is the point $\frac{2}{3}$. If we arbitrarily associate the number 1 to alternative A and 0 to C, then what number should we associate to B to summarize our information about his preferences when this one gamble is allowed? Naturally, $\frac{2}{3}$. If this choice is made, then B is a "fair equivalent" for the gamble with A and C as outcomes with probabilities $\frac{2}{3}$, $\frac{1}{3}$, respectively, in the sense that the utility of B equals the utility expected value of the gamble,

$$1(\tfrac{2}{3}) + 0(\tfrac{1}{3}) = \tfrac{2}{3}.$$

There are triples of numbers other than $(1, \frac{2}{3}, 0)$ which also can serve to summarize the information we have about our subject's tastes, but not nearly so many as when all outcomes were certain. Adding the information of just this one gamble restricts the triples to those of the form

$$a + b, \; \tfrac{2}{3}a + b, \; b,$$

where the number a must be positive.

We note that in all such triples the numerical difference between the utility assignments to B and C is twice that between A and B. Does this permit us to say that going from B to C is twice as (or even just more) desirable than going from A to B? We think not! The number $\frac{2}{3}$ was determined by choices among risky alternatives, and it reflects attitudes toward gambling, not toward the two intervals. Suppose, for example, that, because of his aversion to gambling, our subject reported he would be indifferent between paying out \$9 and having a 50–50 chance of paying out \$10 or nothing. His response could then be summarized by saying that his utilities for \$0, $-\$9$, and $-\$10$ are 1, $\frac{1}{2}$, and 0. We would be unwilling, however, to say that going from $-\$10$ to $-\$9$ is "just as enjoyable" as going from $-\$9$ to \$0.

In this theory it is extremely important to accept the fact that the subject's preferences among alternatives and lotteries came prior to our numerical characterization of them. We do not want to slip into saying that he preferred A to B because A has the higher utility; rather, because A is preferred to B, we assign A the higher utility.

If we add more gambles to the collection and try to assign utilities as we have done, it is clear that to be successful the subject's preferences will have to satisfy some consistency requirements. For example, if he prefers A to B, B to C, and a lottery which yields A with probability $\frac{2}{3}$ and C with probability $\frac{1}{3}$ to a lottery which yields A with probability $\frac{3}{4}$ and C with probability $\frac{1}{4}$, then we are in trouble. Or if he prefers A

to B, B to C, and B to *any* lottery involving A and C as prizes so long as it is a bona fide gamble, i.e., $p \neq 1$, we are again in trouble.

Once one has this idea of utility, then the task is to develop a set of consistency requirements which, on the one hand, seem plausible as an idealized model of human preferences, and which, on the other, allow one to prove that the utility assignments can be made. In the next section we shall present such a set of axioms, but let us first suggest the general nature of these consistency demands by a few descriptive and intuitive words:

i. Any two alternatives shall be comparable, i.e., given any two, the subject will prefer one to the other or he will be indifferent between them.

ii. Both the preference and indifference relations for lotteries are transitive, i.e., given any three lotteries A, B, and C, if he prefers A to B and B to C, then he prefers A to C; and if he is indifferent between A and B and between B and C, then he is indifferent between A and C.

iii. In case a lottery has as one of its alternatives (prizes) another lottery, then the first lottery is decomposable into the more basic alternatives through the use of the probability calculus.

iv. If two lotteries are indifferent to the subject, then they are interchangeable as alternatives in any compound lottery.

v. If two lotteries involve the same two alternatives, then the one in which the more preferred alternative has a higher probability of occurring is itself preferred.

vi. If A is preferred to B and B to C, then there exists a lottery involving A and C (with appropriate probabilities) which is indifferent to B.

2.5. AN AXIOMATIC TREATMENT OF UTILITY

The purpose of this section is to make precise both the consistency requirements and the theorem which we discussed informally in the last section. We shall adopt a set of axioms which are a bit different from those already available in the literature. At some, but relatively unimportant, expense in generality, we can employ axioms which are extremely simple and which lead to the utility numbers quite directly. For other axiom systems the reader is referred to von Neumann and Morgenstern [1947], Herstein and Milnor [1953], and Hausner [1954].

As we present these axioms, it is well to have some interpretation of them in mind. We suggest the following: Suppose that one has to make a choice between a pair of lotteries which are each composed of complicated risky alternatives. Because of their complexity it may be extremely difficult to decide which one is preferable. A natural procedure, then, is to analyze each lottery by decomposing it into simpler alternatives, to make decisions as to preference among these alternatives, and to agree upon some consistency rules which relate the simpler decisions to the more complicated ones. In this way, a consistent pattern is imposed upon the choices between complicated alternatives. Our analysis will follow these lines. At the outset we will not require that a subject choose consistently between all pairs of risky alternatives—just between some of the simpler ones. In the end, we shall show that consistency among the simpler alternatives, plus a commitment to certain rules of composition, implies overall consistency, in the sense that utility numbers can be introduced to summarize choices.

At the same time, as we introduce each assumption (i.e., axiom), we shall view it critically to see just how it will restrict the applicability of the model. Such a model must, inevitably, be a compromise between wider and wider applicability through less restrictive assumptions and richer and more elegant mathematical representation through stronger assumptions.

There is little practical loss of generality if we suppose that all lotteries are built up from a finite set of basic alternatives or prizes, which we denote by $A_1, A_2, \cdots, A_r$. A lottery ticket is a chance mechanism which yields the prizes $A_1, A_2, \cdots, A_r$ as outcomes with certain known probabilities. If the probabilities are $p_1, p_2, \cdots, p_r$, where each $p_i \geqslant 0$ and the sum is 1, then the corresponding lottery is denoted by $(p_1A_1, p_2A_2, \cdots, p_rA_r)$. We interpret this expression to mean only this: one and only one prize will be won and the probability that it will be A_i is p_i. Operationally, one can think of a lottery as the following experiment: A circle having unit circumference is subdivided into arcs of lengths p_1, $p_2, \cdots, p_r$, and a "fair" pointer is spun which if it comes to rest in the arc of length p_i means that prize A_i is the outcome.

The meaning of such a lottery bears some consideration. We are definitely assuming that there is no conceptual difficulty in assigning objective probabilities to the events in question by using symmetries of the experiment and past experience with it. That is to say, we are quite willing to admit a frequency interpretation of probability when assigning probabilities to the events. We do not, however, view the lottery itself from a frequency point of view; it is a single entity that will be conducted *once and only once*, not something to be repeated many times. This restriction to events having known objective probabilities will permit us to deal with most of the conceptual problems of game theory. Those problems arising from experiments having such abstruse events that the probability assignments to them are quite unclear will be deferred until Chapter 13.

We shall now be concerned with an individual's choice between a pair of lottery tickets $L = (p_1A_1, p_2A_2, \cdots, p_rA_r)$ and $L' = (p_1'A_1, p_2'A_2, \cdots, p_r'A_r)$. If L is preferred to L', this means that the individual prefers the experiment associated with L to that associated with L'.

Among the basic prizes, we use the symbolism $A_i \succsim A_j$ to denote that A_j is not preferred to A_i. Equivalently, we say that A_i is preferred or indifferent to A_j.

Assumption 1 (ordering of alternatives). *The "preference or indifference" ordering, $\succsim$, holds between any two prizes, and it is transitive. Formally, for any A_i and A_j, either $A_i \succsim A_j$ or $A_j \succsim A_i$; and if $A_i \succsim A_j$ and $A_j \succsim A_k$ then $A_i \succsim A_k$.*

These assumptions can be criticized on the grounds that they do not correspond to manifest behavior when people are presented with a sequence of paired comparisons. This can happen even over time periods when it is reasonable to suppose individual tastes remain stationary. There are several possible rationalizations for such intransitivities. For one, people have only vague likes and dislikes and they make "mistakes" in reporting them. Often when one is made aware of intransitivities of this kind he is willing to admit inconsistency and to realign his responses to yield a transitive ordering. See Savage [1954, pp. 100–104] for a penetrating discussion of an example due to Allais which traps people, including Savage, into inconsistencies. Once the inconsistency is pointed out, Savage claims that he is grateful to the theory for indicating his inconsistency and he promptly reappraises his evaluations.

A second rationalization asserts that intransitivities often occur when a subject forces choices between inherently incomparable alternatives. The idea is that each alternative invokes "responses" on several different "attribute" scales and that, although each scale itself may be transitive, their amalgamation need not be. This is the sort of thing which psychologists cryptically summarize by terming it a multidimensional phenomenon.

No matter how intransitivities arise, we must recognize that they exist, and we can take only little comfort in the thought that they are an anathema to most of what constitutes theory in the behavioral sciences today. We may say that we are only concerned with behavior which is transitive, adding hopefully that we believe this need not always be a vacuous study. Or we may contend that the transitive description is often a "close" approximation to reality. Or we may limit our interest to "normative" or "idealized" behavior in the hope that such studies will have a metatheoretic impact on more realistic studies. In order to get on, we shall be flexible and accept all of these as possible defenses, and to them add the traditional mathematician's hedge: transitive relations are far more mathematically tractable than intransitive ones.

Since the labeling of the prizes is immaterial, we lose no generality in assuming that they have been numbered so that $A_1 \succsim A_2 \succsim \cdots \succsim A_r$ and that A_1 is strictly preferred to A_r. The latter condition is added only to keep things from being trivial.

Suppose that $L^{(1)}, L^{(2)}, \cdots, L^{(s)}$ are any s lotteries which each involve $A_1, A_2, \cdots, A_r$ as prizes. If $q_1, q_2, \cdots, q_s$ are any s non-negative numbers which sum to 1, then $(q_1 L^{(1)}, q_2 L^{(2)}, \cdots, q_s L^{(s)})$ denotes a compound lottery in the following sense: one and only one of the given s lotteries will be the prize, and the probability that it will be $L^{(i)}$ is q_i.

Assumption 2 (reduction of compound lotteries). *Any compound lottery is indifferent to a simple lottery with $A_1, A_2, \cdots A_r$ as prizes, their probabilities being computed according to the ordinary probability calculus. In particular, if*

$$L^{(i)} = (p_1^{(i)} A_1, p_2^{(i)} A_2, \cdots, p_r^{(i)} A_r), \quad for\ i = 1, 2, \cdots, s,$$

then

$$(q_1 L^{(1)}, q_2 L^{(2)}, \cdots, q_s L^{(s)}) \sim (p_1 A_1, p_2 A_2, \cdots, p_r A_r),$$

where

$$p_i = q_1 p_i^{(1)} + q_2 p_i^{(2)} + \cdots + q_s p_i^{(s)}.$$

This assumption is deceptively simple. It seems to state that any complex lottery can be reduced to a simple one by operating with the probabilities in what appears to be the obvious way. However, consider the lottery $L^{(1)}$, which we have assumed is described by an experiment $\mathbf{p}^{(1)} = (p_1^{(1)}, p_2^{(1)}, \cdots, p_r^{(1)})$, and the more complex lottery which is described by the experiment $\mathbf{q} = (q_1, q_2, \cdots, q_s)$. It is perfectly possible that these two experiments might not be statistically independent; for example, it might happen that, if the first alternative comes up in experiment $\mathbf{q}$, then the third alternative in experiment $\mathbf{p}^{(1)}$ is bound to occur. If so, the reduction given in assumption 2 makes no sense at all. It must, therefore, be interpreted as implicitly requiring one of two things: either that the experiments involved are statistically independent or that such a symbol as $p_j^{(i)}$ actually denotes the conditional probability of prize j in experiment $\mathbf{p}^{(i)}$ given that lottery i arose from experiment $\mathbf{q}$.

Once this interpretation is made, the assumption seems quite plausible. Nonetheless, it is not empty for it abstracts away all "joy in gambling," "atmosphere of the game," "pleasure in suspense," and so on, for it says that a person is indifferent between a multistage lottery and the single stage one which is related to it by the probability calculus. (One neat example of multistage lotteries is found in Paris, as was pointed out to us by Harold Kuhn. Throughout that city are wheels of chance having as prizes tickets in the National Lottery.)

Assumption 3 (continuity). *Each prize A_i is indifferent to some lottery ticket involving just A_1 and A_r. That is to say, there exists a number u_i such that A_i is indifferent to $[u_iA_1, 0A_2, \cdots, 0A_{r-1}, (1 - u_i)A_r]$. For convenience, we write $A_i \sim [u_iA_1, (1 - u_i)A_r] = \tilde{A}_i$, but note well that A_i and $\tilde{A}_i$ are two quite different entities.*

This is a continuity assumption. If $A_1 > A_i > A_r$, it is plausible that $[pA_1, (1 - p)A_r]$ is preferred to A_i if p is near 1, and that the preference is inverted if p is near 0, so it is also plausible that as p is shifted from 1 to 0 there is a point of inversion when the two are indifferent.

Although this assumption seems plausible, at least as a criterion of consistency, there are examples where it does not seem universally applicable. It is safe to suppose that most people prefer \$1 to \$0.01 and that to death. Would, however, one be indifferent between one cent and a lottery, involving \$1 and death, that puts any positive probability on death? When put in such bald form, some, whom we would hesitate to charge with being "irrational," will say No. At the same time, there are others who would argue that the lottery is preferable provided that the chance of death is as low as, say, one in 10^{1000}, for such an event is a virtual impossibility. Even though the universality of the assumption is suspect, two thoughts are consoling. First, in few applications are such extreme alternatives as death present. Second, even if assumption 3 is neither explicitly assumed nor a consequence of other assumptions, a utility calculus can be derived. A single number will no longer suffice; rather, an n-tuple is needed; nonetheless, a good deal of game theory can be constructed on this more complicated utility foundation. We will not describe this theory of n-dimensional utilities; the interested reader can consult Hausner [1954].

Assumption 4 (substitutibility). *In any lottery L, $\tilde{A}_i$ is substitutable for A_i, that is, $(p_1A_1, \cdots, p_iA_i, \cdots, p_rA_r) \sim (p_1A_1, \cdots, p_i\tilde{A}_i, \cdots, p_rA_r)$.*

This assumption, taken with the third, is reminiscent of what is known in other work as the assumption of the *independence of irrelevant alternatives;* this we shall discuss in Chapter 6 and again in Chapters 13 and 14. If one asserts $A_i \sim \tilde{A}_i$, then in view of assumption 4 we also assert that not only are they indifferent when considered alone but also when substituted in any lottery ticket. Thus, the other possible alternatives must be irrelevant to the decision that they are indifferent.

Assumption 5 (transitivity). *Preference and indifference among lottery tickets are transitive relations.*

The comments following assumption 1 apply here even more strongly.

From these first five assumptions it is possible to find for any lottery ticket one to which it is indifferent and which only involves A_1 and A_r. Let $(p_1A_1, p_2A_2, \cdots, p_rA_r)$ be the given ticket. Replace each A_i by $\tilde{A}_i$. Assumption 3 states that these indifferent elements exist, and assumption 4 says they are substitutable. So by using the transitivity of indifference serially,

$$(p_1A_1, \cdots, p_rA_r) \sim (p_1\tilde{A}_1, \cdots, p_r\tilde{A}_r).$$

If now we sequentially apply the probability reduction assumption 2, it is easy to see that we get

$$(p_1A_1, p_2A_2, \cdots, p_rA_r) \sim [pA_1, (1-p)A_r],$$

where

$$p = p_1u_1 + p_2u_2 + \cdots + p_ru_r.$$

A numerical example illustrating this calculation is given at the end of the section.

We now introduce our final assumption:

Assumption 6 (monotonicity). *A lottery $[pA_1, (1-p)A_r]$ is preferred or indifferent to $[p'A_1, (1-p')A_r]$ if and only if $p \geqslant p'$.*

This seems eminently reasonable: between two lotteries involving only the most and least preferred alternatives one should select the one which renders the most preferred alternative more probable. But is it always? A mountain climber certainly prefers the alternative "life" to "death," yet when climbing he prefers some lottery of life and death to life itself, i.e., not climbing. Our trouble here appears to be not so much the assumption but the alternatives we have chosen in the example. A successful climb does not just mean life but also the thrill of the climb, publicity, etc. The real alternative is this "gestalt" which is completely dependent upon their being the risk of death to be attractive.

As this point is important, let us cite another example where the psychological reaction to an outcome of an experiment depends upon the probabilities in the experiment as well as on the actual outcome. Suppose X and Y are two people who are forced to exchange sums of money depending upon the outcome of an experiment. If X is sensitive to Y's feeling, he may prefer that no money be transferred and his preference may decrease with the amount to be transferred (up to some limit, say $100) regardless of who pays. Thus, if

$$A_1 \text{ means } X \text{ pays \$5 to } Y$$

and

$$A_2 \text{ means } Y \text{ pays } \$10 \text{ to } X,$$

X may well exhibit the following preferences:

$$(\tfrac{2}{3}A_1, \tfrac{1}{3}A_2) \succ (1A_1, 0A_2) \succ (0A_1, 1A_2).$$

Such a pattern would violate assumption 6. In other words, X prefers A_2 when it occurs by chance to having it outright.

Although these examples may be a bit strained, they do suggest that, if there is a psychological interaction between the basic alternatives and the probabilities, it may be necessary to use a richer set of basic alternatives in order for assumption 6 to be approximately valid.

With these six assumptions we are done, for if two lotteries L and L' are given, the first five assumptions permit us to reduce them to the form of lotteries in assumption 6, and then we decide between them on the basis of assumption 6. That is, for lotteries $L = (p_1A_1, \cdots, p_rA_r)$ and $L' = (p_1'A_1, \cdots, p_r'A_r)$, we compute

$$p_1u_1 + p_2u_2 + \cdots + p_ru_r \quad \text{and} \quad p_1'u_1 + p_2'u_2 + \cdots + p_r'u_r,$$

and if the former is larger we prefer L to L', if the latter L' to L, and if they are equal L and L' are indifferent. Put as a formal theorem:

If the preference or indifference relation $\succsim$ satisfies assumptions 1 through 6, there are numbers u_i associated with the basic prizes A_i such that for two lotteries L and L' the magnitudes of the expected values

$$p_1u_1 + p_2u_2 + \cdots + p_ru_r \quad \text{and} \quad p_1'u_1 + p_2'u_2 + \cdots + p_r'u_r$$

reflect the preference between the lotteries.

Let us introduce the following terms which will be used in the rest of the book. If a person imposes a transitive preference relation $\succsim$ over a set of lotteries and if to each lottery L there is assigned a number $u(L)$ such that the magnitudes of the numbers reflect the preferences, i.e., $u(L) \geqslant u(L')$ if and only if $L \succsim L'$, then we say there exists a *utility function u* over the lotteries. If, in addition, the utility function has the property that $u[qL, (1 - q)L'] = qu(L) + (1 - q)u(L')$, for all probabilities q and lotteries L and L', then we say the utility function is *linear*.[2] The above

---

[2] Sometimes this property is referred to as the *expected utility hypothesis* since it asserts that the utility of a lottery is equal to the expected utility of its component prizes. Not only is this terminology more explicit (if less brief), but it would help to avoid confusion. The much overworked word "linear" will also arise later with a different meaning. We will sometimes assume that the utility of money is linear with money meaning that a plot of utility versus money forms a straight line.

result can then be stated: if assumptions 1 through 6 are met, then there is a linear utility function over the set of risky alternatives arising from a finite set of basic alternatives.

Specifically, such a utility function u is given by:

$$u(A_1) = 1,$$

$$u(A_i) = u_i, \quad \text{for } 1 < i < r \text{ (see assumption 3)},$$

$$u(A_r) = 0,$$

and

$$u(p_1A_1, \cdots, p_rA_r) = p_1u_1 + p_2u_2 + \cdots + p_ru_r,$$

where $u_1 = 1$ and $u_r = 0$ by definition.

If a and b are any two constants such that $a > 0$, then the function u', where

$$u'(L) = au(L) + b$$

for any lottery L, is also a linear utility function, as is easily shown. Technically, we call u' a positive linear transformation of u. It can also be shown that, if u^* as well as u is a linear utility function representing the ordering $\succsim$, then there exist constants a^* and b^*, $a^* > 0$, such that

$$u^*(L) = a^*u(L) + b^*$$

for all lotteries L. That is, if u^* is a linear utility function, then it is a positive linear transformation of u.

A concrete numerical example may clarify the whole procedure. Consider a person choosing among lotteries involving the four alternatives A_1, A_2, A_3, and A_4, which he prefers in the order given. Of the two lotteries,

$$L = (0.25A_1, 0.25A_2, 0.25A_3, 0.25A_4),$$

$$L' = (0.15A_1, 0.50A_2, 0.15A_3, 0.20A_4),$$

which should he choose? Suppose that we determine that he is indifferent between A_2 and $\tilde{A}_2 = (0.6A_1, 0.4A_4)$ and between A_3 and $\tilde{A}_3 = (0.2A_1, 0.8A_4)$. We know by assumption 3 that some such lotteries involving A_1 and A_4 must exist. Now, by assumptions 4 and 5 it follows that

$$L \sim [0.25A_1, 0.25(0.6A_1, 0.4A_4), 0.25(0.2A_1, 0.8A_4), 0.25A_4],$$

which according to assumption 2 simplifies to

$$L \sim (0.45A_1, 0.55A_4).$$

A similar calculation shows that

$$L' \sim (0.48A_1, 0.52A_4).$$

Thus, if this person is to be consistent with our six assumptions and at the same time have the stated indifferences between A_2 and $\tilde{A}_2$ and between A_3 and $\tilde{A}_3$, then he must prefer L' to L.

Two possible linear utility indicators are given in the following table:

Lottery	A_1	A_2	A_3	A_4	$(p_1A_1, p_2A_2, p_3A_3, p_4A_4)$
u	1.0	0.6	0.2	0.0	$p_1(1) + p_2(0.6) + p_3(0.2) + p_4(0)$
u'	1.6	0.8	0.0	-0.4	$p_1(1.6) + p_2(0.8) + p_3(0) + p_4(-0.4)$

The first of these is the one described above, and the second one is the linear transformation of it obtained by using the constants $a = 2$ and $b = -0.4$.

Given that a subject's preferences can be represented by a linear utility function, then *he behaves as if he were a maximizer of expected values of utility.* It is important to recognize that a subject's manifest behavior may be summarized by a linear utility function without his being consciously aware of making his choices in this manner. About his subconscious awareness we will not comment.

The general theory of utility is not confined to a finite set of basic alternatives nor to cases where a least or a most preferred alternative exists. We have only examined a simple special case, but one with sufficient complexity so that we can see just what is involved when we use utility theory in game theory. If one is interested in the more general theories, which are correspondingly more complicated, see the papers referred to at the beginning of this section.

2.6 SOME COMMON FALLACIES

Newcomers to modern utility theory tend to be critical of the idea, and, to be sure, there are valid reasons, but as criticisms are so often based on a fallacious understanding of the construct we have elected to point out some of the more common misinterpretations.

Fallacy 1. $(p_1A_1, \cdots, p_rA_r)$ *is preferred to* $(p_1'A_1, \cdots, p_r'A_r)$ *because the utility of the former,* $p_1u_1 + \cdots + p_ru_r$, *is larger than the utility of the latter,* $p_1'u_1 + \cdots + p_r'u_r$.

Some care must be taken to see why this is a fallacy, for there are two quite distinct ways of interpreting utility theory. First, we may think of the theory as a description of preference, in which case the causal relationship of the fallacy is the exact opposite of the truth; the preferences among lottery tickets logically precede the introduction of a utility function.   Second, we may think of the theory as a guide to consistent action. Here, again, certain (simple) preferences come first and certain rules of consistency are accepted in order to reach decisions between more complicated choices.   Given these, it turns out that it is possible to summarize both the preferences and the rules of consistency by means of utilities, and this makes it very easy to calculate what decisions to make when the alternatives are complex.   The point is that there is no need to assume, or to philosophize about, the existence of an underlying subjective utility function, for we are not attempting to account for the preferences or the rules of consistency.   We only wish to devise a convenient way *to represent* them.

Fallacy 2. *Suppose that $A \succ B \succ C \succ D$ and that the utilities of these alternatives satisfy $u(A) + u(D) = u(B) + u(C)$, then $(\frac{1}{2}B, \frac{1}{2}C)$ should be preferred to $(\frac{1}{2}A, \frac{1}{2}D)$ because, although they have the same expected utility, the former has the smaller utility variance.*

This is a completely wrong interpretation of the utility notion, and again it results from a failure to accept that preferences precede utilities. It misses the point of utility theory.   The principal result of utility theory for risk is that a linear utility index can be defined which reflects completely a person's preferences among the risky alternatives.   If the fallacy actually made sense, then it would be a beautiful example to show that a utility function is impossible.   This is not to say that, if the prizes are money, we will not find a person preferring the gamble with the smaller money variance when the expected values of money are the same.   We probably will, but this only goes to show that the utility of money (if the concept is meaningful) cannot be linear with money.

Fallacy 3. *Suppose that $A \succ B \succ C \succ D$ and that the utility function has the property that $u(A) - u(B) > u(C) - u(D)$, then the change from B to A is more preferred than the change from D to C.*

Again, if we consider how the utility function is constructed from preferences between pairs of alternatives, not between pairs of pairs of alternatives, it is clear that the above statement is not justified.   Indeed, empirically, it may well be false.   This does not mean that one should not consider constructing a theory of utility which is able to compare utility differences.   We only want to emphasize that the present theory does not permit such comparisons.   The example on p. 22 illustrates this point.

Our fourth fallacy is so important, and in many ways really an unresolved problem, that we shall treat it separately in the next section.

2.7 INTERPERSONAL COMPARISONS OF UTILITY

There is one thing which we stressed when discussing the classical attempts to devise a numerical utility for decision making under certainty which we have not adequately discussed for utility functions when there is risk: the uniqueness of the function. Under certainty, one of the difficulties was the almost complete lack of uniqueness—any order-preserving transformation of the numbers was equally acceptable. It is also true in the risky situations that any order-preserving transformation of a utility function is again a utility function, but such a transformation of a *linear* utility function does not generally result in a *linear* utility function. One must, therefore, keep in mind the class of transformations which take a linear utility function into one of the same type. As we pointed out before, the appropriate class consists of those transformations known as the positive linear ones, i.e., if u is a linear utility function over a set of risky alternatives and if a and b are any constants so long as a is positive, then $u' = au + b$ is again a linear utility function over the set. Conversely, if u and u' are two linear utility functions for a preference relation over the same set of alternatives, then there exist constants a and b, where a is positive, such that $u' = au + b$.

Another way of stating this uniqueness result is that the consistency axioms (such as assumptions 1 through 6 of section 2.5) determine a linear utility function which is unique up to its zero point and its unit. If we choose any two alternatives which are not indifferent, then we can always set the utility of the less preferred to be zero and the utility of the more preferred to be one. As we shall come to see, the non-uniqueness of the zero point is of no real concern in any of the applications of utility theory, but the arbitrary unit of measurement gives trouble. The trouble may be illustrated most easily by a fictitious example in the measurement of distances. Suppose two people are isolated from each other and each is given a measuring stick marked off in certain and possibly different arbitrary units. The one subject is given full-scale plans for an object to be constructed to the same size by the other, and he is permitted to send only messages stating angles and lengths (in his units) of the object. With such limited communication it is clearly possible for the second man to construct a scale model of the object, but it will only be of the correct size if it happens that both measuring rods are marked in the same units. Clearly, once the barriers on communication are dropped, the two men can determine with fair accuracy the relationship between their two units by measuring things they each have and which are known to have about the same size, e.g., the span of a hand, or the width of a trouser leg, etc.

In utility theory, if we should want to compare utilities between two people much the same problem exists: we do not know the relationship between the two units. The big difference between utility and length measurement is that we do not seem to have any "outside thing" which can be measured by both persons to ascertain the relation between the units. Certain proposals for an "outside" standard of unit utility have been offered; for example, it has been suggested that for each person in the situation his most preferred alternative be assigned the value 1 and his least preferred the value 0. Often, however, this seems to fail to capture one's intuitive idea of an interpersonal comparison of utility: in a gamble between a rich man and a poor one which involves money in the range of $-\$1$ to $+\$1$, it is hard to believe that a gain of \$1 should have the same utility for each of them. In some sense, the poor man is far more "sensitive" to a fixed monetary change than is the rich one, and his preference for it is correspondingly more "intense." Just exactly what this means we do not know, but it seems to mean something to each of us. We are forever trying to decide whether one outcome means more or less to another person than a different outcome means to us. For more discussion of these matters in the context of game theory, where they play havoc, see Chapters 6, 7, 8, and 14.

Thus, the fact that a linear utility function is defined only up to a linear transformation leads to the problem of interpersonal comparisons of utility when there is more than one person in the situation. Since it is not solved, one can either assume that such comparisons are possible, knowing that this creates (at least at present) an Achilles' heel in the theory, or one can attempt to devise theories in which comparisons are not made. Both approaches have been taken in game theory.

*2.8 EXPERIMENTAL DETERMINATIONS OF UTILITY

Given such an axiomatic theory as that presented in section 2.5, can we find the utility function for an individual in a given situation? If the question is meant naively, the answer is surely No. If it is refined in various ways, the answer is Maybe—at least several people have tried. We do not propose to examine such work in detail but rather to discuss the general problem very briefly and to guide the reader to the (limited) literature on such problems. For a more detailed guide to this literature, plus some of the purely theoretical work, up to the beginning of 1954, see Edwards [1954 c].

The most obvious difficulty in attempting to confirm the theory is that it rests upon an infinity of paired comparisons, but there are others too. Obviously, in an experiment one will only make a relatively few paired comparisons, so one way or another the verification will have to be based upon these. One procedure is to determine the utilities of a few alternatives experimentally, using the assumption that a utility function exists and that it is linear. Suppose we arbitrarily assign utilities of 0 and 1 to two alternatives and then we determine the utility of a third by finding the lottery of the first two which is indifferent to it. For example, suppose $C \sim [pA, (1 - p)B]$; then by linearity $u(C) = u[pA, (1 - p)B] = pu(A) + (1 - p)u(B) = p$. If this is done for several more points, then one soon knows enough values on the utility scale (assuming it exists) to make predictions. For example, suppose it was found that $u(D) = q$. Then we could predict whether the lottery $[rA, (1 - r)C]$ is preferred or not to the lottery $[sD, (1 - s)B]$ for a particular choice of r and s. If these predictions are confirmed experimentally, we then have some confidence that we have obtained a portion of the utility function. This is the method which was used by Mosteller and Nogee [1951].

A more elegant, if more difficult, alternative to starting with a model having an infinity of comparisons is to devise one in which only a finite number are to be made. Such a model must be quite different from the one we have described if it is to lead to a linear utility function which is unique up to a linear transformation. However, such axiom systems are possible as has been shown by Suppes and Winet [1955] and for this case Davidson, Siegel, and Suppes [1955] have devised an experimental setup in which it can be checked. This work is probably the most experimentally elegant in the area, and the results have been very encouraging.

A second difficulty in attempting to ascertain a utility function is the fact that the reported preferences almost never satisfy the axioms, e.g., there are usually intransitivities. Furthermore, if the same pair is offered several times, then in some cases the subject will not be consistent in his reports. One cannot expect the data to fit the model perfectly, but how does one determine which model they fit most closely and how does one measure how good the agreement is? Such problems pose the following intriguing and important statistical problem: to formulate a model which assumes that a subject is actually (or latently or genotypically) a von Neumann-Morgenstern utilitist in the sense of "having" a linear utility function, but that his responses yield this underlying order confounded by random disturbances and errors. We need not consider the question whether or not a true utility function exists, for we may take the pragmatic approach that such a postulate makes more precise what we shall mean by a "reasonable," or "approximate," or "realistic" fit of the data to theory. To date, little has been published on this problem.

A third, and possibly the most puzzling, difficulty arises from the basic probabilities in the model. We have, in our discussion, identified them with certain physical experiments, and certainly in a normative theory this is what one would want to do. But, if we are trying to describe behavior, it may be unreasonable to suppose that people deal with objective probabilities as if they satisfy the axioms of the calculus of probabilities or that they only cope with situations in which objective probabilities are defined. This leads one to consider introducing the idea of subjective probabilities, on the basis of which people are assumed to act. If such exist, little is known about them—how they combine with one another, how they interact with the utility values, how they are related to the objective probabilities, etc. Edwards (see references [1953, 1954]) has run a series of experiments on this question, and he has considerable evidence to support the view that people react in, shall we say, strange ways to objective probabilities. In their experimental work, Davidson, Siegel, and Suppes had to work with an event having subjective probability $\frac{1}{2}$ and they found that many of the obvious things having objective probability $\frac{1}{2}$ would not do.

While writing this book one of us became intrigued with these last two problems—the probabilistic nature of preferences and the role of subjective probability—and devised a probabilistic theory of utility which closely parallels some models of psychophysical discrimination. For example, subjective probability is axiomatized by properties somewhat similar to those of objective probabilities, but it is then shown that subjective probability possesses the defining property of a subjective (Fechnerian) sensation scale as the term is used in psychophysics. Since this discrimination model of utility seems particularly amenable to experimental verification and is of some conceptual interest, we have formulated the axioms and presented the principal results in Appendix 1.

There can be no question that it is extremely difficult to determine a person's utility function even under the most ideal and idealized experimental conditions; one can almost say that it has yet to be done. Indeed, should it be done? Since we think it should, let us consider it carefully.

If it is so difficult to determine utility functions under the best of conditions, there is certainly no hope at all that it can be done under field conditions for situations of practical interest. Thus, if the theories built upon utility theory really demand such measurements, they are doomed practically; if they can be useful without making such measurements, then why go to the trouble of learning how? As in the physical sciences, we would claim that a theory may very well postulate quantities which cannot be measured in general, and yet that it will be possible to derive some conclusions from them which are of use. To be sure, if the measurements could be made, more could be concluded; but this is not the same as saying that, if the measurements cannot be made, nothing can be concluded. We therefore move on to the second part of our conditional question:

why, then, make any measurements in the laboratory? The main purpose is to see if under any conditions, however limited, the postulates of the model can be confirmed and, if not, to see how they may be modified to accord better at least with those cases. It will still be an act of faith to postulate the general existence of these new constructs, but somehow one feels less cavalier if he knows that there are two or three cases where the postulates have actually been verified.

Every indication now is that the utility model, and possibly therefore the game model, will have to be made more complicated if experimental data are to be handled adequately. Although one such complication of the utility model is discussed in Appendix 1, its domain of applicability is limited and it is completely unclear how it can be utilized in game theory. Furthermore, neither it nor any of the present utility models take into account the intuition, now bolstered by a staggering amount of empirical data for a wide variety of psychological dimensions (see Miller [1956] for a partial survey), that people rarely categorize a single dimension into more than seven or so distinct levels. The major exceptions seem to be cases where the culture provides a simple, fine, and unambigious scale, such as money. Since, however, most decisions, even when money is a factor, are not based entirely on monetary considerations, discrete categorization of preferences may be the basic case to study. As no theoretical work has been carried out on such problems, we can only turn in the following chapters to what has been done to give a model for conflict of interest within the present utility framework.

2.9 SUMMARY

The primary purpose of this chapter was to introduce the central ideas of modern utility theory, which is a cornerstone of much decision theory. As background, we classified decision making according to whether a decision is reached by an individual or a group, and whether it is effected under conditions of certainty, risk, uncertainty, or a mixture of uncertainty and risk in the light of experimental evidence. Using these categories, we described the general structure of the book.

Decision making under certainty encompasses much of formal theory in social science. This problem can be viewed as follows: given a set of possible acts, to choose those which maximize (or minimize) a given index. In many traditional applications, the solution can be formulated in the language of the calculus, but some of the more interesting modern problems require more sophisticated techniques. Although we did not go into decision making under certainty, the example of linear programing was discussed briefly because of its close relation to game theory and its inherent importance in applications.

This led to the question whether any individual decision problem can be represented by a numerical index called utility. This is not only possible but also possible in a great many ways, provided the preference relations are transitive. Historically there have been so many misuses of this representation that it has been totally discredited. However, once decision making under risk became an issue the idea of a numerical utility reappeared. Since numbers would have to be attached to the infinity of possible gambles, it seemed conceivable that there might be sufficient constraints on the index to make it unique, or nearly so, thus avoiding some of the troubles usually associated with utility.

To achieve such a result, it is necessary that the preference relation meet certain more or less plausible consistency requirements. A set of axioms, closely related to those given by von Neumann and Morgenstern, were stated and discussed. Among the more important requirements were these: preference shall be transitive, i.e., if A is preferred to B, and B to C, then A is preferred to C; any gamble shall be decomposed into its basic alternatives according to the rules of the probability calculus; and if A is preferred to B and B to C, then there shall exist a gamble involving A and C which is judged indifferent to B. From these and other axioms it was shown that numbers can be assigned to the basic alternatives in such a fashion that one gamble is preferred to another if and only if the expected utility of the former is larger than the expected utility of the latter. If u is such an index, any other is related to it by a linear transformation, i.e., there is a positive constant a and a constant b such that $au + b$ is the second index. Such an index u is called a *linear utility function*, where "linear" means that the utility of a gamble is the expected value of the utilities of its components.

Certain cautions must be maintained in interpreting this concept: One alternative possesses a larger utility than another because it is more preferred, not the other way round. All the preference information is summarized by the expected value of utility—in particular, utility variance has no meaning. The fact that one utility difference is larger than another does not permit us to say that the one change is subjectively larger than the other, for the utility function was constructed in terms of subjective responses to gambles, not in terms of the subjective evaluation of two different changes. Since neither the zero nor the unit of a utility scale is determined, it is not meaningful in this theory to compare utilities between two people.

The chapter closed with a brief sketch of some of the experimental problems associated with utility theory, and references were given to several experimental studies. It was suggested that a less idealized theory is needed, one more amenable to empirical study.

chapter **3**

EXTENSIVE AND
NORMAL FORMS

3.1 GAME TREES

The mathematical abstraction of a game assumes the three forms we mentioned in the introduction: the extensive, the normal, and the characteristic function forms. The first is our topic throughout most of this chapter; it is an attempt to capture the salient features of certain conflicts of interest, such as those found in a parlor game. The rules of any parlor game specify a series of well-defined *moves*, where each move is a point of decision for a given player from among a set of alternatives. The particular alternative chosen by a player at a given decision point we shall call the *choice*, whereas the totality of choices available to him at the decision point constitutes the *move*. A sequence of choices, one following another until the game is terminated, is called a *play*. These familiar words, particularly the word "move," are being used in somewhat unfamiliar ways. For example, in chess and other board games the word move is used in two ways which differ from our meaning. It is sometimes used to refer to the physical act of moving a piece from one position to another. It is also used in such phrases as "the fifth move." This refers to the set of all possible moves in our sense following any of the possible sequences of four choices from the beginning of the game. Similarly, by a "play" we do not mean the act of participating once in the game, as is intended in ordinary language, but rather a detailed statement of the actual decisions made.

Let us suppose that in one game (at some stage of a play) player 1 has to choose among a king of hearts, a two of spades, and a jack of diamonds, and that in another game a player, also denoted 1, has to choose among passing, calling, and betting. In each case the decision is among three alternatives, which may be represented by a drawing as in Fig. 1.

But how can these two examples be considered equivalent? Certainly it is clear from common experience that one does not deal with every three-choice situation in the same way. One might if they were given

FIG. 1

out of context, for there would be no other considerations to govern the choice; but in a game there have been all the choices preceding the particular move and all of the potential moves following the one under consideration. That is to say, we cannot truly isolate and abstract each move separately, for the significance of each move in the game depends upon some of the other moves. However, if we abstract all the moves of the game in this fashion and indicate which choices lead to which moves, then we shall know the abstract relation of any given move to all other moves which have affected it, or which it may affect.

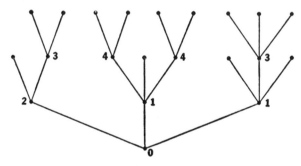

FIG. 2

Such an abstraction leads to a drawing of the type shown in Fig. 2. The number associated with each move indicates which player is to make the choice at the move, and therefore these numbers run from 1 through n when there are n players. In the example of Fig. 2, $n = 4$, and we see that all the moves, save the first, are assigned to one of the players; the first move has 0 attached to it. A move assigned to "player" 0 is a chance move, as, for example, the shuffling of cards prior to a play of poker. To each chance move, which need not be the first move of the game, there must be associated a probability distribution, or weighting, over the several alternative choices.   If a chance move entails the flipping of a fair coin, then there are two alternatives at the move and each will occur with probability ½.

A drawing such as Fig. 2, when considered abstractly, i.e., as a mathematical system, is known as a *connected graph*.   A connected graph consists of a collection of points (called nodes or vertices) and branches between certain pairs of nodes such that a path can be traced out from each point to every other point.   A graph may have closed loops of branches, such as *abca* or *abdeca* in Fig. 3.   A connected graph with no such loops of branches is called a *tree*.   The graph of a game is a tree, which is called the *game tree*.   It may not seem reasonable to assume the graph of a game is a tree, for in such games as chess the same arrangement of pieces on the board can be arrived at by several different routes, which appears to mean that closed loops of branches can exist.   However, in game theory we choose to consider two moves as different if they have different past histories, even if they have exactly the same possible future moves and outcomes. In games like chess this distinction is not really

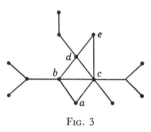

Fig. 3

important and to make it appears arbitrary, but in many ways the whole conceptualization and analysis of games is simplified if it is made.   The tree character of a game is not unrelated to the sinking feeling one often has after making a stupid choice in a game, for, in a sense, each choice is irretrievable, and once it is made there are parts of the total game tree which can never again be attained.

The tree is assumed to be finite in the sense that a finite number of nodes, and hence branches, is involved.   This is the same as saying that there is some finite integer N such that every possible play of the game terminates in no more than N steps.   Such is certainly true of all parlor games, for there is always a "stop" rule, as in chess, to terminate stalemates.   To say the tree is finite is not to say that it is small and easy to work with.   For example, card games often begin with the shuffling of a deck of 52 cards, and so the first 0 move has 52!, i.e., approximately 8.07×10^{67}, branches stemming from it.   Clearly, for such games no one is going to draw the game tree in full detail!

3.2 INFORMATION SETS

The next step in the formalization of the rules of a game is to indicate what each player can know when he makes a choice at any move.   We are not now assuming what sorts of players are to be postulated in game theory but only what is the most that they can possibly know without violating the rules of the game. Clearly, there is the possibility that the rules of the game do not provide a player with knowledge on any particular move of all the choices made prior to that move. This is certainly the situation in most card games which begin with a chance move, or where certain cards are chosen by another player and placed face down on the table, or where the cards in one player's hand are not known to the other players. Indeed, it may be that a player at one move does not know what his domain of choice was at a previous move! The most common example of this is bridge where the two partners must be considered as a single player who alternately forgets and remembers what alternatives he had available on previous moves (see sections 7.4 and 7.5).

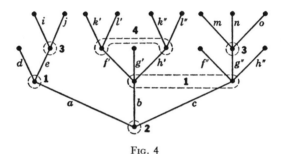

Fɪɢ. 4

To suggest a method to characterize the information available to a player, consider a game whose tree is that shown in Fig. 4. The dotted lines enclosing one or more nodes are something new in our scheme; as we shall see they can be used to characterize the state of information when a player has a move. Let us suppose the rules of this game assert that on move 1 player 2 must choose among three alternatives, denoted a, b, and c. Regardless of player 2's choice, player 1 has the second choice in each play. We shall suppose that the rules of the game permit player 1 to know whether or not player 2 selected choice a. If player 2 chooses b, the rules are such that player 1 can only know that either b or c was chosen but not which. Although verbally this may seem complicated, graphically all we need do is enclose in a dotted line those moves of player 1 which end up on b or c. The dotted line simply means that from the rules of the game the player is unable to decide where he is among the enclosed moves. The single move at the end of choice a is also enclosed, for if that choice is made player 1 knows it. If choice b was in fact made, and if player 1 then makes choice f' (of course, he does not know whether he is making f' or f'') the next move is up to player 4. Note that according to the diagram, the rules of the game make it impossible for him to determine whether he is choosing between k' and l' or between k'' and l''.

In general, the rules of any game must specify in advance which moves are indistinguishable to the players—the sets we have enclosed in dotted lines. Abstractly, there are two obvious necessary features to these sets of moves—which are known as *information sets*. Each of the moves in the set must be assigned to the same player, and each of the moves must have exactly the same number of alternatives. For if one move has r alternatives and another s, where $s \neq r$, then the player would need only count the number of alternatives he actually has in order to eliminate the possibility of being at one move or at the other. A third condition, which may be less obvious, is also assumed, namely: a single information set shall not contain two different moves of the same play of the game tree. The reason for this condition is that it seems to be generally met in practice, and having it makes the theory simpler.

Returning to Fig. 4, consider player 1's information set which has two moves. Since they are indistinguishable, each choice on one move must have a corresponding choice on the other move. It is convenient in these diagrams to pair them systematically, so f' corresponds to f'', g' to g'', and h' to h''. It is clear that this correspondence can be generalized to information sets having more than two moves and other than three alternatives at each move.

When an information set consists of a single move, the player is totally informed in that he knows exactly where he is on the tree. When all the moves are of this type, we say that the game has *perfect information*. Ticktacktoe and chess are examples of games whose rules result in perfect information.

3.3 OUTCOMES

The final ingredient given by the rules of the game is the outcome which occurs at the end of each play of the game. Almost anything may be found to be the outcome of some game; for example, the subjective reward of victory in a friendly game, or the monetary punishment of seeing someone else sweep in the pot, or death in Russian roulette. In any given system of rules for a game there is some fixed set of outcomes from which specific ones are selected by each of the plays. Each of the end points of the game tree is a possible termination point of the game, and it completely characterizes the play of the game which led to that point, for there is only one sequence of choices in a tree leading to a given end point from a fixed first move. We may index these end points and denote a typical one by the symbol α. Now, if Ω is the set of outcomes, the rules of the game associate to each α an outcome from Ω which we may denote by $\omega(\alpha)$. For example, in a game like ticktacktoe the set of outcomes is {player 1 loses and player 2 wins, player 1 wins and player 2 loses, draw}. In this case, and in a wide class of games, it would be sufficient to state the outcomes for only one of the players, but in other situations which are not strictly competitive it is necessary for the elements of the outcome set to describe what happens to each player.

In summary, then, the rules of any game unambiguously prescribe the following:

i. *A finite tree with a distinguished node (the tree describes the relation of each move to all other moves, and the distinguished node is the first move of the game).*

ii. *A partition of the nodes of the tree into $n + 1$ sets (telling which of the n players or chance takes each move).*

iii. *A probability distribution over the branches of each 0, i.e., chance, move.*

iv. *A refinement of the player partititon into information sets (which characterizes for each player the ambiguity of location of the game tree of each of his moves).*

v. *An identification of corresponding branches for each of the moves in each of the information sets.*

vi. *A set Ω of outcomes and an assignment ω of an outcome $\omega(\alpha)$ to each of the end points α (or plays) of the tree.*

3.4 AN EXAMPLE: THE GAME OF GOPS

Gops, which stands for "game of pure strategy" and which also goes under the name of goofspiel, is a two-person card game which, in addition to being easy to describe and illustrative of the preceding formulation, is amusing to play and very revealing of some of the problems and ideas of game theory. A deck of cards is divided into suits, one of which (say clubs) is discarded, a second (spades) is shuffled and placed face down on the table, and each of the two players has, as his hand, a complete suit. In the course of play the spades are turned over one by one and each is captured by one of the players. They are valued: ace = 1, numbered cards their numerical values, jack = 11, queen = 12, and king = 13. The player capturing the larger total value of spades wins. Since the value of the entire suit is 91, the winner must capture 46 or more except when there are ties. The procedure of play is as follows: The first spade is turned over so that the players can see it. Each player then selects whatever value card he wishes from his hand, and these are shown simultaneously; the one having the higher value takes the spade which was showing. But, if they are of equal value, then the victor on the next move takes both spades. If a draw does not occur, the first spade is captured, and the second spade is turned over and the process repeated, the only difference being that each hand is now depleted by the card used for the first spade.   The process continues until the cards are exhausted or until one of the players has spades totaling at least 46 points.   Two variations are possible which from the point of view of game theory are not different, but, given the fallibility of human memory, tend to turn out differently.   Either the cards played from the player's hands are turned face down after each spade is captured, or they remain showing.   The game seems to be more taut with the latter procedure, since the outcome depends entirely upon wit and chance, not upon memory.

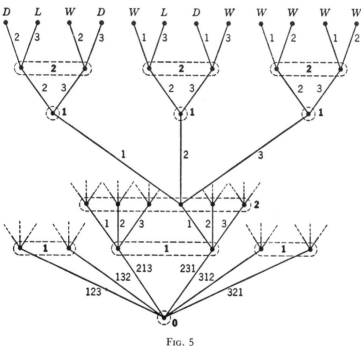

Fig. 5

To get an idea of the interest of the game, suppose the first card turned over is a 10, i.e., a fairly valuable card.   One can be certain of not losing it by playing the king, but this places one in a weakened condition when later the king or queen of spades arises.   If one plays a jack, then, although there is a fair chance of getting the 10, it is an expensive loss if one's opponent plays the queen.   If you play a 6 and if the opponent plays a queen, he wins the 10 at considerable expense to himself and little to you; on the other hand, if he plays a 7 he has won a lot at little cost to himself. Try it to get an idea of the complex reasoning required.

Clearly, there is nothing inherent in these rules requiring suits of 13 cards, so in drawing part of the game tree of Gops let us suppose the suits have only three cards—this makes the game dull, but it illustrates the principles just as well.   The first move of the game is a chance one—the shuffling of the deck—from which there are $3! = 6$ branches, each having a probability of $\frac{1}{6}$ of occurring.   Next, the top card is turned up.   It has one of three values, and the remaining two cards are in one of two possible

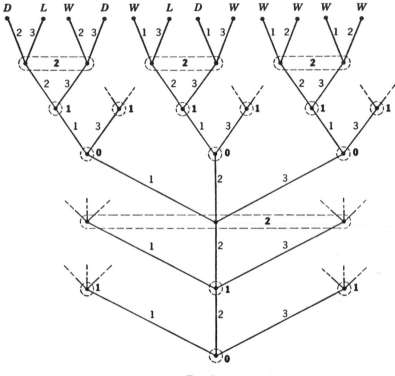

Fig. 6

orders, so for player 1 there are three information sets of two moves each, as seen in Fig. 5.   Rather than fill out the game tree in full, which would be messy, we shall extend it from just one of these information sets.   At this point, player 1 must choose among the three cards in his hand.   Since he does not show his card until player 2 has made a selection, all the possible end points are enclosed in a dotted line.   Player 2 must then select one of his three cards.   When this is done, both players turn over the cards and the next card in the deck is turned over.   Since there are only three cards in the deck this means that there is no longer any ambiguity as to location of the play on the game tree.   The next move is up to player 1 who must select one of two cards, and the final move is up to player 2 who in the absence of knowledge of player 1's choice must choose one out of two cards. Naturally, the outcome of the last card is completely determined by the previous moves. The play is then terminated, and the victor is the player having the largest sum of cards from the deck. Since the sum is 6, there is the possibility of draw as well as win and lose. The numbers alongside the branches denote the card chosen, and W on an end point means player 1 wins, D means a draw, and L that he loses. In other words, the set of outcomes consists of three elements for each player: W, D, and L.

▶ It should be noted that for a given set of rules there is not, in general, a unique tree representing them. For example, it is clearly immaterial whether we put player 1's move before or after player 2's move on the tree, for the information sets are so chosen to make them simultaneous in effect. A more profound change in the tree can be effected by replacing the single chance move by two chance moves. We can either view chance as selecting with probability $\frac{1}{6}$ one of the six permutations of the three cards, or, as is shown in Fig. 6, we may view the first chance move as selecting one of three cards with probability $\frac{1}{3}$, and after players 1 and 2 have each taken a move chance selects one of the two remaining cards with probability $\frac{1}{2}$. The two representations are equivalent to the rules of the game. ◀

3.5 EXTENSIVE FORM

So far we have characterized what we shall mean by the rules of the game, but no players have really been introduced. It is true that we have spoken of players who will make choices at the moves, but they cannot be considered to be introduced until they are characterized by the properties which describe their behavior. It is clear that we may have the same set of rules but very different conflict of interest situations—games—depending upon the nature of the players. In total we shall make three assumptions about our players; these assumptions will characterize them.

The first assumption is that each player has a pattern of preferences over the set of outcomes which satisfies the axioms of utility theory. It will be recalled that we established in Chapter 2 that if this is the case then we can assign a numerical and linear utility function to the outcomes. More formally, we suppose:

vii. *For each player i, there is a numerical and linear utility function M_i defined over the set of outcomes Ω.*

It is easy to see that we can combine conditions (vi) and (vii) into a single condition, for by (vi) an outcome is assigned to each play and by (vii) a utility to each outcome, hence a utility, which we may denote by

$M_i(\alpha)$, is assigned to each play α. The function M_i defined over plays is known as player i's *payoff function*. If we formally combine these conditions we have

vii'. *For each player i there is a numerical and linear utility function M_i defined over the set of end points of the game tree.*

It is important to recognize that M_i is defined over the outcome set and not just over the direct returns received by player i; it is a response to the total situation and may therefore take into account what happens to other players. In this fashion, such notions as altruism and spite are included.

The system consisting of the first five parts of the rules of the game and (vii') is known as a *game in extensive form*. We observe that it is not identical to the rules of the game for it includes the preference patterns of the individuals participating in the game, and so introducing different individuals into a situation having the same rules will, in general, yield a different extensive form.

The original description of a game in extensive form (see von Neumann and Morgenstern [1947]) differs somewhat from and is less compact than this one, which we have paraphrased from the formal definition given by Kuhn [1953 *b*].

In this extensive description of a game all the subtle differences between games are apparent. No matter how intuitively similar two games are, if there is a formal difference as given by the rules or if the rules are the same but there are differences in individual preferences it will show up at this level of abstraction. At such a detailed level the problems of analysis seem overwhelming. This is particularly true for many conflicts of interest arising in economics and the other social sciences, where, at the intuitive level, the moves do not fall into a pattern of well-specified, temporally ordered moves. In these situations, the timing of decisions plays an important role. Now, it is true that such timing problems can be cast into the game mould, either by admitting an infinity of alternatives at each move or by subdividing time into sufficiently small discrete units. In the second case, a finite game in extensive form results if, for example, we suppose that each player decides at each second whether to make one of a set of "positive" choices or not to take any action at that point (which is itself a choice). It is not difficult to see that such a construction reduces any timing problem into an extensive game, but it is also clear that even the simplest such situation leads to an exceedingly complex game tree.

A limited amount of theoretical effort has been spent on games in extensive form. One group of papers has been concerned with the conditions on the extensive form necessary to ensure the existence of a pure strategy equilibrium point (see sections 4.6 and 7.8). Another, which we shall not discuss, has been devoted to defining the notion that two extensive games are equivalent; the reader is referred to Dalkey [1953], Krentel, McKinsey, and Quine [1951], McKinsey [1950 *a*], and F. B. Thompson [1952]. A third group, which is discussed at some length in sections 7.4 and 7.5, introduces an intuitively natural concept of strategy which differs from the one usually used, and the relations between these two strategy concepts are explored. Finally, Berge [1953 *a*, *b*] has proposed a somewhat different concept of two-person extensive games which, among other things, need not have a finite number of alternatives at each move nor a finite number of moves.

3.6 RATIONALITY AND KNOWLEDGE

As we said in the last section, the players in the game are to be characterized by three assumptions, only one of which has so far been given, namely: each player has preferences over the outcomes which meet the axioms of utility theory. The other two assumptions concern what the players know and the basis on which they arrive at decisions.

Let us take up the question of knowledge first. It is assumed that:

viii. *Each player is fully cognizant of the game in extensive form, i.e., he is fully aware of the rules of the game and the utility functions of each of the players.*

It hardly seems necessary to point out that this is a serious idealization which only rarely is met in actual situations. If one systematically examines the several features of a game in extensive form, it is clear that human beings do not generally have the knowledge assumed.

Is there, then, any reason to pursue the theory of games further except as a mathematical topic; can it possibly have any relevance to social science? The answer Yes can be supported by three distinct reasons. First, the theory can be used normatively to tell a person that this is the knowledge he should acquire, and, once he has it, the theory establishes the decisions he should make in order to achieve certain specified ends. Second, the knowledge assumption does not seem nearly so grave when we move to the next level of abstraction—the normal form. Since most of the theory is actually at that level, we should not yet abandon it. Third, it may be possible to weaken the knowledge assumptions of the classical game model. In section 12.4 we shall briefly examine a model where each of the players is supposed to have "misperceptions" concerning the utility functions of the other players; in that model each player has only partial knowledge of the entire postulated structure. We do not want to underestimate the difficulties arising from this knowledge assumption with respect to social science applications, but we do argue that one should examine more of the theory before making a judgment.

Finally, we must consider a "law of behavior" for the players. It should be noted that nowhere in the eight conditions so far stated have we introduced any principle of behavior; rather we have simply described a set of conditions under which any law we postulate must be operative. Recall that for each of the players the word "preference" has already entered as the underlying relation giving rise to the utility functions. We then postulate that

ix. *Of two alternatives which give rise to outcomes, a player will choose the one which yields the more preferred outcome, or, more precisely, in terms of the utility function he will attempt to maximize expected utility.*

The logical quality of this postulate bears some consideration. We shall take it to be entirely tautological in character in the sense that the postulate does not describe behavior but it describes the word "preference." With this interpretation, the problem is not to attempt to verify the postulate but rather to devise suitable empirical techniques to determine individual preferences. It is not the least bit obvious that the "preference" implicitly defined in (ix) will be the same as the "preference" defined by an arbitrarily given set of experimental operations, such as asking a person to state his preferences among the set of outcomes. The alternative to the tautological interpretation is to accept certain experimental operations as defining "preferences" and then to attempt to verify postulate (ix). This is basically much simpler for the experimentalist, but experience indicates that it is not always successful.

The social scientist will immediately recognize that this is not an entirely happy state of affairs, even if we take postulate (ix) to be tautological. To make applications it will be necessary to devise ways to determine those preferences which satisfy (ix). Since exhaustive testing of subjects is impractical outside the laboratory, if even there, a psychological theory is needed which will allow the prediction of utility functions from a knowledge of the objective payoffs and from a relatively few measurements or observations of the subjects. These underpinnings simply do not exist at present, and it remains to be seen whether experimental psychologists working with utility theorists can provide a satisfactory theory.

Postulate (ix) is often described as a postulate of rational behavior. This is all right provided only that one does not impute to rationality more than is contained in postulate (ix); there is a tendency for the word to assume a derogatory character which results from forgetting the tautological character of the postulate. Of course, if one attempts to identify utility with some objective measure of the outcome, such as money, then people are not generally rational in the sense of satisfying postulate (ix). But this is irrelevant; it merely implies that the preference patterns of people are not simply related to expected money returns.

3.7 PURE STRATEGIES AND THE NORMAL FORM

One way to ascertain the outcome of a game in extensive form is to let the players play it and observe the outcome.   Indeed, many would say this is the only way, but they would be wrong, for in principle we could cause each player to state in advance what he would do in each situation which might arise in the play of the game.   From this information for each of the players, an umpire could carry out the play of the game without further aid from the players and thereby determine the payoffs.   Such a prescription of decision for each possible situation is known as a *pure strategy* for a player.

For many games the actual preparation of a pure strategy in a form an umpire could use without ambiguity is a hopeless task; however, certain simple examples of pure strategies are easily given, though in general they would be poor ways to play.   For example, if we suppose that each branch stemming from a move is given a number, 1, 2, $\cdots$, r, where r is the number of branches, then one pure strategy is to take branch 1 at each move.   Another is to take the branch with the largest number.   Indeed, if player i has q different information sets, which we may number 1, 2, $\cdots$, q, then any pure strategy can be represented by a set of q numbers, where the jth number represents the branch chosen when, and if, the play reaches the jth information set.   Thus q-tuples of integers,

$$(y_1, y_2, \cdots, y_q),$$

represent pure strategies.   For example, the strategy in which branch 1 is always taken is represented by

$$(1, 1, \cdots, 1).$$

But each y_j has as its range only a finite number of integers, since each move has only a finite number of branches, and there are only a finite number of y's, namely q, so there are only a finite number of strategies. Without any loss, we may label the strategies by numbers 1, 2, $\cdots$, t, where t is the total number of strategies available to the player.   The number t is finite, but it need not be small.   A game having but 10 information sets for a player and 10 branches at each set is exceedingly simple, but it can have millions of essentially different strategies for the player.

▶ The formulation and use of strategies can be simply illustrated for the game of gops with three card suits (see section 3.4). Let us suppose player 1 employs the following pure strategy which is independent of the behavior of player 2:

Card Appearing in Deck	Card Played
1	2
2	3
3	1

For player 2 we shall suppose a more complex strategy in that it depends upon player 1's choice in the first move: for the first card turned up, player 2 matches it, and for the second card turned up he plays the larger of his remaining two cards if player 1 used card 3 on the first move; otherwise he plays the smaller one. From these we can ascertain what happens for each of the six possible arrangements of the deck:

Deck	Player 1	Player 2	Cards Won by Player 1	Outcome for Player 1
123	231	123	1, 2	Draw
132	213	123	1, draw on 2	Lose
213	321	231	2, draw on 3	Win
231	312	231	2, 1	Draw
312	123	312	1, 2	Draw
321	132	312	2, draw on 1	Lose ◀

Let s_i be the label for player i's typical strategy choice, and S_i be the label for the set of all his possible strategy choices. Now, as we pointed out, when each player has selected a strategy, an umpire is in a position to play the game and to determine the payoffs. That is to say, from the given payoffs of the game in extensive form we may determine a payoff function defined over n different variables, the ith having as its domain the strategies available to player i. First, if there are no chance moves in the extensive game, the selection of the strategies $(s_1, s_2, \cdots, s_n)$, one strategy selected by each player, determines a play α. So we define i's utility for this strategy complex as

$$M_i(s_1, s_2, \cdots, s_n) = M_i(\alpha).$$

If, however, there are chance moves, then the selection of the strategies $(s_1, s_2, \cdots, s_n)$ does not uniquely determine a play, but rather there is a probability distribution over all possible plays (of course, the probability for some plays may be 0). Denote by $p(\alpha)$ the probability of play α occurring. Now, as the payoff associated with the strategies $(s_1, s_2, \cdots, s_n)$ we take the expected value[1] over all the plays, that is,

[1] There has been some misconception that numerical utility is not needed at this point but only when the notion of mixed strategies is introduced, and that as far as pure strategies are concerned we may still deal with orderings of preferences. But, in the case where there are chance moves, a linear utility function is necessary if we are to

$$M_i(s_1, s_2, \cdots, s_n) = \sum_\alpha p(\alpha) M_i(\alpha).$$

We are justified in forming this sum because we have assumed the outcomes are measured in a linear utility, i.e., if outcomes α and β have utilities $M_i(\alpha)$ and $M_i(\beta)$, then the utility of the outcome in which either α or β arises, the former with probability p and the latter with probability $1 - p$, is equal to $pM_i(\alpha) + (1 - p)M_i(\beta)$.

Observe that by means of the strategy notion every game in extensive form has been reduced to a game of the following form: each player has exactly one move (a choice among his several strategies), and he makes his choice in the absence of any certain knowledge about the choices of the other players. The payoff to the players is determined from the functions M_i and the values of s_i. This is a reduction of every game to a simple standard form which is called the *normal form* of a game.

What sleight of hand is this? We began by abstracting parlor games and arrived at the extensive form of a game, which, in general, led to an oppressively complex game tree. For games of any reasonable complexity, the number of possible trees and of variations arising from different information sets immediately led us to believe that there is little hope of finding detailed classifications of games in extensive form or of analyzing player behavior at that level. Then, by introducing the idea of a pure strategy, we suddenly reduce all games to a comparatively simple standard form. Thus, the sleight of hand is to trade the conceptual complexity of a game tree for the numerical tedium of listing all available strategies.

The reduction of any specific game, except the simplest, to normal form is a task defying the patience of man; but, since the normal form of *all possible games* is comparatively simple, one may hope to carry out successfully a mathematical examination of all possible games in normal form. The study of specific games may be close to impossible, but it may now be quite feasible to classify, analyze, and determine the features of all games. For some empirical purposes that may be sufficient.

We see that the normal form of the game is exactly the general problem which was evolved and discussed in the introduction to the book: each player has limited control over the variables which determine what he shall receive, and each wishes to maximize his return.

3.8 SUMMARY

More than half of this chapter was taken up with a detailed description of the concepts required to present a conflict of interest in extended form.

---

assign a payoff to a selection of strategies, and, without such a payoff, the development of the theory would be considerably more complex.

The simplest prototype for the abstraction is a parlor game, but other conflicts extending in time can also be formulated in the same framework. The formal structure of the conflict, the rules of the game, were presented separately from the assumptions about the players involved.   In essence, the rules give all the formal relations among the players: who moves when, what information he has when he moves, what alternatives are available to him, and the ultimate outcome to each sequence of choices.   More formally, the *rules of the game* consist of:

i. A finite tree (whose nodes represent the moves, and whose branches represent the alternatives available at each move) with a distinguished node (which tells which move occurs first).

ii. A labeling of each move into one of $n + 1$ classes, which tells which of the n players, or chance, takes that move.

iii. A probability distribution over the branches of each move assigned to chance.

iv. A partition of each player's moves into subsets (called *information sets*) of moves which he cannot distinguish from one another because of imperfect information about what happened on previous moves.

v. An identification of corresponding branches for all the moves in each of the information sets (such must exist, otherwise the moves in an information set could be distinguished).

vi. An assignment of an outcome from a given set of possible outcomes to each end point of the tree, i.e., to each possible sequence of choices.

We next introduced the three properties which are taken to characterize the players.   First, it was assumed that each player i has a pattern of preference over the outcome set which satisfies the von Neumann and Morgenstern utility axioms.   Hence, his preferences can be represented by a linear utility function M_i.   This assumption, (vii), was combined with (vi) of the rules to give:

vii′. For each player i, there is a linear utility function, known as his *payoff function*, defined over the end points of the game tree.

The first five rules plus assumption (vii′) form what is known as a *game in extensive form*.

The remaining two assumptions were:

viii. Each player is assumed to have full knowledge of the game in extensive form: not only is he assumed to know the rules of the game in full detail, but also the payoff functions of the other players.

It was pointed out that this is a particularly strong assumption.

ix. Each player is assumed to be "rational" in the sense that, given two alternatives, he will always choose the one he prefers, i.e., the one with the larger utility.

Although this assumption of rationality in game theory is often subjected to ridicule, logically it is not really an assumption but a tautology. The knowledge assumption appears to be the real source of unreality in the model; however, treating rationality as tautological does create serious experimental complications.

The remainder of the chapter was devoted to the normal form of a game, which is a radical conceptual simplification of the extensive form. It is achieved by introducing the idea of a *pure strategy:* a detailed prescription of what a player shall do in each eventuality. Given this notion, a *game in normal form* consists of:

i. The set of n players.

ii. n sets of pure strategies S_i, one for each player.

iii. n linear payoff functions M_i, one for each player, whose values depend upon the strategy choices of all the players.

As in the extensive form, part of the description of the players is contained in the payoff functions and the rest is given by a knowledge assumption (now of the game in normal form) and a rationality assumption. Each player attempts to maximize his utility in a situation where his outcome depends not only upon his choice, but upon the choices of each of the other players; in turn, their choices are influenced by the choice they think he is going to make, for they too are attempting to maximize a function over which they do not have full control.

chapter 4

TWO-PERSON
ZERO-SUM GAMES

4.1 INTRODUCTION

Two-person games play a central role in the whole theory of games. One part of the present theory for games with more than two players is devoted to games in which cooperation among the players is prohibited, and the principal notion in that theory is a generalization of the important equilibrium idea which arises very naturally in two-person games. Another part of n-person theory assumes that the players are free to cooperate to their mutual advantage. With this assumption, one examines what happens in the two-person games which arise when a coalition forms and it is opposed by the coalition consisting of all the remaining players. These two reasons alone would be sufficient motivation to study the two-person case, but, of course, it is of interest in its own right, since much conflict of interest involves only two protagonists.

▶ In addition, both historically and in relation to other mathematical theories, the two-person theory has played a more central role than the rest of game theory. First, until Nash's work in 1951 all equilibrium notions for n-person games rested on the two-person theory. Second, mathematics as a whole and two-person theory in particular have received mutual benefit and stimulation from the interplay between the methods of proof of the principal theorem of the two-person-game (the minimax theorem) and the formulation of new variants of the Brouwer fixed-point theorem in topology and the development of the theory of convex bodies. Third, the economic theories of linear programing and activity analysis received much of their initial impetus from their interrelations with two-person games. Fourth, Wald's contributions to the theory of statistical decision functions are intimately related to two-person theory. Indeed, it is quite possible that ultimately the theory of games will be considered important in mathematics mainly because of its historically significant relations to other parts of mathematics rather than for its own sake. Finally, one should not fail to recognize that the mathematical elegance of the two-person theory has served to attract some mathematicians to the study of mathematical applications in the behavioral sciences. Thus, even if the final verdict should be that game theory has nothing substantial to contribute to the social sciences—this we do not believe will be the case, but assuming it so—the by-products of the two-person theory would still be significant for the social sciences. ◀

All that has been said about the role of two-person theory is, in fact, about a special class of two-person games: those which in normalized form have a finite number of pure strategies and have a special property which later we shall call "zero-sum." This chapter is devoted to this class of two-person games.

To find what we mean by a two-person game in normalized form, the reader need only specialize the general definition of section 3.7 to the case $n = 2$. Thus, we have two players 1 and 2 and a game in which each player has only one move—the moves are taken simultaneously or, what amounts to the same thing, they are taken in succession in such a manner that the player who moves second does not know the choice made by the other player. We may denote the two sets of pure strategies as follows:

$$S_1 = A = \{\alpha_1, \alpha_2, \cdots, \alpha_m\},$$

$$S_2 = B = \{\beta_1, \beta_2, \cdots, \beta_n\}.$$

The assumption that the game has a finite number of pure strategies is built into the notation we have used, for the number m denotes the number of strategies available to player 1 and n the number available to player 2.

To a strategy choice for each of the players, α_i for player 1 and β_j for player 2, there is a certain outcome which will be denoted O_{ij}. An outcome O_{ij} can take on a wide variety of possible interpretations, such as:

    i. Player 1 pays player 2 $10.

    ii. Player 2 "wins" the game.

    iii. Player 1 is killed and player 2 is maimed.

    iv. Player 1 can choose any book he wants from player 2's library.

    v. Player 2 receives player 1's record collection.

    vi. Player 1, who is an employer, shuts down production for six weeks and player 2, the set of employees, loses six weeks' wages.

vii. An honest coin is tossed; if heads occurs player 1 receives player 2's library, and if tails occurs player 1 gets player 2's address book.

Three things should be noted: first, the one player does not always get what the other gives up [see (iii) and (vi)]; second, the players need not be individuals but may be aggregates of people [see (vi)]; and, third, the outcome may not be a fixed payment but rather may depend upon chance [see (vii)]. In the last case, the outcome consists of the several possibilities each with a designated probability of materializing.

Since the normalized form of the rules of the game consists of just the two strategy spaces A and B and the set of associate outcomes, we may symbolically represent them as follows:

$$
\begin{array}{c}
\text{Player 2's Pure Strategies}\\
\begin{array}{c}

\end{array}
\begin{array}{cccccc}
\beta_1 & \beta_2 & \cdots & \beta_j & \cdots & \beta_n
\end{array}\\
\begin{array}{c}
\alpha_1\\
\alpha_2\\
\vdots\\
\text{Player 1's}\quad\vdots\\
\text{Pure Strategies }\alpha_i\\
\vdots\\
\vdots\\
\alpha_m
\end{array}
\left[
\begin{array}{cccccc}
O_{11} & O_{12} & \cdots & O_{1j} & \cdots & O_{1n}\\
O_{21} & O_{22} & \cdots & O_{2j} & \cdots & O_{2n}\\
\vdots & \vdots & & \vdots & & \vdots\\
\vdots & \vdots & & \vdots & & \vdots\\
O_{i1} & O_{i2} & \cdots & O_{ij} & \cdots & O_{in}\\
\vdots & \vdots & & \vdots & & \vdots\\
\vdots & \vdots & & \vdots & & \vdots\\
O_{m1} & O_{m2} & \cdots & O_{mj} & \cdots & O_{mn}
\end{array}
\right].
\end{array}
$$

Thus, player 1's choice of strategy α_i and player 2's choice of strategy β_j is equivalent to the choice of row i and column j. The outcome O_{ij} results from these choices.

As in the general theory, the following assumptions are made:

i. *Each player knows the alternatives available both to him and to his opponent, and he knows how the outcome depends upon these choices, i.e., he knows the above table.*

ii. *If the outcome of the game involves a chance mechanism* [case (vii) of the examples of outcomes], *then each player is aware of the different possibilities and their respective probabilities.*

iii. *Each player has a preference ordering over these outcomes*, i.e., for any pair of outcomes, he either prefers one to the other or he is indifferent between them, and these paired comparisons are consistent with regard to their transitivity.[1]

iv. *Each player knows his opponent's preference pattern for the outcomes of the game.*

[1] See section 2.2.

It is obvious that these assumptions are very restrictive; it is equally obvious how to generalize or relax them, provided that one does not worry about creating an interesting theoretical superstructure. These assumptions have the virtue that they do allow a significant theory, and it is felt that they possess sufficient generality to include many interesting cases of interest conflict. Of course, the assumption that each player has only one move is not restrictive, for it will be recalled that in section 3.7 we showed that any n-person extensive game with many moves can be treated as a game with one move per player, i.e., it can be put in normalized form.

Player 1, through his choice of an alternative from A, attempts to control the outcome according to his tastes, and player 2 similarly strives for an outcome he desires by his choice from B. The problem, then, is: *under the given conditions, what choices should the players make?* Or alternatively, if you were player 1, what should you do in order to achieve your desires? The study of two-person games amounts to formulating possible meanings for these questions and giving answers to them. Our first step toward formulating the question is to limit, for the time, the class of games to which it applies—to limit it to those games where the one player has a preference pattern exactly opposite to that of his opponent.

4.2 STRICTLY COMPETITIVE AND NON-STRICTLY COMPETITIVE GAMES

The analysis of some games is trivial and, presumably, inherently different from non-trivial ones. For example, if both players have the same preference pattern over the outcomes, then everything is trivial since both players prefer the same outcome above all others. There is no conflict of interest. It is doubtful that one should try to incorporate the analysis of such games with, say, those of the opposite extreme: if player 1 prefers outcome x to outcome y, then player 2 prefers y to x; and if 1 is indifferent between x and y then so is player 2. In such cases, we say that players 1 and 2 are *strict adversaries* of each other and that they have *strictly opposing* preference patterns for the outcomes of the game. If the players of a game have strictly opposing patterns among the lotteries of outcomes, then we shall call it a *strictly competitive* game.

Are there in fact such games? Or better, are there games for which this category is not too gross a miscarriage of the mathematician's privilege of abstraction? One might be tempted to take war as the most extreme example of interest conflict, but at the global level it is probably not strictly competitive since both factions presumably prefer a draw to mutual annihilation. However, an individual engagement or an air duel can perhaps be considered, as a first approximation, a strictly competitive game. In the realm of less serious examples, many parlor games are strictly competitive—at least, the rules are designed to make them strictly competitive provided each player has a preference pattern that coincides with some natural quantity attached to the outcomes. Actually, win-lose parlor games are sometimes not played in a strictly competitive fashion, for often a player will ignore an almost certain loss in order to take an exciting risk or to impress his opponent with his audacity at bluffing. Often, however, games in which one player pays the other a money payoff will be assumed to be examples of strictly competitive games, which amounts to supposing players whose utility for money is linear with money (see footnote 2, p. 29).

▶ Care must be taken in treating games whose outcomes require monetary exchanges as strictly competitive. Consider the following example: Outcome x requires player 1 to pay player 2 $2, and outcome y requires 1 to pay 2 the sum of $500 with probability $\frac{1}{500}$, and nothing with probability $\frac{499}{500}$. Depending upon the players' attitudes toward gambling we may find both preferring x to y, or both y to x, or one x to y and the other y to x.[2] Nonetheless, our general point remains that there are some games which can be usefully treated as if they are strictly competitive, and equally there are others such as the labor-management game, the duopoly or duopsony game, the game of foreign trade between two countries, etc., which should not be treated as strictly competitive if misleading and erroneous conclusions are to be avoided. It is true that we shall use the strictly competitive theory as a building block to understand these more general cases but not by ignoring the differences.   ◀

4.3 REASONING ABOUT STRICTLY COMPETITIVE GAMES

In this section, and for the remainder of the chapter, we shall restrict our attention to strictly competitive games, i.e., to situations where the lotteries of outcomes are appraised in a strictly opposing manner. A knowledge of the preference pattern of one player implies a knowledge of the pattern of the other.

In the following example we shall suppose that the outcome results in a money payment from player 2 to player 1, where a minus sign (if any existed) would represent a negative payment from 2 to 1, and hence a (positive) payment from 1 to 2. We suppose that the sole guiding principle for 1 is to get as much money from 2 as possible, and that for 2 the motivation is to give 1 as little as possible. (We assume, of course, that both players are required to play the game.)   The example is:

---

[2] The player's attitudes toward gambling are sometimes pertinent even when the outcomes associated with pairs of pure strategies are free of probabilistic elements, since it will be shown that it is often expedient for the players themselves to inject probabilistic considerations into their choices of pure strategies.

$$\begin{array}{c c c c c} & \beta_1 & \beta_2 & \beta_3 & \beta_4 \\ \alpha_1 & \begin{bmatrix} 18 & 3 & 0 & 2 \\ \alpha_2 & 0 & 3 & 8 & 20 \\ \alpha_3 & 5 & 4 & 5 & 5 \\ \alpha_4 & 16 & 4 & 2 & 25 \\ \alpha_5 & 9 & 3 & 0 & 20 \end{bmatrix}. \end{array}$$

If player 1 chooses α_2 and player 2 chooses β_3, then 2 pays 1 \$8, since 8 is the entry of the second row and the third column.

Let us first examine this game from player 1's point of view. If player 1 knew what 2's choice would be, there would be no difficulty in determining his best counterchoice:

If player 2's choice is:	β_1	β_2	β_3	β_4
Player 1's best counterchoice is:	α_1	α_3 or α_4	α_2	α_4
The return to 1 is:	18	4	8	25

Since the counterchoice does in fact depend upon player 2's choice, and since at this level of analysis that choice is not known, it is unclear what 1 should do. It is, therefore, "natural" for player 1 to attempt to see whether his opponent's analysis of the situation would lead to a specific strategy choice for 2 which 1 would counter with the appropriate response, e.g., if it turned out that 2 should clearly take β_3, then 1 should take α_2. Thus, player 1 is led to prepare a similar table for player 2's best counterchoice:

If player 1's choice is:	α_1	α_2	α_3	α_4	α_5
Player 2's best counterchoice is:	β_3	β_1	β_2	β_3	β_3
The return to player 1 is:	0	0	4	2	0

Again, player 2's choice depends upon player 1's choice, which would be unknown to player 2, and therefore at this level of analysis player 1 has no way of deciding upon the best strategy. We have a fully circular argument.

On the other hand, these tabulations do suggest a possible resolution for player 1. From the second tabulation, he sees that if he takes α_1 he is assured an amount 0, similarly for α_2 and α_5, but from α_3 he can be sure of 4 and from α_4 of 2. Let us say that the least amount he can receive from a strategy choice is the *security level* of that choice. Strategy α_3 has the property that it *maximizes* player 1's security level, i.e., by adopting α_3 he can guarantee himself a return of at least 4, and no other strategy can guarantee a return of as much as 4.

Let us repeat this same argument taking player 2's point of view. From the first table it is seen that strategies β_1, β_2, β_3, and β_4 have security levels for player 2 of -18, -4, -8, and -25 respectively; thus, β_2 maximizes player 2's security level.   Strategy β_2 has the property that it limits the payment received by 1 to at most 4, and no other strategy available to player 2 limits 1 to that amount.

Thus, there appear to be good reasons to expect player 1 to use strategy α_3 and player 2 to use β_2; however, these reasons would collapse into nothing if the fact that player 1 could expect 2 to use β_2 would lead him to choose a strategy different from α_3, or if player 2 were induced to use a strategy different from β_2 just because he expected 1 to use α_3.   Such is not the case, for if we return to the two tabulations we see that α_3 is the best counterchoice against β_2, and conversely.   Thus, we are led to say that α_3 and β_2 are in *equilibrium* in the sense that it does not behoove either player to change his choice if the other does not change.

Returning to player 1's point of view, we have singled out α_3 as a likely choice for 1 on the grounds that it has the two properties:

i. It maximizes player 1's security level.

ii. It is the best counterchoice against the maximizer (β_2) of player 2's security level.

Should player 1 choose α_3?   Certainly (ii) is not a very convincing argument if player 1 has any reason to think that player 2 will not choose β_2, which he might expect if he feels the above analysis exceeds 2's capabilities.   Also, (i) implies a rather pessimistic point of view; to be sure, α_3 does guarantee at least 4, but it also yields *at most* 5.   He might reason that player 2 would be tempted to take β_3, for the sum of payments in this column is a minimum.   Furthermore, β_3 has the property that it is the best counterchoice against strategy α_4, and player 2 might anticipate that 1 would adopt α_4 since it yields the maximum row sum.   Thus, player 1 can give a rationalization for 2 choosing β_3, and thus he would select α_2 as the best counterchoice, and so he would expect a return of 8.   One may give player 1 pause by suggesting that player 2 might be aware of this process of reasoning and that, therefore, instead of taking β_3 he will actually take β_1, which is best against α_2, holding 1 down to a payment of 0. So it goes, for nothing prevents us from continuing this sort of "I-think-that-he-thinks-that-I-think-that-he-thinks $\cdots$ " reasoning to the point where all strategy choices appear to be equally reasonable.

What then should player 1 do?   Even though he comes to the realization that α_3 will guarantee him 4, that player 2 can hold him down to that return by playing β_2, and that α_3 and β_2 are in equilibrium, it is still a moot question whether he should play α_3.   If for any of numerous reasons 1 does not believe 2 will play β_2, he would be ill-advised to play α_3. What should he do?   *Game theory does not attempt to prescribe what he should do!*   It does point out that player 1 can guarantee himself 4 by selecting

α_3 and that no other choice has as high a guarantee, but what 1 *should* do the theory is careful to avoid saying.

One might argue: if it is pointed out to both 1 and 2 that α_3 and β_2 are in equilibrium, then 1 *should* choose α_3. Possibly, but nothing in game theory says so. If we were player 1 in this case, we certainly would choose α_3, but we would not call another "irrational" if he did otherwise. Even if we were tempted at first to call an α_3-non-conformist "irrational," we would have to admit that player 2 might be "irrational" in which case it would be "rational" for player 1 to be "irrational"—to be an α_3-non-conformist. We belabor this point because we feel that it is crucial that the social scientist recognize that game theory is not *descriptive*, but rather (conditionally) *normative*. It states neither how people do behave nor how they should behave in an absolute sense, but how they should behave if they wish to achieve certain ends. It prescribes for given assumptions courses of action for the attainment of outcomes having certain formal "optimum" properties. These properties may or may not be deemed pertinent in any given real world conflict of interest. If they are, the theory prescribes the choices which *must* be made to get that optimum.

4.4 AN A PRIORI DEMAND OF THE THEORY

In the preceding section we noted a particular property of the pair of strategies (α_3, β_2) which we may formalize as a demand to be met by any theory of strictly competitive games. It seems plausible that, if a theory offers α_{i_0} and β_{j_0} as suitable strategies, the mere knowledge of the theory should not cause either of the players to change his choice: just because the theory suggests β_{j_0} to player 2 should not be grounds for player 1 to choose a strategy different from α_{i_0}; similarly, the theoretical prescription of α_{i_0} should not lead player 2 to select a strategy different from β_{j_0}. Put in terms of outcomes, if the theory singles out $(\alpha_{i_0}, \beta_{j_0})$, then:

i. No outcome O_{ij_0} should be more preferred by 1 to $O_{i_0j_0}$.
ii. No outcome O_{i_0j} should be more preferred by 2 to $O_{i_0j_0}$.

Any α_{i_0} and β_{j_0} satisfying conditions (i) and (ii) are said to be in *equilibrium*, and the *a priori* demand made on the theory is that the pairs of strategies it singles out shall be in equilibrium.

There is no serious loss of generality if we replace the outcomes O_{ij} by their utilities a_{ij} for player 1 and b_{ij} for player 2. Since the two players have strictly opposed preferences for lotteries of outcomes, the utility of the first person is in effect exactly the negative of the utility of the second one. The only ambiguity that exists is in the value of the units and zero of these functions, which are not determined in the von Neumann-

Morgenstern theory of utility. Since, in the present context of non-cooperative games, we shall have no occasion to compare utilities in any way, it does not matter what choices we make of units and zeros. Thus, we may choose them so that

$$a_{ij} = -b_{ij}.$$

In other words, the strictly competitive game may be represented in such a fashion that the sum of utility payments to the players is zero:

$$a_{ij} + b_{ij} = 0.$$

For this reason, strictly competitive games are known as *zero-sum* games, and we shall use these two terms interchangeably. Clearly, it is sufficient to present the matrix of payments of the first player.

It is not difficult to show that the pair $(\alpha_{i_0}, \beta_{j_0})$ is in equilibrium if and only if

$$a_{i_0 j_0} = \max_i a_{ij_0} = \min_j a_{i_0 j},$$

that is, the entry $a_{i_0 j_0}$ is the maximum of its column j_0 and the minimum of its row i_0. For example, in the game discussed in section 4.3, (α_3, β_2) is in equilibrium since the entry 4 in the third row and the second column is the maximum of its column and the minimum of its row.

This notion of an equilibrium pair, though abstractly arrived at, is not a pure figment of the theoretical mind; it has its counterpart in such practical affairs as battles. Haywood [1950, 1954] has explored the relation between military-decision doctrine and two-person zero-sum game theory.

A military commander may approach decision with either of two philosophies. He may select his course of action on the basis of his estimate of what his enemy *is able to do* to oppose him. Or, he may make his selection on the basis of his estimate of what his enemy *is going to do*. The former is a doctrine of decision based on enemy capabilities; the latter, on enemy intentions.

The doctrine of decision of the armed forces of the United States is a doctrine based on enemy capabilities. A commander is enjoined to select the course of action which offers the greatest promise of success in view of the enemy capabilities. [1954, pp. 365–366.]

If, and this is a big if, a military situation can be viewed as a two-person zero-sum game, then this philosophy can be translated into the rule: maximize one's security level. If both commanders evaluate the situation in the same way and if both adopt this philosophy, then the outcome is an equilibrium pair, provided such a pair exists. These points he illustrates by two examples drawn from World War II; we shall examine one: the Battle of the Bismark Sea.

In the critical stages of the struggle for New Guinea, intelligence reports indicated that the Japanese would move a troop and supply convoy from the port of Rabaul at the eastern tip of New Britain to Lae, which lies just west of New Britain on New Guinea. It could travel either north of New Britain, where poor visibility was almost certain, or south of the island, where the weather would be clear; in either case, the trip would take three days. General Kenney had the choice of concentrating the bulk of his reconnaissance aircraft on one route or the other. Once sighted, the convoy could be bombed until its arrival at Lae. In days of bombing time, Kenney's staff estimated the following outcomes for the various choices:

| | Japanese Strategies | |
	Northern Route	Southern Route
Kenney's Strategies:		
Northern Route	2	2
Southern Route	1	3

It is easily seen that there is one equilibrium point: (northern route, northern route), with an expectation of two days of bombing. These in fact were the choices made; the convoy was sighted about one day after it sailed; and the Japanese suffered severe losses. However, as Haywood emphasizes, "Although the Battle of the Bismark Sea ended in a disastrous defeat for the Japanese, we cannot say the Japanese commander erred in his decision." [1954, p. 369.] Given the total strategic situation, which was not particularly bright for him, his choice was wise in the sense that his northern route strategy was at least as good as his southern route strategy against either one of Kenney's strategies.

4.5 GAMES WITH EQUILIBRIUM PAIRS

There are a number of questions that come to mind about equilibrium pairs: questions of existence, uniqueness, and properties that they may possess. These we must examine.

i. *Do all strictly competitive games have equilibrium pairs?*

The answer is No. This is easily seen by exhibiting an array, $[a_{ij}]$, where there is no entry which is both the minimum of its row and the maximum of its column, for example,

$$\begin{bmatrix} 3 & 1 \\ 2 & 4 \end{bmatrix}.$$

Thus, we are forced to divide the totality of zero-sum games into two classes: those which have an equilibrium pair and those which do not. We postpone discussion of games of the latter type to sections 4.7 and 4.8.

ii. *If a game has an equilibrium pair, is this pair necessarily unique, or may there be several equilibrium pairs?*

No, such pairs are not unique.   For example, in the game

$$\begin{bmatrix} 4 & 5 & 4 \\ 3 & 0 & 1 \end{bmatrix}$$

both (α_1, β_1) and (α_1, β_3) are equilibrium pairs.

iii. *Does the existence of several equilibrium pairs in a game cause any difficulty in the sense of creating a conflict of interest among them?*

Two sources of difficulty seem *a priori* possible.   Suppose $(\alpha_{i_0}, \beta_{j_0})$ and $(\alpha_{i_1}, \beta_{j_1})$ are equilibrium pairs.   If $a_{i_0 j_0}$ were greater than $a_{i_1 j_1}$, then player 1 would prefer the first pair and player 2 the second.   Thus, the solution of some games would be another game in which the conflict centered about the equilibrium points of the first game.   Second, however we might choose to resolve the first difficulty, there certainly is the possibility that player 1 would choose α_{i_0} and player 2 β_{i_1}.   Is $(\alpha_{i_0}, \beta_{j_1})$ also an equilibrium point?

Fortunately, all is well.   If $(\alpha_{i_0}, \beta_{j_0})$ and $(\alpha_{i_1}, \beta_{j_1})$ are equilibrium pairs, then it can be shown[3] that:

1. $(\alpha_{i_0}, \beta_{j_1})$ and $(\alpha_{i_1}, \beta_{j_0})$ are also in equilibrium.
2. $a_{i_0 j_0} = a_{i_1 j_1} = a_{i_0 j_1} = a_{i_1 j_0}$.

Because of these results, it is appropriate to call α_i an *equilibrium strategy* for player 1 if there exists a strategy β_j such that (α_i, β_j) is an equilibrium pair.   The results may then be paraphrased as saying that any pair of equilibrium strategies, one for each player, is an equilibrium pair, and all equilibrium pairs give rise to outcomes with the same utility payment.

To phrase the next two questions we must generalize the definition of security level given in the discussion of the example (section 4.3).   We shall call $\min_j a_{ij}$ the *security level* for player 1 *of strategy* α_i, since if he chooses α_i he cannot receive less than $\min_j a_{ij}$.

---

[3] The proofs are simple: We know that $(\alpha_{i_0}, \beta_{j_0})$ is an equilibrium pair if and only if $a_{i j_0} \leqslant a_{i_0 j_0} \leqslant a_{i_0 j}$, for all i and j.   Similarly, the fact that $(\alpha_{i_1}, \beta_{j_1})$ is also an equilibrium pair implies $a_{i j_1} \leqslant a_{i_1 j_1} \leqslant a_{i_1 j}$, so

$$a_{i_0 j_0} \leqslant a_{i_0 j_1} \leqslant a_{i_1 j_1} \leqslant a_{i_1 j_0} \leqslant a_{i_0 j_0}.$$

Since the same number appears both at the left and the right, all the inequalities must be equalities, which proves 2.   To show $(\alpha_{i_0}, \beta_{j_1})$ is an equilibrium pair, we note

$$a_{i j_1} \leqslant a_{i_1 j_1} = a_{i_0 j_1} = a_{i_0 j_0} \leqslant a_{i_0 j},$$

for all i and j.

iv. *Does an equilibrium strategy maximize a player's security level?*

By this question we mean: if α_{i_0} is one of player 1's equilibrium strategies, then does it have the property that the security level for any other strategy α_i is not greater than that for α_{i_0}? The answer is Yes.[4] Since there is an equilibrium pair $(\alpha_{i_0}, \beta_{j_0})$, the security level of α_{i_0} is

$$\min_{j} a_{i_0 j} = a_{i_0 j_0},$$

so

$$\max_{i} (\min_{j} a_{ij}) = a_{i_0 j_0}.$$

The strategy α_{i_0} is said to be a *maximin* strategy since i_0 has the property that it is a maximizer over i of the expression $\min_{j} a_{ij}$.

For player 2, an analogous situation obtains. Let $\max_{i} a_{ij}$ be called the security level of strategy β_j, and so, to obtain the best security level, player 2 should choose β_j to minimize $\max_{i} a_{ij}$. It follows readily that 2 can achieve this goal by choosing an equilibrium strategy β_{j_0} and that this choice gives a security level of $\max_{i} a_{i j_0} = a_{i_0 j_0}$. Thus,

$$\min_{j} (\max_{i} a_{ij}) = a_{i_0 j_0}.$$

The strategy β_{j_0} is said to be a *minimax* strategy.

We have thus shown that if there exists an equilibrium pair $(\alpha_{i_0}, \beta_{j_0})$, then

$$\max_{i} a_{i j_0} = \min_{j} (\max_{i} a_{ij}) = \max_{i} (\min_{j} a_{ij}) = \min_{j} a_{i_0 j} = a_{i_0 j_0}.^{[5]}$$

To summarize, an equilibrium strategy not only attains the best security level for player 1 but it is also good against that strategy of player 2 which attains his best security level.

v. *If a strategy maximizes a player's security level, is it an equilibrium strategy?*

[4] To see this assertion, we note that

(Security level of α_{i_0}) $= a_{i_0 j_0} \geqslant a_{i j_0} \geqslant$ (Security level for α_i, any i).

[5] The existence of an equilibrium pair is seen to imply

$$\min_{j} (\max_{i} a_{ij}) = \max_{i} (\min_{j} a_{ij}),$$

i.e., the operators min and max are commutative. From the above argument it can also be seen that the converse holds: that if the operators are commutative, then an equilibrium pair exists. Incidentally, it is not customary to include the parentheses, as we have done for greater clarity, but rather to write $\min_{j} \max_{i} a_{ij} = \max_{i} \min_{j} a_{ij}$.

The answer cannot be an unqualified Yes, since a strategy yielding the best security level is always defined, whereas some games do not have equilibrium pairs. But, if a game has an equilibrium pair, then the answer is Yes. Operationally, therefore, one can search for equilibrium strategies by first finding those strategies which attain the best security level. Unless it is known that the game has an equilibrium pair, it is necessary to verify that these maximizers of security level do result in equilibrium pairs.

*4.6 EQUILIBRIUM PAIRS IN EXTENSIVE GAMES

Since, as we have seen, not all games in normalized form have equilibrium pairs, the following question is natural: If a game is given in extensive form, is it possible to tell directly from the game tree structure (without computing the normal form) whether or not optimal—minimax and maximin—strategies exist? It turns out that it is not too difficult to show that they exist for such games as chess, checkers, and ticktacktoe, and indeed for any games with perfect information (von Neumann and Morgenstern [1947]). It will be recalled (section 3.2) that a game has perfect information if at any move the player has complete and unambiguous information concerning the choices made at previous moves. This condition is sufficient, but it is by no means necessary; for a necessary and sufficient condition see Dalkey [1953].

To see intuitively that perfect information is sufficient, we may argue as follows: At a terminal choice point—we are assuming that all games have a stopping rule, and this enables us to work backwards—the player whose move it is will naturally adopt the choice which suits him best. Thus, since the last choice is determinate, we may as well delete it and place the appropriate payoff directly at the terminal move position. If this is done for each terminal move, the penultimate moves now play the role of terminal moves, and so the process may be carried backward to the starting point. The crux of the argument, which incidentally indicates where the assumption of perfect information is needed, depends upon the fact that in any partial play of the extensive game the players are precisely aware of their position in the game tree.

4.7 GAMES WITHOUT EQUILIBRIUM PAIRS

For games with equilibrium pairs, we have seen that we can demand that the theory lead to such pairs. There remains, however, the nontrivial case of games without equilibrium pairs. One tool that we know exists for all games are the maximin and minimax strategies—the ones which maximize a player's security level. Let us examine how they might be used in the game

$$
\begin{array}{cc}
 & \beta_1 \ \ \beta_2 \\
\begin{array}{c} \alpha_1 \\ \alpha_2 \end{array} & \begin{bmatrix} 3 & 1 \\ 2 & 4 \end{bmatrix},
\end{array}
$$

which does not have an equilibrium point. By choosing α_2 player 1 obtains his maximum security level of 2, and player 2 attains his, 3, by choosing β_1. Not only do α_2 and β_1 attain the maximum security levels, but β_1 is good against α_2. These two arguments tend to reinforce player 1's belief that player 2 will choose β_1, in which case 1 is better off taking α_1 than α_2. But, if player 2 follows this argument, then he clearly will take β_2 rather than β_1. That being the case, player 1 should take α_2, etc. This cyclic effect is all the more reason why both players should stick to the maximin and minimax strategies, in which case player 1 should defect to α_1, etc., and we go around in circles again.

Such an argument seems to force us to assert that player 1 should be indifferent between α_1 and α_2. If so, then he should be willing to toss a coin to decide between them. Clearly, if player 2 chooses β_1, then tossing a coin is preferable to α_2; if 2 chooses β_2, tossing the coin is preferable to α_1. But if 1 does not know whether β_1 or β_2 will be chosen, should he prefer a toss of the coin in preference to either α_1 or α_2? After all, we have argued ourselves into the position where we are indifferent between α_1 and α_2, so what help is tossing a coin to decide? The answer is that it raises the security level. Let us see how.

If β_1 is used, then player 1 receives 3 with probability $\frac{1}{2}$ and 2 with probability $\frac{1}{2}$, which is certainly preferable to a certainty of 2. If β_2 is chosen, then player 1 receives 1 with probability $\frac{1}{2}$ and 4 with probability $\frac{1}{2}$. In this case the direction of preference is less clear, but we note that the expected value of the gamble is $\frac{5}{2}$. Can one say this is preferred to 2? Certainly there is no such assurance if we are talking about money, but this difficulty is avoided by assuming that these numbers are utilities, i.e., numbers which arose from the individual's preference patterns over outcomes which may have been money or other stimuli. As we pointed out in Chapter 2, if the preference pattern over gambles satisfies certain consistency requirements, then the expected value of the utility function represents the preferences. Thus, the gamble with the expected value $\frac{5}{2}$ is preferred to the sure thing of 2, even though in the gamble player 1 may get as little as 1. If this were not so, we would have to conclude that the numbers 1, 2, and 4 are inappropriate numerical indices of the subjective worth—utility—of the outcomes they purport to reflect.

Perhaps the argument that a chance event such as the toss of a coin may raise the security level can be reinforced by noting that it enables player 1 to hedge against the 1 entry with the 4 entry in the second column, just as one may raise one's "security level" in betting on the horses by not placing all of one's money on a single horse.

The technique of considering a probability mixture of pure strategies in order to raise a player's security level turns out to have some startling and elegant consequences. Von Neumann exploited this notion to create a beautiful order among those zero-sum games without equilibrium points —an order quite analogous to that exhibited by games with equilibrium points.

Let us denote the strategy of choosing α_1 and α_2 each with probability $\frac{1}{2}$ by $(\frac{1}{2}\alpha_1, \frac{1}{2}\alpha_2)$, or in the general case where the probabilities are x_1 and x_2 by $(x_1\alpha_1, x_2\alpha_2)$. Since the numbers x_1 and x_2 form a probability distribution over the two alternatives we know that $x_1 \geqslant 0$, $x_2 \geqslant 0$, and $x_1 + x_2 = 1$. The strategy of choosing α_1 with probability x_1 and α_2 with probability x_2, i.e., $(x_1\alpha_1, x_2\alpha_2)$, is called a *mixed* (or randomized) *strategy* to contrast it with a pure strategy such as choosing α_1. It is a "mixture" of pure strategies. Of course, each pure strategy is a special case of a mixed strategy in which all the weight is on one component.

If player 2 chooses the pure strategy β_1, then player 1's expected value from the mixed strategy $(x_1\alpha_1, x_2\alpha_2)$, i.e., his utility for the lottery (x_1O_{11}, x_2O_{12}), is $3x_1 + 2x_2$, whereas if 2 chooses β_2 it is $1x_1 + 4x_2$. The security level of the mixed strategy is then the minimum of these two quantities. Since $(\frac{1}{2}\alpha_1, \frac{1}{2}\alpha_2)$ gives rise to an expected value of $\frac{5}{2}$ for both β_1 and β_2, its security level is $\frac{5}{2}$. It can be shown that no other mixed strategy $(x_1\alpha_1, x_2\alpha_2)$ has as high a security level, so $(\frac{1}{2}\alpha_1, \frac{1}{2}\alpha_2)$ is said to be player 1's *maximin strategy in the mixed strategy sense.*

For player 2 we have similar concepts. If y_1 and y_2 form a probability distribution, $(y_1\beta_1, y_2\beta_2)$ represents the mixed strategy which selects β_1 with probability y_1 and β_2 with probability $y_2 = 1 - y_1$. Player 1's return when he selects α_1 against this mixed strategy is $3y_1 + 1y_2$ and when he selects α_2 it is $2y_1 + 4y_2$. Player 2's security level with the given mixed strategy is the maximum of $3y_1 + 1y_2$ and $2y_1 + 4y_2$. It can be shown that 2 attains his best security level with $(\frac{3}{4}\beta_1, \frac{1}{4}\beta_2)$, and so this is called 2's *minimax strategy in the mixed strategy sense.* For this strategy, the return to player 1 when he chooses α_1 is $(\frac{3}{4})3 + (\frac{1}{4})1 = \frac{5}{2}$ and when he chooses α_2 it is $(\frac{3}{4})2 + (\frac{1}{4})4 = \frac{5}{2}$.

In summary, then, there exists a number $\frac{5}{2}$, a mixed strategy $(\frac{1}{2}\alpha_1, \frac{1}{2}\alpha_2)$ for player 1, and a mixed strategy $(\frac{3}{4}\beta_1, \frac{1}{4}\beta_2)$ for player 2 such that

i. $(\frac{1}{2}\alpha_1, \frac{1}{2}\alpha_2)$ gives player 1 a security level of $\frac{5}{2}$, and no other mixed strategy has a better security level for him.

ii. ($\frac{3}{4}\beta_1$, $\frac{1}{4}\beta_2$) gives player 2 a security level of $\frac{5}{2}$, and no other mixed strategy has a better security level for him.

iii. These two mixed strategies are in equilibrium in the sense that there is no advantage for one player to change from his strategy if the other player holds his strategy choice fixed.

The first statement says that by playing his mixed strategy player 1 can be certain of an expected (utility) return of at least $\frac{5}{2}$, and the second one says that player 2 can hold him down to at worst that expectation by playing his mixed strategy. The final statement says that neither can improve his expectations by changing his choice of mixed strategies.

4.8 THE MINIMAX THEOREM

The gist of this section is an informal statement of the central theorem of two-person zero-sum theory; in effect, it says that the analysis given in the last section of a specific example did not rest on any peculiar features of that example, but that the same properties hold for any zero-sum (strictly competitive) game. In Appendix 2 a self-contained and rigorous statement of this theorem is given, and if the reader desires more formality at this point he should turn to that appendix.

Let us first recall the result that holds for strictly competitive games which have an equilibrium pair among pure strategies (sections 4.4 and 4.5):

There exists a number v, a pure strategy (a maximin strategy) for player 1 which guarantees him at least v, and a pure strategy (a minimax strategy) for player 2 which guarantees that player 1 gets at most v. These pure strategies are in equilibrium, and any pair of pure strategies which are in equilibrium yield a maximin and a minimax strategy for 1 and 2 respectively.

Now let us consider all strictly competitive games with a finite number of pure strategies, whether or not they have equilibrium pairs. It is clear that we shall want to generalize the notion of a mixed (or randomized) strategy as given in the last section. If player 1 has m pure strategies α_1, α_2, $\cdots$, α_m, then a mixed strategy is a probability distribution over m points, i.e., it is a set of m numbers x_1, x_2, $\cdots$, x_m such that

$$x_i \geqslant 0 \quad \text{for } i = 1, 2, \cdots, m, \quad \text{and} \quad x_1 + x_2 + \cdots + x_m = 1.$$

The mixed strategy can be symbolized by ($x_1\alpha_1$, $x_2\alpha_2$, $\cdots$, $x_m\alpha_m$). A similar definition holds for the mixed strategies of player 2, except that the probability distribution is over n points.

The principal theorem—known as the minimax theorem—asserts that the italicized assertion for games with equilibrium pairs is valid for all zero-sum games with a finite number of pure strategies *provided that the domains of choice of the players is enlarged from their sets of pure strategies to their sets of mixed strategies.*

The mixed strategies isolated by the theorem are called *maximin* and *minimax* strategies, and, extending the previous terminology, they are said to form an *equilibrium pair*. As with games having pure strategy equilibrium pairs, the value v in the theorem is unique and any maximin strategy together with any minimax strategy forms an equilibrium pair. Thus, there does not result any conflict of interest among the equilibrium mixed strategy pairs. This unique number v is called the *value of the two-person zero-sum game*.

At about the time von Neumann proved that every two-person zero-sum game has an equilibrium point when randomized strategies are permitted, at least two other authors had also come to realize the importance of randomization in analyzing games, but they failed to achieve this central theorem. As we mentioned in Chapter 1, Borel in the early '20's published several papers on games in which randomization played a crucial role. Later, but independently of both Borel and von Neumann, R. A. Fisher, who introduced randomization so effectively in experimental design, came up with the same idea to resolve an old enigma in the game "le Her" which had earlier attracted some scientific attention through an exchange between Nicolas Bernoulli and Montmort. Both men had agreed that in reality each player had only two reasonable (technically, undominated) strategies and that the resulting two by two game lacked a pure strategy equilibrium point. Fisher [1934] showed that by introducing randomization "the chances of the game are stabilized at the saddle," which is to say that he stated and proved the minimax theorem for the two by two cases, but apparently he was unaware of its generalization.

▶ One interesting result on the utility of money rests upon the preceeding analysis of strictly competitive games. Suppose that a money payoff matrix is given; call it G. From player 1's point of view, it is not this game he is playing but rather the one obtained by substituting his utility for money for the monetary payoffs in G. If u denotes his utility function, then we let $u(G)$ denote his utility payoff matrix for the given monetary game. Now, observe that if we add the same constant amount of money, h, to each entry in G we leave the objective strategic problem unchanged. If we let E be a payoff matrix which is the same size as G but has 1 for every entry, then the payoff obtained by adding h to each entry of G is given by $G + hE$. The corresponding utility payoff for player 1 is, of course, $u(G + hE)$. The question raised by Kemeny *et al.* [1955] is: What possible form may u take if it is strictly increasing and differentiable function of money (both very plausible conditions) and if the set of maximin strategies in $u(G)$ is the same as that in $u(G + hE)$? The latter condition is assumed to hold for all possible monetary payoff matrices G and for all values of h. They show that u must be either linear or exponential with money, or, to be more exact, if x denotes the monetary variable, then either

$$u(x) = ax + b, \quad \text{where } a > 0,$$

or

$$u(x) = ae^{bx} + c, \quad \text{where } ab > 0.$$

The primary assumption leading to these conclusions is that the two sets of maximin strategies should be the same.   Put in another and more intuitive way, it requires that the absolute level of a person's wealth—which is what changes when a constant amount is added to each payoff entry—shall not result in a utility payoff which alters his strategic considerations.   As a normative condition, this is acceptable and interesting; as a description of behavior, it is very doubtful. Absolute levels of wealth do appear to influence behavior.   ◄

4.9 COMPATIBILITY OF THE PURE AND MIXED STRATEGY THEORIES

Although there is no conflict of interest among the mixed strategy equilibrium pairs nor among the pure strategy equilibrium pairs when they exist, one might fear that there would be between the two classes of equilibrium pairs.   In other words, one might fear that the mixed strategy theory and the pure strategy theory would differ for those games having a pure strategy equilibrium pair, that the optimal security level using mixed strategies would differ from that using pure strategies.   It can be shown that this is not the case.   Furthermore, the pure strategy equilibrium pairs are also mixed strategy equilibrium pairs; hence, with respect to the best security levels and equilibrium pairs, no complications result by introducing mixed strategies into games with pure strategy equilibrium pairs.

It remains true, however, that even though a player has a pure equilibrium strategy and realizes it, he may have reasons to play differently if he thinks that his opponent may not play an equilibrium strategy.   One may depart from an equilibrium strategy to exploit an opponent's supposed ignorance, but it may be sensible to use a mixed strategy as a hedge against extremely unfavorable situations and against the possibility that one's opponent has more insight into one's behavior than anticipated.

It will be recalled that we have gotten to this point by the following chain of steps.   In section 4.4 we set up an *a priori* demand of any theory leading to pure strategy selections, and we showed that this meant that an acceptable theory would have to single out equilibrium pairs.   In section 4.5 we established, first, that not all strictly competitive games have pure strategy equilibrium pairs, and, second, that, in those games which do, they are substantially interchangeable, since an equilibrium strategy from each player results in an equilibrium pair and all equilibrium pairs have the same security level. For games without equilibrium pairs, we were led in section 4.7 to use mixed strategies, and all of the above statements—including the *a priori* demand of section 4.4—hold in the general case with the mere substitution of mixed for pure strategies. Furthermore, the two notions are in agreement when there is a pure strategy equilibrium pair. Thus, we are led to say that any adequate theory of strictly competitive two-person games should single out mixed strategy equilibrium pairs, and, since they are all substantially equivalent in the sense of leading to the same security level, we are led to accept the equilibrium pairs as *the* theory of such games.

4.10 ON THE INTERPRETATION OF A MIXED STRATEGY

What we have just related is certainly mathematically above reproach, but the skeptical reader may question whether it possesses any meaning conceptually. What does it mean to select a mixed strategy and would one ever really choose one?

As to the meaning, this depends very greatly upon one's interpretation of probability. We shall take the point of view that the selection of a pure strategy by means of a mixed strategy is equivalent to performing an experiment. Let us suppose player 1 has the pure strategies α_1, α_2, $\cdots$, α_m, and let $\mathbf{x}$ be the symbolic representation of the mixed strategy where player 1 adopts one and only one pure strategy, but where the probability of adopting α_i is x_i, i.e.,

$$\mathbf{x} = (x_1\alpha_1, \ x_2\alpha_2, \ \cdots \ , \ x_m\alpha_m).$$

To employ $\mathbf{x}$ we say that player 1 must perform an experiment in which he partitions a set of possible outcomes into m mutually independent and exhaustive events whose probabilities are x_1, x_2, $\cdots$, x_m and to which he associates the strategies α_1, α_2, $\cdots$, α_m, respectively. In practice one might use a table of random numbers for this purpose; however, for didactic purposes it is more convenient to take the following "physical" experiment as typical. The player is given a "fair" spinner centered at a disk of unit circumference, and he partitions the circumference into m arcs of lengths x_1, x_2, $\cdots$, x_m. If the spinner comes to rest in the ith arc of length x_i (some arcs may be of zero length), then strategy α_i is adopted. We shall refer to this experiment as experiment $\mathbf{x}$.

Let $\mathbf{x}^{(0)}$ denote one of player 1's maximin mixed strategies, and let us suppose that at time t_0 he sets up but does not perform the experiment $\mathbf{x}^{(0)}$. At time $t_1 > t_0$ experiment $\mathbf{x}^{(0)}$ is performed, yielding the pure strategy he will use later at time t_2 when the game is played. Let $v(\mathbf{x})$ denote the security level associated with the mixed strategy $\mathbf{x}$, i.e., it is the minimum utility expected by player 1 against all the mixed strategies available to player 2. Clearly, in the interval from t_0 to t_1 (prior to conducting experiment $\mathbf{x}^{(0)}$) player 1's security level is $v(\mathbf{x}^{(0)})$. At time t_1 the spinner yields some pure strategy, say α_6; thus, from t_1 to t_2 player 1's security level is changed to $v(\alpha_6)$.

Now, if we are motivated solely by security levels, as might be the case, we may be quite unhappy with the outcome of the spinner—it may be that some other strategy, say α_9, has a much higher security level than α_6—and so we may be tempted at t_2 to adopt a pure strategy different from the one dictated by the spinner! But, if we are going to tamper with our fate as given by the experiment, why be so silly as to construct and perform the experiment in the first place? After all, if we are cognizant that we will not abide by the experiment $\mathbf{x}^{(0)}$, then our security level is *not* $v(\mathbf{x}^{(0)})$ from t_0 to t_1.

Indeed, let us suppose that the player is offered one of two options: In option 1 he does *not* have the privilege of ignoring the dictate of the spinner, and in option 2 he need not follow its dictates. Since option 2 includes all the possibilities available to him under option 1, plus others, it would seem that it should be preferred. In that case, a critic of the randomized strategy concept can argue that between t_1 and t_2 the player knows what pure strategy the spinner has selected, and so the mixed strategy $\mathbf{x}^{(0)}$ used to arrive at it is strategically irrelevant. One must compare that pure strategy on its own merits with the other pure strategies. Consequently, the concept of a mixed strategy is a convenient mathematical tool but it completely fails to be realistic.

The most common counterargument in defense of mixed strategies is the observation that they withhold from our opponents knowledge of the pure strategy we will use. This, it is contended, is important, for such knowledge can be exploited. Although granting that this is often a pertinent point, some authors feel that it certainly cannot be a complete defense, for mixed strategies are appropriate even when we know that the other player is *not in any way concerned with one's behavior*. This may occur if the other player is not aware of the game payoffs, or in games against nature. But, if the defense of mixed strategies should not be confined solely to a secrecy argument, what should it be based on? The supplementary defense is that, psychologically, option 1 should be preferred to option 2, contrary to our assumption above, just because it does not permit us to fall prey to our human frailty. It is not unlike the person who wants to go on a diet. He announces his intention, or accepts a wager that he will not break his diet, so that later he will *not* be free to change his mind and to optimize his actions according to his tastes at *that* time—e.g., to eat an ice-cream sundae.

There are some who feel strongly that this argument is spurious and who contend that the only valid argument for randomization in games rests on the secrecy aspect. When we discuss games against nature and statistical inference in Chapter 13, we shall have occasion to deal again with the conceptual pros and cons of randomization. For example, we shall describe Chernoff's [1954] eloquent defense for using pure strategies in statistical decisions, but we must point out that many statisticians remain unconvinced.

A strategy which is good in the total context of the conflict of interest may appear to be poor in a limited context. In evaluating strategies this distinction between contexts is, of course, important, but it is often difficult to maintain when considering particular cases. This is particularly true after the outcome of the game is known and the wisdom of the choices is under consideration. The problem can be vividly illustrated by two military examples. Compare the role of an aerial strategist who selects one of several strategies for fighter pilots in dog fights with the roles of the pilots themselves. Suppose the strategist has arrived at a mixed strategy x, which he tells to a briefing officer who in turn determines a pure strategy for each pilot. This the briefing officer does by performing the experiment x (presumably in private) and then instructing each pilot in the strategy he is to assume. The conflict of point of view becomes apparent if we suppose the experiment has led to an unhappy strategy for a pilot with whom the briefing officer is friendly. From the strategic point of view the mixed strategy is profitable. From the individual point of view it does not seem so. This example may seem special in that there will probably be many simultaneous occurrences of the same game, and so a mixed strategy would mean, roughly, the fraction of games in which each strategy is employed. Actually, however, the argument is just as valid for a *single* foray. A mixed strategy will still be considered best (in the sense of security level) to the military strategist—but perhaps not to the pilot who has to adopt the specific pure strategy dictated by the outcome of the experiment.

As the second example, imagine a congressional investigation of a military commander, or an agency chief, who has adopted a specific pure strategy which has been ruinous. What would be the reaction if his defense hinged on the fact that he adopted this pure strategy by a throw of dice? Or equally, imagine a commander so ill advised that on being commended for a brilliant successful strategic move disclosed the fact that his choice had resulted from the toss of a coin. Unfortunately, the strategist is often evaluated in terms of the outcome of the adopted choice rather than in terms of its strategic desirability in the whole risky situation.

4.11 EXPLOITATION OF OPPONENT'S WEAKNESSES

Let us suppose that the value of a game to player 1 is v, and that he can achieve this by playing the optimal (maximin) mixed strategy $\mathbf{x}^{(0)}$. *Even if player 2 does not employ his optimal strategy,* there are many games where 1's return will be v or only slightly more than v if he plays $\mathbf{x}^{(0)}$; hence we are led to consider how a player may exploit his opponent's "mistakes." We shall still assume a conservative philosophy by concentrating on security levels.

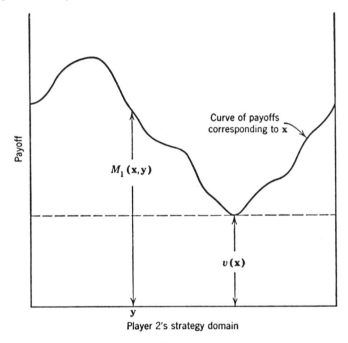

FIG. 1

It will be helpful to employ the pictorial representation shown in Fig. 1. Let each point on the horizontal axis represent symbolically a pure or mixed strategy for player 2, and let the vertical axis represent the payoffs to player 1 when he employs the strategy $\mathbf{x}$. Thus, to each $\mathbf{y}$ for player 2 there is a value $M_1(\mathbf{x}, \mathbf{y})$ for player 1; therefore for each $\mathbf{x}$ we can draw a curve representing the payoff to 1 as the strategy for 2 varies over its domain. Each such curve is associated with a specific strategy $\mathbf{x}$, so when we wish to consider more than one strategy for player 1, we shall have to deal with several curves, each labeled by its associated strategy.

In Fig. 2, strategy $\mathbf{x}^{(1)}$ is clearly as good as $\mathbf{x}^{(2)}$ for all values of $\mathbf{y}$, and

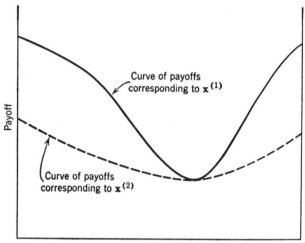

Player 2's strategy domain

Fig. 2

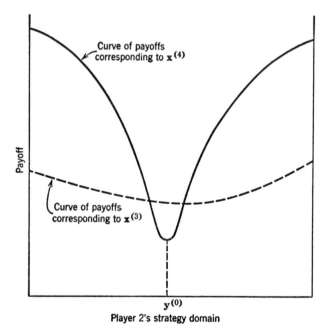

Player 2's strategy domain

Fig. 3

for some **y**'s it is better.   In such a case we say that $\mathbf{x}^{(2)}$ is *inadmissible* and that it is *dominated by* $\mathbf{x}^{(1)}$.   If the curves are as in Fig. 3 we cannot conclude that either strategy dominates the other.   It is true that $\mathbf{x}^{(4)}$ is better than $\mathbf{x}^{(3)}$, for all **y**'s not in the neighborhood of the strategy $\mathbf{y}^{(0)}$,

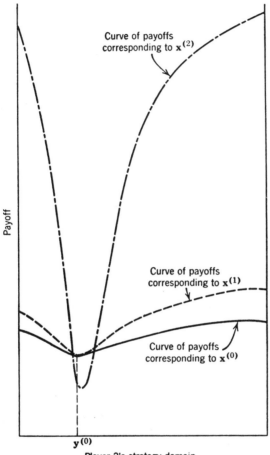

Fig. 4

but it is possible that player 2 is aware of this and so will adopt a strategy in that neighborhood.

Even if both $\mathbf{x}^{(0)}$ and $\mathbf{x}^{(1)}$ are optimal strategies for player 1, one may dominate the other, as $\mathbf{x}^{(1)}$ dominates $\mathbf{x}^{(0)}$ in Fig. 4.   Against an optimal strategy $\mathbf{y}^{(0)}$ for player 2 they both yield the same payoff v, but not in other cases.   It seems obvious that 1 should prefer $\mathbf{x}^{(1)}$ to $\mathbf{x}^{(0)}$.   Indeed, if 1 has reason to suspect that 2 will not play $\mathbf{y}^{(0)}$, then he might use a strategy such as $\mathbf{x}^{(2)}$, which has a lower security level but which exploits deviations from 2's optimal strategy much better than a maximin strategy. Just how great a reduction of security level 1 should risk depends upon the value v, upon 1's subjective appraisal of 2's intellectual capacity, etc. In short, the answer lies outside the present mathematical model.

Some of the above points are most sharply illustrated by considering games in extensive form. There is no loss of generality in supposing the value of the game is zero. Again we shall take player 1's point of view, and we shall suppose that he is fully aware of the game-theoretic analysis and that he is capable of making all the calculations needed–the latter is no small assumption. Suppose after players 1 and 2 have each taken some moves, 1 takes stock of the situation. From his current partial knowledge of the play of the game, which is equivalent to his knowledge of the information set he is in, he can treat the remainder of the game as a new game with a restricted set of strategies for both 2 and himself. Suppose when he does so he finds that, because of the previous moves made, he is able to guarantee himself a security level larger than 0, say $\frac{1}{2}$. We may suppose that he governs his choice at the following move to squeeze out this advantage. Similarly, at each later stage he can calculate his security level relative to the information set he is in and he can act to guarantee that higher level. Now the following question arises: Suppose during the course of play, 1 finds that his opponent has repeatedly made serious mistakes thereby allowing 1 to increase his security level, should 1 still play conservatively in the sense of guaranteeing himself his latest security level, or, anticipating 2's future mistakes, capitalize on them by playing less conservatively?

An example will make 1's problem vivid. Suppose they are playing a game, Γ_1, with perfect information and with the game tree shown in Fig. 5. The value of this game is 0 since 2 can obviously hold 1 down to 0 by taking choice b at move 1, and, if 2 chooses a on move 1, player 1 can be certain of 0 by choosing d on move 2. Although the choice of d following 2's (non-optimal) choice of a is a maximin strategy for 1, it is *not admissible*, for at move 2 the security level for 1 is $\frac{1}{2}$ and selecting c achieves it. If at move 3 player 2 adopts e, player 1 is in a position at move 4 to get $\frac{3}{4}$ by choosing h. However, at move 4, player 1 is well aware of the two serious mistakes 2 has already made and he might be tempted to choose g rather than h depending upon the outcomes denoted by Γ_2 and Γ_3. Let us suppose that the choice of j at move 5 results in a complicated game tree Γ_2 whose value is $-\frac{1}{2}$ and that the choice of i leads to a complicated game tree Γ_3 whose value is 3. If 1 chooses g instead of h at move 4, then he gives up a sure $\frac{3}{4}$ to play, possibly, a game with a security level of $-\frac{1}{2}$. Player 1 can argue that this makes sense, for he already knows that 2 has made some stupid mistakes at elementary moves, so there appears to be a good chance that 2 will take choice i at move 5.   This argument might be particularly forceful if several of the outcomes of game Γ_3 with overall value 3 are particularly attractive to player 2, say with returns of -10, whereas none of the outcomes of Γ_2 are less than -1.

In summary, a strategy which dictates choices c and g on moves 2 and 4, respectively, cannot be maximin for this game, but it can very well be admissible (non-dominated) and perhaps psychologically desirable—but this is a metatheoretic statement, for such considerations are not formally encompassed by the theory.   The difficulty of including them is illustrated by the case where player 2 plays 1 for a sucker: By making mistakes

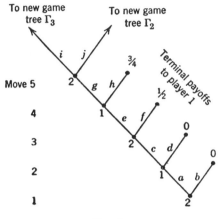

Fig. 5

2 induces player 1 to choose g at move 4 and then responds by selecting j and playing the game Γ_2 impeccably.   We shall return from time to time to the psychological problems which beset the shrewd player pitted against an antagonist who is shrouded in some mystery; however, this topic is best resumed under such headings as game learning in the non-strictly competitive case and games against nature.

It should be pointed out that our using the extensive rather than the normal form for this discussion was not accidental.   It is true that the reduction to the normal form leaves the strategic aspects of the game intact, but often it does not seem best suited to dealing with pertinent psychological, i.e., descriptive, information which we may have.

*4.12 A GUIDE TO THE APPENDICES ON
TWO-PERSON ZERO-SUM GAMES

Several books have been written concerned with the mathematical details of the two-person strictly competitive game, yet, for the most part, we have not given these details any consideration.   The reason is that our purpose in this book is to give a critique of the conceptual aspects of game theory, not a presentation of its mathematical intricacies; however, the two are not unrelated.   For this reason, and in order to give a more adequate survey of the field, we have included in the appendices a survey of this mathematical material.

Appendix 2 presents, in a relatively precise and self-contained manner, the principal theorem—the minimax theorem—of the two-person strictly competitive (zero-sum) theory.   A topological proof is given which rests on the Brouwer fixed-point theorem.

Appendices 3 and 4 give two different geometrical insights into the algebraic structure of games.   The geometry employed in Appendix 3 is very intuitive, and it makes plausible the truth of the principal theorem, although making the argument rigorous is a bit messy.   This geometrical model provides the basis for the double description method of solving games (see Appendix 6).   The geometry described in Appendix 4 illustrates clearly the mathematical relationship between the theory of games and the theory of convex bodies and their supporting hyperplanes.   The proof of the principal theorem follows readily from an understanding of this geometry.

Appendix 5 describes the relationship between two-person zero-sum game theory and the theory of linear programing (see section 2.3).   The reduction of a game to a linear-programing problem is examined, and this relationship is utilized to suggest the duality theory of linear programing. That theory is then stated precisely, and the major theorem, which is (not quite) proved, is then employed to give the converse reduction of a linear-programing problem to a game problem.

Appendix 6 is concerned with finding solutions to games with a finite number of strategies.   This is important, for the principal theorem establishes the existence of solutions but it does not indicate how to find them. Here, at best, the story is quite discouraging.   Although several methods are known and discussed for solving games, these algorithms usually require a fantastic amount of work, at least for games which purport to be realistic replicas of actual conflicts of interest.   The realism is achieved only at the expense of introducing a fabulous number of pure strategies. One might hope that cases involving such a large number of strategies could be idealized by a continuous model and that refined analytical methods could be brought to bear on the idealization.   In part this is possible, as we shall see in our discussion of infinite games in Appendix 7. However, in all honesty, we must admit that the number of existing techniques for the infinite case is small, and even in examples which have their "natural" description in the infinite case the usual hope is to reduce them to approximate finite games (cf. discussion of polynomial and polynomial-like games in Appendix 7).

There are two saving features in the solution of many games which arise in practice. First, although a game may involve a huge number of strategies, one can sometimes use the practical context to help reduce the model to its bare essentials by discarding many of the inadmissible strategies. Second, the context of the game often leads one to shrewd guesses about solutions or, in iterative procedures, about intelligent starting points.

We believe, and probably most of our colleagues would agree, that many important and interesting games will *never* be solved. This does not imply that game theory will never contribute anything to the understanding of these realistic games. Often, a *modus operandi* for a complicated case is to consider an auxiliary game which is motivated by and related to the original one in such a way that many of the important phenomena of the original are retained while the auxiliary remains solvable. From the solution of the auxiliary game one speculates informally how the results are modified in the original game. Thus, for example, there are simplified variants of both poker and bridge in the literature (see Bellman and Blackwell [1949], Bellman [1952 *b*], Gillies, Mayberry, and von Neumann [1953], Kuhn [1950 *b*], and Nash and Shapley [1950]). Such studies are in much the same spirit as economic analyses of idealized *Robinson Crusoe* and *Swiss Family Robinson* economies which, by means of a lot of hand waving, are used "to explain" economic phenomena and to reach policy decisions concerning the economy at large. This is dangerous, yes! Yet it is quite stimulating to our creative intuitions and often helpful in purely literary, pseudological (not said deprecatingly, but rather pragmatically) theorizing.

Contentwise, Appendix 6 begins with a discussion of the trial-and-error technique. A few guide posts are suggested for those who have occasion to indulge in this guessing game. Next, a rough geometrical (Appendix 3) explication of the Shapley-Snow algorithm is given and its relation to the double description method is explained. The simplex method—the most common way to solve linear-programing and game problems—is described, and its relation to the dual simplex method is illustrated geometrically. Finally, iterative techniques for finding solutions are given, e.g., a differential equations approach to equilibrium is examined. An iterative solution of games by fictitious play, due to Brown and von Neumann [1950], is of particular interest here since it has conceptual overtones for a descriptive theory of games. Brown [1951] states: "The iterative method in question can be loosely characterized by the fact that it rests on the traditional statistician's philosophy of basing future decisions on the relevant past history.   Visualize two statisticians, perhaps ignorant of min-max theory playing many plays of the same discrete zero-sum game.   One might naturally expect a statistician to keep track of the opponent's past plays and, in the absence of a more sophisticated calculation, perhaps to choose at each play the optimum pure strategy against the mixture represented by all the opponent's past plays."   The punch line is that such iterations generate a solution of the game.   The exact sense in which we mean "generate" is discussed in Appendix 6. Consequently, if one wants to find the solution of a game which is to be played but once, he can set up two fictitious players, generate a fictitious iteration of games with the players behaving as naive statisticians, and observe the outcomes which generate a solution.

Appendix 7 is devoted to two-person strictly-competitive games with an infinite number of pure strategies.   An extension of the minimax (equilibrium) theorem to a special class of these games was first accomplished by Ville [1938].   During the late 1940's a great deal of research at the RAND Corporation was devoted to these games.   This work was largely motivated by games of timing (when to shoot in a duel) and by games of partitioning (what proportion of one's resources should be allotted to a given endeavor).   In many of these games it was suitable to identify each pure strategy with a real number on the unit interval; hence there now exists a sizable (unclassified) literature on games over the (unit) square, a point (α, β) of the square corresponding to strategy α for the first player and a strategy β for the second.   It is only fair to remark that a partitioning-like game—the deployment of military forces—with an infinite number of pure strategies was extensively and elegantly treated by Borel [1938] before Ville's systematic treatment of a special class of infinite games.

At much the same time, but independent of Ville, Wald published a series of works [1945 a, 1945 b, 1947 b, 1950 a] on statistical decision theory in which he developed an extensive theory of two-person games with an infinity of pure strategies.[6]   Much that Wald did in statistical decision theory, using the theory of games that he so aptly developed, can perhaps be accomplished now more elegantly without game theory, but probably today's refinements would not have been achieved so readily without the hindsight of Wald's pioneering game-theoretic framework.

Once we allow an infinity of pure strategies the neat little story of games with a finite number of pure strategies—the existence of a value and optimal equilibrium strategies—can be violated in several different ways. Some of these complications are presented in Appendix 7.

[6] The interrelations between game theory and statistical inference are discussed at some length in Chapter 13.

In the final appendix we survey the literature of an important class of two-person games in extensive form, which, at the time of writing, constitutes one of the most active areas of research. Their major features are:

i. A succession of stages or trials.

ii. At each trial the players have complete information about past choices.

iii. At each trial the players make simultaneous choices.

iv. The number of trials, finite or infinite, may or may not be fixed in advance, but usually the length of play is determined by chance and by the actual sequence of player's choices.

The strategic problem at each trial can be visualized as a two-person game in normal form, which itself is treated as a single component of the dynamic supergame.

In the class of games known as *stochastic*, the strategy choices in a component game not only determine the positional payoff—an exchange of money or goods—but also control the probabilities governing which component game is to be played at the next trial, if any at all. The structure is so restricted that the play is almost certain to terminate in an undetermined but finite number of trials. In a *recursive* game payoffs are not made during the play; rather they occur at termination if the play is finite. For plays of infinite duration a convention is adopted for assigning payoffs. Such "real world" problems as games of (military) survival, attrition games, economic ruin games, dynamic programing, or compound statistical decision problems (e.g., classifying a stream of subjects) fall—admittedly a little unnaturally at times—into these categories of dynamic games.

4.13 SUMMARY

In this chapter we have covered what is probably the most central aspect of game theory: the existence of pairs of equilibrium strategies in two-person zero-sum games. A *strictly competitive* (or, equivalently, *zero-sum*) two-person game is one in which the two players have precisely opposite preferences. It is, therefore, a game in which cooperation and collusion can be of no value. Any improvement for one player necessitates a corresponding loss for the other. The term "zero-sum" is used because it is possible to choose the zeros and units of the two utility functions so that they always sum to zero. Such games are most compactly represented by matrices: the rows representing player 1's pure strategy choices; the columns, player 2's strategies; and the entries, the utility payoffs (by convention, those to player 1). The central problem was to devise a suitable theory for strategy choices in such games.

The theory was arrived at by the following metatheoretical condition: If the theory states that α_{i_0} and β_{j_0} are suitable (pure) strategy choices for players 1 and 2 respectively, this knowledge should not be sufficient reason for either player to make a different choice. In terms of the payoffs, this means that

$$a_{i_0 j_0} \geqslant a_{i j_0} \qquad \text{for every } i,$$

and

$$a_{i_0 j_0} \leqslant a_{i_0 j} \qquad \text{for every } j.$$

In other words, the strategies selected by the theory must have the property that the resulting utility is the maximum entry in its column and the minimum entry in its row. Such strategies are said to be in *equilibrium*, and each is called an *equilibrium strategy*.

It was shown that not every zero-sum two-person game has a (pure strategy) equilibrium pair, and that when one does exist it is not necessarily unique. The non-uniqueness does not, however, generate a new conflict of interest, for equilibrium pairs are *equivalent* in the sense of having the same utility payoff, and equilibrium strategies are *interchangeable* in the sense that any pair, one from each player, forms an equilibrium pair. Finally, any equilibrium strategy maximizes a player's security level, and, provided that an equilibrium pair exists, a strategy which maximizes his security level is an equilibrium strategy. Because of these properties, there does not seem to be any reason to restrict the theory to a subset of the equilibrium pairs.

This does not provide a theory for the games lacking pure strategy equilibrium pairs. By examining ways of raising security levels in such games, we were led to the idea of a *mixed* (or *randomized*) *strategy:* a probability distribution over a player's pure strategies. This resulted in a unified theory for all zero-sum two-person games, which is summarized in the famed *minimax theorem:* In the domain of mixed strategies, every zero-sum two-person game has at least one equilibrium pair, and when there are several they are equivalent and the equilibrium strategies are interchangeable. The common utility v of the equilibrium pairs is known as the *value* of the game.

The pure strategy and the mixed strategy theories are not in conflict: Every pure equilibrium strategy is also a mixed equilibrium strategy, and so the two theories yield a common value for the game.

How should a mixed strategy be interpreted, and of what value is it to a player? We advocated that it be interpreted as the selection of a pure strategy by a simple physical experiment. But this casts doubt on its worth, for, if in the final analysis a pure strategy is to be used, shouldn't the "best" pure strategy be selected in the first place? Keeping the pure strategy choice secret is often suggested as the reason for using mixed strategies. Although this is perhaps the most compelling reason, for some it cannot be the only reason since they advocate mixed strategies in games where it is immaterial whether the pure strategy choice is known. It was argued that they present a more flexible and profitable hedge against the total strategic situation than any of the pure strategies.

Equilibrium strategies possess certain optimal properties if both players use them, but what if one fails to? It was pointed out that the other player might profit by also deviating from an equilibrium strategy, but that this carries with it the risk of a more serious loss than could occur with an equilibrium strategy.

chapter 5

Two-person non-zero-sum non-cooperative games

5.1 INTRODUCTION

A non-strictly competitive game is exactly what it says: a game which fails to be strictly competitive because there is at least one pair of lotteries L and L' over the outcomes of the game such that one player prefers L to L' and the other does not prefer L' to L. For such games it is impossible to choose the utility functions of the players so that they sum to zero; hence we may use the terms "non-strictly competitive" and "non-zero-sum" interchangeably. Most economic, political, and military conflicts of interest can be realistically abstracted into game form only if their non-strictly competitive nature is acknowledged.

Naively, one would suspect that the element of agreement between the players would simplify the analysis; certainly in the extreme case where there is perfect agreement the analysis is trivial. In general, however, it is not simplified! We shall see that partial agreement confounds the issue to such an extent that there is neither as elegant nor as cohesive a theory as has been constructed for the strictly competitive game. In practice, an adequate discussion of non-zero-sum games seems possible only in terms of special cases, and, even so, one is often forced into extra-theoretic questions such as the "bargaining psychologies of the individuals," "interpersonal comparisons of utility," etc. The extent and complexity of this penumbra of indeterminateness, even in an idealized mathematical model, should invite speculation and experimentation among economists, sociologists, and psychologists, and at the same time it should give them pause when formulating a hasty verbal generalization or explanation.

In strictly competitive games it is impossible for the players to achieve mutual benefit by any form of cooperation; however, in non-strictly competitive games such mutual gain is always a possibility. Thus, we are forced to consider explicitly whether or not the players are permitted to cooperate. We shall only examine the two most extreme assumptions. By a *cooperative game* is meant a game in which the players have complete freedom of preplay communication to make joint *binding* agreements. In a *non-cooperative game* absolutely no preplay communication is permitted between the players. The latter games with two players are our present topic; the two-person cooperative games will be taken up in the next chapter. However, even in this chapter we shall from time to time attempt to indicate the differences preplay communication introduces.

As before, we denote the players by 1 and 2, their respective strategy sets by $A = \{\alpha_1, \cdots, \alpha_m\}$ and $B = \{\beta_1, \cdots, \beta_n\}$, and the outcome associated with (α_i, β_j) by O_{ij}. We assume that each player has preferences among mixtures of outcomes which lead to a linear utility function; let a_{ij} denote the utility of outcome O_{ij} for player 1, b_{ij} for player 2. This leads to a table of the form:

$$
\begin{array}{c}
\begin{array}{ccccc}
\beta_1 & \beta_2 \cdots & \beta_j & \cdots & \beta_n
\end{array} \\
\begin{array}{c}
\alpha_1 \\
\alpha_2 \\
\vdots \\
\alpha_i \\
\vdots \\
\alpha_m
\end{array}
\left[
\begin{array}{ccccc}
 & & & & \\
 & & \cdot & & \\
 & & \cdot & & \\
\cdots & (a_{ij}, b_{ij}) & \cdots & & \\
 & & \cdot & & \\
 & & \cdot & & \\
 & & & &
\end{array}
\right].
\end{array}
$$

Unless otherwise stated, we shall assume that both players are aware of all the data contained in the above table.

5.2 REVIEW OF THE SALIENT ASPECTS OF ZERO-SUM GAMES

It will be useful once again to summarize some of the important properties of the strictly competitive game, for these will be found not to hold in some non-zero-sum games.

If player 1 uses the randomized strategy $\mathbf{x} = (x_1\alpha_1, \cdots, x_m\alpha_m)$ and 2 uses $\mathbf{y} = (y_1\beta_1, \cdots, y_n\beta_n)$, the outcome O_{ij} occurs with probability x_iy_j. The utilities associated with it are a_{ij} and b_{ij}, so the expected utility of the choice $(\mathbf{x}, \mathbf{y})$ is

$$M_1(\mathbf{x}, \mathbf{y}) = \sum_{i, j} x_i a_{ij} y_j$$

for player 1 and

$$M_2(\mathbf{x}, \mathbf{y}) = \sum_{i, j} x_i b_{ij} y_j$$

for player 2. The motivation of 1 is to choose $\mathbf{x}$ so as to maximize M_1 and of player 2 to choose $\mathbf{y}$ to maximize M_2. In the strictly competitive game we selected the units and zeros of the utility functions so that

$$M_2(\mathbf{x}, \mathbf{y}) = -M_1(\mathbf{x}, \mathbf{y}),$$

which led to the term zero-sum.

We noted the following properties of zero-sum games:

i. It is never advantageous to inform your opponent of the (pure or mixed) strategy you plan to employ. (Of course, if a player plans to use an equilibrium strategy, his security level is not diminished by disclosing his intentions—but nothing is gained by the disclosure.)

ii. It never benefits the players to communicate prior to the play and to decide upon a *joint* plan of action.

iii. If $(\mathbf{x}, \mathbf{y})$ and $(\mathbf{x}', \mathbf{y}')$ are both in equilibrium, then:

(1) $(\mathbf{x}, \mathbf{y}')$ and $(\mathbf{x}', \mathbf{y})$ are both in equilibrium,

and

(2) $M_1(\mathbf{x}, \mathbf{y}) = M_1(\mathbf{x}', \mathbf{y}') = M_1(\mathbf{x}, \mathbf{y}') = M_1(\mathbf{x}', \mathbf{y}),$

and

$$M_2(\mathbf{x}, \mathbf{y}) = M_2(\mathbf{x}', \mathbf{y}') = M_2(\mathbf{x}, \mathbf{y}') = M_2(\mathbf{x}', \mathbf{y}).$$

iv. If $\mathbf{x}$ is a maximin strategy and $\mathbf{y}$ a minimax strategy, then $(\mathbf{x}, \mathbf{y})$ is an equilibrium pair, and conversely.

The next section is devoted to an analysis of a very simple two-person non-zero-sum game for which all four of these properties are violated.

5.3 AN EXAMPLE: BATTLE OF THE SEXES

The game we shall discuss is the following:

$$
\begin{array}{c@{}c@{}c}
 & \beta_1 & \beta_2 \\
\begin{array}{c}\alpha_1 \\ \alpha_2\end{array} &
\left[\begin{array}{cc}
(2, 1) & (-1, -1) \\
(-1, -1) & (1, 2)
\end{array}\right].
\end{array}
$$

Various interpretations are possible, but one seems most familiar; we may call it the "battle of the sexes": A man, player 1, and a woman, player 2, each have two choices for an evening's entertainment. Each can either go to a prize fight (α_1 and β_1) or to a ballet (α_2 and β_2). Following the usual cultural stereotype, the man much prefers the fight and the woman the ballet; however, to both it is more important that they go out together than that each see the preferred entertainment. Let us see whether this game possesses any of the four characteristics of zero-sum games.

The power of disclosing one's strategy. Player 1 would be most content with (α_1, β_1) whereas 2 prefers (α_2, β_2). If 1 announces that he plans to choose α_1 and that no arguments will alter his choice, and if 2 has faith in 1's stubbornness in sticking to his announced intentions, then she has no alternative but to choose β_1. A similar argument holds if 2 announces her intentions first. Thus, we see that it is advantageous in such a situation to disclose one's strategy first and to have a reputation for inflexibility. It is the familiar power strategy: "This is what I'm going to do; make up your mind and do what you want." If the second person acts in his own best interests, it works to the first person's advantage.

Preplay jockeying and its effect on utilities. In connection with the preceding point, we should like to recognize another phenomenon which will play havoc in much of the subsequent discussion; however, having raised it, we wish to de-emphasize it for now and to return to deal with it more fully later.

If, in the preplay discussion, the man says he is already committed to the prize fight and demonstrates his intention of going by producing the ticket he has already purchased, this may cause the woman to submit to his will, as argued above. But, to some spirited females, such an offhand dictatorial procedure is resented with sufficient ferocity to alter drastically the utilities involved in the payoff matrix. Preplay communication is considered outside the game structure of the payoff matrices, yet in some cases it may result in a radical change of one player's preference pattern and therefore of the payoff matrix. In such cases we could, perhaps, enlarge the space of strategies and complicate the game to include the preplay negotiations. Later we shall return to such points, but for now we shall suppose the payoff matrix remains invariant during the negotiations.

Some complications in the equilibrium concept. Continuing with the same payoff matrix, we note that both (α_1, β_1) and (α_2, β_2) are equilibrium pairs, since each strategy in one of the pairs is best against the other in the same pair. However, neither (α_1, β_2) nor (α_2, β_1) is an equilibrium pair. Furthermore, (α_1, β_1) and (α_2, β_2) do not yield the same returns to the players. Note well how completely these observations contrast with the properties of equilibrium pairs in zero-sum games (property iii, 1 and 2, of section 5.2).

Let us suppose that the players have no preplay communication and that they must make their choices simultaneously, i.e., they are to play the non-cooperative version of the game. Player 1 might reason as follows: "I want (α_1, β_1) and clearly my opponent wants (α_2, β_2), but if I take α_1 and she takes β_2, then we both lose out. Suppose, then, that I give in and take α_2—I still will do pretty well. But player 2 may reason the same way and give in to me, and again we would both lose with the (α_2, β_1) pair. Indeed, whatever rationalization I give for either α_1 or α_2 there is, by the symmetry of the situation, a similar rationalization for player 2, and so it seems inevitable that we both lose. This approach is none too promising, so let me consider maximizing my security level. I want to choose a mixed strategy $\mathbf{x}^{(0)} = (x_1^{(0)}\alpha_1, x_2^{(0)}\alpha_2)$ such that $\mathbf{x}^{(0)}$ maximizes the minimum of the two quantities

$$M_1(\mathbf{x}, \beta_1) \text{ and } M_1(\mathbf{x}, \beta_2)$$

associated with each $\mathbf{x}$." After some calculation, 1 finds that his maximin strategy is $(\tfrac{2}{5}\alpha_1, \tfrac{3}{5}\alpha_2)$ and that this results in a security level of $\tfrac{1}{5}$. Furthermore, he sees that if 2 selects β_1 the returns are $(\tfrac{1}{5}, -\tfrac{1}{5})$ and with β_2 they are $(\tfrac{1}{5}, \tfrac{4}{5})$.

Player 1 continues his monologue: "Hmm, by taking my safe strategy $\mathbf{x}^{(0)}$ I can guarantee myself at least $\tfrac{1}{5}$, but if 2 has any idea I'm going to play it safe she will play β_2 and get $\tfrac{4}{5}$. That is, if I can rationalize $\mathbf{x}^{(0)}$ for myself, then I can rationalize β_2 for player 2, in which case it would be best for me to choose α_2, and here we go again."

Similarly, 1 can compute 2's maximin strategy, which is $\mathbf{y}^{(0)} = (\tfrac{3}{5}\beta_1, \tfrac{2}{5}\beta_2)$, and the resulting returns, which are $(\tfrac{4}{5}, \tfrac{1}{5})$ if 1 plays α_1 and $(-\tfrac{1}{5}, \tfrac{1}{5})$ if he plays α_2. Thus, if he expects 2 to play her maximin strategy, then he should play α_1 for the return of $\tfrac{4}{5}$. But, if both take the "double cross" strategies (α_1, β_2), then the return is $(-1, -1)$, which is all the more reason to play the "safe" maximin strategies, which is all the more reason to "double cross," etc.

The difficulty, of course, is that the pair of maximin strategies $(\mathbf{x}^{(0)}, \mathbf{y}^{(0)})$ is not in equilibrium.

With this, we see that this single simple example fails to have all four properties of the zero-sum game; and it is for these reasons that the analysis of non-zero-sum games is so much wilder and (depending upon one's viewpoint) so much more interesting than is the zero-sum case.

The nature of the difficulties is very clearly illustrated by the following "giveaway program." Let two contestants play the following game: Each has a "safe" and a "double cross" strategy. The safe yields a return of

$1 to the player; double cross yields $1000 to a player provided the other player selects safe; and a pair of double crosses causes each player to give up $.05 to the house. The players are incommunicado, and they must announce their choices simultaneously. They are explicitly told that the house wants both to use the double cross strategy but that, after all, it is a fair game to the contestants since they are sure to get $1 if they play safe. The game is played only once by contestants who have never met.   It should be cheap advertising!

A geometrical representation of the game. Another way to see the complexities of the non-cooperative version of the game we have been discussing is to make a geometrical plot of the possible payoffs. Along the

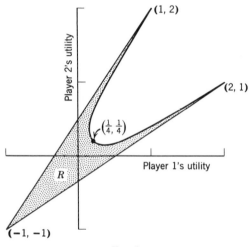

Fig. 1

horizontal axis we plot player 1's utility, and along the vertical, player 2's. Only certain combinations can arise; these are shown as the shaded area of Fig. 1. To each pair of mixed strategies $(\mathbf{x}, \mathbf{y})$ there corresponds a payoff which is one of the points in the shaded region; conversely, to each point in the region there corresponds at least one pair of strategies with this point as payoffs.

▶ It is worth noting that the pair of mixed strategies $[(\frac{3}{5}\alpha_1, \frac{2}{5}\alpha_2), (\frac{2}{5}\beta_1, \frac{3}{5}\beta_2)]$ is a (symmetric) equilibrium pair with a payoff of $(\frac{1}{5}, \frac{1}{5})$. There is little reason, however, to expect the players to choose these strategies, for if 2 were to choose $(\frac{2}{5}\beta_1, \frac{3}{5}\beta_2)$, then all of 1's strategies are equally good against it in the sense that they all have an expected return of $\frac{1}{5}$, and player 1's maximin strategy $(\frac{3}{5}\alpha_1, \frac{2}{5}\alpha_2)$ guarantees him $\frac{1}{5}$ against *any* strategy 2 might select. Thus, the maximin strategy seems preferable to the strategy of the symmetric equilibrium pair; yet it produces the complications we discussed above. Although this seemingly innocuous game possesses some symmetries, it is difficult to see how to exploit them.   ◄

On using the cooperative game to solve the non-cooperative one. A player of this non-cooperative game might wonder what he and his opponent would do if the game were actually cooperative, for if their roles are clear in the cooperative context then each might act as if he were in collusion with the other player even though no preplay communication is possible. Although this suggestion is natural, as we shall see it is not profitable, and, in fact, a much more useful idea is to reverse the theme: to analyze a cooperative game by constructing an appropriate auxiliary non-cooperative game.

Cooperatively, it is clear that the players would try to arrive at (α_1, β_1) or (α_2, β_2), and an "equitable" solution is for them to toss a coin, heads meaning that (α_1, β_1) is jointly chosen, and tails that (α_2, β_2) is jointly chosen. In the language of our interpretation: heads, the couple goes to the fight; tails, to the ballet. The utility for this *jointly* arranged randomized strategy is $(\frac{1}{2})2 + (\frac{1}{2})1 = \frac{3}{2}$ for each player. Note, however, that in the non-cooperative context a return of $(\frac{3}{2}, \frac{3}{2})$ is never possible— it lies outside the shaded region of Fig. 1. The strategy which randomizes between (α_1, β_1) and (α_2, β_2) can never be achieved if each player randomizes his strategies *independently*—which is exactly what must occur in the non-cooperative context. Thus, contemplating• the cooperative situation is merely frustrating for those prohibited from preplay communication.

Temporal collusion. Before we turn to a different example, let us briefly consider what might happen if this game were played not once but repeatedly in time and the payoffs were made after each game prior to the next play. Even when no preplay communication is permitted, there is, nonetheless, a form of involuntary communication. The players signal to each other via their choice patterns on previous plays. Introspectively, we would suspect that, after some preliminary jockeying, the players would settle on a pattern of alternation between (α_1, β_1) and (α_2, β_2). In this case the players can be thought of as playing $(\frac{1}{2}\alpha_1, \frac{1}{2}\alpha_2)$ and $(\frac{1}{2}\beta_1, \frac{1}{2}\beta_2)$, but their choices cannot be independent—they "correlate" their mixed strategies. We shall return again (in sections 5.5 and 5.6) to this notion of temporal collusion in games without preplay communication.

5.4 AN EXAMPLE: THE PRISONER'S DILEMMA

We turn now to a different example of a non-zero-sum game. This one is attributed to A. W. Tucker, and it has received considerable attention by game theorists. The payoff matrix is:

$$G: \begin{matrix} & \beta_1 & \beta_2 \\ \alpha_1 & \begin{bmatrix} (0.9,\ 0.9) & (0,\ 1) \\ \alpha_2 & (1,\ 0) & (0.1,\ 0.1) \end{bmatrix} \end{matrix}.$$

The following interpretation, known as the prisoner's dilemma, is popular: Two suspects are taken into custody and separated. The district attorney is certain that they are guilty of a specific crime, but he does not have adequate evidence to convict them at a trial. He points out to each prisoner that each has two alternatives: to confess to the crime the police are sure they have done, or not to confess. If they both do not confess, then the district attorney states he will book them on some very minor trumped-up charge such as petty larceny and illegal possession of a weapon, and they will both receive minor punishment; if they both confess they will be prosecuted, but he will recommend less than the most severe sentence; but if one confesses and the other does not, then the confessor will receive lenient treatment for turning state's evidence whereas the latter will get "the book" slapped at him. In terms of years in a penitentiary, the strategic problem might reduce to:

	Prisoner 2	
Prisoner 1:	Not Confess	Confess
Not Confess	1 year each	10 years for 1 and 3 months for 2
Confess	3 months for 1 and 10 years for 2	8 years each

If we identify α_1 and β_1 with not confessing and α_2 and β_2 with confessing, then—providing neither suspect has moral qualms about or fear of squealing—the above payoff matrix in utilities has the right character for the prisoner's dilemma. The problem for each prisoner is to decide whether to confess or not. The game the district attorney presents to the prisoners is of the non-cooperative variety.

Another version of this payoff matrix which will be intuitively more useful in some of the following discussion is

$$H: \begin{matrix} & \beta_1 & \beta_2 \\ \alpha_1 & \begin{bmatrix} (5,\ 5) & (-4,\ 6) \\ \alpha_2 & (6,\ -4) & (-3,\ -3) \end{bmatrix} \end{matrix}.$$

This will be given the interpretation that an entry $(-4, 6)$ means player 1 loses \$4 and player 2 receives \$6, and we shall suppose that each player wishes to maximize his monetary return. Note that if we take the utility of money to be linear with money and set the utility of \$6 to be 1 and of $-\$4$ to be 0, then the game G results from H.

Dominated and equilibrium strategies. Let us examine G first from 1's point of view. If 2 chooses β_1 or β_2, 1's second strategy is preferred to his first, since 1 is larger than 0.9 in the first case and 0.1 is larger than 0 in the second. Thus, α_2 strictly dominates α_1. Similarly, β_2 strictly dominates β_1. Since the players each want to maximize utility, α_2 and β_2 are their "rational" choices. Of course, it is slightly uncomfortable that two so-called irrational players will both fare much better than two so-called rational ones. Nonetheless, it remains true that a rational player (an α_2 or β_2 conformist) is always better off than an irrational player. In further support of these strategy choices, we may point out that (α_2, β_2) is the unique equilibrium pair of the game and α_2 and β_2 are also the unique maximin strategies for 1 and 2 respectively. However, the really important fact is that α_2 strictly dominates α_1 and β_2 strictly dominates β_1.

One might try to argue that the differences between 1 and 0.9 and between 0.1 and 0 are so small that even a criminal's ethics would make him select the first strategy so that they would not both be caught in the "stupid" (0.1, 0.1) trap. Such an argument is inadmissible since the numerical utility values are supposed to reflect all such "ethical" considerations. No, there appears to be no way around this dilemma. We do not believe there is anything irrational or perverse about the choice of α_2 and β_2, and we must admit that if we were actually in this position we would make these choices.

Cooperative aspects. Let us suppose that the players of the game G could cooperate; then it is clear that they would enter into a binding pact to stick to (α_1, β_1), since the only real alternative is (α_2, β_2) which neither prefers. But (α_1, β_1) is not in equilibrium, which is but a formal way of saying that there is good reason for each of them to defect on the bargain. The good reason is that, if a player defects and his opponent does not, then he profits; whereas, if he fails to defect and his opponent does, he loses more than he would if they were both to defect. Within the criminal context, such a "double cross" may engender serious reprisals and so it might be argued that it would not be worth while. This seems, however, to deny the utility interpretation of the given numbers. If we have ignored such considerations in abstracting a game from reality, we had better include the breaking of a binding agreement as an integral aspect of an enlarged game purporting to summarize the conflict of interest. Alternatively, we may suppose that the effect of breaking a binding agreement is so disastrous that it is not considered.

If we accept the second alternative, then clearly (α_1, β_1) is the choice. Does this knowledge help to resolve the non-cooperative game in any way different from (α_2, β_2)? We think not! The hopelessness that one feels in such a game as this cannot be overcome by a play on the words "rational" and "irrational"; it is inherent in the situation. "There should be a law against such games!" Indeed, some hold the view that one essential role of government is to declare that the rules of certain social "games" must be changed whenever it is inherent in the game situation that the players, in pursuing their own ends, will be forced into a socially undesirable position. That such social and economic games exist is illustrated in the next paragraph.

An alternative interpretation. As an n-person analogy to the prisoner's dilemma, consider the case of many wheat farmers where each farmer has, as an idealization, two strategies: "restricted production" and "full production." If all farmers use restricted production the price is high and individually they fare rather well; if all use full production the price is low and individually they fare rather poorly. The strategy of a given farmer, however, does not significantly affect the price level—this is the assumption of a competitive market—so that regardless of the strategies of the other farmers, he is better off in all circumstances with full production. Thus full production dominates restricted production; yet if each acts rationally they all fare poorly.

In practice the equilibrium may not occur since the farmers can, and sometimes do, enter into some form of weak collusion. In addition, a farmer does not play this game just once. Rather it is repeated each year and this introduces, as we shall see in the next section, an element of collusion. Finally, sometimes the government feels as we do, steps in, and passes a law against such games. Of course, in this analysis we have neglected the consumer. When he is included collusion may not be socially desirable even if it is desirable for the farmer.

5.5 TEMPORAL REPETITION OF THE PRISONER'S DILEMMA

Let us consider a game which is analogous to the prisoner's dilemma in the sense that it has the same strategic one-play aspects but which can be meaningfully repeated in time. Let it be of the form

$$H: \begin{array}{c} \\ \alpha_1 \\ \alpha_2 \end{array} \begin{array}{c} \beta_1 \qquad\quad \beta_2 \\ \left[\begin{array}{cc} (5,\,5) & (-4,\,6) \\ (6,\,-4) & (-3,\,-3) \end{array} \right], \end{array}$$

and suppose that it is repeated successively in time. We shall suppose that at each repetition (trial) the players make their choices simultaneously and that after each trial they receive the payoff resulting from that trial. This amounts to assuming that one's utility for the sequence of outcomes is the simple sum of the utilities in each component game. There is, in fact, nothing in utility theory to justify forming this sum, for it may or may not reflect one's overall preference.   What we are assuming explicitly in the following discussion is that it does reflect their preferences.   If this seems too gross an abuse of the utility notion, consider players who are *only* interested in the maximization of their *own* expected monetary return, and let the numbers in the payoff matrix represent money returns.

Let us suppose that the players have in one way or another arrived at a pattern of selecting (α_1, β_1).   Since player 1 then has reason to suppose that his opponent will "probably" choose β_1, he may be tempted to squeeze a bit more out of the next game by choosing α_2.   However, he may—and probably should—anticipate that the occurrence of (α_2, β_1) will ensure β_2 in the next game, in which case he is driven to play α_2 in that game, and so in total he will lose more than is compensated for by a 6 instead of 5. Thus, we may argue that his contemplation of the resulting chaos tends to keep 1 in line, and, if he is unable to reason so clearly about the future, a little experience should soon set him straight.   From these arguments, we see that in the repeated game the repeated selection of (α_1, β_1) is in a sort of quasi-equilibrium: it is not to the advantage of either player to initiate the chaos that results from not conforming, even though the non-conforming strategy is profitable in the short run (one trial).

It is intuitively clear that this quasi-equilibrium pair is extremely unstable; any loss of "faith" in one's opponent sets up the chain which leads to loss for both players.   Let us examine this in more detail.   Suppose the players are told that game H is to be played exactly twice, and suppose that each player is shrewd enough to see that his second strategy strictly dominates his first one in a single play of the game.   Thus, before making their first move, each realizes that in the second game the result is bound to be (α_2, β_2), for, after the first game is played, the second one must be treated as if H is going to be played once and only once.[1]   The second play being perfectly determined, the first play of the game can be construed as H being played once and only once.   Thus, it appears that (α_2, β_2) must arise on both trials.   The argument generalizes: Suppose they know that H is to be played exactly 100 times.   Things are clear on the last trial, the (α_2, β_2) response is assured; hence the penultimate trial, the 99th, is now in strategic reality the last, so it also evokes (α_2, β_2); hence the 98th is in strategic reality the last, so it evokes (α_2, β_2), etc.

[1] One might object that the particular choices made on the first trial will alter the utilities on the second, since these arise not only from the "physical" outcomes but also from one's attitude toward the opposing player.   Although this is sometimes true, it need not always be.   For example, if the outcomes are sufficiently important the changes in one's attitude will not alter the qualitative nature of the component payoff matrix. · Or we can assume that the outcomes are changed in such a manner as to compensate for the attitudinal changes.   In any case, the abstraction is clear enough!

This argument leads to (α_2, β_2) on all 100 trials. Indeed, if player 2 is a β_2-conformist on all trials, then player 1 is best off choosing α_2 on all trials, and conversely, i.e., (α_2, β_2) *on all trials* is an equilibrium pair.

This series of 100 trials of the component game H can easily be conceived of as a supergame in extensive form involving 100 simultaneous moves, where at each move each player has a pair of choices. A play of the supergame is the selection of a particular sequence of 100 pairs of choices, and the payoff is merely the sum of payoffs generated by the component games.[1] A strategy for player 1 in this new game is complex since it must prescribe what choice 1 is to make on each move and it may explicitly take into account the past history of all preceding moves. Complicated or not, it is certainly possible to consider the supergame in normalized form where each player merely chooses one overall strategy which dictates his full behavior pattern. It turns out that *any equilibrium pair of strategies will result in the repeated selection of α_2 and β_2 on each of the* 100 *moves*. Furthermore, repeated use of α_2 and β_2 are maximin strategies. Therefore, if one wishes to maximize his security level he should play the second strategy in each component game, and this strategy is reinforced by being in equilibrium. Although *any* equilibrium pair will result in the repeated use of α_2 and β_2, this does not mean that each player has a unique equilibrium strategy. For example, an equilibrium pair results when each player adopts the strategy of playing α_2 or β_2 for the first 99 moves and then changes to α_1 or β_1 on the 100th move if his adversary has used his first strategy for the first 99 moves. There are many other such pairs. The point, however, is that, even though there is not a unique equilibrium pair, there is a unique equilibrium outcome.

An indication of the technique used to prove this result will prove useful, for the method applies to other situations. Let us list the overall strategies of the supergame as:

$$A_1 = \{r_1, r_2, \cdot \cdot \cdot, r_{N_1}\} \qquad B_1 = \{s_1, s_2, \cdot \cdot \cdot, s_{N_1}\},$$

where N_1 is a finite but fantastically large number. One first notes that some of 1's strategies may be strictly dominated, i.e., there is at least one pair r_i and r_j such that r_i is never worse than r_j no matter what strategy 2 employs and for some of 2's strategies it is better. These dominated strategies may be thrown away with no loss to 1, leaving a new pure strategy set A_2 with $N_2 \leqslant N_1$ strategies. Similarly, player 2 has some strategies which are strictly dominated relative to A_1 and these may be thrown away leaving a set B_2 with (because of the symmetry of the supergame) N_2 strategies. Now, as long as 1 knows that 2 will confine himself

---

[1] See footnote on opposite page.

to B_2 (2 would be stupid not to!), he may be able to throw away more strategies that are strictly dominated relative to any choice 2 may make from B_2 (note: not from B_1). This results in a set A_3 with N_3 pure strategies, where $N_3 \leqslant N_2 \leqslant N_1$. Similarly, we define B_3. In this manner we go back and forth throwing away more and more strategies for 1 and 2. A strategy which is thrown away at any stage of this process is said to be *dominated in the wide sense*. For the supergame generated from H, one shows that any pair of strategies not dominated in the wide sense requires that the second strategy be used at every move. Furthermore, it can be shown that a strategy which is dominated in the wide sense cannot be part of an equilibrium pair.

If we were to play this game we would *not* take the second strategy at every move!

Let us see why. Denote by $\alpha_2^{(i)}$ the strategy in which α_2 is used exclusively from trial i on to trial 100, and for trial $j < i$, α_1 is used if and only if player 2 has used β_1 on trials 1, 2, $\cdots$, $j - 1$. In other words, we use α_1 so long as 2 uses β_1 or until trial i, whichever comes first, and after that we use α_2. Clearly, $\alpha_2^{(1)}$ is the equilibrium strategy of playing α_2 on each of the 100 trials. In a similar way, $\beta_2^{(i)}$ is defined. Now, if player 1 knows that 2 plans to use $\beta_2^{(k)}$ for $k > 1$, then his best response is $\alpha_2^{(k-1)}$, and if $k > 2$ his worst response is the equilibrium strategy $\alpha_2^{(1)}$, assuming that he must select one of these 100 strategies. If we were player 1 and were confined to these strategies, probably we would select a strategy $\alpha_2^{(i)}$, where i is some number in the nineties. Our particular choice would depend upon our subjective probability distribution over the set of strategy choices available to our adversary.

In practice, of course, one is not restricted to these strategies and probably would not choose from among them. A strategy like $\alpha_2^{(96)}$ seems too harsh in that an early aberration by player 2 is punished with an unrelenting ferocity—so much so that we would be "cutting off our nose to spite our collective face." Rather, we want to punish 2 for each choice of β_2 early in the game by using a series of α_2's, but at the same time we want to give him another chance to get into the (α_1, β_1) routine until we are ready to defect to α_2. Thus, after a series of punishing trials we would play α_1 again to see if he will accept it—a sort of game teaching, if you please. For this payoff matrix, we might reasonably allow 20 trials, say, for teaching purposes during which we would punish a β_2 choice against an α_1 choice by countering with an α_2 on the next trial. The total return to 2 for these two trials is less than the 10 he would have received had he played β_1 both times. If in that time he did not learn, then we would resort to playing α_2. Of course, it must not be forgotten that it is actually penalizing to 1 if he attempts to teach an opponent who is committed to the equilibrium strategy, but the opponent can benefit himself if he will cooperate a bit.

It should be mentioned that a dangerous but enticing strategy principle exists for this supergame: play dumb initially, and let your opponent teach you. In this way you are able to cash in a few times on your second strategy before settling down to (α_1, β_1). This becomes particularly tempting if, instead of playing H 100 times, the following are the payoffs:

$$\begin{bmatrix} (5, 5) & (-50, 50) \\ (50, -50) & (-3, -3) \end{bmatrix}.$$

This game has the same qualitative characteristics as H in that the second strategy dominates the first for each player; however, there is a much stronger temptation to play the second strategy. In this case a strategy like $\alpha_2^{(75)}$, literally interpreted, seems "reasonable."

Let us recapitulate. If H is played but once, we feel that it is "reasonable" to single out (α_2, β_2) as the "solution" of the game provided that there is no preplay communication. By reasonable we mean that we predict that intelligent players will play accordingly and, furthermore, that they will *still* do so even after a full airing of the "theory" of such a game. True, players will find themselves completely frustrated; nonetheless they have no real alternative. (Incidentally, in some contexts, if the two players are frustrated, it may be beneficial to society.) In contrast, we do not think it "reasonable" to single out $(\alpha_2^{(1)}, \beta_2^{(1)})$ as the "solution" when H is iterated n times, even though an equilibrium point results from this type of behavior and even though each of these strategies is undominated in the wide sense. It is not "reasonable" in the sense that we predict that most intelligent people would not play accordingly. Unfortunately, we know some individuals who, although brilliant in other ways, insist they would choose $\alpha_2^{(1)}$. Yet, we feel that, as long as our "subjective a priori probability" for our opponent selecting $\beta_2^{(1)}$ is less than 1, we should not single out $\alpha_2^{(1)}$. We feel that in most cases an unarticulated collusion between the players will develop, much in the same way as a mature economic market often exhibits a marked degree of collusion without any communication among the participants. This arises from the knowledge that the situation will be repeated and that reprisals are possible. One cannot help reflecting that, unfortunately, in the military and political spheres the participants all too often have a *single-play orientation*.

Flood [1952] has performed some empirical work on the 100-fold iteration of the above game. His running dialogue of the players' move-by-move reactions is amusing as well as instructive.

The iterations we have been discussing are of a somewhat special nature, namely: we assumed that both players knew exactly how many iterations would occur. Does the analysis alter markedly if we suppose this number is not known exactly, or not known at all? Intuitively, it seems plausible that the strategy of playing α_1 so long as 2 uses β_1 and a change to α_2 if 2 uses β_2 is an equilibrium strategy if H is repeated indefinitely. Actually, there is some difficulty in showing this to be so, since it is difficult to make precise the interpretation of the overall supergame. What is the payoff? There are an infinity of moves (games H) each of which has a finite payoff, so the sum of payoffs does not exist. Three methods for dealing with this have been suggested: First, to introduce a constant discount factor and to let all payments in the future be discounted back to the present (this serves to make the payoff sum convergent). Second, to let the number of trials be a random variable with a known distribution of the exponential type, i.e., the conditional probability of exactly n more trials, given that k trials have already occurred, is independent of k. An equivalent formulation of this stopping rule is to terminate the play at each trial with probability $1 - p$ and to repeat the game again with probability $p \neq 1$. Games of this general type, which are known as stochastic games, are studied in Appendix 8. If, in the first case, the discount rate is not too small and, in the second, the "repeat" probability p is not too small, then it is possible to show an equilibrium pair exists which results in the repeated use of the first strategy by both players. The third method assumes that the number of trials is fixed but unknown to the players. This last suggestion can be catalogued under the heading of a two-person, non-zero-sum, non-cooperative game against nature. In the extensive form, nature has the first move, so to speak, a choice of a number, and the other players cannot in any objective fashion assign a probability distribution over the possible choices of nature. This game, so far as we know, has not been analyzed, but we conjecture that a reasonable optimality criteria could be concocted so that repeated choice of the first strategy is optimal.

5.6 ITERATIONS OF ZERO-SUM GAMES

In our preceding discussion the numbers in the payoff matrices reflected the player's preferences; however, one might argue that a player is far more concerned with his relative advantage. Thus, we are led to the question: how does the analysis develop if we suppose that the game is interpreted in terms of a monetary payoff and one is not interested in the amount of money one obtains but in one's relative monetary advantage? It is clear that in such cases each component trial is a strictly competitive (zero-sum) game, so we restrict our attention to that case.

As a special case of the results in Appendix 8 we can conclude that, in a supergame comprising iterations of a zero-sum game, an overall strategy of choosing a maximin strategy at each trial is itself maximin in the super-game.   Furthermore, since the supergame is zero-sum, any such pair of strategies is an equilibrium pair.   This is an important difference between zero-sum and non-zero-sum games.   For zero-sum games, the repeated use of strategies which are optimal in the small constitutes a strategy which is optimal in the large; for non-zero-sum games, the repeated use of strategies which are optimal in the small might be unrealistic in the large, although if both players choose these strategies they are in equilibrium.

Actually, when a zero-sum game is iterated one is in a position to gain and exploit statistical information about one's opponent.   Clearly, this is important in life, and it emphasizes the difference between a descriptive and normative theory.   An example may point this up.   Consider the zero-sum game of matching pennies in which each player has two pure strategies: heads (α_1, and β_1) and tails (α_2, and β_2).   If they both choose the same strategy, 1 receives \$1 from 2; if they differ, 2 receives \$1 from 1. The money payoff matrix for player 1, therefore, is:

$$\begin{array}{c} & \beta_1 & \beta_2 \\ \alpha_1 & \begin{bmatrix} 1 & -1 \\ \alpha_2 & -1 & 1 \end{bmatrix}. \end{array}$$

Let us assume that each player attempts to maximize his expected monetary return.   Player 1's maximin strategy places a probability of $\frac{1}{2}$ on each strategy; this yields an expected return of zero no matter what 2 does. Thus, even if 2 fails to play his minimax strategy this deviation is not exploited by 1's maximin strategy.   This is the story we outlined in sec-4.11.   If the game is iterated, 1 can obtain sequential information about his opponent's strategy and, if it appears not to be optimal, then he can attempt to exploit the deviation.   To take an extreme example, suppose 2 is playing heads with probability $\frac{3}{4}$; then 1 should play heads with a probability greater than $\frac{1}{2}$.   Mathematically, at least for a single trial, his best choice is heads with probability 1; however, in an iterated game 2 would soon spot this strategy.   Player 1 must try to exploit 2's blunders without, however, teaching him the error of his ways.   If 1 judges his opponent to be shrewd, but not as shrewd as he, a more subtle tactic might be used: 1 departs slightly from optimality, and he lies in wait knowing that ultimately 2 will notice it and attempt to exploit it.   When 2 alters his strategy, 1 detects the change quickly since he is anticipating it and, since this strategy cannot be optimal, 1 subtly changes his strategy accordingly.   When 2 again catches on, 1 is ahead.

Such considerations as these seem realistic and would have to be encompassed in a fully descriptive theory, but the present theory of games cannot cope with such non-normative aspects in any formal manner.

5.7 THE ROLE OF EQUILIBRIUM PAIRS IN NON-ZERO-SUM GAMES

The game

$$\begin{bmatrix} (2, 1) & (-1, -1) \\ (-1, -1) & (1, 2) \end{bmatrix}$$

discussed in section 5.3 illustrates the complexities involved in constructing a normative theory for the non-cooperative non-zero-sum case. If there is to be a non-cooperative theory for this game, the least we can expect it to do is to suggest a strategy or class of strategies for each of the players; yet if the pair of strategies chosen is not in equilibrium there are reasons for the players not to act in conformity with the theory (see section 4.4). But since in the example (α_1, β_1) and (α_2, β_2) are both in equilibrium, yet (α_1, β_2) and (α_2, β_1) are not, what can the theory suggest to the players?

One might expect, however, no difficulty if each player had but one equilibrium strategy, as in the game

$$\begin{bmatrix} (0.9, 0.9) & (0, 1) \\ (1, 0) & (0.1, 0.1) \end{bmatrix}.$$

The pair (α_2, β_2) is the unique equilibrium pair, and, although we can sympathize with the frustrated player of this game, we are willing to subscribe to (α_2, β_2) as the "solution." On the other hand, a 100-fold iteration yields a non-zero-sum game with equilibrium behavior in which both players adopt their second strategies on all moves. We are not willing to subscribe to this as the "solution" of the iterated game.

Do these examples sound the death knell for the equilibrium concept as the principal ingredient of a theory of non-cooperative non-zero-sum games? In our opinion, the answer must be Yes if one demands a realistic theory for all possible non-cooperative non-zero-sum games, but it is No if one is willing to restrict the set of games for which it is asserted that "a solution exists." We shall ultimately make this precise.

It is unfortunate (or fortunate, depending upon your viewpoint) that a unified theory for all non-cooperative games does not seem possible. The only alternative seems to be to complicate the problem by introducing more initial information in the form of boundary and initial conditions—information referring to personality traits, psychologies of the players, etc.

On the other hand, there are some workers in the field who are not particularly disturbed by this state of affairs since they hold that almost no important realistic games are played in the non-cooperative context. We cannot help feeling that the realistic cases actually lie in the hiatus between strict non-cooperation and full cooperation, but that one should first attack these polar extremes.

We may also add that, even if it is possible to produce pathological examples which throw doubt upon the universality of a concept, this does not necessarily undermine its importance. It merely establishes that care must be exerted to check whether the concept is plausible in the specific cases to which it is applied. Ideally, one should attempt to investigate the mathematical restrictions which should be placed on the domain of admissible games so that the concept is plausible. In the case of the equilibrium point concept for non-cooperative games, we know that several major difficulties exist; nonetheless, it is an exceptionally important tool for the analysis of wide classes of economic games.

Even if we were to decide to reject equilibrium points as a normative theory for non-cooperative games (and remember there is no real alternative) it may still be that the notion is relevant in a description of behavior. Although not "all life is a game," at least not in our sense, we cannot fail to recognize that people are constantly jockeying to better their lot in a manner which is quite analogous to playing in an extremely complicated many-person game. For a given society, a set of mores and patterns of behavior gradually build up and then remain stationary for long periods of time; yet another society, with approximately similar initial conditions, will evolve to a quite distinct pattern of cultural norms. Loosely speaking, we may regard these as two possible equilibrium "solutions" to this "game." They are equilibria in the sense that an individual usually finds it disadvantageous to buck the tide of society's opinion. It is our impression that players of a game do, in some sense, evolve to an equilibrium position—not necessarily a unique one—and we can say that, from a descriptive, though not a normative, point of view, the set of equilibrium points of a game do constitute a "characterization of the solution" of the game. Although the following is not grounded in any specific empirical studies, we can imagine the participants of a complicated game floundering about using trial and error methods to arrive at a suitable mode of behavior and finally settling down to a pattern which is not in any sense a "social optimum," but which nevertheless is in equilibrium, since it does not profit any player individually to pioneer in new directions. Much of the n-person theory we shall discuss, if it is to be interpreted descriptively, must be interpreted in this manner.

*5.8 EXISTENCE OF EQUILIBRIUM PAIRS

In an extremely elegant proof, Nash [1951] has shown that *every non-cooperative game with finite sets of pure strategies has at least one mixed strategy equilibrium pair*. The idea involved in the proof (see Appendix 2) is delightfully simple: to each pair of mixed strategies $(\mathbf{x}, \mathbf{y})$ there is associated, by means of a mapping T, a new pair $(\mathbf{x}', \mathbf{y}')$ in such a manner that $(\mathbf{x}, \mathbf{y})$ is in equilibrium if and only if $(\mathbf{x}, \mathbf{y}) = (\mathbf{x}', \mathbf{y}')$. The nature of the mapping T is such that from a general existence theorem (the Brouwer fixed-point theorem) one can conclude that there is at least one element which remains fixed under T, i.e., there is at least one equilibrium pair.

Considering our previous discussion, it is reasonable that we should want to distinguish games in which the equilibrium pairs are equivalent and those in which they are interchangeable. So we give the following formal definitions: Two equilibrium pairs $(\mathbf{x}, \mathbf{y})$ and $(\mathbf{x}', \mathbf{y}')$ are *equivalent* if the returns to each player are the same, i.e.,

$$M_1(\mathbf{x}, \mathbf{y}) = M_1(\mathbf{x}', \mathbf{y}') \quad \text{and} \quad M_2(\mathbf{x}, \mathbf{y}) = M_2(\mathbf{x}', \mathbf{y}').$$

They are said to be *interchangeable* if $(\mathbf{x}, \mathbf{y}')$ and $(\mathbf{x}', \mathbf{y})$ are also in equilibrium.

Property iii of section 5.2 is equivalent to saying that for a zero-sum game any two equilibrium pairs are equivalent and interchangeable.

*5.9 DEFINITIONS OF "SOLUTION" FOR NON-COOPERATIVE GAMES[2]

A non-cooperative game is said to be *solvable in the sense of Nash* if every pair of equilibrium pairs are interchangeable.

Thus, the prisoner's dilemma is solvable in the sense of Nash as is the supergame generated from it by a fixed number of iterations, but the battle of the sexes

$$\begin{bmatrix} (2, 1) & (-1, -1) \\ (-1, -1) & (1, 2) \end{bmatrix}$$

is not since (α_1, β_1) and (α_2, β_2) are in equilibrium and are not interchangeable.

The *solution* of a game that is solvable in the sense of Nash is its set of equilibrium pairs.

A Nash-solvable game need not have equivalent equilibrium pairs, so Nash was led to define the *upper value* for a player as the most he can get

---

[2] This section tends to be more technical than most and not correspondingly more rewarding, so many readers may prefer to skip it.

from some equilibrium pair and the *lower value* as the least he can possibly get.

In the game

$$\begin{bmatrix} (1, 3) & (2, 3) \\ (1, 1) & (2, 1) \end{bmatrix}$$

every pair of pure or mixed strategies is in equilibrium and so it is automatically Nash solvable. Observe that a player's strategy choice has absolutely no influence on his return; this is entirely governed by his opponent's choice. Such a game is particularly frustrating since all it amounts to is this: 1 can give 2 from one to three (utility) units of satisfaction while remaining completely indifferent himself among his choices; similarly, 2 can give 1 from one to two units of satisfaction while remaining completely indifferent among his choices. Player 1's upper and lower values are 2 and 1, 2's are 3 and 1. Naturally, if the players could communicate they would make a binding agreement to adopt (α_1, β_2) yielding (2, 3), but in the non-cooperative context all strategies are indifferent for each of the players. In cases where the equilibrium concept does not lead to a unique mode of behavior, the players probably do well to contemplate the cooperative game.

A strategy pair $(\mathbf{x}, \mathbf{y})$ is said to be *jointly inadmissible* if there exists a strategy pair $(\mathbf{x}', \mathbf{y}')$ such that each prefers the latter to the former, i.e.,

$$M_1(\mathbf{x}', \mathbf{y}') > M_1(\mathbf{x}, \mathbf{y}) \qquad \text{and} \qquad M_2(\mathbf{x}', \mathbf{y}') > M_2(\mathbf{x}, \mathbf{y}).$$

In this case, $(\mathbf{x}', \mathbf{y}')$ is said to *jointly dominate* $(\mathbf{x}, \mathbf{y})$. A pair $(\mathbf{x}, \mathbf{y})$ is *jointly admissible* if and only if it is not jointly dominated by another pair.

A non-cooperative game is said to have a *solution in the strict sense* if:

i. There exists an equilibrium pair among the jointly admissible strategy pairs.

ii. All jointly admissible equilibrium pairs are both interchangeable and equivalent.

The second condition prohibits confusion in the case of non-unique jointly admissible equilibrium pairs.

The pairs (α_1, β_1) and (α_2, β_2) are in equilibrium in the game

$$\begin{bmatrix} (1, 1) & (0, 0) \\ (0, 0) & (2, 2) \end{bmatrix},$$

but they are not interchangeable, so the game is not Nash solvable; however, (α_1, β_1) is jointly dominated by (α_2, β_2), so the latter is the solution in the strict sense. The prisoner's dilemma has no jointly admissible equilibrium pair, hence it is not solvable in the strict sense, but it is solvable in the Nash sense.   Similarly, when that game is iterated a fixed number of times it is not solvable in the strict sense but again it is solvable in the Nash sense.

We now wish to weaken this last concept in such a way that there is a "solution" when the prisoner's dilemma is played only once but not when it is iterated.

The first thing we must do is to introduce a suitable definition of domination for mixed strategies.   Let X and Y be arbitrary subsets of the sets of mixed strategies for players 1 and 2, respectively.   If we suppose that player 1 knows that he will restrict his attention to X and that 2 will restrict his attention to Y, then it is actually possible that 1 need only attend to a subset X^* of X.   This would be so if every strategy which is in X but which is not in X^* were dominated by some mixed strategy in X^*. Clearly, we would want to consider the smallest such set.   Thus, we are led to the following definition: Given X and Y, the subset X^* of X is said to be a *minimal complete class* of strategies of X relative to Y if:

i. For any $\mathbf{x}$ in X but not in X^* there is at least one $\mathbf{x}^*$ in X^* such that $M_1(\mathbf{x}^*, \mathbf{y}) \geqslant M_1(\mathbf{x}, \mathbf{y})$ for all $\mathbf{y}$ in Y, and greater than holds for at least one $\mathbf{y}$ in Y.

ii. No proper subset of X^* has property i.

An analogous definition holds for player 2.   It can be shown that minimal complete classes of mixed strategies over a finite number of pure strategies exist and are unique.

Now, if we have a non-cooperative game with spaces X and Y of mixed strategies each based on a finite number of pure strategies, then player 1 might just as well confine himself to the minimal complete class for X relative to Y, call it $X^{(1)}$, and 2 to the minimal complete class for Y relative to X, call it $Y^{(1)}$.   In general, this will effect some reduction of the number of strategies under consideration.   If $(X, Y; M_1, M_2)$ denotes a non-cooperative game with mixed strategy spaces X and Y and payoff functions M_i, then the *associated reduced game* is defined to be $(X^{(1)}, Y^{(1)}; M_1, M_2)$.   In other words, it is the same game except that the players confine themselves to their minimal complete classes.   For example, the associated reduced game of the prisoner's dilemma is trivial: there is only the one strategy pair (α_2, β_2).

We shall say that a game is *solvable in the weak sense* if its associated reduced game is *solvable in the strict sense*.

The prisoner's dilemma is solvable in the weak sense, but n-fold iterates of it are not.

Since the players have nothing to lose by confining themselves to the sets of strategies $X^{(1)}$ and $Y^{(1)}$, one is tempted to define $X^{(2)}$ as the minimal complete class for $X^{(1)}$ relative to $Y^{(1)}$, and $Y^{(2)}$ as the minimal complete class for $Y^{(1)}$ relative to $X^{(1)}$, and to argue that they should attend only to $X^{(2)}$ and $Y^{(2)}$. But there are dangers, for although player 1 has nothing to lose in considering only $X^{(1)}$, we can say the same for $X^{(2)}$ only provided (and this is the rub) that 2 confines himself to $Y^{(1)}$. If 2 does not confine himself to $Y^{(1)}$ (which seems silly, but empirically it is at least plausible) then 1 might suffer a disadvantage by restricting himself to $X^{(2)}$.

In like manner, we may define sets $X^{(3)}$ and $Y^{(3)}$, $X^{(4)}$ and $Y^{(4)}$, etc., where $X^{(n)}$ is the minimal complete class for $X^{(n-1)}$ relative to $Y^{(n-1)}$ and $Y^{(n)}$ is a minimal complete class for $Y^{(n-1)}$ relative to $X^{(n-1)}$. As far as 1 is concerned, a reduction from $X^{(n-1)}$ to $X^{(n)}$ is only safe if he feels confident that 2 will confine himself to $Y^{(n-1)}$; if 2 does not and 1 keeps on reducing the set of strategies he will consider, he may be asking for trouble.

If 1 and 2 both keep reducing their strategy spaces, the process must eventually terminate in the sense that there is an integer N such that $X^{(N)} = X^{(N+1)}$ and $Y^{(N)} = Y^{(N+1)}$. $X^{(N)}$ and $Y^{(N)}$ are called the *completely reduced* strategy spaces and $(X^{(N)}, Y^{(N)}; M_1, M_2)$ the *completely reduced game* associated with $(X, Y; M_1, M_2)$. A non-cooperative game is said to be *solvable in the complete weak sense* if the associated completely reduced game is solvable in the strict sense.

Any pair of strategies in the completely reduced strategy spaces of the n-fold iterate of a prisoner's-dilemma-type game reduces to the choices of α_2 and β_2 from the first move on, so it is solvable in the complete weak sense.

5.10 SOME PSYCHOLOGICAL FEATURES

Although we have considered many special definitions of what constitutes a "solution" of a non-cooperative game, the above analysis is pitifully incomplete. One might be tempted to argue that, in the absence of any adequate theory, the players should, as a last resort, choose the maximizer of their security level; however, it can be shown that the resulting maximin value (optimal security level) never exceeds that of any equilibrium pair. Thus, even from a very conservative point of view, the equilibrium pairs are worthy of a great deal of consideration.

Within the realm of equilibrium pairs, one might hope to extend the domain of analysis further by introducing a more subtle partial ordering of the equilibrium points by taking into consideration psychological factors. For example, in the game

$$\begin{array}{c} \quad\quad \beta_1 \quad\quad\quad \beta_2 \\ \begin{array}{c} \alpha_1 \\ \alpha_2 \end{array} \left[\begin{array}{cc} (4, \ -30) & (10, \ 6) \\ (8, \ 8) & (5, \ 4) \end{array} \right] \end{array}$$

the pairs (α_2, β_1) and (α_1, β_2) are jointly admissible equilibrium pairs, but they are neither interchangeable nor equivalent. Thus, there is no solution to this game in the strict, weak, complete, or Nash sense. Nonetheless, if 2 has any reason to fear that 1 will take α_1, then he dare not take β_1 for fear of getting -30 (change this to -300 or -3000 if the point is not clear), but 1, knowing this, has every reason to take α_1, which gives him his best return. But now the argument is cyclic, for 2, having some rationalization for 1's adoption of α_1, has all the more reason to avoid β_1. Thus, the equilibrium pair (α_1, β_2) "psychologically dominates" (α_2, β_1). In this analysis it is not only the qualitative ordering of the numbers which counts; the quantitative aspects are extremely significant.

In the game

$$\left[\begin{array}{cc} (4, \ -3000) & (10, \ 6) \\ (12, \ 8) & (5, \ 4) \end{array} \right]$$

the equilibrium pair (α_1, β_2) is jointly dominated by (α_2, β_1) since the former yields a return of $(10, 6)$ versus $(12, 8)$ for the latter. Thus, the latter point is a solution in the strict sense. Yet if we were player 1 we would hesitate to use α_2 on the grounds that player 2 would argue that β_2 psychologically dominates β_1, and so long as 2 can give any rationale for 1's choosing α_1, 2 does not dare choose β_1. The argument is cyclic and it reinforces (α_1, β_2) even though it is jointly dominated by (α_2, β_1).

No doubt there are other psychological aspects of a similar sort, and at present the theory seems inadequate to cope with them. Although admitting that the structure developed in these sections does not adequately reflect many of the psychological factors of the non-cooperative games, we feel that there is some element of realism in most of these concepts—enough, at least, to demand scrutiny of these ideas before coming to a decision in a non-cooperative game.

5.11 DESIRABILITY OF PREPLAY COMMUNICATION

One might hope that some of the many difficulties in analyzing non-cooperative games would evaporate once preplay communication and the existence of binding agreements were assumed. At least this seems to have been the case in all the examples up to now. In actuality, although some of the difficulties are ameliorated, preplay communication generates a new crop of puzzling features which demand clarification. The following chapter is concerned with them, but before that we can consider the question whether the members of a non-cooperative game should always elect to have preplay communication if it is offered to them.

If there is no preplay communication, the analysis of the game

$$\begin{bmatrix} (1,\ 2) & (3,\ 1) \\ (0,\ -200) & (2,\ -300) \end{bmatrix}$$

is simple because α_1 strictly dominates α_2, and β_1 strictly dominates β_2. Furthermore, the pair $(\alpha_1,\ \beta_1)$ is the unique equilibrium point which is jointly admissible, so it is the solution (in any sense) of this non-cooperative game. Now suppose the players were forced into preplay communication. Player 1 can demand that they enter into a binding agreement to choose $(\alpha_1,\ \beta_2)$ by the threat to choose α_2 if 2 does not agree. To be sure, 1 does not want to take α_2, which would give him only 0, but if he does 2 is faced with a loss of 200 (which cannot be said to give 1 any satisfaction beyond 0 since we are already dealing with utilities). It is reasonable to suppose 2 will succumb to the "threat" if the same numbers for players 1 and 2 somehow denote changes of comparable importance. Regardless of some of the potential pitfalls of the above analysis, it is to 2's advantage to refuse to come to a conference table, for to confer would only allow 1 to browbeat him into an agreement.

If this non-cooperative game is iterated, then it might as well be cooperative, for 1 can force his desires on 2 by taking α_2 a few times until 2 learns the "score." This is a vivid example of "collusion through iteration."

In this example we have some preview of coming attractions in the cooperative game: the threat powers of players and their attendant interpersonal comparisons implied in such phrases as "this will hurt you more than it does me."

5.12 SUMMARY

The analysis of non-strictly competitive, i.e., non-zero-sum games, is inherently different from that of strictly competitive ones. In the zero-sum case, it is never advantageous to disclose one's strategy, equilibrium pairs are equivalent, equilibrium strategies are interchangeable, and maximin and minimax strategies are in equilibrium. An example served to show that all these assertions are false, in general, for non-zero-sum games. Further examples exhibited other pathologies. In the prisoner's dilemma, each player has one undominated pure strategy, but the payoff to that pair, even though it is in "equilibrium," is not jointly desirable. Thus, although a "rational" player cannot do better than play his undominated strategy (assuming a single-shot game without preplay communication), two "irrational" players will always fare better than two "rational" ones.

When a non-zero-sum non-cooperative game is repeated many times, certain of the strategic aspects change. For example, even without any formal preplay communication, the players can develop some form of temporal collusion. This can assume the form of a temporal patterning of their choices of pure strategies, resulting in a correlated joint strategy (e.g., the pair (α_1, β_1) is chosen on odd trials and (α_2, β_2) on even trials); or it can involve threat strategies to police an informal status quo agreement. For example, in the prisoner's dilemma, the players may, after some experience, each repeatedly select his first pure strategy—the dominated strategy when the game is viewed as a single-shot affair. But, should one player succumb to the temptation of a short range gain, the other can resort to punitive action by also defecting to his undominated strategy for the next few trials. This same game, when repeated a known fixed number of times, can be analyzed as a supergame from an equilibrium point of view. The unique overall equilibrium behavior demands that each player employ his undominated strategy on each trial, which seems contrary to ordinary wisdom. Though we conclude from such examples that the equilibrium notion is not universally applicable to non-zero-sum games, it still remains an important analytic tool for a wide class of games. In addition, even if it is rejected as the basic tool for a normative theory, we argued that it may be of pragmatic importance in descriptive studies.

In two starred sections, 5.8 and 5.9, we stated the important theorem that mixed-strategy equilibrium pairs always exist, and we gave alternative definitions for what might reasonably be meant by a "solution" of a non-zero-sum game. In section 5.10, the possibility was raised that a subtle partial ordering of the several equilibrium points of a game could be effected by considering psychological factors. Although this was illustrated by several examples, the analysis is pitifully incomplete, and such considerations remain in the realm of "artful judgments."

We assumed at the outset that in this chapter preplay communication would be prohibited by the rules of the game. However, in most of our examples, it appeared as if life would be considerably simpler were preplay communication allowed. To avoid the illusion that this is always so, an example was given (section 5.11) where one player would definitely prefer not to have the privilege of preplay communication, for with it would come the realistic possibility of his opponent threatening to hurt him badly, even at some expense to the opponent, if he did not agree to a strategy particularly beneficial to his opponent. More of this in the next chapter.

Some authors are not disturbed by the inadequacies of the non-cooperative theory, since they feel that preplay communication is usually possible in realistic contexts.   Other authors feel, on the contrary, that the non-cooperative theory is of paramount importance, even were preplay communication invariably permitted.   Their goal is to formalize the preplay communication as precise moves, thus yielding an extensive game which, so to speak, is grafted onto the beginning of the normalized game.   This enlarged game is then to be treated as non-cooperative in character. Again, more of this in the next chapter.

chapter 6

TWO-PERSON
COOPERATIVE GAMES

6.1 INTRODUCTION

In the preceding chapters we have prohibited both preplay discussion and binding agreements between the players, except to the extent of considering them as possible (and not too successful) intuitive aids to finding solutions of non-cooperative games. In this chapter we turn full attention to the other extreme: cooperation in two-person games. Explicitly, we assume that:

i. All preplay messages formulated by one player are transmitted without distortion to the other player.

ii. All agreements are binding, and they are enforceable by the rules of the game.

iii. A player's evaluations of the outcomes of the game are not disturbed by these preplay negotiations.

Of these assumptions, the third is the least palatable for many applications. Should it be an unreasonable assumption for a particular abstraction of some reality, then an alternative abstraction *must be* effected which includes the negotiations as an integral aspect of the strategic possibilities. In this way the outcomes can be made to depend upon the negotiations.

Since most of our examples will suggest that cooperation is valuable to all concerned, it is worth recalling that games can be constructed (see section 5.11) such that one player can be expected to resist coming to a conference table, because his willingness to cooperate would subject him to realistic threats without, at the same time, benefiting him. This is to say, there may well be strategic considerations in agreeing to negotiate; these we shall ignore, and we shall assume that negotiation is compulsory.

Most authors feel that, if such economic problems as duopoly, labor-management disputes, trade regulations between two countries, etc., can be treated as games at all, then it will have to be in the cooperative context. In like manner, one may hope that it will prove possible to formulate cooperative game models which reflect limited aspects of the diplomatic relations between two countries or of the political conflict between two parties within a single country. To be sure, such small parcels of a complex social or economic problem can be realistic only to the extent that the utility functions chosen do reflect the subtle interrelations between the game-in-the-small and the overall problem. Given the present state of game theory, we are indeed skeptical that many such problems can be given a realistic formal analysis; rather, we would contend that a case can be made for studying simplified models which are suggested by and related to the problem of interest. The hope is that, by analogy, their analysis will shed light—however dim and unreliable—on the strategic and communication aspects of the real problem.

6.2 THE VON NEUMANN-MORGENSTERN SOLUTION

Consider again (see section 5.3) the payoff matrix

$$
\begin{array}{c}
\quad\quad\quad \beta_1 \quad\quad\quad \beta_2 \\
\begin{array}{c} \alpha_1 \\ \alpha_2 \end{array}
\left[
\begin{array}{cc}
(2, 1) & (-1, -1) \\
(-1, -1) & (1, 2)
\end{array}
\right].
\end{array}
$$

It will be recalled that the set of possible payoffs can be given in a drawing as shown in Fig. 1. To any point in the shaded area R there is a pair of mixed strategies $(\mathbf{x}, \mathbf{y})$ such that the payoffs $[M_1(\mathbf{x}, \mathbf{y}), M_2(\mathbf{x}, \mathbf{y})]$ are the coordinates of that point; and, conversely, to every pair of mixed strategies, the corresponding pair of payoffs constitutes a point in the shaded area.

Recall that if this game is repeated in time, it is reasonable for the players to alternate, in phase, between their first and second strategies. This yields (2, 1) and (1, 2) as alternate payoffs, and the average payoff per trial is $(\tfrac{3}{2}, \tfrac{3}{2})$. This expected payoff cannot be achieved in a single trial if the players randomize without any preplay communication; how-

ever, if they can communicate, then they can achieve the single trial
expectation of $(\tfrac{3}{2}, \tfrac{3}{2})$ by tossing a fair coin to decide whether to choose
(α_1, β_1) or (α_2, β_2). Thus, by correlating their mixed strategies, which
is possible with preplay communication, the players are able to enlarge
their potential payoff set in this game. Let us call the region generated
by considering all correlated mixed strategies R'.

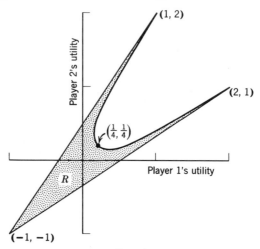

Fig. 1

▶ In general, not every joint randomization which yields a pure strategy pair
(α_i, β_j) with probability z_{ij} is realizable from the independent selection of mixed
strategies. For example, in the above example the randomization of probability
$\tfrac{1}{2}$ on (α_1, β_1) and $\tfrac{1}{2}$ on (α_2, β_2), i.e.,

$$[\tfrac{1}{2}(\alpha_1, \beta_1),\ 0(\alpha_1, \beta_2),\ 0(\alpha_2, \beta_1),\ \tfrac{1}{2}(\alpha_2, \beta_2)],$$

which yields the expected payoff $(\tfrac{3}{2}, \tfrac{3}{2})$ is not realizable if the players use independent random strategies. It is, of course, possible if the players *act cooperatively
through preplay communication*.

Any randomized strategy agreed upon by both players will be termed a *joint
randomized strategy*. Typically, the symbol **z** will be used for a joint randomized
strategy, and the set of all such joint randomized strategies will be denoted by Z.
The expected payoffs from the choice of a joint mixed strategy **z** are, of course,

$$M_1(\mathbf{z}) = \sum_{i,j} a_{ij} z_{ij},$$

$$M_2(\mathbf{z}) = \sum_{i,j} b_{ij} z_{ij},$$

where, it will be recalled,

$$a_{ij} = M_1(\alpha_i, \beta_j), \qquad b_{ij} = M_2(\alpha_i, \beta_j), \qquad \cdot$$

and z_{ij} is the probability assignment of the pair (α_i, β_j) for mixed strategy **z**.

As in the non-cooperative case, when z runs over all its possible values in Z, the set of points $[M_1(z), M_2(z)]$ generates a region in the plane, which will be denoted by R'. For the game given at the beginning of the section, the associated region is shown in Fig. 2.

In general, the region R' can be described in the following way: Plot the points in the plane associated to all (a_{ij}, b_{ij}) pairs (e.g., the points $(1, 2)$, $(2, 1)$ and $(-1, -1)$ in Fig. 2); then R' is the smallest (polygonal) convex body containing these points. This definition is clear except for the words "convex body." A set of points in the plane (or in any Euclidean space, for that matter) is called *convex*

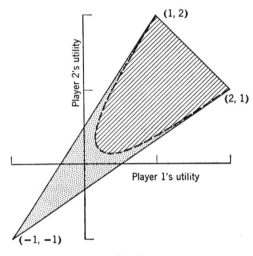

FIG. 2

if whenever two points lie in the set the line segment joining them also lies in the set. Thus, for example, all the points interior to a circle comprise a convex set—but not a convex body. A *convex body* is a convex set which also contains its boundary[1] and which is bounded in the sense that there is a circle about the origin which includes the whole set if the radius is chosen to be sufficiently large. Another way to describe R' is as the "convexification" of R: the least addition of points to R which results in a convex body.   ◄

Suppose that the region R' of some game is of the form shown in Fig. 3. By acting jointly, the players can achieve any point of R' as their payoff. A point (u, v) of R' is said to be *jointly dominated* by a *different* point (u', v') of R' if both $u' \geqslant u$ and $v' \geqslant v$. Clearly, the players need not consider any point which is jointly dominated by another one of R'. Thus, after a little preliminary negotiation, they can be expected, if they are rational, to confine their attention to the jointly *un*dominated outcomes, which in

[1] Formal mathematical definitions of such words as "boundary," "interior," and "exterior" can be given, but it is hardly appropriate to do so here.

this case form the darker line a, b, c, d of R' (see Fig. 3).   These undominated outcomes are called the *joint maximal set* of R' (the term *Pareto optimal set* is also used).   Let it be clear that the knowledge assumptions of the underlying game model mean that the region R' and the joint maximal set are known to each of the players; there is no element of "hidden strategies" or of bluffing permitted in the present analysis.

Clearly, player 1 desires the point d most and player 2 the point a. Moreover, on the joint maximal set the player's preferences are strictly opposing; hence, once they confine themselves to consideration of the joint maximal set it is not possible for them to cooperate further for mutual benefit.   Nonetheless, although each player may prefer an end point of the joint maximal set, it is easy to see that such desires are generally totally unrealistic.   For example,

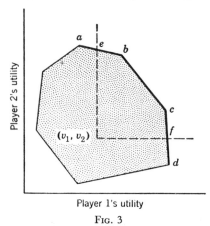

Player 1's utility

Fig. 3

by treating the game in a non-cooperative manner, player 1 can guarantee himself an amount v_1 and player 2 an amount v_2, the maximin values,[2] and it is unreasonable to suppose either player will accept less than his maximin value in the negotiations.   These acceptable points—the points of the joint maximal set of R' which yield each player at least as much as he can secure independently by playing his maximin strategies— form what is called the *negotiation set* of the game; this set is denoted by $\mathfrak{N}$.   In Fig. 3 the negotiation set is the part of the dark line marked e, b, c, f.

The cooperative two-person theory of von Neumann and Morgenstern [1947] singles out the negotiation set as the "cooperative solution" of the game.   In words, the players act jointly to discard all jointly dominated payoff pairs and all undominated payoffs which fail to give each of them at least the amount he could be sure of without cooperating.   They have argued that the actual selection of an outcome from the multiplicity of points in the negotiation set $\mathfrak{N}$ depends upon certain psychological aspects of the players which are revelant to the bargaining context.   They acknowledge that the actual selection of a point from $\mathfrak{N}$ is a most intriguing problem, but they contend that further speculation in this direction is not

---

[2] Of course, knowing the region R' is not sufficient to calculate these values; the payoff matrix is required.   Thus, their location in Fig. 3 depends upon the payoffs, and these we have not specified.   So our choice of their location is arbitrary.

of a mathematical nature—at least, not with the present mathematical abstraction.

6.3 SOLUTIONS—IN WHAT SENSE?

Although von Neumann and Morgenstern do not feel that further restrictions can be placed upon the outcome of the cooperative game, several other authors have attempted to single out a unique payoff from $\mathfrak{N}$ which they put forward as a "realistic" solution to the abstract problem. For example, in section 6.9 we shall discuss a scheme due to Nash which formalizes in a precise, but particular, fashion the preplay communication and bargaining. The resulting extensive game is treated as a non-cooperative game without preplay communication, and one of its equilibrium points which has a special mathematical property is taken to be the solution to the cooperative game. This will be preceded by a discussion of Nash's bargaining model (section 6.5) which is a stepping stone for his treatment of the general cooperative game.

In what sense can any such unique point be considered a solution? For example, in the game

$$\begin{bmatrix} (2,\ 1) & (-1,\ -1) \\ (-1,\ -1) & (1,\ 2) \end{bmatrix}$$

there are many who would object to choosing $(\frac{3}{2},\ \frac{3}{2})$ as the cooperative solution. To be sure, they would agree that the symmetry of the matrix affords some argument for choosing the symmetric outcome in $\mathfrak{N}$, but they would question whether a possible asymmetry of the roles of the players has not been overlooked. For example, player 1 may refer to a whole labor union and 2 to a single entrepreneur; or 1 may be the government serving the interests of the nation, whereas 2 is a monopolist. But if such role asymmetry exists, then surely "solution" is not meant in an ethical sense—as a concept of a *just* outcome. Furthermore, it surely cannot be meant in a descriptive sense, since the players may also be asymmetric in bargaining ability, one being a tough bargainer and the other indifferent. So, in what sense is it a solution?

Indeed, one can argue that the whole of $\mathfrak{N}$ is probably not suitable as a solution in a descriptive sense. Consider the game

$$\begin{bmatrix} (0,\ 100) & (100,\ 0) \\ (-1,\ -200) & (-40,\ -300) \end{bmatrix}.$$

Let us suppose that the payoffs are in dollars and that the two players are in roughly equivalent financial positions. In the non-cooperative game

α_1 dominates α_2 and β_1 dominates β_2, and therefore the pair (α_1, β_1), which gives rise to the payoff $(0, 100)$, is clearly the solution; however, in the cooperative game player 1 can reasonably demand the $(100, 0)$ payoff by threatening to take α_2 if 2 does not comply with his wish. Note that the threat strategy α_2 is inadmissible from 1's selfish viewpoint in the non-cooperative game, but in the cooperative game it is a possible threat since, presumably, 2 will be hurt more than 1 if it is used. If 2 does not agree to the $(100, 0)$ payoff, player 1 cannot very well merely threaten to play the non-cooperative game; either he must continue to bargain or he must be willing to play out his threat α_2. In turn, player 2 might threaten to take β_2 if 1 does not agree to $(0, 100)$; but this does not seem very convincing since it punishes 2 "more" than it does 1. At this point the bargaining personalities of the two players will begin to enter: Each can hurt the other in differing degrees and so they will dicker, threaten, withdraw, bluff, etc., until an outcome is reached.

To predict what will in fact happen without first having a complete psychological and economic analysis of the players seems foolish indeed. We would claim no descriptive limits: anything is liable to happen—including the $(-40, -300)$ payoff! Specifically, *we would not claim that an actual observed outcome will lie in the negotiation set.*

What then is the point and interpretation of a solution? Our views are given in the next section where we discuss arbitration schemes.

Before we turn to that, it will be useful to catalogue three possible and extreme interpretations of payoffs of a game. How a cooperative game is treated depends significantly on which of these, or other, possibilities is assumed:

i. Payoffs are in utility terms, no interpersonal comparisons of utility are permitted, and no side payments are allowed.

ii. Payoffs are in utility terms, interpersonal comparisons are meaningful (or partially meaningful), and no side payments are allowed.

iii. Payoffs are in monetary terms, utility is linear in money, interpersonal comparisons are meaningful, and monetary side payments are allowed.

A fourth case, not discussed in this chapter, is partially explored in section 10.4. There the payoffs are assumed to be physical commodities and interpersonal comparisons are not assumed to be meaningful, but side payments in physical goods are permitted.

In all cases it is assumed that the payoffs are fully known to the players and that the preplay communication does not alter the utilities. We agree that such extreme assumptions are probably not quite realistic, yet, in some special contexts, one or another may serve as a reasonable abstraction.

6.4 ARBITRATION SCHEMES

Let us suppose that the players of a specific cooperative game have restricted their attention to the negotiation set and that they are bitterly bargaining over which point to select. The harder each bargains, the more he will probably get—except, of course, if he is forced to carry out a threat! This is the rub about cooperative games. It is well known that in such situations so-called "rational" people (and countries) frequently have failed to reach an agreement, and the threats have had to be carried out, to their mutual discomfort. For these reasons players are often willing to submit their conflict to an arbiter, an impartial outsider who will resolve the conflict by suggesting a solution.

We may suppose that the arbiter sincerely envisages his mission to be "fairness" to both players; however, there are not, as yet, any simple and obvious criteria of "fairness," so, in effect, he is being asked to express a part of his ethical standards when resolving the game. The arbiter can be assumed to want to suggest a solution which will seem "reasonable," both because he is sincere and because he may wish to be hired for such tasks in the future. Thus, for example, he would be mistaken to suggest a solution having an obvious alternative which is preferred by both players. Or suppose there are two different conflict situations and that everyone agrees player 1 is strategically better off in the first than the second; then the arbiter should not give player 1 less in the first than the second. In short, an arbiter will (or should) try to satisfy some consistency requirements. In addition, as with most adjudicators, he will be anxious to defend his suggested solutions with some fairly good rationalization. All of this means that he should be prepared to formulate and to defend the basic principles which lie behind his suggested compromises—they should not be completely arbitrary!

Whatever principles may be involved, the net effect is to associate to each possible conflict of interest a single outcome. Thus, we define an *arbitration scheme* to be a function, i.e., rule, which associates to each conflict, i.e., two-person non-strictly competitive game, a unique payoff to the players. This payoff is interpreted as the arbitrated or compromised solution of the game. Without further specification, there are clearly an infinity of such functions. We could try to select one on an intuitive basis and attempt to defend it; however, we should always fear that someone might concoct a hypothetical situation for which the arbitrated solution is at variance with our intuitive "ethical norms." Rather than dream up a multitude of arbitration schemes and determine whether or not each withstands the test of plausibility in a host of special cases, let us invert the procedure. Let us examine our subjective intuition of "fairness" and formulate this as a set of precise desiderata that any acceptable arbitration scheme must fulfill.   Once these desiderata are formalized as axioms, then the problem is reduced to a mathematical investigation of the existence of and characterization of arbitration schemes which satisfy the axioms.

It may turn out that no arbitration scheme can exist which satisfies all the requirements—that is, the desiderata as formulated may be self-contradictory.   If so, and it is not uncommon, then we are forced to an "agonizing reappraisal" of our intuitive norms.   Let there be no mistake that one need not worry about this happening: the inner inconsistency need not be the least bit obvious.   Such a possibility should give many social scientists pause, for think how futile is a search for a subjectively "reasonable" arbitration scheme when our notions of what is reasonable are inconsistent.

Or, it may turn out that there is exactly one arbitration scheme compatible with our desiderata.   This is the ideal: the desiderata can then be considered a full characterization of that arbitration scheme, and, of course, they help in defending the reasonableness of the scheme.

Or, finally, it may turn out that many schemes are compatible with the axioms.   In that case, the desiderata are a partial characterization of the set of compatible schemes.   Any scheme of this set may serve as an acceptable arbitrator of conflict situations.   If, however, certain of them still yield arbitrated results which we deem "unfair," then one must search for further stipulations as to the meaning of "fairness" to add to our axioms and thereby eliminate those we consider undesirable.

The problem of multiple arbitration schemes may appear to give rise to a dilemma, since each scheme will compromise a particular game in a specific manner.   This situation seems hopeless, for each player will clearly prefer that scheme which yields him the largest payoff.   Thus, in terms of a particular game each player will rank the schemes according to their desirability to him, and these rankings are strictly opposed; this substitutes another game over the arbitration schemes for the original bargaining conflict over payoffs of a game.   Not much is accomplished.

But is this the problem?   Or is it to confront the same pair of people with the same set of arbitration schemes before these individuals are concerned with the arbitration of a specific game?   Put more precisely, they should evaluate the schemes with respect to a wide class of possible games.   It is now more difficult to rank the schemes, to be sure, but also there is much more hope that they may agree as to the most desirable or the "fairest" one.   In a sense, this is the problem we all face when we vote for one among a slate of judges.   Each offers himself as an arbiter for an unknown group of legal contests, and in principle he is required to offer a platform, i.e., a set of reasonable principles, which characterizes the nature of the decisions he will make if elected.   By and large, a common cultural background results in considerable unanimity about basic legal principles which renders such elections perfunctory, except in those occasional cases when there is a burning legal issue which forces the electorate to evaluate the principles of the several candidates not in general terms but in terms of this specific issue.

The power of the axiomatic method is this: By means of a (small) finite number of axioms we are able "to examine" the infinity of possible schemes, to throw away those which are unfair, and to characterize those which are acceptable.   The only alternative—to examine in detail each of the infinity of schemes for each of the infinity of possible conflicts it is supposed to arbitrate—is not practical.

What are some reasonable principles for an arbitration function to fulfill?   At this time we do not wish to go into great detail, and so the reader should be tolerant of the following formulations; they are stated in a very rough manner—ambiguities exist and misinterpretations may be anticipated.   However, such ambiguities will be eliminated once we make the conditions mathematically precise.

i. The arbitrated solution of a specific conflict situation (two-person non-strictly competitive game) should be an element of the negotiation set of the game.   In other words, the arbitrated solution should give each player as much as he could be expected to gain if he played non-cooperatively, and there should not be any other feasible payoff preferred by both players to the solution.

ii. The arbitrated solution, as seen in terms of the real underlying conflict, should not depend upon the particular utility units used in abstracting the problem into a formal framework.

iii. The arbitration scheme should be egalitarian in the sense that it is independent of the names or labels of the individuals in the conflict.

iv. If two games are "close to each other" in some strategic sense, the arbitrated solutions should also be close.   Put another way, slight perturbations or errors of measurement should not drastically alter the arbitrated solution.

v. The arbitrated solution should reflect the threat capabilities of the players in the conflict situation.   (This condition, in particular, needs considerable clarification.)

It is all too easy to say "Ah yes, these are very reasonable conditions," so let us append a word of caution.   It is often difficult to assess how reasonable an axiom actually is in its abstract setting; we must seek its meaning in concrete contexts, looking particularly for cases where it leads to peculiar outcomes.   In this way its real limitations become apparent. Furthermore, if we accept a set of conditions as reasonable, we also accept their consequences.   It is perfectly possible to cite examples of seemingly reasonable conditions, each appearing to be innocuous in itself, which collectively yield unpalatable consequences.   In this case, by going back to the original conditions, we can usually see why one or more are not as reasonable as they first seemed.   Consequently, before committing ourselves on the acceptability of a set of axioms or conditions, we are well advised to investigate some of the consequences with an eye to ferreting out the hidden jokers.

Other principles, some of which will appear in later sections, could be listed.[3]   But, rather than do this, our strategy will be, first, to consider a special class of conflicts, called bargaining games, in which we can illustrate in detail the mathematical counterparts of some of the above verbal conditions.   Then, second, we will turn to the more complex case of the general two-person non-strictly competitive game; however, we shall not study it in the same detail as the bargaining games.

6.5 NASH'S BARGAINING PROBLEM

"The economic situations of monopoly versus monopsony, of state trading between two nations, and of negotiation between employer and labor union may be regarded as bargaining problems."   [Nash, 1950 b.] The task is to give a formal definition of a bargaining problem and to solve it; this Nash has attempted, and we shall discuss his work on bargaining in this and the following section.

Consider two individuals, 1 and 2, who "are in a position where they may barter goods but have no money to facilitate exchange," but who do have the facilities to perform randomized experiments.   Each comes to the market with an initial bundle of goods, and a trade takes place if and only if each consents to it.   By a trade we mean an actual reapportionment of the joint bundle of goods held by them.   Let T, T', etc., denote different possible trades.   In the class of all possible trades there is one which is distinguished, namely, when no trade actually occurs—the status quo. This we denote by T^*.

---

[3] For example, we might impose the condition that an arbitration scheme should not hedge in the following sense: In one situation, the compromise is grossly unfair to player 2; whereas, in another, it is grossly unjust to 1.   Its unreasonableness in both situations is rationalized on the grounds that the overall expected payoffs are reasonable since it is not known which situation, if either, will arise.   To defend such a scheme it really would be necessary to make an *a priori* probability assignment to potential conflict situations, and, although it may be possible to do so in some contexts, we will restrict our attention to applications where such hedging would be unacceptable.

We shall suppose that each player's preferences over randomized outcomes are consistent in the sense of Chapter 2, and so they may be mirrored by a numerical utility index. Thus, associated to each trade T is a pair of utilities (u, v) representing, respectively, the utility of T for 1 and 2. Denote the utility pair of T^* by (u^*, v^*). In this way, each trade T can be represented as a point in the plane. If T is represented by (u, v) and T' by (u', v'), a randomization between T and T' is represented by a specific point on the line segment joining (u, v) and (u', v'); and, conversely, any such point represents a randomization between these trades.

Let R denote the set of all points representing trades and randomizations between trades. R is bounded, convex, and closed (i.e., it contains

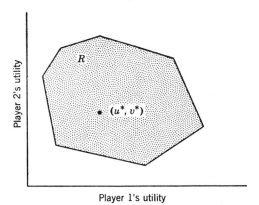

Player 2's utility

R

$\bullet$ (u^*, v^*)

Player 1's utility

Fig. 4

its boundary). In Fig. 4 a typical R is shown. The region will be polygonal, as shown, if the underlying set of trades is finite, i.e., if no commodities are infinitely divisible; otherwise it need not be.

The choice of utility functions is not unique, as has been stressed, but, if we had made other choices having different units of scale and different origins, the new region would be merely a translation and stretching of the old one. Thus, the same underlying bargaining problem has an infinity of simply related abstract versions.

In summary, then, a bargaining problem is characterized by a region R of the plane and a special point (u^*, v^*) of R. We denote such a bargain by $[R, (u^*, v^*)]$. The following interpretation holds: If no trade occurs, the payoff is (u^*, v^*), and a trade takes place if and only if both players agree upon a unique point of R which then constitutes the payoff. Naturally, 1 desires a trade represented by a point as far to the right as possible in R, and 2 wishes to obtain a point as high in R as possible. In general, of course, these are incompatible desires; nonetheless, it behooves them to perform some trade so long as there are any points in R both above and to the right of (u^*, v^*). *The existence of such points will be assumed throughout.*

Now, for each such game $[R, (u^*, v^*)]$, we want to single out a unique payoff (u_0, v_0) which is a "fair" outcome for the players. In the terms of the preceding section, we wish to prescribe a precise formula which will arbitrate any such bargaining problem. Symbolically, we want a function F which operates on $[R, (u^*, v^*)]$ to give a point (u_0, v_0) of R. We shall first present such a function, one due to Nash, and then present the desiderata which lead to it and which are its defense. The "formula" is:

i. Change the origin of measurement of utility for each player so that the point (u^*, v^*) is transformed into $(0, 0)$, and let the resulting transformation of R be denoted by R'. In words, we choose utility functions such that the status quo gives zero payoff.

ii. In the region R' find the unique[4] point (u_0', v_0') such that $u_0'v_0'$ is the maximum of all products uv, where (u, v) is in R', i.e.,

a. (u_0', v_0') is a point of R', $u_0' > 0$, $v_0' > 0$.

b. $u_0'v_0' \geqslant uv$, for all (u, v) belonging to R' such that $u \geqslant 0$ and $v \geqslant 0$.

The point (u_0', v_0') is the "Nash solution" to the bargaining game $[R', (0, 0)]$. The solution to $[R, (u^*, v^*)]$ is obtained by inverting the utility transformations on (u_0', v_0'). This point can be characterized more directly—though perhaps less suggestively for the proof below—as the unique point (u_0, v_0) of R such that

$$(u_0 - u^*)(v_0 - v^*) \geqslant (u - u^*)(v - v^*),$$

for all (u, v) belonging to R, and such that $u \geqslant u^*$ and $v \geqslant v^*$.

But why this scheme rather than any number of other possible choices? The primary reason is that it is the only arbitration scheme which satisfies the four properties given below, and it is mainly in the light of their "reasonableness," plus experience with the scheme for specific examples, that we can judge the plausibility of the formula.

Let F denote an arbitration scheme which maps a typical bargaining game $[R, (u^*, v^*)]$ into an arbitrated outcome. We demand the following:

Assumption 1 (invariance with respect to utility transformations). *Let $[R_1, (u_1^*, v_1^*)]$ and $[R_2, (u_2^*, v_2^*)]$ be two versions of the same bargaining game, i.e., they differ only in the units and origins of the utility functions. Then the arbitrated values $F[R_1, (u_1^*, v_1^*)]$ and $F[R_2, (u_2^*, v_2^*)]$ shall be related by the same utility transformations.*

---

[4] Such a point exists because R' is bounded and closed. It is unique because R' is convex and because we are assuming the existence of a point (u, v) in R' such that $u > 0$ and $v > 0$.

Assumption 2 (Pareto optimality). *An arbitrated outcome* (u_0, v_0) *shall have the following properties:*

(i) $u_0 \geqslant u^*$ *and* $v_0 \geqslant v^*$,
(ii) (u_0, v_0) *is a point of* R,
(iii) *There is no* (u, v) *in* R, *different from* (u_0, v_0), *such that* $u \geqslant u_0$ *and* $v \geqslant v_0$.

In words, the arbitrated value must be: (*i*) *at least as good as the status quo,* (*ii*) *feasible, and* (*iii*) *not bettered by any other feasible point.*

Assumption 3 (independence of irrelevant alternatives). *Suppose that two different bargaining games have the same status quo point and that the trading possibilities of the one are all included in the other. If the arbitrated value of the game with the larger set of alternatives is actually a feasible trade in the game with the smaller set, then it shall also be the arbitrated value of the latter game. Put another way, if certain new feasible trades are added to a bargaining problem in such a manner that the status quo remains unchanged, either the arbitrated solution is also unchanged or it becomes one of the new trades. In symbols, suppose* $[R_1, (u^*, v^*)]$ *and* $[R_2, (u^*, v^*)]$ *are the two games and that*

(i) R_1 *is a subset of* R_2,
(ii) $F[R_2, (u^*, v^*)]$ *is in* R_1,

then

$$F[R_1, (u^*, v^*)] = F[R_2, (u^*, v^*)].$$

Assumption 4 (symmetry). *Suppose the version* $[R, (u^*, v^*)]$ *of a bargaining game has the following properties:*

(i) $u^* = v^*$,
(ii)   *If* (u, v) *is in* R, *then* (v, u) *is in* R,
(iii) $(u_0, v_0) = F[R, (u^*, v^*)]$,

then

$$u_0 = v_0.$$

In words, if an abstract version of a bargaining game places the players in completely symmetric roles, the arbitrated value shall yield them equal utility payoffs, where utility is measured in the units which made the game symmetric.

Now we repeat the punch line: *The Nash formula described above not only satisfies these four assumptions, it is the only function which does so.* Put another way, these desiderata implicitly define a unique arbitration scheme for bargaining games.

▶ The proof is simple, so we will outline it for the mathematically oriented. Consider a version $[R, (u^*, v^*)]$ of a bargaining game, and let $[R', (0, 0)]$ be the translation of it which puts the status quo at the origin. Find the point (u_0', v_0') in R' such that $u_0'v_0' \geqslant uv$ for all (u, v) in R'. Next, change the scales of utility measurement in such a way that (u_0', v_0') maps into $(1, 1)$, and call this representation of the game $[R'', (0, 0)]$. We now show that $(1, 1)$ is the solution of $[R'', (0, 0)]$, from which it will follow by assumption 1 that (u_0', v_0') is the solution of $[R', (0, 0)]$ and so $(u_0' + u^*, v_0' + v^*)$ is the solution of $[R, (u^*, v^*)]$. There exists a symmetric set R''' which contains R'', and so $(1, 1)$, but no other point (x, y) such that $x \geqslant 1$ and $y \geqslant 1$. According to assumptions 2 and 4, $(1, 1)$ is a solution of $[R''', (0, 0)]$; hence by assumption 3 it is also a solution of $[R'', (0, 0)]$. Thus, the Nash formula is a solution to the problem. To show that it is the unique solution is easy, for suppose F' is a function distinct from F which is also a solution. If it disagrees with F at some version $[R, (u^*, v^*)]$, then the above construction is contradicted. This shows that if a solution exists to Nash's axioms it must be this one; it is straightforward to show that in fact it does satisfy them. ◀

6.6 CRITICISMS OF NASH'S MODEL OF THE BARGAINING PROBLEM

A number of criticisms can and have been made of Nash's model; some that we shall examine do not appear to us to be relevant, but others seem more serious. Some confusion has resulted from Nash's presentation of the problem. He has always confined himself to formal versions of the bargaining situation in which the status quo trade is at the origin, or, in other words, the zeros of the individual utility functions were arbitrarily chosen to be the case of no trade. "In a bargaining situation one anticipation is especially distinguished; this is the anticipation of no cooperation between the bargainers. It is natural, therefore, to use utility functions for the two individuals which assign the number zero to this anticipation. This still leaves each individual's utility function determined only up to multiplication by a positive real number." [1950 *b*, p. 157.] This has caused confusion, for many readers have falsely interpreted this to mean a definite loss of generality. They have argued that not only is this choice unnatural but that it implicitly establishes an interpersonal comparison of utility. It has been held that it is a serious flaw in an otherwise ingenious argument. In our presentation we purposely departed from Nash's discussion so as to give a simple refutation of these contentions.

A second group of criticisms surround the meaning of this "solution" concept. Certainly, it is not a prediction of what actually happens in bargains—it is easy to cite empirical cases where an agreement is not reached and the players end up at the status quo point. Nash contends that his solution is a "fair" division which purportedly should reflect the "reasonable expectancies" of "rational bargainers." "Now, since our solution should consist of *rational* expectations of gain by the two bargainers, these expectations should be realizable by an appropriate agreement between the two. Hence there should be an available anticipation which gives each the amount of satisfaction he should expect to get.   It is reasonable to assume that the two, being rational, would simply agree to that anticipation or to an equivalent one.   Hence we may think of one point in the set of the graph as representing the solution and also representing all anticipations that the two might agree upon as *fair bargains.*" [1950 *b*, p. 158.]   It is not easy to make these statements more precise,

other than to say that assumptions 1 and 4 stipulate the principle of "fairness"; nonetheless, the spirit of the argument is clear.

If this is accepted, then most remaining criticism takes the form of examples where it is contended that the Nash solution is not fair. We feel that often these criticisms are not just to the arbitration scheme, since the critics seem to be demanding that a "fair" arbitration scheme yield a "fair" solution when applied to an "unfair" situation.   Consider the game where two players are to divide $100

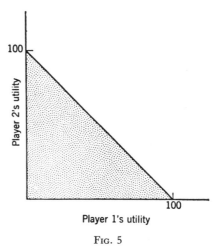

Fig. 5

between them if they can reach an agreement, or where they will receive nothing if they do not agree.   The region of possible outcomes is shown in Fig. 5.   If they can agree upon a point in this region, bargainer 1 will receive the value of the first coordinate and bargainer 2 the value of the second one; if not, they each get zero.   If utility is assumed to be linear with money, then the Nash solution is (50, 50).   However, suppose there is an asymmetry in the economic roles of the players.   For example, suppose 1 is rich and selfish and 2 is poor.   Player 1 may then make a good case for (75, 25), let us say, on the grounds that the utility increment to 2 of $25 is at least as great as the gain to 1 of $75.   He would argue that an ethically fair division of the proceeds is a fair division of utility, not of money.   For example, suppose that (part of) the utility functions were these:

Monetary Payoff		Utility		
1	2	1	2	Product of Utilities
$0	$100	0.00	1.00	0.000
25	75	0.25	0.98	0.245
50	50	0.50	0.90	0.450
75	25	0.75	0.73	0.548
100	0	1.00	0.00	0.000

Recall that we give these numbers the following interpretation. The values of zero and one are chosen to be the extreme values, and there is no loss of generality since any linear transformation of a utility function is acceptable. Any other value represents an outcome which is indifferent to a lottery between the two extremes, e.g., for player 1, the rich man, ($75, $25) is indifferent to the lottery which gives him $100 (and player 1 $0) with probability 0.75, and gives him $0 (and player 1 $100) with probability 0.25; for player 2, the same outcome is indifferent to ($0, $100) with probability 0.73 and ($100, $0) with probability 0.27. For the points indicated, the outcome ($75, $25) has the maximum utility product, and it is clear that we can fill in all other values of utility in such a way that this remains true, in which case this "unfair" division is the Nash solution to the bargaining problem.

Note well how the asymmetry in the roles of the players enters into the solution. The model does not equate the status quo positions of the players, but the asymmetry of economic roles is partially reflected in the shape of their utility functions. Ethically, this may be unfortunate, since the economic asymmetry works to the detriment of the poor man and to the benefit of the rich one. This has been cited as an example to show that the Nash axioms are not "ethically fair." We must point out, however, that Nash's solution only purports to give a "fair" arbitrated value of the bargain when the *strategic aspects* are taken into account. These are captured by the utility functions, and, had the utilities of the rich man been more "socially ethical," the poor man would have received a better break. As an illustration, keep the same utilities for the poor man and change those of the rich one to:

Monetary payoff	$0	25	50	75	100
Utility	0.00	0.30	0.85	1.00	0.90

Then it is possible to fill in the rest of the values so that the Nash solution is ($50, $50). We observe that this function is chosen so that it takes player 2's preferences into account in that, for example, the ($75, $25) split is preferred by 1 to ($100, $0).

Having rejected these general objections, let us turn to a critical evaluation of each of the axioms. With respect to assumption 1, we hold that the applicability of the model is restricted just because it does *not* allow for some interpersonal comparison. Contrast this with the fact that Nash's solution has been falsely accused of establishing an implicit interpersonal comparison of utility. An example will make this difficulty clear. Suppose that the players must, without using any side payments, agree upon a point in the region shown in Fig. 6. If we suppose that each player's utility is linear in money, then Nash's solution is (5, 50). Is this fair?

Suppose, first, that the players are in roughly equivalent economic positions, then player 1 can make a good case for the point on the line joining (10, 0) to (0, 100) in which both players get the same rewards, approximately (9.09, 9.09). His argument can be based on two grounds: First, the threat that, if 2 holds out for ($5, 50), then 1 will not agree to it. This he can afford to do since he will lose much less than 2 will. Second, the ethical argument that the reference point really should be (0, 0) and that each should be made to gain equally. Player 2 would surely argue that if they move from (5, 50) to (9.09, 9.09), then he will have given up $40.91 and 1 will have only added $4.09. Again, 1 would question why (5, 50) is the reference point rather than (0, 0), claiming that the asymmetry of the problem is far more apparent than real. "For instance," he might argue, "suppose that there were two games under consideration: the one we are playing, which I shall call A, and another, B, which has the payoff region shown in Fig. 7. In B it is reasonable that we should each get $5. This is how much you say that I should receive from

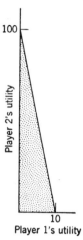

FIG. 6

game A; however, I personally would much prefer to play in A than in B. That I do presumably means only that I expect to get more out of A than out of B, and not that it gives you a much better return with no benefit to me. I am not reluctant to let you have more in A than in B so long as I get something out of it. You should not get false aspirations because of the asymmetry of the region of payoffs in A—after all, I control the outcome as well as you and (0, 0) looks awfully symmetric to me!"

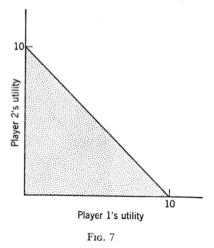

FIG. 7

Such an argument implicitly assumes the existence of an interpersonal comparison of utility. Nash feels that such comparisons are not meaningful and that such an argument cannot, therefore, be made sound. We would agree that interpersonal comparisons have not been given a rigorous meaning, but we feel that an abstraction completely omitting such considerations is perhaps departing too far from reality *in certain contexts*. In many bargaining parleys, reference is made—vaguely, to be sure—to interpersonal comparisons of values, and presumably mathematical abstractions of such situations should incorporate comparisons.

The second assumption, Pareto optimality, seems quite innocuous from a normative point of view, and since we are unaware of any serious misgivings on this score we shall not discuss it normatively. However, it is most clearly open to doubt as a description of behavior. This is worth some consideration. One reason that the players may not reach a Pareto optimal point in practice, even if they achieve agreement and do not resort to implementing threat strategies, is that they hide their true utilities. This point will be discussed later, but we may mention here that such behavior violates the basic knowledge assumption of game theory and so is not really relevant in the present context. More to the point is the fact that the dynamic process usually takes the form of a series of small changes commencing at the status quo point and ultimately terminating in a local optimum. It can happen that this is a Pareto optimum; however, often it is only a local optimum in the sense that there exist *drastically different* outcomes which would be preferred by both players if brought to their attention. This appears to be a phenomenon subject to laboratory investigation. For example, let two subjects bring to a bargaining table a set of objects, such as books, records, etc., which do not have an obvious price tag. Prior to the bargaining, obtain from each of the subjects a confidential statement of his preferences. Conjecture: in many cases it will be possible to suggest to the players a trade which each prefers to the one upon which they have agreed. In cases where it is suspected that the preferences have changed as a result of the bargaining, these can be checked again; the nature of such changes should be interesting. L. Hurwicz has suggested an ingenious modification which may make this experiment easier to carry out. Instead of using books, each player can be given marked tags which will be redeemable for money and a key which discloses the monetary worth *to him* of each tag and set of tags (certain complementarities can be introduced). He may be given only partial information as to their worth to his adversary, or in a variant he may be given full information. It is plausible that in such a bargain individuals will falsify certain aspects of their true preferences. If so, will they achieve a Pareto optimal solution?

The independence of irrelevant alternatives assumption is the source of considerable contention. In social welfare economics, a stormy controversy centers about an analogous assumption (see Chapters 13 and 14), and many, but not all, of the criticisms raised there can be carried over into the bargaining context. Consider the two bargaining games with the payoff regions shown in Fig. 8. In both games let us suppose that the payoffs are in money, that utility is linear in money for both players, and, if they do not agree upon a common point, neither player receives anything. In each game the Nash solution is (5, 50). Let us not question whether (5, 50) is a reasonable solution for game *A*, but rather whether it should be for game *B* *if* it is for game *A*, as follows from the assumption of the independence of irrelevant alternatives.

In judging the plausibility of a potential solution for a game, one may try to evaluate the "levels of aspiration" of the players, for that is certainly one of the psychological factors often involved in bargaining temperaments. The critics of the third assumption argue that whatever a "fair" solution to game *A* may be, a "fair" solution of game *B* should yield less to 2 than the solution of game *A*, for in game *A* his potentialities are far greater. Put another way, let us suppose the players have agreed upon a solution to game *B*, and then they are told that actually the game is *A*; it is reasonable for player 2 to argue that he now deserves more. If so, assumption 3 is violated.

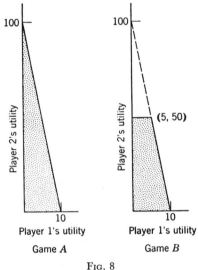

Fig. 8

We feel at this time—the implication being that we have changed our minds in the past—that this argument against assuming independence of irrelevant alternatives loses its appeal when applied to bargaining problems; the reason is that the naturally distinguished trade, the status quo, serves to point out that certain aspirations are merely empty dreams.

A word of caution about this assumption: although it may itself be reasonable, there are numerous related assumptions which appear to be equally plausible at first glance but which are not. For example, consider this one: If new trades are admitted in a bargaining game and the status quo point is held fixed, this should either not affect the solution or the new solution should be one of the new trades. Furthermore, if there are new trades which both players prefer to the old solution, the solution to the new game should be one of these preferred new trades. This assumption, while apparently reasonable, is contradictory if Pareto optimality is assumed. We show this by considering three games—*A*, *B*, and *C*—whose associated regions R_A, R_B, and R_C are shown in Fig. 9. The region R_A is simply the straight line from the origin to the point *a*,

R_B the straight line from the origin to the point b, and R_C a triangle with the two axes as sides and including the points a and b. In game A, Pareto optimality requires that a be the solution.   In game B it requires that b be the solution.   Now, according to the principle we have stated, any solution of C must dominate the solution of A, so it must be in the little triangle with origin at a and with the dotted sides.   Equally well, any solution of the game C must dominate the solution of B, so it must be in the little triangle with origin at b and with the dotted sides.   But these two triangles have no points in common, so we are led to an inconsistency.

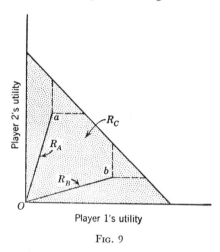

Symmetry, assumption 4, is open to major ethical criticisms in certain contexts. For example, player 1 might be a single individual and player 2 a whole community, then whether or not we wish to assume symmetry—say, equality before the law—very much depends upon the context. Nash asserted [1950 *b*] that this assumption "expresses equality of bargaining skill," but later [1953] he disavows this interpretation.

Player 2's utility

Player 1's utility

Fig. 9

We would agree with his second stand, for otherwise one has the impression that the solution describes what players of "equal bargaining skill" will do or what they can "rationally" expect from the situation.

In closing this section of criticisms, we should point out (again) that one of the basic assumptions of game theory, and of the bargaining model in particular, is extremely doubtful, namely: that each player knows the true tastes—the utility functions—of the others.   For example, suppose that in a bargaining situation the players agree to submit to an arbiter who is committed to Nash's assumptions.   To resolve the conflict, the arbiter must first ascertain their utility functions; hence the situation deteriorates into a game of strategy where each player tries to solve the problem of how best to falsify or exaggerate his true tastes.   In most situations, a player's preferences are only partially known to his adversary, and falsification of one's true feelings is an inherent and important bargaining strategy.   An arbiter, to be successful, must skillfully ferret out at least a part of the truth.   This reality is seriously idealized in game theory, and thereby the theory is severely restricted.   This is not to say it is useless in all situations, but only that there is always the fear that the real problem may have been abstracted away.

6.7 ALTERNATIVE APPROACHES TO THE BARGAINING PROBLEM

Harsanyi [1956] notes that Zeuthen's [1930] solution to the bargaining problem is mathematically equivalent to Nash's product maximization solution, and he feels that Zeuthen's formulation has the added merit of "supplying a plausible psychological model for the actual bargaining process." In examining this formulation we shall continue to use our former terminology, but the spirit of the Harsanyi-Zeuthen remarks will be conveyed.

Let a bargain be given with the region R of possible payoffs and the status quo at the origin. Suppose that player 1 is holding out for a trade with utility payoffs (u_1', u_2') and 2 is demanding (u_1'', u_2''), where the two points are different and each is Pareto optimal. Who should make a concession? The argument is, and we shall examine it in detail later, that player 1 "should" make a concession if and only if

$$\frac{u_1' - u_1''}{u_1'} \leqslant \frac{u_2'' - u_2'}{u_2''},$$

and player 2 should make it if the inequality is reversed. It is easy to see that this inequality is equivalent to

$$u_1'u_2' \leqslant u_1''u_2''.$$

Concession need not necessarily mean accepting the opponent's demand; rather, the conceding player can suggest an alternative trade which will not require him to make a further concession in the next round of negotiations. But, for this to be so, he must propose some (u_1''', u_2''') having a component product $u_1'''u_2'''$ at least as large as the component product of his opponent's demand, and larger if possible. Clearly, this procedure raises the component product at each stage, and so it inexorably leads to the point for which the component product is a maximum—Nash's solution.

As presented, the concession principle is totally arbitrary, but Harsanyi and Zeuthen have attempted to provide some rational underpinning for it. When the two demands are (u_1', u_2') and (u_1'', u_2''), then very crudely $(u_1' - u_1'')/u_1'$ and $(u_2'' - u_2')/u_2''$ measure, respectively, the relative losses incurred when players 1 and 2 concede. The assumption, then, is that the player whose relative loss is the smaller will concede.

A further derivation of the concession principle is presented by Harsanyi which is based on postulated human behavior. He assumes, among other things, that each player knows his opponent's subjective probability of conceding, in which case it becomes meaningful to consider the (expected) utility of conceding or not conceding. This assumption, plus the usual ones about symmetry and Pareto optimality, leads (by the tautological utility maximization principle) to the same concession principle.

It is again difficult to interpret such results either as a descriptive model of human behavior or as a piece of (conditionally) normative counseling. We can, however, see merit in the concession principle as a negotiation scheme which two players might agree is "fair in the abstract" and which they would use to resolve any specific conflict. In this sense, the Harsanyi-Zeuthen result seems to help one accept the Nash solution—or vice versa.

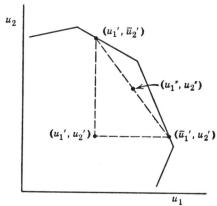

Fig. 10

Raiffa [1951] also envisaged using such "fair" negotiation models to motivate more abstract arbitration schemes. 'Beginning at the status quo point, which need not be the origin, the negotiation model effects step by step improvements in the player's positions until a Pareto optimal point is reached. This point is offered as a "reasonable" arbitrated value. For example, suppose that at some stage of the process the players are at the point (u_1', u_2') shown in Fig. 10. Player 1's most selfish demand would be the Pareto optimal point $(\bar{u}_1', u_2')$, and player 2's would be $(u_1', \bar{u}_2')$. As a compromise, one "reasonable" negotiation model would suggest a simple average of the utilities for each of the players, i.e., the point (u_1'', u_2''), where

$$u_1'' = (u_1' + \bar{u}_1')/2 \quad \text{and} \quad u_2'' = (u_2' + \bar{u}_2')/2.$$

In a variation of this model, Raiffa suggests not moving in discrete jumps as above, but in a continuous motion—think of it as a limiting case of making more and more discrete jumps of smaller and smaller size—such that the slope of the movement at (u_1', u_2') is the same as the line joining that point and (u_1'', u_2''). This idealization leads, in general, to a non-linear motion from the status quo along what may be called the "negotiation curve" to the arbitrated value.

Both of these schemes satisfy all of Nash's axioms save the independence of irrelevant alternatives, and depending upon one's point of view this may or may not be a fault.

It should be pointed out in conclusion that, in the continuous motion model, the slopes of the negotiation curve and of the Pareto optimal curve are of the same magnitude at their point of intersection, but of opposite sign. If one "linearizes" this model by demanding that the negotiation curve be a straight line having this same relation between its slope and that of the Pareto optimal curve at their point of intersection, then the arbitrated point is Nash's point where the product is a maximum.

6.8 ARBITRATION SCHEMES FOR NON-STRICTLY COMPETITIVE GAMES: THE SHAPLEY VALUE

Having digressed into the philosophy of arbitration and discussed simple bargaining games, we return in this and the following two sections to the general case of non-strictly competitive games. It will be recalled that we denote by R the set of all possible pairs of payoffs when the two players use uncorrelated mixed strategies and by R' all possible pairs of payoffs when the players cooperate and adopt joint randomized strategies. Every point in R is in R', but the converse is not necessarily true. By (v_1, v_2) we denote the payoffs representing the security levels of the players, i.e., player 1 can guarantee himself an expected return of v_1 by a suitable strategy choice, and v_2 is similarly defined by changing the roles of 1 and 2. The negotiation set $\mathfrak{N}$ is the part of the northeast boundary of R' which dominates (v_1, v_2). The problem is to single out a point of the negotiation set of each game as its "solution."

One simple way to do it is to treat the cooperative game as the bargaining game $[R', (v_1, v_2)]$. That is, a given non-strictly competitive game induces a bargaining game where the status quo is taken to be the security level of the players, and the solution of the game is taken to be the Nash solution of the bargain associated with the game. This procedure has some nice properties: it picks out a unique point of the negotiation set, it is invariant with respect to the origin and the scale of utility measurement, and it is symmetric or egalitarian in that it does not look at the labels of the players.

We may call this the *Shapley procedure* on the grounds that it is a slight generalization of a very special case of the Shapley value of an n-person game. The latter notion will not be discussed until section 11.4, but it will suffice here to say that it gives the *a priori* payoff expectation of each player in cooperative *n*-person games where side payments are allowed. Consider, then, the two-person game in which bribes, or side payments, are allowed, and let us suppose that the possible outcomes are any pair that does not sum to more than some number c.   The region of possible payoffs is shown in Fig. 11, and suppose (v_1, v_2) is the point shown.   The negotiation set of this game consists of all points on the line segment from

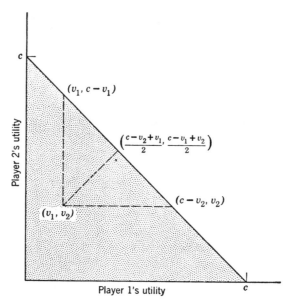

FIG. 11

$(c - v_2,\ v_2)$ to $(v_1,\ c - v_1)$.   The induced bargaining game has the solution

$$[(c - v_2 + v_1)/2,\ (c - v_1 + v_2)/2)],$$

which is also the Shapley value of this game.   The payoff to player 1 is the average of the following two payoffs: v_1, the amount he can definitely secure for himself, and $c - v_2$, the marginal amount that he contributes when he joins player 2 in a coalition.   This interpretation of the Shapley value generalizes in an appropriate manner for cooperative games with more than two players.   Alternatively, the payoff may be written as $(c/2) + (v_1 - v_2)/2$, that is, player 1 receives half the maximum total c plus half of the relative (dis)advantage in optimum security levels (maximin values) of the non-cooperative game.

We shall now offer an example which casts some doubt on this procedure for solving two-person cooperative games. Let the game have only two pure strategies for each player and let the payoff matrix be:

$$\begin{array}{c} & \beta_1 & \beta_2 \\ \alpha_1 \\ \alpha_2 \end{array} \left[\begin{array}{cc} (1, 4) & (-1, -4) \\ (-4, -1) & (4, 1) \end{array} \right].$$

The bargaining region R' is that shown in Fig. 12, and the security levels are $(0, 0)$. The maximin strategies are $(\frac{4}{5}\alpha_1, \frac{1}{5}\alpha_2)$ and $(\frac{1}{2}\beta_1, \frac{1}{2}\beta_2)$.

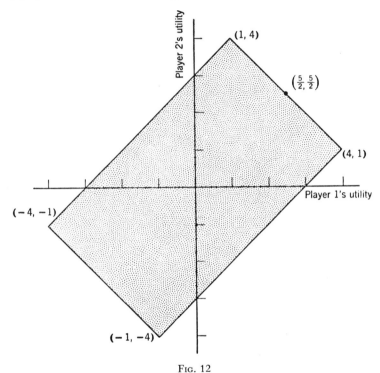

FIG. 12

Since the induced bargaining situation for this game, $[R', (0, 0)]$, is symmetric, it is easily seen that the Nash solution is $(\frac{5}{2}, \frac{5}{2})$, which is the midpoint of the negotiation set, i.e., the midpoint of the line segment from $(1, 4)$ to $(4, 1)$.

Observe that this induced bargaining game, and the solution, are perfectly symmetric between the two players, but that the game itself is not —player 2 has a distinct advantage. For suppose 2 threatens to play strategy β_1, then what alternatives has 1? If he plays α_1, then the payoff is $(1, 4)$—the best possible for 2 in the negotiation set of the game. Therefore 1's only realistic counter threat is α_2, leading to the payoff $(-4, -1)$.

Hence the status quo for the bargaining game should not be $(0, 0)$ but rather $(-4, -1)$. With this as the status quo, the bargaining game $[R', (-4, -1)]$ is not symmetric and the solution of the bargain is easily seen to be $(1, 4)$. The asymmetry is even clearer if we are willing to make an assumption of interpersonal comparisons, for then we see that strategy β_1 commands a relative advantage of three units for player 2 regardless of 1's choice.

To summarize: the Shapley "solution" is not sensitive to the obvious advantage that player 2 has in this game as reflected in his effective threat strategy β_1. Thus, one argument against the Shapley value (for two-person games!) is the inappropriateness of the maximin pair (v_1, v_2) as the basis for bargaining.

6.9 ARBITRATION SCHEMES FOR NON-STRICTLY COMPETITIVE GAMES: NASH'S EXTENDED BARGAINING MODEL

By selecting a different status quo point than in the Shapley procedure, Nash [1953] has extended his analysis of the bargaining model in much the same way to yield an analysis of all two-person non-strictly competitive games. Roughly, his idea boils down to this: Each player adopts a mixed strategy as a "threat"; the pair of threats establishes a payoff, which, in turn, acts as the status quo point for future bargaining; and the bargaining problem is resolved in the manner discussed in section 6.5. Therefore, the problem is reduced to selecting the threat strategies so as to influence the status quo—which controls the ultimate payoff—in the most favorable manner. Thus, the given game G induces a new non-cooperative game G^*, and it has a well-behaved solution in the following sense: There is a payoff $(v_1{}^*, v_2{}^*)$ such that:

i. Player 1 can guarantee himself at least $v_1{}^*$ by a suitable strategy choice $\mathbf{x}^*$.

ii. Player 2 can guarantee himself at least $v_2{}^*$ by a suitable strategy choice $\mathbf{y}^*$.

iii. $\mathbf{x}^*$ and $\mathbf{y}^*$ are in equilibrium, i.e., each is good against the other.

iv. If $\mathbf{x}'$ and $\mathbf{y}'$ are in equilibrium, then $\mathbf{x}'$ will guarantee 1 at least $v_1{}^*$ and $\mathbf{y}'$ will guarantee 2 at least $v_2{}^*$.

It can be shown that $(v_1{}^*, v_2{}^*)$ is Pareto optimal and that it lies on the negotiation set of G. The pair $(v_1{}^*, v_2{}^*)$ is called the *Nash solution* of G, and the optimal strategies $\mathbf{x}^*$ and $\mathbf{y}^*$ of G^*, which need not be unique, are called the *optimal threat strategies* for the game G.

Nash attempts to defend this solution in two ways: by constructing a non-cooperative game of which it is an equilibrium point and by axiomatizing the solution of a general cooperative game.   Let us examine each of these.

The first procedure reflects Nash's belief that the non-cooperative games are more basic than cooperative ones; that one should extend any cooperative game to a non-cooperative game having as formal well-defined moves any negotiation and bargaining moves that are admissible. In this case, he proposes that it be done in the following manner:

Move 1: Player 1 chooses a mixed strategy $\mathbf{x}$.

Move 2: Player 2, with no information about the choice made at move 1, chooses a mixed strategy $\mathbf{y}$.

Move 3: Player 1, knowing the choices at moves 1 and 2, makes a demand d_1, i.e., chooses a number d_1.

Move 4: Player 2, knowing the choices at moves 1 and 2, but not at move 3, chooses a number d_2.

Moves 1 and 2 and moves 3 and 4 can be thought of as pairs of simultaneous moves.   Any play of this extensive game can be described as a 4-tuple $(\mathbf{x}, \mathbf{y}, d_1, d_2)$.   The payoff associated with such a play is defined as follows:

i. If (d_1, d_2) is a point of R', i.e., if it is a feasible payoff in the game G, then players 1 and 2 receive d_1 and d_2, respectively.

ii. If (d_1, d_2) is not a point of R', then the payoff is $[M_1(\mathbf{x}, \mathbf{y}), M_2(\mathbf{x}, \mathbf{y})]$, where M_1 and M_2 are the payoff functions in the game G.

A strategy for player 1 in this extensive game is a pair $(\mathbf{x}, d_1)$, where d_1 may depend upon $\mathbf{x}$ and $\mathbf{y}$; for player 2, a strategy is a pair $(\mathbf{y}, d_2)$, where d_2 may depend upon $\mathbf{x}$ and $\mathbf{y}$.   Nash asserts that optimal strategies for the players are $(\mathbf{x}^*, v_1^*)$ and $(\mathbf{y}^*, v_2^*)$, where these quantities are defined above.

It is true that these strategies are in equilibrium and that they yield a Pareto optimal payoff; however, as Nash is well aware, there is in general a continuum of other inequivalent equilibrium pairs.   The weak link in the argument is to single out this particular pair.   Nash offers an ingenious and mathematically sound argument for doing so, but we fail to see why it is relevant.

Thus the equilibrium points do not lead us immediately to a solution of the game.   But if we discriminate between them by studying their relative stabilities we can escape from this troublesome non-uniqueness.

To do this we "smooth" the game to obtain a continuous payoff function and then study the limiting behavior of the equilibrium points of the smoothed game as the amount of smoothing approaches zero. [1953, p. 131.]

Nash then shows that his "solution" is "the only necessary limit of the equilibrium points of smoothed games." Indeed, this is true, but isn't it a completely artificial mathematical "escape from this troublesome non-uniqueness"? Would it have any relevance to the players?

Thus, we take exception to Nash, because we feel that the extensive form of the negotiation game associated with a given cooperative game does not have a meaningful unique "non-cooperative solution." We would claim that a solution of the negotiation game can only be said to exist if the players are committed to Nash's solution— which we interpret as an arbitrated solution.

As an alternative defense, Nash has offered an axiomatic definition consisting of seven axioms which are satisfied only by his solution of the negotiation model. As could be expected, five of these constitute an "appropriate modification" of the principles used in the bargaining model, namely: (1) feasibility, i.e., the solution should be in R'; (2) Pareto optimality; (3) invariance with respect to utility scales; (4) symmetry, i.e., independence of labels of players; and (5) independence of irrelevant alternatives. The remaining two axioms describe the behavior of the "solution" when the domains of available strategies are modified, but the payoffs remain fixed. Axiom 6 requires, roughly, that, if a player's choice of strategies is restricted while at the same time the other player's strategy space is unchanged and the payoffs are held fixed, then his return from the solution cannot increase. "A player's position in the game is not improved by restricting the class of threats available to him." Axiom 7 requires that if, say, player 1 is restricted to a single strategy, there exists a way of restricting 2 to a single strategy without increasing the return to 1 above that given by the solution. Nash states that: "The need for Axiom 7 is not immediately obvious. Its effect is to remove the possibility that the value to a player of his space of threats should be dependent on collective or mutual reinforcement properties of the threats." [1953, p. 138.]

The real problem with this axiom system is how to rationalize axioms 6 and 7. In this connection, Nash makes the following puzzling remarks:

[These] axioms · · · lead to the same solution that the negotiation model gave us; yet the concepts of demand or threat do not appear in them. Their concern is solely with the relationship between the solution (interpreted here as value) of the game and the basic spaces and functions which give the mathematical description of the game.

It is rather significant that this quite different approach yields the same solution. This indicates that the solution is appropriate for a wider variety of situations than those which satisfy the assumptions we made in the approach via the model. [1953, p. 136.]

What puzzles us is how one can rationalize the last two axioms other than by contemplating a negotiation model similar to Nash's and by employing his "threat" and "demand" notions. We feel that the negotiation model and the axiomatic approach are quite similar in spirit and that *they serve to complement each other very well.*

6.10 ARBITRATION SCHEMES FOR NON-STRICTLY COMPETITIVE GAMES: THE CASE OF MEANINGFUL INTERPERSONAL COMPARISONS OF UTILITY

In the preceding sections we have been concerned with arbitration in situations where interpersonal comparisons of utility are assumed to be meaningless; in this section we shall suppose that they can be given meaning. Raiffa [1953] considered both cases in his work. For the case where they are meaningless, he suggested a class of procedures out of which he 'singled one for special attention. Even though the two authors worked independently and devised quite different rationales for the procedure, Raiffa's special scheme is operationally identical to Nash's extended bargaining model (section 6.9). Of these two rationalizations, Nash's is the less *ad hoc*. In the context of meaningful interpersonal comparisons, Raiffa has offered an arbitration scheme which deals directly with the cooperative game and which is independent of Nash's solution of the bargaining problem. This we shall now discuss.

Depending upon the situation, we may or may not wish to permit side payments, which simply means we choose to deal with different regions of payoff pairs; the analysis is the same in both cases. For ease of discussion, let us suppose that the individual security levels are both zero, and so the negotiation set is the northeast boundary of R' from a to b, as shown in Fig. 13. If we take any point (u_1, u_2) of R', then the relative advantage to player 1 is $u_1 - u_2$. All points of R' having the same relative advantage obviously lie on the 45-degree line passing through (u_1, u_2), so the contour lines of constant relative advantage are all the 45-degree lines.

Suppose, for the moment, that player 1 has a strategy $\mathbf{x}^*$ such that, regardless of player 2's choice, his payoff in the non-cooperative game is on or below the contour line passing through some point c in the negotiation set. Similarly, suppose 2 has a strategy $\mathbf{y}^*$ such that, independent of 1's choice, his payoff is on or above the same contour line. If so, then we submit that c is a "reasonable" candidate for the arbitrated solution. The reason is this: If 2, for example, wishes to move from c, then 1 can threaten to use $\mathbf{x}^*$ which will maintain or increase the relative advantage

1 has with respect to 2 at the point c.   A similar argument holds for 1. Let it be clear that, if such a threat is carried out, each player may suffer an absolute loss, but the relative advantage of the player not agreeing to c will not be increased.   For example, if 2 demands b then 1's threat $\mathbf{x}^*$ is cogent, for, relative to b, player 2 has more to lose than does 1 if the threat is carried out.

To find the point c, one proceeds as follows: For each strategy pair $(\mathbf{x}, \mathbf{y})$ the relative advantage to player 1, $R(\mathbf{x}, \mathbf{y}) = M_1(\mathbf{x}, \mathbf{y}) - M_2(\mathbf{x}, \mathbf{y})$,

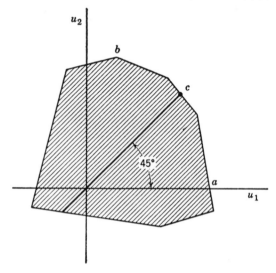

FIG. 13

can be taken to be the payoff function of a two-person zero-sum game. By the results of section 4.8 we know there exists a value v and optimal strategies $\mathbf{x}^*$ and $\mathbf{y}^*$, and

(i) $M_1(\mathbf{x}^*, \mathbf{y}) - M_2(\mathbf{x}^*, \mathbf{y}) \geqslant v$, all $\mathbf{y}$,

(ii) $M_1(\mathbf{x}, \mathbf{y}^*) - M_2(\mathbf{x}, \mathbf{y}^*) \leqslant v$, all $\mathbf{x}$.

On the contour line of constant relative advantage v, i.e., the locus of points (u_1, u_2) such that $u_1 - u_2 = v$, find the point of intersection with the boundary of R' in the northeast region.   This amounts to adding a constant to the minimax payoff for each of the players.   If this boundary point is Pareto optimal, as c is in Fig. 13, then this is the arbitrated solution; if it is not Pareto optimal, proceed along the boundary in a northeasterly direction from this point of intersection until a Pareto optimal point is reached, then this is the arbitrated solution.   In other words, the players first engage in a strictly competitive game of relative advantage, and after it is resolved they cooperate fully to increase their payoffs as much as possible while preserving the relative advantage.

The procedure may be illustrated for the game matrix

$$\begin{bmatrix} (1, 4) & (-1, -4) \\ (-4, -1) & (4, 1) \end{bmatrix}$$

studied in section 6.8. It is easy to compute that the induced zero-sum game of relative advantage yields player 1 the following payoffs:

$$\begin{array}{c} \quad\quad \beta_1 \quad \beta_2 \\ \begin{array}{c} \alpha_1 \\ \alpha_2 \end{array} \begin{bmatrix} -3 & 3 \\ -3 & 3 \end{bmatrix}. \end{array}$$

The value to 1 of this game is clearly -3. Since the point $(1, 4)$ of the game yields a relative (dis)advantage of -3 to 1 and is also on the north-east boundary of the region (see Fig. 12), we concluded that it is the arbitrated solution. This, it will be recalled, was the solution we obtained before when we admitted that the game is not symmetric in the sense that $(-4, -1)$ should be the status quo point, not $(0, 0)$.

Such a procedure does not depend upon the origins of the utility functions, but it is sensitive to changes in units of measurement. Thus, when we speak of relative advantage, we are assuming that there is a common unit of measurement. For many situations money serves this purpose. In some contexts where money is not appropriate, it may be possible to determine a common unit by choosing two stimuli to serve as reference points for equating tastes. In still other cases, interpersonal comparisons may be deemed entirely inappropriate, in which case this procedure is useless. It would be desirable to have a method for establishing interpersonal comparisons within the framework of the game itself, equating certain distinguished values of the game. Two *ad hoc* methods for doing so are discussed in the next section, and they are illustrated by a specific example.

6.11 TWO DEFINITIONS OF INTERPERSONAL COMPARISONS IN TWO-PERSON GAMES

To illustrate the definitions which will be presented, consider the following amusing conflict situation introduced by R. B. Braithwaite in *Theory of Games as a Tool for the Moral Philosopher* [1955]:

Suppose that Luke and Matthew are both bachelors, and occupy flats in a house which has been converted into two flats by an architect who had ignored all considerations of acoustics. Suppose that Luke can hear everything louder than a conversation that takes place in Matthew's flat, and vice versa; but that sounds in the two flats do not penetrate outside the house. Suppose that it is legally impossible for either to prevent the other from making as much noise as he wishes, and economically or sociologically impossible for either to move elsewhere. Suppose further that each of them has only the hour from 9 to 10 in the evening for recreation, and that it is impossible for either to change to another time. Suppose that Luke's form of recreation is to play classical music on the piano for an hour at a time, and that Matthew's amusement is to improvise jazz on the trumpet for an hour at once. And suppose that whether or not either of them performs on one evening has no influence, one way or the other, upon the desires of either of them to perform on any other evening; so that each evening's happenings can be treated independently. Suppose that the satisfaction each derives from playing his instrument for the hour is affected, one way or the other, by whether or not the other is also playing; in radio language, there is "interference" between them, positive or negative. Suppose that they put to me the problem: Can any plausible principle be devised stating how they should divide the proportion of days on which both of them play, Luke alone plays, Matthew alone plays, neither play, so as to obtain maximum production of satisfaction compatible with fair distribution?   [1955, pp. 8–9.]

Let us suppose that we have ascertained their utility functions for this situation and that the strategy matrix is

$$\text{Player 2 (Matthew)}$$

$$\text{Player 1 (Luke)} \quad \begin{matrix} & \beta_1 \text{ (play)} & \beta_2 \text{ (not play)} \\ \alpha_1 \text{ (play)} & \left[\begin{matrix} (1, 2) \\ \alpha_2 \text{ (not play)} & (4, 10) \end{matrix}\right. & \left.\begin{matrix} (7, 3) \\ (2, 1) \end{matrix}\right] \end{matrix}.$$

Luke, the pianist, we observe, most prefers that he play alone, next that Matthew play alone, then that neither play, and finally that both play; whereas the trumpeter would rather he play alone, that Luke play alone, that they both play, and finally that neither play. Of course, the numbers chosen are not unique: those associated to each player can have a constant added and can be multiplied by a positive constant without altering the structure of the game.

Raiffa [1953] suggests one such transformation, namely, set the utility of the worst payoff to be 0 and of the best to be 1 for each player. In these scales, the matrix becomes

$$M$$

$$L \quad \begin{matrix} & \beta_1 & \beta_2 \\ \alpha_1 & \left[\begin{matrix} (0, \frac{1}{9}) \\ \alpha_2 & (\frac{1}{2}, 1) \end{matrix}\right. & \left.\begin{matrix} (1, \frac{2}{9}) \\ (\frac{1}{6}, 0) \end{matrix}\right] \end{matrix}.$$

Although there is no adequate rationale for doing so, one can assume that this choice establishes an interpersonal comparison of utility (for the purposes of this game!) and can solve the game using the arbitration scheme discussed in the preceding section. The diagram of payoffs is shown in Fig. 14. One can easily verify that Luke, by using strategy

α_1, can keep the payoff on or below the dotted line, and that Matthew, by playing β_1, can keep it on or above the line. Thus, the arbitrated solution yields a payoff of 0.652 utiles, i.e., Luke is indifferent between the arbitrated solution and a lottery which weights his most preferred alternative (α_1, β_2) with probability 0.652 and his least preferred alternative (α_1, β_1) with probability $1 - 0.652 = 0.348$. Similarly, Matthew's payoff is 0.763 utiles. To achieve this arbitrated result, Matthew should play while Luke remains silent, (α_2, β_1), on 16 out of every 23 nights, while Luke should play and Matthew be silent, (α_1, β_2), 7 out of 23 nights.

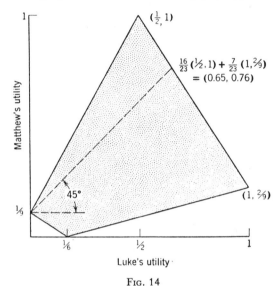

FIG. 14

The overall procedure, then, is to take the arbitrary payoff matrix, reduce it to canonical form by transforming the utility scales so that the most preferred outcome has utility 1 and the least preferred 0, find the solution as given in section 6.10, and then convert the answer back to utiles in the original measurement scales. This procedure has two interesting features: First, it yields an arbitrated solution which is *invariant with respect to origins and units of utility measurement*—even though as a technical device a specific pair of scales were singled out as an integral part of the analysis. Second, it satisfies *all save one* of the axioms demanded by Nash for "any reasonable value." The exception is the independence of irrelevant alternatives. But for this reason, the procedure is open to serious criticism. For example, one can add irrelevant strategy alternatives, any pair of which will certainly not be adopted by the players, which alter the least favorable outcomes, and hence may alter the arbitrated solution. The counterargument is that one should first delete such extraneous strategies before arbitrating the game. The counter-counter-argument: "Ah, but how?"

Braithwaite [1955] has proposed a related, but more sophisticated, scheme for normalizing the utility functions, which, unfortunately, possesses the same weakness as the above procedure. Let us also illustrate it in terms of the Matthew-Luke problem. Consider the following four (distinguished) strategies:

i. x_1 is player 1's (Luke's) maximin strategy $(\frac{1}{4}\alpha_1, \frac{3}{4}\alpha_2)$, which yields him a security level of 3.25.

ii. y_1 is player 2's maximin strategy $(\frac{1}{5}\beta_1, \frac{4}{5}\beta_2)$, which yields him a security level of 2.80.

iii. x_2 is player 1's minimax strategy $(\frac{9}{10}\alpha_1, \frac{1}{10}\alpha_2)$ against 2, which holds 2 to at most 2.80.

iv. y_2 is player 2's minimax strategy $(\frac{5}{8}\beta_1, \frac{3}{8}\beta_2)$ against 1, which holds 1 to at most 3.25.

In tabular form, the payoffs corresponding to these strategy choices are:

$$M$$

$$L: \begin{array}{c} x_1 \\ x_2 \end{array} \begin{array}{c} y_1 \qquad\qquad y_2 \\ \left[\begin{array}{cc} (3.25,\ 2.80) & (3.25,\ 5.56) \\ (5.46,\ 2.80) & (3.25,\ 2.80) \end{array} \right]. \end{array}$$

Braithwaite claims that a natural method to obtain a common unit of utility measurement is to assume that *each player benefits equally by a change from his maximin strategy, which guarantees an optimal security level for himself, to his minimax strategy, which guarantees a minimal security level for his adversary, assuming that the adversary is holding to his maximin strategy.* In this example, if M holds to his maximin strategy y_1, then the increment to L is 5.46 − 3.25 = 2.21 utiles as he changes from x_1 to x_2. Similarly, if L holds to x_1, the increment to M is 5.56 − 2.80 = 2.76. The ratio is 4 to 5, so the payoffs in the original game are normalized by dividing L and M's utilities by 4 and 5, respectively, yielding

$$\begin{array}{c} \alpha_1 \\ \alpha_2 \end{array} \begin{array}{c} \beta_1 \qquad\qquad \beta_2 \\ \left[\begin{array}{cc} (\frac{1}{4},\ \frac{2}{5}) & (\frac{7}{4},\ \frac{3}{5}) \\ (1,\ 2) & (\frac{1}{2},\ \frac{1}{5}) \end{array} \right]. \end{array}$$

The corresponding payoff regions R and R' are shown in Fig. 15.

If L plays α_1, he can hold M down to the 45-degree line passing through $(\frac{1}{4}, \frac{2}{5})$; and M can hold L on or above this line by playing β_1. Thus, using Braithwaite's procedure of interpersonal comparison, the arbitrated solution is (1.29, 1.45), which amounts to letting M play 26 out of 43 nights while L remains silent, and letting L play 17 out of 43 nights while M remains silent.

This solution is somewhat less favorable (about 60% versus 70% of the nights) to Matthew than the one given before. Nevertheless, "This

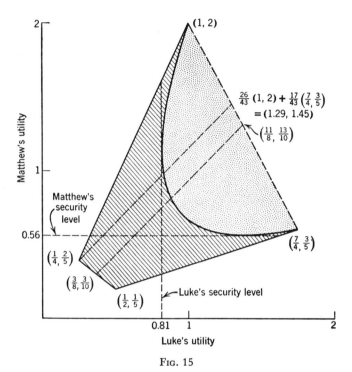

Fig. 15

arrangement is relatively advantageous to Matthew, and it is worth examining why this is so. Matthew's advantage arises purely from the fact that Matthew, the trumpeter, prefers both of them playing at once to neither of them playing, whereas Luke, the pianist, prefers silence to cacophony." [1955, p. 37.] Matthew has the threat advantage.

Braithwaite makes two other observations which seem, to him, to bolster his arbitrary normalization of the utility scales. First, the northeast boundary of the region R is actually a parabola whose axis makes a 45-degree angle with the horizontal when his procedure is used to normalize the payoffs. In general, then, he chooses contours of constant relative advantage which are parallel to the axis of the parabola which forms the northeast boundary of R. Second, his choice of scales also means that the line segment joining (*a*) the midpoint of the line segment between the two least favorable outcomes to (*b*) the midpoint of the line segment between the two most favorable outcomes in parallel to the axis of the parabola, and hence has a 45-degree slope.   [In this case it is the line joining $(\frac{3}{8}, \frac{3}{10})$ to $(1\frac{1}{8}, 1\frac{3}{10})$.]   Thus, there is a "somewhat sophisticated sense of 'symmetry' · · · and this symmetry may, to some people, be intuitively obvious as a criterion for fairness."   [1955, p. 42.]   He is eloquent in defense of his procedure, and his small book can be recommended as the best way to decide for one's self.   Among other things in its favor is the fact that it is a procedure which is invariant up to arbitrary origins and units of utility measurement, and that it satisfies most of Nash's axioms.   Again, the one which fails is the independence of irrelevant alternatives.

Nonetheless, we consider his procedure, in particular the rationalizations of his definition of the common unit of measurement, arbitrary.   This is not necessarily to be interpreted as an unfavorable criticism of the "reasonableness" of the procedure, but rather as a recognition that a clinching argument, showing how this method is better than others, has not yet been produced.   With this, we are sure Braithwaite would agree.

▶ We have considered two apparently quite different methods of arbitration: In sections 6.8 and 6.9 the analysis depended upon a negotiation game which in turn rested upon a prior analysis of bargaining games.   In section 6.10 and this one no explicit mention was made of either negotiation models or bargaining games; rather, the analysis hinged upon contour lines in R' of "equal relative advantage" (or "isorrhopes" to use Braithwaite's term).   Actually, these two procedures are technically very similar.

To see this, consider first the case where we have the contour lines, and let us give a negotiation-bargain interpretation.   If the status quo point (u_1, u_2) in R' is given, then define a solution of this bargaining problem to be the point where the contour through (u_1, u_2) intersects the Pareto optimal set of R'.   Given this as the solution to the bargaining problem, then the negotiation problem simply entails that the players play a non-cooperative threat game to determine the status quo position.

Conversely, we want to give contour line interpretation of the negotiation-bargain solution.   We need not restrict ourselves to Nash's solution of the bargaining problem, but we do suppose that a specific solution criterion is given.   We now define contour lines in R' as follows: two points of R' lie on the same contour line if and only if they yield the same arbitrated solution (according to the given criterion) when the two points are taken to be the status quo points.   With these contour lines we define an arbitrated solution as before (section 6.10), which it will be recalled is a point c with the following property: each player has a strategy ($\mathbf{x}^{(0)}$ and $\mathbf{y}^{(0)}$, such that, no matter what the other does in the non-cooperative game, the payoff to the first will fall on a contour which yields him returns at least as good as those on the contour passing through c.   This point c will also be the solution of the Nash negotiation-bargaining game, and the strategies $\mathbf{x}^{(0)}$ and $\mathbf{y}^{(0)}$ will be the optimal threat strategies for that game.   ◀

*6.12 STABILITY OF ARBITRATION SCHEMES

When the bargaining model, or any other game theoretic mechanism for that matter, is applied to an empirical problem, the utilities used must be determined by experimental techniques. They are, therefore, certainly going to be in error, and so it would be most unfortunate if small perturbations in the utilities could produce drastic changes in the arbitrated solution. In other words, we should demand of an arbitration scheme that the arbitrated solution be a continuous function of changes in the utilities.

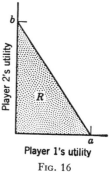

Player 1's utility

Fig. 16

We shall say that an arbitration scheme is *stable* (or, equally well, that the arbitrated solution is continuous) if it possesses the following property. Let $G^{(n)}$, where $n = 1, 2, \cdots$, be a sequence of games, with the nth game having the joint payoffs $(a_{ij}^{(n)}, b_{ij}^{(n)})$ and arbitrated values $(v_1^{(n)}, v_2^{(n)})$. Suppose that the sequence approaches the game G in the sense that the numbers $a_{ij}^{(n)}$ and $b_{ij}^{(n)}$ approach, respectively, the corresponding payoffs a_{ij} and b_{ij} of G as n approaches infinity. The scheme is stable if for each such convergent sequence it is also true that the numbers $v_1^{(n)}$ and $v_2^{(n)}$ approach, respectively, the arbitrated values v_1 and v_2 of G.

It is easy to give examples of unstable schemes. Consider bargains where the status quo is at the origin and where utilities are interpersonally comparable. Let the arbitrated value be that point on the negotiation set having the largest utility component, provided this point is unique; otherwise, let it be the point of the negotiation set which gives each bargainer the same utility. In the problem shown in Fig. 16, the arbitrated value is:

$$(a, 0), \qquad \text{if } a > b,$$
$$(0, b), \qquad \text{if } a < b,$$
$$(a/2, a/2), \qquad \text{if } a = b.$$

Obviously, in the neighborhood of $a = b$, slight perturbations in the utility values can alter the arbitrated solution drastically—from $(a, 0)$ or thereabouts to $(0, a)$ or thereabouts.

None of the schemes which have been considered seriously either here or in the literature exhibit such a trivial pathology; however, they universally possess a more subtle instability. We may illustrate it in a representative special case. Consider the two bargaining regions shown in Fig. 17. In this example we shall not suppose that utilities are interpersonally comparable. With any "reasonable" scheme, the arbitrated values of R_n should be $(1/n, 1)$ and of R_n' should be $(1/2n, \frac{1}{2})$. As n is increased, both bargains approach a region R consisting of the line segment from $(0, 0)$ to $(0, 1)$. But $(1/n, 1)$ approaches $(0, 1)$, and $(1/2n, \frac{1}{2})$ approaches $(0, \frac{1}{2})$. So, what should be the arbitrated value of R? Obviously, such a scheme cannot at the same time be stable and yield a unique solution for R.

There are two obvious ways out of the impasse, and which is adopted seems mainly a matter of convenience or taste. We can retain the stability condition unchanged and deny that R has a unique solution. This

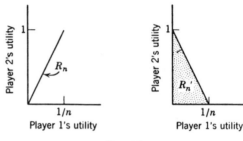

FIG. 17

we tend to favor, for among other things the region R fails to meet the condition, which was needed earlier to obtain Nash's result, that there be a point in R which is better than the status quo for both players. The alternative is to weaken the stability requirement to the extent of not demanding that it hold for a limiting game which contains either a vertical or horizontal line segment in its northeast boundary.

6.13 SUMMARY

In a cooperative two-person game the players are permitted to communicate prior to playing the game, to make binding agreements, to use joint (or correlated) randomized strategies, and, in some variations, to make side payments.

The *negotiation set* consists of all undominated payoffs (the Pareto optimal set) for which each player gets at least his maximin value. Von Neumann and Morgenstern feel that, within the framework of game theory, further restrictions are not possible and that in any bargaining context the actual selection of an outcome from the multiplicity of points in the negotiation set depends upon certain psychological characteristics of the players. Other authors, noting that in many realistic examples players rarely agree on a jointly acceptable point of the negotiation set, have attempted to restrict the "solution" to a single point. Attention was given to what possibly can be meant by such a solution, and we concluded that it can be neither a descriptive nor an ethical statement. We elected to interpret such solutions from the point of view of an arbiter attempting to devise a "fair" scheme to arbitrate all such games, where "fair" is meant in the tautological sense that the scheme satisfies certain reasonable desiderata.

For the two-person cooperative games, it was suggested that reasonable axioms might specify that the point lie in the negotiation set; require that the point be independent of the utility units used and of the labeling of the players; require stability in the sense that slight perturbations of the payoff entries not drastically affect the arbitrated value; and reflect the threat capabilities of the two players. The rest of the chapter was devoted to more precise statements of specific axiom schemes.

The first of these was restricted to the class of cooperative games where only exchanges of goods occur: bargaining problems. Nash assumes that solutions to this problem should satisfy: invariance with respect to utility transformations, Pareto optimality, independence of irrelevant alternatives, and symmetry. From these it follows that there is a unique solution which may be obtained as follows: translate the utility scales so that the status quo point is at the origin, find the point for which the product of the two coordinates is a maximum, and then invert the utility transformations. A number of criticisms of the axioms were presented and discussed.

Several approaches to unrestricted cooperative games were described. The first, which we called the Shapley procedure because it is a slight extension of the Shapley value from n-person theory, takes the maximin values as the basis of bargaining. An asymmetric game which is symmetrically resolved by the procedure was offered as a criticism.

Second, we described Nash's extension of his solution to the bargaining problem. The easiest way to present it is as a reduction of the cooperative game to a non-cooperative one; however, this is completely *ad hoc*. Alternatively, an axiomatic method for obtaining the same result was sketched. In essence, two axioms were added to those for the bargaining problem, but their rationalization did not seem adequate. Although we were critical of both of Nash's separate approaches to this problem, we felt that each helped to support the other, and that collectively they have much merit. The Nash solution was independently arrived at by Raiffa who used a different type of rationalization.

The next procedure, one of several suggested by Raiffa, rests upon an assumed intercomparability of utility. The cooperative game is transformed into a zero-sum game of relative advantages, its value is obtained, and the corresponding contour of relative advantage is found in the cooperative game. Ignoring some exceptional cases, the intersection of this contour with the Pareto optimal set is taken to be the solution. This solution is not independent of changes in utility units.

In cases where interpersonal comparisons are not initially meaningful, Raiffa presents the following scheme: transform the utility scales of the players in the game under consideration so that they satisfy some specific requirement for that game; assuming that this choice establishes an inter-personal comparison of utility (for the purposes of this game!), find the arbitrated value; and then transform the solution back into one for the original game by inverting the utility transformations. As an illustration of the procedure, the utility scales can be transformed so that the most preferred outcome has utility 1 and the least preferred has 0. Braithwaite suggests an alternative transformation in which the utility interval from a player's maximin strategy (based upon his payoffs) to his minimax strategy (based upon his opponent's payoffs), under the condition that his opponent uses his maximin strategy, is taken as the unit. Both pro-cedures satisfy all of Nash's axioms save the independence of irrelevant alternatives. Note that the solutions are independent of separate changes in utility units for the players.

A strong technical similarity was established between these last pro-cedures, which rest upon contours of relative advantage, and Nash's exten-sion of his bargaining solution.

Finally, the stability of an arbitration scheme was defined, roughly, to mean that the arbitrated solution is continuous in changes in the utility scales. All schemes exhibit an instability for regions where, for example, there is no point better for both players than the status quo, but this does not seem to be a serious difficulty.

chapter 7

THEORIES OF n-PERSON GAMES
IN NORMAL FORM

7.1 INTRODUCTION

The theory of games would be a very incomplete edifice, both esthetically and practically, if it were restricted to the two-person case. It is not. In this and the following five chapters we examine the general theory which is, in the main, very different from the two-person theory and, we are forced to admit, less satisfactory.

Intuitively, it is reasonable to suppose that the two most significant notions of the two-person theory—mixed strategies and equilibrium points —can be extended to games with more than two players, and this extension we shall discuss in the present chapter. Were these generalizations and the resulting theorems the totality of n-person theory, we should have presented it in a unified manner for all $n \geqslant 2$. However, it has long been recognized in sociology, and in practical affairs, that between two-person situations and those involving three or more persons there is a qualitative difference which is not as simple as the difference between 2 and 3. Georg Simmel writes, "The essential point is that within a dyad, there can be no majority which could outvote the individual. This majority, however, is made possible by the mere addition of a third member." [1950, p. 137.] Again, "The typical difference in sociological constellation,

thus, always remains that of two, as over against three, chief parties."
[1950, p. 144.] The recognition of this feature—of the possibility of
coalitions in the language of von Neumann and Morgenstern—has resulted
in an *n*-person theory markedly different from two-person theory.

A major obstacle to developing a satisfactory theory of coalition formation is that in the present formalizations of a game no explicit provisions
are made about communication and collusion among the players (see
section 7.6 for more discussion of this point). Thus any theory of collusion, i.e., of coalition formation, has a distinctly *ad hoc* flavor. The difficulties in making explicit assumptions about communication appear, at
least superficially, to stem from the variety of rules which are found in
empirical situations. Collusion in parlor games is prohibited by social
sanctions and by a sense of sportsmanship; that the rules are well heeded
is, one supposes, a reflection of how little is usually at stake. Of course,
there are known exceptions in the history of gambling. In the economy
one finds the whole gamut from no rules at all, through moral sanctions,
to elaborate legal codes as in the antitrust laws. In international affairs,
coalitions and their disruption bulk large throughout Western history;
the rules obeyed seem to have been few.

In addition to the conceptual complications of collusion, there are
inherent practical complications as *n* gets larger, for the number of possible coalitions increases at a fantastic rate; the difficulty of a detailed analysis of a two-person game such as chess is minor compared to a similar
analysis of most *n*-person games. One of the principal features of the
current theory is to bypass such a detailed analysis. That we can successfully avoid combinatorial problems at the conceptual level does not necessarily mean that we can do so when dealing with empirical situations.
Naively, it appears that in an empirical study one must deal with specific
games in all their complexity; however, this presumption does not seem to
cover the issue entirely, for ways have been proposed to avoid some of
these difficulties in empirical work, but we must postpone more discussion
to Chapter 12.

Before digging into the conceptual problems of *n*-person theory, let us
re-emphasize that for the most part we are following the framework set
up by von Neumann and Morgenstern—a framework which they and
others have criticized but not replaced.[1] To some extent this may be due
to inattention, for the vast majority of work in the dozen years since the
first printing of their book has been devoted to the finite two-person

---

[1] Von Neumann and Morgenstern raised objections to the two distinct theories to
which they were forced, and they suggested that when the theory is more mature it
may be unified for all $n \geqslant 2$ and the now important characteristic function will appear
only as an unnecessary technicality. [1947, p. 606–608, particularly p. 608.]

theory, to extensions of it to infinite games and sequentially compounded games, and to related topics such as linear programing and statistical decision theory; the published papers on n-person games number few more than a score. Several facts may be mentioned which seem relevant to this phenomenon: the relation of the two-person game to linear programing and to statistics has attracted considerable attention because of the known importance of the latter two subjects; mathematicians have been intrigued by the two-person theory because it draws on more advanced mathematics than does n-person theory; and many workers have felt dissatisfied with the present formalization of n-person theory and rather than meet the conceptual challenge they have, for the most part, withdrawn to other issues.

Nonetheless, it is the n-person theory which must be of greater interest in sociology and economics. It is here, more than in two-person theory, that game theory as a part of social science, though not as a part of mathematics, will stand or fall.

7.2 MIXED STRATEGIES AND THE NORMAL FORM

Back in section 3.7 we arrived at the normal form of an n-person game in pure strategies; it will be recalled that it consists of:

(i)   The set I_n of n players,

(ii)   The n strategy sets $S_1, S_2, \cdots, S_n$,

and

(iii) The n real-valued payoff functions $M_1, M_2, \cdots, M_n$, where $M_i(s_1, s_2, \cdots, s_n)$ is the utility payoff to player i when player 1 uses strategy s_1, 2 uses strategy s_2, $\cdots$, and player n uses s_n.

In addition it was assumed that each of the players knows the entire structure of the game in normal form and that each is governed in his behavior by an inflexible desire to maximize expected utility. Beginning with this structure and specializing it for $n = 2$, the discussion of two-person games forced us to introduce the concept of a mixed strategy, i.e., of a probability distribution over the set of pure strategies. It seems reasonable that if this concept was needed there it will also be needed for $n > 2$. The generalization is practically obvious, but for the sake of completeness we shall present it here.

▶ If s_i is a typical pure strategy in S_i, then a mixed strategy σ_i for player i assigns a probability to each s_i. If we denote this probability by $\sigma_i(s_i)$, then we must have $\sigma_i(s_i) \geq 0$ and the sum of all these quantities over all s_i in S_i must be 1. Let us suppose that player i chooses the mixed strategy σ_i and that each of the players chooses a pure strategy, say $s_j^{(0)}$ for player j, $j \neq i$, then the outcome resulting from this n-tuple of strategies $(s_1^{(0)}, s_2^{(0)}, \cdots, \sigma_i, \cdots, s_n^{(0)})$ is a lottery among the outcomes associated to n-tuples of pure strategies. Specifically, the outcome associated with the pure strategy n-tuple $(s_1^{(0)}, s_2^{(0)}, \cdots, s_i, \cdots, s_n^{(0)})$ occurs with probability $\sigma_i(s_i)$. The utility evaluation of this lottery for player j is simply the expected utility of the outcomes associated with the pure strategy n-tuples, i.e.,

$$M_j(s_1^{(0)}, s_2^{(0)}, \cdots, \sigma_i, \cdots, s_n^{(0)})$$
$$= \sum_{s_i \text{ in } S_i} \sigma_i(s_i) M_j(s_1^{(0)}, s_2^{(0)}, \cdots, s_i, \cdots, s_n^{(0)}).$$

If we proceed in this fashion to each of the other players, it is clear that the payoff functions can be extended to the spaces of mixed strategies. ◀

7.3 CONSTANT-SUM AND ZERO-SUM GAMES

In the theory of two-person games a strictly competitive game was called zero-sum because it is always possible to choose the zeros and units of the player's utility functions in such a manner that the sum of the two utility functions for any strategy choices is zero. This only reflected the fact that the interests of the players were strictly opposing; these choices of units did not make an assumption that utility is comparable between the players. For games with more than two players this notion can be directly generalized: an n-person game is *zero-sum* if there is a choice of utility unit and origin for each of the players such that the sum of the utility numbers associated to each n-tuple of strategies is zero. Formally, the units and zeros of the utility functions M_i can be so chosen that, for every n-tuple of strategies $(s_1, s_2, \cdots, s_n)$,

$$\sum_{i=1}^{n} M_i(s_1, s_2, \cdots, s_n) = 0.$$

If this is possible, it is also always possible to choose the zeros of the functions so that they add up to any arbitrary constant, and conversely. This has led to the introduction of the term *constant-sum*, which is widely used in n-person theory, but it is well to keep in mind that this means nothing more nor less than zero-sum.

If we have a game, such as a parlor game, where the payoffs to the players are money and they always sum to a constant, and if each player's utility is linear in money, then the game is zero-sum. Such an assertion does not require any interpersonal comparison of utility. In general, however, utility functions are not linear in money, so the resulting game is not zero-sum. Most economic processes, even if they are games, cannot be treated as zero-sum games.

*7.4 BEHAVIORAL STRATEGIES AND PERFECT RECALL

It has surely occurred to the reader that, although these notions of strategies, both pure and mixed, may be fine tricks for the mathematical development of game theory, people almost never pick a strategy on such a grand scale. Even for most parlor games, the domain of strategies is just too large ever to have been completely given; in all the years that chess has been played and analyzed, only a small fraction of partial strategies has ever been discussed and listed. Thus, one might wonder about a theory of games with a more modest view of the strategy notion. One of a somewhat special and limited nature has been examined and the results are interesting, for in a certain important class of games they justify a theory based on mixed strategies.

Instead of giving a mixed strategy to the umpire, a player might specify for each of his information sets a probability distribution over the alternatives of the set. Such a class of distributions—one for each information set—is known as a *behavioral strategy* for the player. Now, although it is still a monumental task to list behavioral strategies for most games, it may be felt that in effect a player has such a distribution in his mind when he makes decisions during a play of the game, and that by making him play it many times (after learning has occurred) and observing his choices we could get experimental estimates of these distributions.

A neat way of viewing the difference between mixed and behavioral strategies has been suggested to us by Harold Kuhn. One can think of each pure strategy as a book of instructions where each page refers to just one information set and states exactly what should be done at that information set. The strategy set is a library of such books. A mixed strategy chooses one book out of the library by means of a chance device having the probability distribution of the mixed strategy. A behavioral strategy, on the other hand, is a book of a different sort. Although each page still refers to a single information set, it states a probability distribution over the alternatives at that set, not a specific choice.

It will help in understanding the several points to be made in this section to have a specific example in mind. Consider the game tree shown in Fig. 1. Player 1 has four pure strategies (a, c) (a, d), (b, c), and (b, d) which we may denote by α_1, α_2, α_3, and α_4, respectively. As we proceed we will illustrate the several concepts in terms of this game.

It is reasonably clear—and it can be shown—that each mixed strategy for a player induces a unique behavioral strategy for him, namely, the induced probability distribution at each information set. Less obvious, but also true, is the fact that if we are given a behavioral strategy for a player there always exists a (not necessarily unique) mixed strategy whose induced behavioral strategy is the given one.

As an illustration, consider the mixed strategies

$$\sigma_1' = (\tfrac{1}{2}\alpha_1, 0\alpha_2, 0\alpha_3, \tfrac{1}{2}\alpha_4) \quad \text{and} \quad \sigma_1'' = (\tfrac{1}{4}\alpha_1, \tfrac{1}{4}\alpha_2, \tfrac{1}{4}\alpha_3, \tfrac{1}{4}\alpha_4)$$

in the game of Fig. 1. Both of these induce the same behavioral strategy, namely: use a and b, each with probability $\tfrac{1}{2}$, on move 1 and c and d, each with probability $\tfrac{1}{2}$, on move 3.

From a collection of behavioral strategies, one for each player, one can compute the probability distribution over the end points of the game tree, and thus one can compute the expected payoff to each of the players.

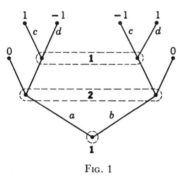

Fig. 1

Two *n*-tuples of mixed strategies $(\sigma_1', \sigma_2', \cdots, \sigma_n')$ and $(\sigma_1'', \sigma_2'', \cdots, \sigma_n'')$ will be said to be *behaviorally equivalent* if both σ_i' and σ_i'' induce an identical behavior strategy for $i = 1, 2, \cdots, n$. For example, in the game of Fig. 1, (σ_1', σ_2) and (σ_1'', σ_2) are behaviorally equivalent for any of player 2's mixed strategies σ_2. It is of interest to classify those games for which every pair of behaviorally equivalent mixed strategy *n*-tuples results in an identical pair of payoff *n*-tuples to the players. For in games of this type a player does not suffer any loss of strategic versatility by confining himself to his behavioral strategies. Stated in a slightly different form, in games of this type a player's set of behavioral strategies are *sufficient* for strategic purposes in the sense that any mixed strategy chosen by him can be strategically matched as far as payoffs are concerned by its induced behavioral strategy—regardless of the strategies chosen by the other players. (The role played by behavioral strategies is conceptually analogous to the role played by a sufficient statistic or a sufficient partition in statistical inference.) This problem was first posed by Kuhn [1953 *b*], and he gave the solution we shall discuss.

Roughly, this class of games is characterized by the property that each player remembers at each of his moves everything he did prior to it, as in a game with perfect information (section 3.2), but unlike these games a player may not always know what choices the other players have made. Most parlor games satisfy this condition, but bridge is a notable exception because the pairs of partners are single players who alternately "forget" their choices. To illustrate the idea more concretely, again consider the game tree shown in Fig. 1. It fails to have perfect recall because when player 1 is in his second information set he is unable to recall the choice he made at the first move. For a fixed strategy choice by player 2, the behavioral strategy which uses a and b, each with probability $\frac{1}{2}$, and c and d, each with probability $\frac{1}{2}$, and the mixed strategy $\sigma_1'' = (\frac{1}{4}\alpha_1, \frac{1}{4}\alpha_2, \frac{1}{4}\alpha_3, \frac{1}{4}\alpha_4)$ result in the same distribution over end points; but replacing σ_1'' by the behaviorally equivalent mixed strategy $\sigma_1' = (\frac{1}{2}\alpha_1, 0\alpha_2, 0\alpha_3, \frac{1}{2}\alpha_4)$ results in a different distribution. The principal feature of the mixed strategy σ_1' is that it produces a correlation between the choices in the first and the second information sets; this is not possible with behavioral strategies. The possibility of correlating the choices on different information sets is related to the signaling phenomena discussed below. With the payoffs to player 1 indicated in Fig. 1, it is clear that pure strategies α_2 and α_3 are dominated by α_1 and α_4 respectively. The mixed strategy is maximin for 1 guaranteeing him a security level of $\frac{1}{2}$. But its induced behavior strategy results in a constant payoff to 1 of 0 units.

Intuitively it is clear what we are trying to say. Now the only problem is to give a suitable general definition. There are several ways this can be done, but possibly the simplest is to introduce a concept due to Thompson [1953 a] which we shall also need in the next section. He defines an information set U of player i to be a *signaling information set*[2] if we can find some other information set, say V, later on in the game tree which also belongs to player i and a branch numbered r leaving the set U such that:

i. There is at least one move in V which can be reached by a path starting with the rth branch of a move in U.

ii. There is at least one move in V which cannot be reached by any path starting with the rth branch of a move in U.

We see that, if U is a signaling information set for player i, when he is at V he finds it impossible to know whether or not he chose the rth alternative at U. For example, the first information set of player 1 in Fig. 1, i.e., the first move, is a signaling information set since one of the moves in his second information set can be reached by a path beginning with the left branch of the first move and the other cannot.

A game is said to have *perfect recall* if there are no signaling information

---

[2] The term "signaling" used here arises, presumably, from a consideration of bridge, which has to be treated as a two-person game with a pair of partners constituting a single player. The choice(s) of one partner often serves to signal considerable information to the other (and the term "signal" is part of the vocabulary of bridge), but rarely, if ever, is the ambiguity totally removed by such signals.

sets.   As we mentioned earlier, it is easy to see that a game with perfect information has perfect recall, but the converse is not true.

The theorem proved by Kuhn [1953 b] is that, if a game has perfect recall and if $(\beta_1, \beta_2, \cdot \cdot \cdot, \beta_n)$ are the behavioral strategies induced by the mixed strategies $(\sigma_1, \sigma_2, \cdot \cdot \cdot, \sigma_n)$, then for each player i

$$M_i(\beta_1, \beta_2, \cdot \cdot \cdot, \beta_n) = M_i(\sigma_1, \sigma_2, \cdot \cdot \cdot, \sigma_n);$$

furthermore, this equality holds in general only in games with perfect recall.   In interpreting this result, remember that for each behavioral strategy there may be many mixed strategies which induce it.   This result means that for such games it does not matter to the players whether they take the global view of mixed strategies or the more restricted (and plausible) view of behavioral strategies.

*7.5   COMPOSITE STRATEGIES

Given that in games with perfect recall the analysis can be at the level either of behavioral or mixed strategies without affecting the expectations of the players, the question arises whether anything more can be said for games without perfect recall.   Thompson [1953 a] attacked this problem, and he has given an intuitively satisfactory solution.

Let Δ_i denote the set of all signaling information sets for player i (see the preceding section for the definition of a signaling information set). It will be recalled that, if $\Delta_i = \phi$ (= the empty set) for all i, the game is said to have perfect recall and behavioral strategies are as good as mixed strategies.   Presumably in games without perfect recall we must focus our attention on those information sets—the signaling ones—which prevent perfect recall.

Earlier we defined the notion of a pure strategy over the set of all information sets; we can, of course, do the same thing over the set Δ_i, and we call this a *pure signaling strategy* for player i.   In like manner, a probability distribution over the pure signaling strategies of player i is called a *mixed signaling strategy*.   These notions are exactly the same as those given in sections 3.7 and 7.2 except that the domain of definition is restricted to the signaling information sets rather than to all information sets.

An *associated behavioral strategy* for player i is a behavioral strategy over the information sets of player i which are not signaling information sets. A *composite strategy* is the pair consisting of a mixed signaling strategy (over the signaling information sets) and an associated behavioral strategy (over the remaining information sets).   That is, over the information sets having perfect recall we continue to use behavioral strategies, and over all other information sets we use mixed strategies.   It is easy to see that for games with perfect recall composite strategies are the same as behavioral strategies.

To each mixed strategy there is a naturally induced (unique) composite strategy, and to each composite strategy one can find a (not necessarily unique) mixed strategy whose induced composite strategy is the given one. Again, as in games with perfect recall, the payoff to player i using a composite strategy is defined in the natural way, and Thompson [1953 a] has shown that the payoffs associated with an n-tuple of mixed strategies are identical to the payoffs of the induced composite strategies. Thus " · · · any payoff which players can obtain by means of mixtures of pure strategies, they can also obtain by means of composite strategies." [1953 a, p. 275.]

This result is of considerable importance in the examination of specific games without perfect recall. Thompson remarks, "This theorem, together with the fact that the normalized form of the game obscures signaling strategies, explains one reason why the normalized form of the game is not always the best form in which to solve actual games." [1953 a, p. 275.]

In another paper, which we shall not go into here, Thompson [1953 b] uses the notion of signaling strategies to examine a simplified form of bridge.

7.6 COMMUNICATION BOUNDARY CONDITIONS

With the last section we have completed the discussion we shall give of games in extensive form. From now on we shall deal only with games in normal form and with an abstraction derived from them, the characteristic function.

Under the assumption that individuals are interested in maximizing their expected utility, the aim of present-day game theory is to construct a notion of equilibrium social behavior and to investigate the properties of such a concept. The word "equilibrium" indicates that the theory is in some sense a static one, and for, at least, practical purposes this must be considered a serious limitation of the present theory. Once one has fixed upon the normal form, then there can be no consideration of the dynamic process whereby the equilibrium states are achieved nor can any learning be admitted at that level. The individual preference patterns, as given by the payoff functions, are assumed to be invariant both in time and with the unfolding experience of participating in the extensive form of the conflict of interest. Furthermore, the strategy spaces are assumed invariant, which is certainly false for the many economic processes which are subject to modification by technological developments and research.

All in all, it is very doubtful that much social behavior is actually equilibrium behavior for any *naive* characterization of the situation as a game in extensive or normal form.

The above points are not theoretically correct, or, at least, they can be misleading. It is always possible to conceive the extensive form of a situation in a sufficiently complicated way to include many of the dynamic features. Or, sometimes it is possible to embody the dynamics as a repetition of games in normal form, which creates a supergame. In these ways the dynamics of learning, of invention, and so on can all be included formally in the static game models. In practice, however, our initial comments are at present largely correct. Usually, it is most difficult to describe an extensive form of a game which takes into account the dynamics, and, in general, very few results are known which will help us to understand social dynamics via an extensive game model. This is even true if we restrict ourselves to the construction of supergames from simpler ones in normal form. However, some exceptions exist (see the discussion of section 5.5, Shubik [1957], and Appendix 8). Moreover, the theory of repeated normalized games, especially two-person ones, is one of the most active topics of mathematical research at the present time, and so we suspect and hope that these comments may soon be dated.

As we mentioned in the introduction to this chapter, as soon as the number of players exceeds two there is the possibility for communication and collusion among them, and in most games it is possible for a player to increase his expected payoff by collaborating with other players in the choice of strategies. We feel that it is exactly this feature of *n*-person theory which is (or should be) of particular interest to the social scientist. Clearly, coalition formation is a sociological phenomenon and one which has received considerable descriptive attention. This literature, and common observation, suggests that one important aspect of the phenomenon are the restrictions society places upon coalition formation and coalition changes. The exact nature of these restrictions and sanctions seems to depend upon the specific situation, its history, the general mores of the society, the legal structure, etc. It is perfectly clear that the characterization of a game in normal form—the strategy spaces and the payoff functions—plus the postulates of complete knowledge and of rational individual behavior (see Chapter 3)—includes no features corresponding to sanctions or restrictions upon coalition formation.

We should judge this omission of sociological assumptions at the level of the normal form to be one of the two major practical faults of present-day *n*-person theory, the other being the previously mentioned static character. The task of formalizing the communication process, especially the prohibitions against communication among the players, is far from trivial.

It appears that to include it in a generalization of game theory will be an exciting major theoretical step. Lacking such a generalization, several tacks have been taken, each of which is unhappily special and arbitrary. It is somewhat consoling to observe, however, that we can find an analogous situation in the physical theories of continuous processes, but more of that later. We shall cite three different approaches to the problem of restricted collusion.

First, there is the one extreme in which any collusion logically possible is allowed to occur. This is characteristic of the von Neumann and Morgenstern theory of solutions, which will be discussed in detail in Chapter 9. In their theory such freedom to cooperate leads to vast numbers of "solutions" with no criteria to select among them. They are forced, as we shall see, to the *ad hoc* assumption that in practice there exist social standards which determine *the* solution which actually occurs, but no attempt is made to exhibit a theory of these standards.

Second, there is the other extreme which prohibits any collusion at all. Such a condition may not be nearly so limiting as it first seems. Certain authors, notably Nash [1951], hold that non-cooperative games are theoretically basic and that cooperative games can and should be subsumed under that theory by making communication and bargaining formal moves in a non-cooperative extensive game. The resulting normalized game would simply enlarge the domain of the various strategies, and the payoff functions could be extended to these larger sets in the natural manner. Were it possible to give an explicit and intuitively acceptable way of enlarging an extensive game so as to include communication the argument would be very convincing. Specific criticism is hard to make since this view has never been fully elaborated, but McKinsey [1952 *a*, p. 359] has pointed out, "It is extremely difficult in practice to introduce into the cooperative games the moves corresponding to negotiations in a way which will reflect all the infinite variety permissible in the cooperative game, and to do this without giving one player an artificial advantage (because of his having the first chance to make an offer, let us say)."

In a way, this conceptual solution to the formalization of preplay communication simply buries some of the most interesting aspects of the problem. One is interested in understanding the forces which lead groups to cooperate, in the cohesiveness of coalitions over repeated plays of the game, and so on, and we do not want to prejudge these problems by entering them into the extensive form in some special manner.

But whether or not we accept the belief that all games can or should be recast in terms of non-cooperative games, one part of the theory certainly should be devoted to such games, and presumably it should be a "natural" extension of the (non-cooperative) two-person zero-sum theory.   Nash has offered such a theory, and we shall discuss it in section 7.8.

The third tack attempts to characterize, in what is surely an oversimplified manner, *some* types of restrictions on collusion.   We shall describe it fully here, for it will be useful in the discussion of section 7.9 and it is necessary in Chapter 10; for other discussions of the idea see Luce [1954, 1955 *a*].   It seems plausible to suppose, at least as an approximation to reality, that the collusion among the players results in non-overlapping coalitions within which there is perfect cooperation and among which there is ruthless competition.   Such a partition of the players into coalitions will be termed a *coalition structure*.   For example, the partition $(\{1\}, \{2\}, \cdot \cdot \cdot , \{n\})$ represents the case where there are no non-trivial coalitions, i.e., there is no cooperation.   The partition $(\{1, 2, \cdot \cdot \cdot , n - 1\}, \{n\})$ represents the case where the first $n - 1$ players have joined forces in a coalition against player n.

Let us now suppose that at some stage of the bargaining in a given conflict of interest the players are arranged in coalitions according to a partition, which we shall denote by τ.   Since each player is a ruthless rationalist, we must assume that each is considering various potential changes of alliance in an attempt to better his own outcome.   These changes may be thought of as all occurring prior to the play of the game, or, if it is repeated many times, they can occur one at a time prior to each play. Some of these alliances are undoubtedly rejected because they are not profitable, but, if our intuition is correct, sometimes there are alliances which are simply never considered even when they are profitable.   This may occur because of sanctions against them, or it may be that the change is simply too complicated to be considered practical.   One presumes that in economic situations communication and changes of alliance are expensive, and so many logically possible changes are ignored as too costly. Another, and possibly more important reason, is the ever present fear that a radical change will upset the status quo, and who knows where he will end up following a "revolution."   The complication of the other players not reacting passively to the changes made by one, or by a few players, will return to plague us again (sections 7.8, 7.9, and 10.2).   On the other hand, the addition of one player, or the expulsion of one, from a coalition and other similar simple changes do occur, and, of course, a long sequence of such changes can radically alter the original coalition structure.

A *rule of admissible coalition changes*, denoted ψ, states for each coalition structure τ precisely those coalitions of players which, for whatever reasons, possess the freedom of unrestricted preplay communication and can potentially act jointly, e.g., by choosing a joint strategy.   For any coalition structure τ, the class of such coalitions—which are said to be admissible changes according to the rule ψ from the structure τ—is labeled $\psi(\tau)$. *It will be convenient to make the convention that all the coalitions in the structure τ are always members of the list $\psi(\tau)$.* An example may be illuminating. For three players, suppose we explicitly write down the rule which permits as an admissible change any coalition formed simply by the addition of a single player to an already existing coalition:

τ	$\psi(\tau)$
$(\{1\}, \{2\}, \{3\})$	$\{1\}, \{2\}, \{3\}, \{1, 2\}, \{1, 3\}, \{2, 3\}$
$(\{1, 2\}, \{3\})$	$\{1, 2\}, \{3\}, \{1, 2, 3\}, \{1, 3\}, \{2, 3\}$
$(\{1, 3\}, \{2\})$	$\{1, 3\}, \{2\}, \{1, 2, 3\}, \{1, 2\}, \{2, 3\}$
$(\{2, 3\}, \{1\})$	$\{2, 3\}, \{1\}, \{1, 2, 3\}, \{1, 2\}, \{1, 3\}$
$(\{1, 2, 3\})$	$\{1, 2, 3\}$

Mathematically, a rule ψ can be identified with the table composed of the lists $\psi(\tau)$ for each τ. That is, ψ can be thought of as a function which maps each τ into a class of coalitions.

It may be worth appending that this third notion includes the other two (i.e., no changes permitted and all possible changes permitted) as special cases: Define the function ψ' to have the property that $\psi'(\tau)$ is the class of all coalitions in a structure τ for each τ; then no changes are permitted. Define the function ψ'' to consist of all subsets of I_n for all values of τ, then all logically possible changes are permitted.

There seem to be at least two major objections to postulating such a function ψ: first, we have no theory to justify it; and, second, we have no idea how it is to be determined in particular cases. Without attempting to dispute these points, it may be worth observing that such functions seem to play a role not unlike boundary conditions in some of the continuous flow theories of physics, whereas the given coalition structure τ can be thought of as an initial condition. The boundary conditions of, say, a heat or air flow problem are not given by the physics of the flow process but rather are supposed to represent certain salient facts about the particular physical configuration under study. The form of the boundary conditions is in part determined by the equations representing the flow process, but their detailed selection is arbitrarily given by the scientist performing the analysis, who calls upon one's intuition to accept his choice. There is clearly an art to making such choices, an art which in physics and engineering has gradually become highly sophisticated as the result of past failures and successes in relating theory to data.

The role of the function ψ may also be viewed in another way: In the normal form of the game the payoff functions represent, in a sense, the economics of the model, i.e., they prescribe the returns accruing to certain choices. The rationality postulate—the desire of individuals to maximize utility—amounts to the psychology built into the model. But where is the sociology of the model? Nothing within the basic structure of the model describes any limitations on the interactions between the participants. So far we have not directly established the need for any sociological assumptions in the model, and it is conceivable that none are needed; however, our interpretation of results to be presented in Chapters 9 and 10 is that there does seem to be such a need. In any case, the functions ψ just described fulfill such a role, and, although it is clear that other sociological postulates are possible, they are the only ones which have been investigated in the literature.

7.7 CLASSIFICATION OF CONTEXTS FOR *n*-PERSON GAMES

Once one leaves two-person zero-sum games, there are serious questions of extra-game-theoretic assumptions. One—the limitations on collusion —was discussed in the preceding section, but it is by no means the only one which has been considered in the literature. It seems appropriate to specify these here and then to classify the several theories (and suggestions for theories) in terms of these. This will permit the reader to find his way around the complexities of *n*-person theory a little more simply.

Most of the past work in *n*-person theory has supposed that, in addition to receiving the payoffs prescribed by the rules of the game, the players are permitted to make additional transfers—side payments in the delicate language of the theory, considerations or bribes in more direct vocabularies (see section 8.1). Indeed, a far stronger assumption is made which is generally subsumed under the phrase that utility is "unrestrictedly transferable." Of course, it is never utility as such that is transferred, for utility is a derivative concept, but commodities to which utility can indirectly be attached by the players. To make any sense of the elliptic concept of unrestricted transferability and of the mathematics employed, one must suppose that there exists an infinitely divisible, real, and desirable commodity (which for all the world behaves like money) such that any reapportionment of it among the players results in increments and decrements of individual utilities which sum to zero according to some specific set of utility scales for the players. This can happen if money exists, provided that each player's utility for money is linear and that the zero and unit of each utility function is so chosen that the conservation of money implies conservation of utility. When else it can realistically happen is obscure.

It might be thought that this selection of a zero and a unit, in essence setting the utility of money to be equal to money, constitutes a decision as to the interpersonal comparison of utility. Such is not the case, provided that nowhere in the mathematics these scale choices are compared one with another, as for example in making threats. Since, as we shall see, direct threats are one of the missing components of n-person theory, no trouble arises on the score of interpersonal comparisons. While we are mentioning this concept, it should be pointed out that assuming inter-personal comparisons plus unrestricted side payments in an infinitely divisible and desirable commodity does _not_ imply transferability of utility as it is presently meant in n-person theory, for it does not imply conserva-tion of utility under reapportionments. The question of side payments will be dealt with more fully in section 8.1.

If the assumption of unrestricted transferability is dropped, then it is clear that there is a whole complex of possible cases ranging from perfect transferability to none at all. It seems hopeless to try to develop a theory covering all cases, and it is much too tedious to examine many of the intermediate assumptions, so, as is often done in mathematics, only the polar extremes are studied. Thus, a theory will assume either that per-fect transferability is possible using some infinitely divisible and desirable commodity which is conserved, or that no transferability is possible at all. The latter assumption is not lacking in interest, for in many situations the mores and legal codes prohibit bribing.

Still another question of context is whether, when preplay communica-tion can occur, the players are able to employ correlated (i.e., joint) strategies, or whether they must only agree upon individual strategies which, although possibly coordinated choices, are not correlated, i.e., independent in the statistical sense. To correlate strategies when a game is not temporally repeated necessitates preplay communication.

Only certain combinations of these contexts have received much atten-tion in the literature. In a way, the cases of neglect are curious, for it is among them that one finds some of the natural generalizations of the central notions of two-person theory. The following table presents all these possible combinations of contexts, lists the sections where the corre-sponding theory is discussed, and states the name of the theory (if one exists). Under the column Preplay Communication, the word "partial" includes, as a special case, no limitations on communication whereas "all" refers only to total freedom to communicate.

Possibly the most surprising omission in the literature is the case of no side payments and partial preplay communication. In section 7.9 we discuss some suggestions toward such a theory, under the assumption that communication enables the players of a coalition to choose joint strategies but does not allow intercoalition communication. In particu-lar, one coalition is not allowed to threaten another. Were this per-mitted, then, in all likelihood, interpersonal comparisons of utility would be necessary. Thus, when side payments are prohibited we still do not have a full generalization of two-person theory, for in that theory there were two aspects to preplay communication: threat and collusion. In the theories admitting side payments, the threat phenomenon is partially included, but in an indirect manner so as to bypass direct confrontations of the form "you will be hurt more than I will" and hence to bypass interpersonal comparisons.

Side Payment	Preplay Communication	Correlation of Strategies in Coalitions	Transferable Utility	Section	Name
	None	Irrelevant	Irrelevant	7.8	Equilibrium Points
No	Partial	Yes	Irrelevant	7.9	$\cdots$
		No	Irrelevant	7.9	$\cdots$
	All	Yes	Yes	9.1–9.7	Solutions
			No	10.4	$\cdots$
Yes	Partial	Yes	Yes	10.1, 10.2	ψ-stability
			No	10.4	$\cdots$
		No	$\cdots$	$\cdots$	$\cdots$

7.8 NON-COOPERATIVE GAMES: EQUILIBRIUM POINTS

This section is devoted to what may be described as an extension of non-cooperative two-person games to non-cooperative *n*-person games. Nash [1951] first introduced the notion of an equilibrium point, and he showed that every game possesses such a point in mixed strategies and that for two-person zero-sum games the definition is identical to the maximin concept. This was an important step, for, previously, no one had seen how to extend the maximin notion beyond $n = 2$.

Let us suppose we have a game in normal form with payoff functions M_i, and let us also assume that no cooperation among the players can occur, i.e., several players cannot get together and agree upon a joint selection of strategies. Now suppose through a normative analysis of the game, or from experience in it, or whatever, the strategy *n*-tuple $(s_1, s_2, \cdots , s_n)$ is singled out for consideration by the players. These strategies will be in equilibrium provided that no player finds it is to his advantage to change to a different strategy so long as he believes that the other players will not change. Thus, if we look at player i it must be that he cannot expect to benefit by employing strategy r_i instead of s_i. Now, it may well be that, if player i could communicate to player j and they agreed upon some joint change of strategies, they could both benefit, but we have assumed that no collusion is permitted. Thus, player i can only consider changes which are under his direct control, i.e., changes in his own strategy choice, and it is argued that, if none of these changes benefit him, he will not change. If $(s_1, s_2, \cdot \cdot \cdot, s_n)$ should be chosen so that what is true for player i is also true for all the other players, then there are no resulting forces to change the given system of strategies, and hence the n-tuple of strategies is in equilibrium. To say that player i does not benefit by changing his strategy choice simply means that his payoff does not increase by any other choice, i.e.,

$$M_i(s_1, s_2, \cdot \cdot \cdot, s_i, \cdot \cdot \cdot, s_n) \geqq M_i(s_1, s_2, \cdot \cdot \cdot, r_i, \cdot \cdot \cdot, s_n).$$

Thus, one is led to the following definition: An n-tuple of pure strategies $(s_1, s_2, \cdot \cdot \cdot, s_n)$ is an *equilibrium point* in pure strategies if the above inequality holds for every i and for every choice of r_i in the set S_i of pure strategies for player i. It is not difficult to show that when $n = 2$ this definition is the same as that given earlier in the two-person theory (sections 4.8 and 5.7).

As in the two-person case, there is no assurance in general that an equilibrium point exists in pure strategies. It is known that a sufficient condition for games to have equilibrium points in pure strategies is that they have perfect information [Kuhn, 1953 b], but this is not a necessary condition. Dalkey [1953] and Otter and Dunne [1953] have given necessary and sufficient conditions, but the statement of these results is too complicated to warrant inclusion in this book. Birch [1955] has extended these results by giving a sufficient condition for a game to have a mixed strategy equilibrium point (see below) in which certain players use pure strategies.

Fortunately, the parallel with the two-person case extends to the point where mixed strategies again suffice to establish existence. The above definition of an equilibrium point in pure strategies can obviously be reworded to give a definition of an equilibrium point in mixed strategies by the simple formal substitution of "mixed strategies σ_i" for "pure strategies s_i." Nash's principal theorem [1951] shows that over the domain of mixed strategies every finite game has at least one equilibrium point. This result shows that Nash's definition has one extremely desirable property of an equilibrium notion: existence. (The proof given in Appendix 2 generalizes to any n.)

However, just as with the two-person non-zero-sum theory, the properties of interchangeability and equivalence, which held for two-person zero-sum games, do not hold in general. To be more specific, suppose that $(\sigma_1, \sigma_2, \cdots, \sigma_n)$ and $(\lambda_1, \lambda_2, \cdots, \lambda_n)$ are both equilibrium points of a general game, then, first, there is no assurance that an intermixture of strategy choices, such as $(\lambda_1, \sigma_2, \sigma_3, \cdots, \lambda_i, \cdots, \sigma_j, \cdots, \lambda_n)$ is also an equilibrium point; and, second, there is no assurance that the payoff to a player is the same for two different equilibrium points, i.e., in general:

$$M_i(\sigma_1, \sigma_2, \cdots, \sigma_n) \neq M_i(\lambda_1, \lambda_2, \cdots, \lambda_n).$$

The failure of the general equilibrium notion to have these two properties raises much more serious questions as to its merits than could be raised against the minimax concept. These points have already been made when we talked about non-zero-sum two-person games, but we shall repeat them here. First, if each player were to confine his strategy choice to those which are a part of some equilibrium *n*-tuple, the resulting problem faced by each player is again a game. It is a contraction of the old game, but it may be just as difficult to analyze conceptually as the original game. Indeed, in some sense it may be more difficult for a player to analyze it because it crystallizes the difficulties involved. Thus, the equilibrium notion does not serve in general as a guide to action.

Second, one may look upon the notion as possibly descriptive in nature. If a game is repeated many times, one may hope that ultimately an equilibrium point will be found. There are, however, various doubts that can be raised. It will be recalled that in our discussion of the prisoner's dilemma (sections 5.4 and 5.5) we argued that temporal collusion was a very likely possibility, and this can also be expected for more general games. On the other hand, as the number of players increases the chance of successful inarticulated collusion seems more remote. Another point that is descriptively relevant for repeated games is just how far ahead does a player see. If not at all, the equilibrium notion is reasonable, but, if he considers several steps ahead, it is doubtful. For example, a player whose expected payoff is low in a given equilibrium point may be expected to argue that, by selecting a strategy different from the equilibrium one, he can create a situation of flux which after several repetitions of the game will terminate in a different equilibrium point. This new point may be more advantageous to him. Intuitively, it is perfectly clear that people do take actions whose immediate effect is disadvantageous but whose ultimate effect is thought or expected to be beneficial; the part of the puritan ethic which leads to saving and reinvestment in the economy is of this nature.

Nonetheless, we continue to have one very strong argument for equilibrium points: if our non-cooperative theory is to lead to an n-tuple of strategy choices and if it is to have the property that knowledge of the theory does not lead one to make a choice different from that dictated by the theory, then the strategies isolated by the theory must be equilibrium points.

The complications of non-equivalence and non-interchangeability of equilibrium points lead one to ask whether there is not some plausible condition which may be added to isolate a single equilibrium point as more acceptable than the others. For games with perfect information (see section 3.2) Gale [1953] has presented the following idea. He calls two strategies *equivalent* if they yield the same payoff to a player against all combinations of strategies for the other players, and he defines one strategy *to dominate* another if the former never results in a smaller payoff than the latter and yet they are not equivalent. Within a given game, two operations are defined: first, *simultaneous* averaging of all equivalent strategies for each of the players; and, second, *simultaneous* deletion of all dominated strategies for each of the players. Gale's theorem then states that, if we begin with a game having perfect information and if these operations are applied recursively in the order given, after a finite number of applications a game results which has but one strategy for each player. These unique strategies are in fact mixed strategies of the original game, and they constitute an equilibrium point in that game; this Gale calls the "solution."

Against this idea, at least as a plausible descriptive notion, one can argue that there is no compelling reason why a player should put equal probability weights over all equivalent strategies. Why not put all the weight on the strategy in the equivalence set which yields the best average return to his opponents? Or why not a host of other possibilities? Also, can one expect players to go through the averaging and deleting process as described? What about other alternatives? For example, in a game with three players, 1 might carry out the process conceptually with only player 2, holding 3's strategy space fixed. When that reduction is completed, suppose he engages in the mental process of averaging and deleting with player 3, then back to 2, and then back to 3, etc. It is possible that this too might lead to an equilibrium point, and, indeed, it might be preferred by the players to the one given by Gale's procedure. In other words, we object to the arbitrary and implausible nature of the process assumed. Nonetheless, it does have the distinct merit of being completely symmetric among the players, and it has the property that, if $n - 1$ players are committed to this suggestion, the nth player might as well go along. It will be recalled that in connection with the temporal repetition of the prisoner's dilemma (section 5.5) we described an analogous procedure.

7.9 COOPERATIVE GAMES WITHOUT SIDE PAYMENTS

Let us suppose that we have an *n*-person game being played in a social context which prohibits side payments, which allows (at least) some cooperation, and which does not prohibit players who have agreed to cooperate from correlating their strategy choices. If $n = 2$, then we have the theory that was discussed in section 6.2. It will be recalled that our typical example was the game with the payoff matrix

$$\begin{bmatrix} (2, 1) & (-1, -1) \\ (-1, -1) & (1, 2) \end{bmatrix}$$

and that cooperation between the players allowed them to correlate their strategies so as to randomize between (2, 1) and (1, 2). It will also be remembered that, if we plot the set of feasible payoff pairs, we argued that the players should restrict themselves to the northeast corner of the area—to the Pareto optimal set. Furthermore, it was argued that neither player should accept less in the cooperative context than he can guarantee for himself in the non-cooperative version of the game by using his maximin strategy. The set of Pareto optimal points which exceed the maximin values for each player was called the negotiation set, and this was taken to be the "solution" of the cooperative game by von Neumann and Morgenstern. We shall now describe one possible general definition of a solution to an *n*-person cooperative game in which side payments are prohibited, but in which correlated strategies and partial preplay communication are possible. When specialized to $n = 2$ and when preplay communication is not restricted, this definition is the same as that of the negotiation set.

As in Nash's theory of equilibrium points, we shall look for equilibrium behavior when it is assumed that each of the players is attempting to maximize his own expected utility, but we shall allow for some measure of cooperation. The Nash theory offered *n*-tuples $(\sigma_1, \sigma_2, \cdots, \sigma_n)$ of mixed strategies which were in equilibrium. Actually, however, one can argue that he found pairs consisting of an *n*-tuple of strategies and a coalition structure, i.e., pairs of the form

$$[(\sigma_1, \sigma_2, \cdots, \sigma_n), \tau],$$

which were in equilibrium. But the assumption of non-cooperation meant that only the coalition structure having no non-trivial coalitions, i.e.,

$$\tau = (\{1\}, \{2\}, \cdots, \{n\}),$$

could arise, and, since it remained fixed, there was no necessity to specify it explicitly. In a generalization to cooperative games such is not the case, so one might be led to search for pairs $[(\sigma_1, \sigma_2, \cdots, \sigma_n), \tau]$ which are in equilibrium when cooperation is allowed. But this will not do, for we have assumed correlated strategies are possible, and such a strategy n-tuple does not allow for that. To symbolize the correlation of strategies within a coalition, let $\tau = (T_1, T_2, \cdots, T_t)$ and let $\sigma(T_i)$ denote a typical correlated mixed strategy jointly chosen by the players in the coalition T_i. Then, the aim of the theory will be to characterize those pairs $[(\sigma(T_1), \sigma(T_2), \cdots, \sigma(T_t)), (T_1, T_2, \cdots, T_t)]$ which are in equilibrium when cooperation is allowed.

The next question to consider is: what cooperation? Our assumption will be that there are limitations upon the contemplated changes of coalitions from a given coalition structure τ. Or to put it in another fashion, given a coalition structure, not all the possible subsets of players are allowed to have preplay communication and to adopt joint mixed strategies. That is, we shall suppose that there are given sociological restrictions which can be summarized as a function ψ of the type discussed in section 7.6.

Consider the following argument for the equilibrium of a pair $[(\sigma(T_1), \cdots, \sigma(T_t)), \tau]$. If S is an element of $\psi(\tau)$, i.e., S is a possible coalition change when the players are arranged according to the coalition structure τ, then S may be expected to form and to disrupt the given pair if each of the players in the set S can be made to gain by the change. Thus, a necessary condition for the pair to be in equilibrium is that, for each S in $\psi(\tau)$ and for each selection of a cooperative strategy $\sigma(S)$, there shall be at least one member of S who does not profit by the change. Since we have assumed that there are no side payments, this simply means that the payoff as given by the rules of the game is no greater than it was in the equilibrium state.

Let us try to formalize this: The pair $[(\sigma(T_1), \cdots, \sigma(T_t)), \tau]$ is in equilibrium if, for each S in $\psi(\tau)$ and for each correlated mixed strategy $\sigma(S)$, there is at least one player j in S such that

$$M_j[\sigma(T_1), \sigma(T_2), \cdots, \sigma(T_t)] > M_j[\sigma(S), \cdots].$$

The question is how to fill in the dots on the right. If S were to form, τ is disrupted, and so, in general, the original strategies $\sigma(T_i)$ become meaningless since some of the coalitions T_i no longer exist. It is surely not reasonable to suppose that the remaining groups of players will carry on as if the change had not occurred, and, indeed, one may expect the change involving the formation of S to cause those players not in S to reappraise completely their collusive arrangements. So S will not know the reaction to its contemplated actions, but if the players in S are conservative they will expect the worst—they will expect all the remaining players, $-S$, to form a coalition against S. Furthermore, the most effective way that $-S$ could go after the coalition S would be to seek out the weakest member of S and to attack him with unrelenting pressure. In other words, the dots will be filled by a mixed strategy $\sigma(-S)$ which is so chosen as to minimize the payoff to j. Let us emphasize that this is an extremely conservative definition. It fails to take into account that $-S$ very likely might not form, and that, even if it did, it might find it distinctly to its disadvantage to try to disrupt S. However, if these conservative conditions are met, then we claim that the resulting pair is certainly in equilibrium. In summary, then, the definition reads: A pair $[(\sigma(T_1), \cdots, \sigma(T_t)), \tau]$ is ψ-stable (for a cooperative game without side payments but with correlated strategies) if, for each S in $\psi(\tau)$ and for each mixed strategy $\sigma(S)$, there is at least one player j in S vulnerable to an attack by $-S$, i.e., $-S$ has a strategy $\sigma(-S)$, say, such that

$$M_j[\sigma(T_1), \cdots, \sigma(T_t)] > M_j[\sigma(S), \sigma(-S)].$$

When we reach Chapter 10, we will see that this notion is conceptually similar to ψ-stability theory for cooperative games with side payments. In addition, it is a generalization of the negotiation set, as we now show.

▶ Since we are considering cooperation, $\tau = (\{1, 2\})$. Therefore, the definition requires that, if $[\sigma(\{1, 2\}), \tau]$ is ψ-stable, then for every other joint mixed strategy $\sigma'(\{1, 2\})$ either

$$M_1[\sigma(\{1, 2\})] \geqslant M_1[\sigma'(\{1, 2\})]$$

or

$$M_2[\sigma(\{1, 2\})] \geqslant M_2(\sigma'(\{1, 2\})],$$

which simply says that the joint strategy $\sigma(\{1, 2\})$ must be Pareto optimal. If no changes from the coalition are possible, that is all one can say. If, however, defections are possible, then we have, for *every* mixed strategy σ_1 for player 1,

$$M_1[\sigma(\{1, 2\})] \geqslant \min_{\sigma_2} M_1(\sigma_1, \sigma_2).$$

Thus,

$$M_1[\sigma(\{1, 2\})] \geqslant \max_{\sigma_1} \min_{\sigma_2} M_1(\sigma_1, \sigma_2).$$

Similarly, for player 2,

$$M_2[\sigma(\{1, 2\})] \geqslant \max_{\sigma_2} \min_{\sigma_1} M_2(\sigma_1, \sigma_2).$$

The right-hand terms are the maximin values. Thus, when changes are permitted, the definition yields the negotiation set, as it should.   ◀

For more than two players, this generalization of the negotiation set becomes more complicated since it depends upon the coalition structure and upon the admissible changes from the coalition structure. No work has been done on this concept for n larger than two, and so no comments can be offered as to its properties, faults, or merits.

For cooperative games without side payments and in which correlated strategies are prohibited, a similar definition is possible. The idea, of course, is that among the admissible coalition changes coordinated (but uncorrelated) changes of strategies will always fail to yield a profit to at least one member of the coalition. If we let $\mathfrak{d} = (\sigma_1, \sigma_2, \cdots, \sigma_n)$ denote an uncorrelated mixed strategy n-tuple, τ a coalition structure, and $\psi(\tau)$ the class of coalitions capable of disrupting τ, then $[\mathfrak{d}, \tau]$ is *not* ψ-stable if and only if there exists a coalition S in $\psi(\tau)$ whose members can so coordinate their choices of mixed strategies (without correlating them) that each improves by the defection when the remaining players hold fixed to their original strategy choices. Such a pair is ψ-stable if this condition does not hold, i.e., if for every coordinated choice of mixed strategies by S there is at least one member of S who fails to improve by the defection. It is easily seen that, if $\tau = (\{1\}, \{2\}, \cdots, \{n\})$ and if ψ admits no changes from τ (which describes the non-cooperative case), then a necessary and sufficient condition that $[\mathfrak{d}, \tau]$ be ψ-stable is that $\mathfrak{d}$ be a Nash equilibrium point; in other words, the definition is a generalization of Nash's notion. One might wish to generalize this notion to allow the members of a disrupting coalition S to correlate their strategies.

Farquharson [1955] and Shubik [1957] have given this definition for the special cases where $\tau = (\{1\}, \{2\}, \cdots, \{n\})$, and where $\psi(\{1\}, \{2\}, \cdots, \{n\})$ consists of all coalitions with k or fewer players. In other words, they consider simultaneous changes of k or fewer players, whereas Nash considered only the case where $k = 1$. Shubik used the term k-stability for this notion (the same term will be used in section 10.2 for a somewhat different but related idea), and Farquharson speaks of equilibrium points of order k.

7.10 SUMMARY

For the most part, this chapter has been concerned with the extension of some concepts from two-person theory to games with more players and with some new distinctions which seem necessary. Among the ideas extended were: mixed strategy, zero-sum game, and equilibrium point. The new distinctions were largely in the realm of extra-game-theoretic concepts including side payments, coalition structures, limitations on collusion (communication boundary conditions), correlated strategies, transferable utilities, and interpersonal comparisons of utility.

As in two-person theory, a *mixed strategy* is simply a probability distribution over a player's set of pure strategies. His payoff function is extended to the domain of mixed strategies by taking the appropriate expected values of pure strategy payoffs—a procedure justified by the central theorem of utility theory.   Once we had this concept, it was possible (in the two starred sections 7.4 and 7.5) to formulate and present the final extensive game problem to be studied.   A *behavioral strategy*—a more modest concept than a mixed strategy—is simply a collection of probability distributions over the branches of each of a player's information sets. It differs from a mixed strategy in that choices in two information sets cannot be correlated; therefore, although a mixed strategy always induces a behavior strategy on the information sets, the payoffs in the two cases need not be the same.   Kuhn has shown that they are the same when and only when a player remembers (in the sense of information sets) his previous choices; he need not, however, recall the choices made by the other players.   Thompson has extended this result to general games by showing over which information sets mixed strategies must be used and those for which behavioral strategies suffice to achieve the same expectation as with mixed strategies.

A second central notion from two-person theory which was generalized is an *equilibrium point:* a collection of strategies, one for each of the *n* players, such that no player is able to increase his payoff by changing his strategy choice when the others hold theirs fixed.   Nash has shown that every game has at least one mixed strategy equilibrium point, but not necessarily a pure strategy one.   In general, equilibrium points are neither equivalent (yield the same payoffs) nor interchangeable (yield an equilibrium point when strategies are intermixed).   The same questions of interpretation therefore exist as for two-person non-zero-sum games.   To bypass this problem, one could wish for a way of selecting a unique one. Gale's method for doing so in games with perfect information was described and criticized.

It was made clear that Nash's theory is very special in that it only makes sense if the players are not permitted to cooperate.   Once the possibility of collusion is admitted, then a variety of extra-game-theoretic assumptions had to be discussed.   We interpreted the need for these assumptions as reflecting the fact that the game model, although embodying economic and psychological assumptions, fails to include any sociological ones. One proposed assumption postulates limitations on the changes of collusive arrangements from any given arrangement.   In particular, it was assumed that such arrangements always result in a partitioning of the players, called a *coalition structure*, such that there is cooperation within the subsets and competition among them.   For each such structure τ, a set of admissible changes, $\psi(\tau)$, was postulated.   At the one extreme, no changes from τ are permitted, which is part of what Nash means by no cooperation; and, at the other extreme, all possible changes are admitted, which is the assumption made by von Neumann and Morgenstern in their theory of solutions (chapter 9).

In addition to these communication boundary conditions, several other distinctions are needed in the cooperative case. Can the players within a coalition use *correlated* (joint) *strategies* or not? Can utility *side payments* among the players be effected or not? If so, does utility possess that special feature, known as *unrestricted transferability*, which makes it behave like money, i.e., is it infinitely divisible and is it conserved when transferred? Depending upon which choice is made in each case, different theories arise; these were charted in a table in section 7.7.

Of the eight cases, only three have been extensively studied in the literature. Of the omissions, the cooperative case in which side payments are prohibited seemed most surprising, for that appears to be a natural generalization of two-person theory. In the final section, two definitions (depending upon whether correlated strategies are permitted or not) for the no side payment case were proposed but not explored. These are both based upon an assumed boundary condition ψ. The general idea is that a pair consisting of strategies (correlated or not, as the case may be) and a coalition structure will be in "equilibrium" provided that at least one person in each of the admissible coalition changes (as given by ψ) is placed in danger of suffering a loss by participating in the change. It was shown that the uncorrelated case is a generalization of Nash's equilibrium point theory and that the correlated case is a generalization of the concept of a negotiation set.

chapter 8

CHARACTERISTIC FUNCTIONS

8.1 SIDE PAYMENTS

What we have studied in the last chapter is not the mainstream of work on n-person games. It is true that the extensive form and, more especially, the normal form of a game are necessary backgrounds to further work, but little of the effort expended has been on the study of cooperative behavior when the players are assumed to receive only the payments prescribed by the rules of the game and are not allowed to pay and to receive "bribes." Most of the work, which will be discussed in this and the following three chapters, assumes side payments among the players, i.e., exchanges among the members of a coalition to equalize any inequities arising from their cooperation.

The possibility of side payments introduces serious practical, if not conceptual, complications. However, considerable simplification can be achieved if we assume that side payments may be made in terms of a "transferable utility." We will make this assumption in the remaining chapters on game theory, and we shall see that it leads to an extremely compact representation of a game. In section 10.4 some tentative suggestions are made for the case where side payments of real commodities are allowed but where no assumption of a "transferable utility" is made.

In section 7.7 the distinction between assuming that utility is interpersonally comparable and assuming that it may be conserved in transfers from one player to another was pointed out. We emphasized that it is possible to make the one assumption without implying the other.

We again stress (see section 7.7) that in making side payments it is goods or money which are transferred, not utility, which is a subjective quantity. When we speak elliptically of a "transferable utility" we mean that there exist a set of utility scales for the players and an infinitely divisible homogeneous commodity such that the changes in individual utilities which result when this commodity is transferred conserve the total utility sum. In terms of these utility scales it is meaningful to talk about the total utility accruing to a coalition and about a partition of this utility sum among the participants of the coalition. Of course if utilities are based on *other* scales (i.e., different origins and units), the utility changes which will occur with transfers of physical goods need not necessarily conserve the utility sum.

The assumption of a transferable utility is reasonable if monetary side payments are allowable and if each player's utility for money is approximately linear in the range of potential payoffs of the game. Although these conditions are sufficient they are not necessary. Money is not the only commodity which can serve the purpose of transfering utility. For example, labor may play a similar role in some contexts (e.g., side payments in the husband-wife coalition may consist of the transfer of household chores). Also, if monetary side payments are allowable, it is mathematically possible— but not very probable—that money transfers may be made which conserve utility even though each of the players' utility for money is non-linear.

To each outcome of a game—an outcome which might involve monetary as well as non-monetary consequences—each player can associate an equivalent pure monetary return in the sense that he is indifferent between this outcome and this amount of money. Monetary equivalences for outcomes do not constitute a suitable utility indicator *unless* the players' utility functions for money are linear in money. In this case, the set of utility scales which agree with the scales of monetary equivalences are especially convenient, for in terms of these scales we can speak elliptically about transferring utility with conservation. We emphasize again that this in no way implies an interpersonal comparison of utility.

We suggest that, if the above remarks are kept in mind, the reader will not go astray in the sequel by interpreting all payoffs in terms of monetary equivalences and assuming each player's utility for money is linear in money. One does not have to assume that money has the same "worth" —whatever that means—for each player.

8.2 DEFINITION OF CHARACTERISTIC FUNCTION

From now on we shall take for granted the existence of a transferable utility. Let us restrict our attention initially to zero-sum games, i.e., to those for which the sum of utility payments to the several players is always zero no matter what strategies are selected. Suppose that S is a subset of the players who have decided to form a coalition in the sense that they shall decide as a group upon individual courses of action which, together, cause the group to do as well as possible. How the individual payments come out does not, for the moment, matter, as long as the summation of these over the members of S is, in a sense to be specified, as good as possible. Still, one might object that, if each time the coalition did its best one of its players did no better, or even worse, than he could have alone, it might indeed be difficult to persuade him to remain in the coalition. As long as the payoff is in some sort of transferable commodity which results in the transferability of utility, this is no problem. The other members of the coalition may simply extend him side payments in order to keep him in the coalition. The extent of the side payment is a difficult problem of prediction, but it presumably depends, in part, upon his contribution to the total strength of the coalition and upon the damage he can cause to the coalition if he defects to another coalition. This suggests that it may be sufficient in developing a theory to look at the total payments which may be expected by coalitions.

The worst possible strategic situation met by a coalition is for the remaining set of players, $-S$, also to form a coalition. The strategic effects of any other possible system of coalitions opposing S can always be achieved by coalition $-S$, and in general it can achieve some outcomes not possible were it divided into several subcoalitions. Coalition S versus coalition $-S$ is, of course, a two-person zero-sum game—a type of game we have already examined in Chapter 4 and for which there is a unique (conservative) value given by the minimax theorem. Let this value for the coalition S be denoted by $v(S)$. That is, if coalition S forms, then the members of S can employ a joint mixed strategy which will guarantee a return (expected value) of $v(S)$ regardless of the joint strategy employed by the strictly competing coalition $-S$. At the same time the coalition $-S$ can employ a joint strategy which will hold S down to, at most, $v(S)$. In other words, S has a strategy which insures $v(S)$ but it has no strategy which can insure it more than $v(S)$.

Since $v(S)$ may be computed for each possible coalition, i.e., subset of players, we have therefore obtained a function v with domain the subsets of $I_n = \{1, 2, \cdots, n\}$ and range the real numbers, i.e., a real-valued set function. If the strategy spaces are at all rich in possibilities, the calculations involved in determining v are generally very complicated. This however, does not weaken the power of the theory to study all such functions!

The function v is not without certain restrictions; it may be shown that in the zero-sum case it satisfies

(i) $v(I_n) = 0$,

(ii) $v(S) = -v(-S)$, for all subsets S of I_n,

(iii) $v(\phi) = 0$, where ϕ is the empty set,

and

(iv)   If R and S are any two disjoint subsets of players,

$$v(R \cup S) \geqslant v(R) + v(S).^1$$

(Note that, given condition ii, conditions i and iii are equivalent.) The first two conditions simply reflect the zero-sum character of the game. The third is a formal statement of the "obvious fact" that the subset involving no players neither loses nor wins anything. The important condition is iv, which, on reflection, is extremely plausible. It says that the whole does not obtain less than the sum of its parts, or, put another way, a coalition composed of the disjoint sets R and S can do anything R and S can do as separate coalitions, and possibly more.

The function v has been named the *characteristic function of the zero-sum game* from which it was derived.

It is interesting and important that any real-valued set function v satisfying conditions i through iv is the characteristic function of a zero-sum game. That is, given any such v, it is possible to construct a game in normal form which has as its characteristic function the given function v.

The extension of the concept of the characteristic function v to non-zero-sum games is completely straightforward. For any coalition S we let $v(S)$ be the maximin value (optimal security level) of the coalition S in the non-cooperative two-person non-zero-sum game S versus $-S$ (see Chapter 5). Of course, the function v no longer has two of the properties of the zero-sum case, $v(I_n) = 0$ and $v(S) = -v(-S)$, but it does satisfy the other two, namely:

(i) $v(\phi) = 0$,

(ii)   If R and S are any two disjoint subsets of players,

$$v(R \cup S) \geqslant v(R) + v(S).$$

▶ Property (i) is essentially a convenient definition, but (ii) has to be established. By a mathematical trick its proof is reduced to proving the corresponding state-

---

[1] If R and S are sets, $R \cup S$ denotes the set of elements which are in R, or in S, or in both R and S.

ment for the zero-sum case. To the non-zero-sum game with n players we add a fictitious player, who is not truly a free agent in the game. He has no strategy choices and does not play a significant role in coalition formation, but his payoff is so chosen that the $(n + 1)$-person game is zero-sum. The characteristic function of the augmented zero-sum game satisfies property (ii), and when it is restricted to the subsets of the original n players it agrees with the definition of v for that game.   ◀

If, in addition, the game is constant-sum (not necessarily zero-sum, but not excluding that case) then:

(iii) $v(S) = v(I_n) - v(-S)$, for all subsets S of I_n.

Of course, if we assume the game is zero-sum, then $v(I_n) = 0$, and (iii) becomes the old condition (ii).

When we use the term *characteristic function* we shall mean any real-valued set function satisfying (i) and (ii), for it is again true that for each such function v there is a game in normal form (no longer zero-sum) which has as its characteristic function the given function v.

As in the case of zero-sum games, the first condition reflects the strategic inconsequence of the null set, and the second that any coalition is at least as potent as any pair of disjoint subcoalitions. Now, these conditions have been derived from game considerations, first, of a game in extensive form, then in normal form, and then using the two-person theory. But, it must be admitted that we could not require less were we to think about coalition formation among rational agents removed from any specific theory of games. That is, were we to suppose the potency or strength of a coalition could be measured by a single number,[2] then we should, at the very least, require that conditions (i) and (ii) be met—indeed, we would probably try to specify more. It is surprising that by restricting our analysis to a game we do not obtain further requirements to be met by characteristic functions. Thus, although we shall make certain criticisms of the characteristic function as an interpretation of the structure of a game, the abstract notion appears to be very well suited to a simple numerical representation of the "power" of coalitions in human situations. Of course, the numerical values obtained from a game analysis may well differ from those we might assign by some other considerations. This suggests—and it is easy to confirm—that the study of characteristic functions, which is completely related to game theory, is also more general, since situations which are not games in normal form can give rise to such functions. For example, let there be a given set of people and a prescribed function v which associates to each possible coalition of players S a monetary value $v(S)$ which is paid to the coalition when, say, a certain

---

[2] This is not a weak supposition; for further discussion see section 8.5.

period of time has elapsed and the members of S have in fact agreed to be a coalition. Certainly this is a conflict of interest which will lead to jockeying for the most advantageous agreements. It is true that, if v satisfies conditions i and ii above, we can produce games in normal form which have v as their characteristic function. Yet it is artificial to say that the given situation is a game in normal form. Nonetheless, in conformity to present usage, we shall refer to a set of players and a characteristic function defined over the subsets of players as a *game*, and whenever it is necessary to avoid ambiguity we shall add "in characteristic function form."

This last discussion suggests that one interesting and comparatively easy way to study people's responses to a characteristic function experimentally is to present the game in terms of the payoffs to coalitions, i.e., in the form of the characteristic function. One can then observe the steps of coalition formation, the resulting coalitions, and how the spoils accruing to a coalition are divided among its members. Exactly such an experiment has been performed; it is discussed in section 12.3.

Our next step is to divide characteristic functions into two classes. It is conceivable that there are games in which no coalition of players is more effective than the several players of the coalition operating alone, in other words, that for every disjoint R and S,

$$v(R \cup S) = v(R) + v(S).$$

Such games are called *inessential;* any game which is not inessential is called *essential.* It is not difficult to show that a game is inessential if and only if the total payment to the set of all players is exactly the same as the sum of payments to all the individual players, i.e.,

$$v(I_n) = \sum_{i=1}^{n} v(\{i\}).$$

Since nothing is gained by forming coalitions in inessential games, it is clear that we cannot expect any theory of coalition formation in that case, and so we shall be concerned only with essential games from now on.

8.3 *S*-EQUIVALENCE AND NORMALIZATION OF CHARACTERISTIC FUNCTIONS

Frequently in mathematics, and in its applications to science, a large class of objects is defined, all of which satisfy certain conditions; characteristic functions form such a class. Commonly, such a class can be partitioned into a number of non-overlapping subclasses, the elements of each subclass being equivalent insofar as theories about them are concerned. When such a partitioning exists, a representative is selected from each class and the theory is developed in terms of those representatives. It is always necessary to show that the theory is invariant under the equivalence concept which originally allowed the partitioning. We must turn to this problem for characteristic functions. The intuitive idea of equivalence which we want to isolate may be described as "strategic equivalence," i.e., we want to consider as equivalent two characteristic functions which lead to the same strategic considerations on the part of the players.

Suppose that one characteristic function v differs from another v' only by a multiplicative positive constant c, i.e.,

$$v(S) = cv'(S), \qquad \text{for all subsets } S \text{ of } I_n;$$

then the two characteristic functions differ only in the unit whereby we measure the utility. One example would be to transform a characteristic function originally in dollars to one in cents. It is clear that such a change of unit cannot possibly affect the strategic character of the game to rational players.

Next, suppose that we have a game with characteristic function v and suppose that, in one way or another, each player i is paid (or is caused to pay, depending upon the sign) an amount a_i prior to the play of the game. Certainly these payments do not alter the strategic considerations of the game, and so they should not have an effect upon the rational selection of strategies nor on the outcomes of the game.[3] But, if this is done, then the total payment to a coalition S is not just the $v(S)$ of the game, for that ignores the payments a_i. The fixed payments to (or from) the coalition S are $\sum_{i \text{ in } S} a_i$, so the total payment to S is:

$$v(S) + \sum_{i \text{ in } S} a_i.$$

It is easy to show that this last function is a characteristic function, and our argument that the payments a_i should not alter the strategic considerations shows that we should want to treat this function as strategically equivalent to v.

If we combine these two conditions into a single one, we may make the

---

[3] Empirically, it is doubtful that behavior is independent of total wealth, as we have assumed in this argument. Gambling experiments and observations strongly suggest that many people become more conservative after major losses and more rash after large winnings, which, of course, can be interpreted as meaning that a person's preferences (utility function) are changed by changes in wealth.

following definition: Two n-person games with characteristic functions v and v' defined over the same set of players are *S-equivalent* if it is possible to find n constants $a_1, a_2, \cdot \cdot \cdot, a_n$ and a positive constant c such that

$$v'(S) = cv(S) + \sum_{i \text{ in } S} a_i$$

for every subset S of I_n.

It may not be obvious, at this point, that this definition of equivalence is a suitable one, and that no further grouping is needed; but the results we shall cite near the end of section 9.1 show that it is adequate, at least for the von Neumann-Morgenstern theory of solutions.[4]

The relation of S-equivalence, defined over the set of all characteristic functions on n players, can be easily shown to satisfy the conditions of an *equivalence relation;* that is, one can show:

(i)   It is reflexive: for all v, v is S-equivalent to v,

(ii)   It is symmetric: if v is S-equivalent to v', then v' is S-equivalent to v,

and

(iii)   It is transitive: if v is S-equivalent to v' and if v' is S-equivalent to v'', then v is S-equivalent to v''.

When a relation is reflexive, symmetric, and transitive it behaves in very much the same way as the notion of equality (or sameness), and in particular it divides the elements of the set over which it is defined into non-overlapping subsets such that the elements within any one set are all equivalent and any two elements from different subsets are not equivalent. These subsets defined by an equivalence relation are called equivalence classes. Within each set the characteristic functions entail the same strategic considerations, and those in different sets require different considerations.

Assuming that S-equivalence is the appropriate grouping of characteristic functions, we must next confront the task of selecting one representative from each class in terms of which we shall construct our theories and examples. Two suggestions have been offered, each of which has certain advantages, primarily in the simplicity of stating certain games and certain definitions. The principle behind both of them is the same: it is possible to require that part of the representative characteristic function

---

[4] Experimental data will be cited in section 12.3 in which two S-equivalent games appear not to have been subjected to the same strategic considerations by the players. This suggests that the arguments leading to the definition of S-equivalence are not valid, but another interpretation is possible if the underlying knowledge postulate of game theory is weakened (see section 12.4).

be the same in all equivalence classes. Ignoring the (single) class of inessential games, von Neumann and Morgenstern [1947] have shown that there is one, and only one, characteristic function in each of the equivalence classes which satisfies

$$v(\{i\}) = -1 \quad \text{for every } i \text{ in } I_n,$$

and

$$v(I_n) = 0.$$

This they called the *reduced form* of an equivalence class of characteristic functions; we shall use the more specific and more popular term, -1, 0 *normalization*.

A second normalization, which is known as the 0, 1 *normalization*, results from stipulating that

$$v(\{i\}) = 0 \quad \text{for every } i \text{ in } I_n,$$

and

$$v(I_n) = 1.$$

As in the other normalization, there is one and only one characteristic function in each equivalence class meeting this condition. It is our impression that, by and large, the 0, 1 normalization results in greater simplicity of statement so we shall use it throughout.

It is instructive to consider the characteristic function in the two-person non-zero-sum case. The 0, 1 normalization conditions show that there is but one such game, namely:

$$v(\{1\}) = v(\{2\}) = 0, \quad v(\{1, 2\}) = 1.$$

Thus, the players can be looked on as engaging in the following bargain: they have one unit to share between themselves provided that they can come to an agreement; otherwise, each receives nothing. This is exactly the form of the two-person bargain discussed in section 6.5.

▶ If we are given the characteristic function of an essential game, the question arises how to find those constants c and a_i which transform it into either of the normalizations. Let v'' denote the given characteristic function, then it is not difficult to show that

$$v'(S) = \frac{v''(S) - \sum_{i \text{ in } S} v''(\{i\})}{v''(I_n) - \sum_{i \text{ in } I_n} v''(\{i\})}$$

is the 0, 1 normalization corresponding to v''. The further transformation

$$v(S) = nv'(S) - |S|,$$

where $|S|$ denotes the number of players in S, yields the -1, 0 normalization.

The first of these two transformations makes clear the necessity of restricting our attention to essential games, for only then is the denominator of the fraction different from zero. ◀

*8.4 SET FUNCTIONS

One of the first advantages—possibly the most important—of the 0, 1 normalization is its emphasis on the relation between characteristic function theory and the concept of a probability measure over the subsets of a finite set. Let us place side by side the conditions for a characteristic function in 0, 1 normalization and for a probability measure:

0, 1 Normalization	Probability Measure
i. v is a non-negative real-valued set function.	p is a non-negative real-valued set function.
ii. $v(I_n) = 1$.	$p(I_n) = 1$.
iii. $v(\phi) = 0$.	$p(\phi) = 0$.
iv. If R and S are disjoint subsets of I_n, $v(R \cup S) \geqslant v(R) + v(S)$.	If R and S are disjoint subsets of I_n, $p(R \cup S) = p(R) + p(S)$.
v. $v(\{i\}) = 0$.	
vi. if the game is constant-sum, $v(S) = 1 - v(-S)$, for all subsets S.	It follows from (ii) and (iv) above that $p(S) = 1 - p(-S)$, for all subsets S.

Although the resemblance between v and p is marked, there are differences, the most important being the inequality in the former and the equality in the latter for (iv), and the lack of the fifth condition for p. We cannot have $p(\{i\}) = 0$ for all i, for, if this were the case, by repeated application of (iv) we could conclude $p(I_n) = 0$, which contradicts condition (ii). We shall return to this correspondence again when we try to characterize the principal problem of n-person game theory (section 8.6).

This comparison suggests that the study of general games by means of characteristic functions could have been entitled the study of "finite superadditive measures." Conditions (i), (iii), and (iv) above suggest the name "superadditive measure," and condition (ii) simply means that the measures are normalized, as in the theory of probability. However, condition (v), $v(\{i\}) = 0$, is most unusual in measure theory. It is worth pointing out, at least for the mathematician, that we may drop this condition when we are studying theories invariant under S-equivalence, since under the transformation

$$v'(S) = \frac{v(S) - \sum_{i \text{ in } S} v(\{i\})}{v(I_n) - \sum_{i \text{ in } I_n} v(\{i\})},$$

v' satisfies $v'(\{i\}) = 0$ even if v does not.

These remarks serve to place the study of characteristic functions in a more general mathematical framework, namely, in the study of arbitrary, finite, normalized, real-valued set functions.   If R and S are any disjoint subsets of a finite set and if the difference

$$v(R \cup S) - v(R) - v(S)$$

is always equal to zero, then the measure is additive and the theory is that of discrete probabilities.   If the quantity is always less than or equal to zero, then the measure is called subadditive.   Some work has been done on these functions in conjunction with the theory of additive measures. Now in game theory the study of the other extreme is begun, i.e., of finite superadditive measures.   This work has, so far, resulted in a theory very different from the subadditive or additive one.   Probably this is an inherent difference and not simply a reflection of the game terminology and motivation.

Shapley in his thesis [1953 c] has undertaken an elegant study of arbitrary finite set functions; he has obtained results which show that under certain conditions the general study can be reduced to a study of these three special types of functions.   His important work is beyond the scope of our book, but the reader interested in research in characteristic function theory should be familiar with it.

8.5   CRITICISM

Before turning to theories based on the characteristic function idea, we should, in all honesty, point out that the simplification in passing from the normal form of a game to the characteristic function form presents its own difficulties.   Indeed, it would be surprising if such a radical simplification could be made of the theory of all n-person games with side payments without overlooking some significant aspects.   Possibly the simplest way of illustrating these difficulties is via an example presented by McKinsey [1952 a, p. 351] which shows that they exist even when $n = 2$. In this game player 1 has but one strategy and player 2 has two, and the payoff matrix is

$$[(0, -1000) \qquad (10, 0)].$$

Even though player 1 has no strategy choice, it is clear that he is in a much more advantageous position than player 2 (assuming interpersonal comparability of utility) for he is almost certain of getting 10 even though player 2 can hold him down to 0.   The cost of this action to player 2, namely -1000, is so great that we must anticipate his choosing the second strategy. Compare this analysis with a calculation of the characteristic function:

$$v(\phi) = 0, \qquad v(\{1\}) = v(\{2\}) = 0, \qquad v(\{1, 2\}) = 10.$$

Thus, although the normal form is distinctly asymmetric, the characteristic function is perfectly symmetric, reflecting no difference between the two players. It would be difficult to deny that an analysis of this situation based upon its characteristic function would necessarily miss the governing aspect of the normal form.

This example is merely a special case of the more general observation that characteristic functions tend to be inadequate representations of non-constant-sum games. There is considerable question whether the numbers so assigned to coalitions can usefully be conceived as the "strength" of the coalitions. Certainly, the above example suggests that they cannot.

Such remarks do not invalidate our earlier comment that, if one wishes to represent coalition strength in a conflict of interest by a single valued numerical function, it is intuitively plausible that it should satisfy the two conditions of a characteristic function. They do, however, raise serious doubts as to the formal procedure, now current, of obtaining these numbers for non-constant-sum games.

Unfortunately, it is also a question whether characteristic functions ever adequately represent a game, at least insofar as a descriptive theory is concerned. Let us restrict our attention to zero-sum games. It will be recalled that the characteristic function is derived by supposing S and $-S$ form opposing coalitions and $v(S)$ is the value to the "player" S of the two-person zero-sum game. A theory based on this characteristic function seems reasonable if it is supposed that, whenever S forms, $-S$ also forms. But does that not prejudge the theory by demanding that all conflicts of interest always reduce to two opposing coalitions? Certainly, this is not observably true, and, to the extent that the formulation of the game situation demands it, the formulation is probably inadequate for social science. In actual fact, the several theories now current do not necessarily stipulate that the game reduce to two opposed factions. Yet, the calculations are based on the characteristic function, which means that the model assumes each coalition takes the most pessimistic view of the opposition it will face. The players of these theories are conservative in the extreme. Presumably, an adequate descriptive theory will incorporate the expectation of each potential coalition as to the reaction of the remaining players if that coalition should actually form. How these expectations should be calculated is far from clear.

We shall not deny that we feel these two groups of criticisms are very serious indeed, and as a consequence we have limited faith in the ability of n-person theory, as it now stands, to deal with the sociological phenomena of coalition formation. At the same time we would urge the social scientist to continue exploring what theory exists. Granted that its grave inadequacies limit—but do not eliminate—the possible applications, still, one may obtain some insights into coalition formation and learn further how "qualitative" ideas are given mathematical form. It is, we would insist, a merit of the mathematization that the flaws and weaknesses of the theory can be made so apparent; to be sure, it would not have been difficult to have slurred over them, but at least it is not necessary to do so.

8.6 IMPUTATIONS AND THE CORE

So far we have dealt with only one ingredient of the n-person game with side payments: the strength of the different coalition possibilities as measured by the characteristic function. Distinct from this, though presumably influenced by it, are the payments that the individual players finally receive. Since we have assumed payments in terms of unrestrictedly transferable numerical utilities, the direct payments as prescribed by the normal form of the game and any side payments resulting from agreements arrived at during coalition formation can all be added up for each player. Let x_i denote this summary payment for player i; then the total set of payments to the n players forms an n-tuple of real numbers which we may write as $\mathbf{x} = (x_1, x_2, \cdots, x_n)$.

We may look upon the task of n-person (characteristic function) theory to characterize which of these n-tuples may arise in an equilibrium state, i.e., we may expect the theory to take the form of a series of defensible restrictions on the $\mathbf{x}$'s. The development of these restrictions traditionally falls into two stages: In the first, which we present in this section, two fairly weak conditions are stipulated. These are common to the several equilibrium theories to be discussed, and they are conditions which seem to be more readily accepted than those embodied in what are called the equilibrium theories. The second stage involves the conditions of the theories themselves, which, as we have indicated, are subject to more controversy. The following three chapters are devoted to these conditions and their discussion.

Of the two restrictions imposed in the first stage, there is little objection to one. Whether a player is in a coalition or not, it is difficult to imagine, if he is rational, that he will accept a final payment less than the least he can expect to receive if he were to play alone against a coalition of all other players. Indeed, such an occurrence would be a direct violation of our principle of individual behavior [postulate (ix), section 3.6]. We are therefore led to impose the condition of individual rationality:

(i) $v(\{i\}) \leqslant x_i$, for every i in I_n.

The second condition results if we use an analogous argument for the set of all players, but this argument is far less acceptable, as we shall see. It runs that rational players, no matter how they constitute themselves into coalitions, should not accept a total payment $\sum\limits_{i \text{ in } I_n} x_i$ less than $v(I_n)$, for, if they did, each player could be made to gain without loss to the others. For example, each could be made to gain the amount

$$\left[v(I_n) - \sum_{i \text{ in } I_n} x_i \right] / n.$$

If this argument is accepted, we must then require that $v(I_n) \leqslant \sum\limits_{i \text{ in } I_n} x_i$. Since $v(I_n)$ represents the most that the players can get from the game by forming one grand coalition, it is impossible for $\sum\limits_{i \text{ in } I_n} x_i$ to exceed $v(I_n)$, so this condition, Pareto optimality, amounts to:

(ii) $\sum\limits_{i \text{ in } I_n} x_i = v(I_n)$.

Any n-tuple $\mathbf{x}$ of real numbers satisfying (i) and (ii) is called an *imputation* of the game with characteristic function v, and it is held that any equilibrium payment must be selected from among the imputations.

The controversies about imputations are entirely restricted to the second condition. First of all, it is clear that if $\sum\limits_{i \text{ in } I_n} x_i$ is less than $v(I_n)$ each of the players can be made to profit without loss to the others. Yet, it is by no means clear that players will be able to reach agreements effecting this. The argument leading to group rationality, i.e., Pareto optimality, is an attempt to extend the postulate of individual rationality to groups of players; however, the notion of group rationality is neither a postulate of the model nor does it appear to follow as a logical consequence of individual rationality. One might attempt to argue it as follows: Any imputation which does not add up to $v(I_n)$ will not be in equilibrium because the individuals will see that their own position can be bettered, and, being rational, they will certainly refuse to stay put.[5]   But, if we accept this, why then should the argument have force only for the whole set I_n? Does it not equally apply to each coalition S?   That is to say, should we not also impose the condition on admissible n-tuples $\mathbf{x}$ that:

(iii) $v(S) \leqslant \displaystyle\sum_{i \text{ in } S} x_i$ for every subset S of I_n?

[Observe that (iii) includes both (i) and (ii) as special cases.]

It is hard not to say Yes; however, as we shall see, this leads to trouble. So one is led to look hard for a defense to keep condition (ii) while dropping (iii).   The following is a weak, but possible, argument: Suppose $\mathbf{x}$ is an imputation such that the sum of payments to players in S is less than $v(S)$. The above argument was that the members of S would not be content with $\mathbf{x}$ because they can command $v(S)$ and therefore each can receive more than x_i.   But suppose that the coalition $-S$ forms and threatens to disrupt S through attractive offers to one or more of its members if S does not accept the imputation $\mathbf{x}$.   Such pressures seem at least a possibility for all coalitions save I_n, for it can be threatened only by the empty set. Thus (ii), but not necessarily (iii), should hold, according to this argument.

The set of n-tuples satisfying condition (iii) has been termed the *core* by Gillies [1953 b] for the very good reason that these n-tuples should be included in any definition of equilibrium we propose.   The difficulty in setting up the core as the equilibrium definition for characteristic function theory is that for very many games it is empty, i.e., no n-tuple meets condition (iii).   For example, suppose v is the characteristic function of a constant-sum game and suppose $\mathbf{x}$ meets condition (iii), then for any S

$$v(S) \leqslant \sum_{i \text{ in } S} x_i,$$

$$v(-S) \leqslant \sum_{i \text{ in } -S} x_i.$$

Adding, we have

$$v(I_n) = v(S) + v(-S) \leqslant \sum_{i \text{ in } I_n} x_i.$$

[5] This argument would have considerable force were there but one imputation, but there are many, and so a problem arises of selecting among them.   Failure to reach agreement, which is quite possible since individuals will differ in their preferences among them, may result in an outcome which is not an imputation.   This is not a new problem; it is similar to the discussion in Chapter 6 about Pareto optimal points, where, it will be recalled, disagreements about which optimal point should be chosen could result in the players ending up at a point which is far from Pareto optimal. Certainly, this has been observed to happen in practice.

But from our previous argument, the inequality on the right must in fact be an equality, hence it follows that for every S

$$v(S) = \sum_{i \text{ in } S} x_i.$$

If we choose S to be a single player, then we see $v(\{i\}) = x_i$, so we have shown that for every S

$$v(S) = \sum_{i \text{ in } S} v(\{i\}).$$

That is, we have shown that, if the core of a constant-sum game is non-empty, the game must be inessential.   Therefore, if the core is taken as the definition of equilibrium, all essential constant-sum games lack equilibrium n-tuples of payments.   We may fairly conclude that we are in serious trouble if we accept the full consequences of the argument leading to condition (ii); up to the present we can find in the literature three ways to avoid or bypass these troubles.

The most obvious way to avoid them is not to impose condition (ii) and therefore not condition (iii).   This tack has not been fully explored, but as we shall see in section 9.7 Shapley and Gillies have examined the effect of dropping condition (ii) within the framework of the solution theory of von Neumann and Morgenstern, and they have established that for solution theory it makes no difference.   Nonetheless, since the solution theory was presumably devised as a way of keeping condition (ii) and bypassing condition (iii), one can raise the question whether an entirely new equilibrium theory different from solutions should not be devised when condition (ii) is omitted.

The resolution offered by von Neumann and Morgenstern, which retains condition (ii) but not condition (iii), involves the idea that it is not an imputation which is in equilibrium but rather a set of them.   These sets of imputations—which are called solutions—possess certain properties of inner stability which we shall discuss in the next chapter.

The third major approach to bypass the difficulty of condition (iii) is to demand that it hold only for certain coalitions S.   Luce [1954, 1955 a] has argued that, if the game model included appropriate sociological assumptions, in general the contemplated changes from an equilibrium state would be restricted (see section 7.6) and that, therefore, condition (iii) would not have to hold for all coalitions S.   Milnor [1952] in his suggested reasonable outcomes has, in effect, allowed condition (iii) to be violated for certain coalitions S.   These ideas are discussed in detail in Chapters 10 and 11.   It must be emphasized that these resolutions of the difficulty say, in effect, that there are some significant restrictions in actual conflicts of interest which are not embodied in game theory. It may well be that no satisfying theory of n-person games can be developed until these intuitions are made a formal part of the underlying structure of the game. As we noted before, general game theory seems to be in part a sociological theory which does not include any sociological assumptions, and, although one might hope one day to derive sociological theory from individual psychology, it may be too much to ask that any sociology be derived from the single assumption of individual rationality.

▶ For those readers who did not skip section 8.4, we may phrase, in a simple way, the problem of characteristic function theory when condition (ii) is accepted. If we substitute the condition of 0, 1 normalization into the definition of an imputation, we find that for an n-tuple to be an imputation it is necessary that

(i) $x_i \geqslant 0$, for all i in I_n,

and

(ii) $\displaystyle\sum_{i \text{ in } I_n} x_i = 1.$

In other words, the set of imputations corresponding to the 0, 1 normalization is identical to the set of all probability distributions over the elements of I_n, i.e., an imputation is a distribution to the individual players of the total payments received by all of the players.

If $\mathbf{x} = (x_1, x_2, \cdots, x_n)$ is a probability distribution over I_n, then it is easy to show that the set function

$$x(S) = \sum_{i \text{ in } S} x_i$$

is a probability measure over I_n (see section 8.4). $x(S)$ is simply the sum of payments to the members of S, whereas we interpreted $v(S)$ as characterizing the strength of S. It is, of course, by the interplay of coalitions and possible coalitions, by threats to form new coalitions or to defect from old ones if certain agreements are not accepted, that the final payment $\mathbf{x}$ must be determined. The aim of any theory is to determine this outcome payment by formalizing what the threats must be, but it is clear that any of these theories must have the property that if $v(S)$ is much larger than $x(S)$ there will be strong forces for the coalition S to form and to demand a new outcome, say $\mathbf{x}'$, such that $x'(S)$ is close to $v(S)$. Thus, the equilibrium problem of game theory involves finding a probability measure x which in some sense approximates the normalized superadditive measure v. The heart of each theory is the specification of the intuitive idea behind the phrase "in some sense approximates."   ◀

8.7 SUMMARY

We confined our attention in this chapter, as we will in the next three, to the special class of n-person games where utility side payments among the players are allowed and where utility acts like money—that is, it is freely transferable in any amount up to the maximum held and conserves its numerical value whenever a transfer occurs. For such games, two central concepts are introduced: characteristic functions and imputations.

If a subset S of players forms a coalition, its worst strategic prospect is for the remaining players, $-S$, also to form a coalition. In that case, a two-person game, S versus $-S$, results. The maximin value for "player" S can be computed; let it be denoted $v(S)$. These numbers, which are defined for every subset of players, form what is called the *characteristic function* of the game. It can be shown to satisfy:

(i) $v(\phi) = 0$, where ϕ is the empty set,

and

(ii)   If R and S are disjoint coalitions, $v(R \cup S) \geqslant v(R) + v(S)$.

By considering the effect of changing the unit of the characteristic function and of making preplay payments to or from the players, we were led to a concept of strategic equivalence. Formally, two characteristic functions v and v' on the same set of players are said to be *S-equivalent* if there exist a positive constant c and constants a_i, $i = 1, 2, \cdots, n$, such that

$$v'(R) = cv(R) + \sum_{i \text{ in } R} a_i,$$

for every coalition R. This relation partitions the set of all characteristic functions on n players into non-overlapping subsets of equivalent characteristic functions. It is held that any equilibrium theory should yield corresponding results for all members of the same subset; thus, it is sufficient to examine just one representative from each. The one chosen for essential games satisfies $v(\{i\}) = 0$ and $v(I_n) = 1$, and it is called the 0, 1 *normalization* of the members of its subset.

These ideas capture something of the strategic potentialities of the several coalitions, but they fail to deal with the returns to individual players. It was argued that the special assumptions about utility allow us to summarize the total payment to player i by a single number x_i. From the individual rationality assumption, it follows that

(i) $x_i \geqslant v(\{i\})$, for each player i.

It was also argued that for the same reason the set of all players cannot be expected to accept less than the total payment available, i.e., the group as well as the individual is rational:

(ii) $\displaystyle\sum_{i \text{ in } I_n} x_i = v(I_n).$

An n-tuple satisfying these two conditions is called an *imputation*.

Actually, the group rationality (Pareto optimality) assumption as embodied in requirement (ii) does not appear to be a logical consequence of the underlying assumptions of game theory. But, should the argument for the reasonableness of (ii) be accepted, a slight modification of it leads to the requirement that every subset of players be rational, i.e.,

(iii) $\displaystyle\sum_{i \text{ in } S} x_i \geq v(S)$, for all coalitions S.

The set of imputations satisfying (iii) is called the *core* of the game v. Many games, including all essential constant-sum ones, have a vacuous core; thus, serious doubt is cast on the appropriateness of (ii). What happens to the several theories when (ii) is not assumed will be discussed as they are presented.

chapter 9

SOLUTIONS

9.1 THE VON NEUMANN-MORGENSTERN
DEFINITION OF A SOLUTION

In the published literature of n-person games one definition, based on characteristic functions and imputations, has received primary attention; this definition, introduced at length by von Neumann and Morgenstern [1947], was offered as the "solution" to the n-person cooperative game—indeed, it was given the name solution. Following their exposition, we may first suggest the idea by an example. It is not difficult to see that the 0, 1 normalization of an essential constant-sum three-person game is unique, and that it is:

(i) $v(\{1\}) = v(\{2\}) = v(\{3\}) = 0,$

(ii) $v(\{1, 2\}) = v(\{1, 3\}) = v(\{2, 3\}) = 1,$

(iii) $v(\{1, 2, 3\}) = 1.$

[*Note:* properties (i) and (iii) are specifically required by the 0, 1 normalization (section 8.3), and (ii) follows from these and the constant-sum requirement.] Suppose, for the moment, that the coalition $\{1, 2\}$ forms. According to the characteristic function it may command a payment of 1, and since players 1 and 2 have symmetric roles (in the sense that if we

were to change their labeling so that 1 were 2 and 2 were 1 the characteristic function would be unchanged), it is not unreasonable to suppose that
they would divide the payment equally.   But, arguing by symmetry
again, there is no reason to single out the coalition $\{1, 2\}$ as superior to
$\{1, 3\}$ or to $\{2, 3\}$, and so any of the three imputations

$$(\tfrac{1}{2}, \tfrac{1}{2}, 0), \ (\tfrac{1}{2}, 0, \tfrac{1}{2}), \ (0, \tfrac{1}{2}, \tfrac{1}{2})$$

seem to be reasonable outcomes.   Let us call this set of imputations F.

Now, return to the case where players 1 and 2 have formed a coalition
and have agreed upon the imputation $(\tfrac{1}{2}, \tfrac{1}{2}, 0)$.   Player 3 can, for
example, propose to player 2 that they form the coalition $\{2, 3\}$ which can
command the payment 1.   Since, of course, player 2 is already assured a
payment of $\tfrac{1}{2}$, it would be silly of 3 to propose the imputation $(0, \tfrac{1}{2}, \tfrac{1}{2})$
of F; rather he must offer something like $(0, \tfrac{3}{4}, \tfrac{1}{4})$ in which both members of the coalition receive equal incremental benefits.   Since both would
benefit, the change should occur.   We say that $(0, \tfrac{3}{4}, \tfrac{1}{4})$ "dominates"
$(\tfrac{1}{2}, \tfrac{1}{2}, 0)$ with respect to the coalition $\{2, 3\}$ because both 2 and 3 benefit and the imputation can be enforced.   It is easy to see that for any
imputation in F there exist imputations not in F which are better for two
individuals and which can be enforced by them if they form a coalition, i.e.,
each imputation in F is dominated by one or more imputations not in F.
On the other hand, it is also true that for any imputation not in F there is
one in F which dominates it.   This latter statement seems unlikely, for
there are only three imputations in F, whereas there are an infinity not in
F; however, it is not difficult to show that this is the case.

▶ Suppose $\mathbf{x} = (x_1, x_2, x_3)$ is not in F, and suppose that the largest entry is x_i. We
show that both of the other two entries are less than $\tfrac{1}{2}$. If $x_i > \tfrac{1}{2}$, then the other
two are each less than $\tfrac{1}{2}$ because $x_1 + x_2 + x_3 = 1$ and $x_j \geqslant 0, j = 1, 2, 3$. If
$x_i \leqslant \tfrac{1}{2}$, then since x_i is the largest entry, the other two are $\leqslant \tfrac{1}{2}$. If one of
these $= \tfrac{1}{2}$, then $x_i = \tfrac{1}{2}$, which means $\mathbf{x}$ is in F, contrary to assumption. So
both are $< \tfrac{1}{2}$. Thus the element in F with its ith entry 0 dominates $\mathbf{x}$.   ◀

Continuing our special example, if players 1 and 3 form a coalition,
then $(\tfrac{1}{2}, 0, \tfrac{1}{2})$ can be enforced, and it dominates $(0, \tfrac{3}{4}, \tfrac{1}{4})$ with respect
to $\{1, 3\}$.   Finally, it should be observed that none of the imputations in
F dominate either of the other two.

We are thus led to suspect that a set of imputations like F has a peculiar
inner stability: every imputation outside the set is dominated by one in it,
and none in the set dominates any of the others in the set.   There is a
further property of F suggested by our example which should insure its
stability.   Player 2, who initially had a payment of $\tfrac{1}{2}$, not only failed to
achieve $\tfrac{3}{4}$ permanently by his coalition with player 3, but ultimately he
lost everything.   Such a possibility would seem to reduce the chance that

he would ever get involved with player 3 to begin with. The question arises whether these notions of stability can be generalized. Actually, von Neumann and Morgenstern's generalization ignores the second property of F—player 2's ultimate loss due to his defection from F—and deals only with the first. The work of Vickrey discussed in section 9.6 takes the second property into account.

Clearly the argument which led to calling F stable is dynamic in nature, and yet we conceive of game situations as one-shot affairs. How, then, can we possibly formalize such reasoning? The answer is that we do not; we only characterize those sets of imputations which ultimately possess the inner stability relative to the notion of domination. It should be remembered that, although the game itself is played but once, there is nothing which prevents the pregame negotiations among players (which are not now formalized in game theory) from having such a dynamic quality.

Von Neumann and Morgenstern give the following definitions which generalize those of the example. Let the game have characteristic function v; then an imputation $\mathbf{y} = (y_1, y_2, \cdots, y_n)$ *dominates* an imputation $\mathbf{x} = (x_1, x_2, \cdots, x_n)$ *with respect to the coalition* T provided that T is not the empty set and the following conditions are met:

(i) $v(T) \geqslant \sum_{i \text{ in } T} y_i,$

and

(ii) $y_i > x_i$ for every i in T.

The first condition admits $\mathbf{y}$ as dominating $\mathbf{x}$ only if $\mathbf{y}$ is feasible in the sense that the members of T can expect to have the amount prescribed by $\mathbf{y}$ to distribute among themselves. The second condition says that everyone in T strictly prefers $\mathbf{y}$ to $\mathbf{x}$. The set T in such a domination is called an *effective* set.

Observe that exactly this condition was met when, in the example, we said that $(0, \frac{3}{4}, \frac{1}{4})$ dominates $(\frac{1}{2}, \frac{1}{2}, 0)$ with respect to the set $\{2, 3\}$.

We say that an imputation $\mathbf{y}$ *dominates* $\mathbf{x}$ (without specific reference to the effective set) if there is at least one effective set T such that $\mathbf{y}$ dominates $\mathbf{x}$ with respect to T. It is not difficult to see that all the logically possible cases can arise, namely:

  i. $\mathbf{y}$ dominates $\mathbf{x}$, but $\mathbf{x}$ does not dominate $\mathbf{y}$.

  ii. $\mathbf{y}$ dominates $\mathbf{x}$, and $\mathbf{x}$ dominates $\mathbf{y}$ (with respect to non-overlapping coalitions, of course); e.g., in any five-person game with $v(\{1, 2\}) =$

$v(\{3, 4\}) = \frac{1}{2}$, $\mathbf{x} = (\frac{1}{4}, \frac{1}{4}, \frac{1}{6}, \frac{1}{6}, \frac{1}{6})$ dominates $\mathbf{y} = (\frac{1}{6}, \frac{1}{6}, \frac{1}{4}, \frac{1}{4}, \frac{1}{6})$ with respect to $\{1, 2\}$ and $\mathbf{y}$ dominates $\mathbf{x}$ with respect to $\{3, 4\}$.

iii. Neither $\mathbf{y}$ nor $\mathbf{x}$ dominates the other.

Furthermore, dominance is not in general a transitive relation. For example, in the three-person constant-sum game, $(\frac{1}{2}, \frac{1}{2}, 0)$ dominates $(\frac{1}{4}, \frac{1}{4}, \frac{1}{2})$ (with respect to $\{1, 2\}$) and $(\frac{1}{4}, \frac{1}{4}, \frac{1}{2})$ dominates $(0, \frac{3}{4}, \frac{1}{4})$ (with respect to $\{1, 3\}$), but $(\frac{1}{2}, \frac{1}{2}, 0)$ does not dominate $(0, \frac{3}{4}, \frac{1}{4})$.

Having now a general notion of domination, we may simply substitute this into our characterization of the set F in order to get a general notion of a stable set of imputations: A *solution* of a game in characteristic function form is defined to be any set A of imputations such that:

i. If $\mathbf{x}$ and $\mathbf{y}$ are imputations in A, then neither $\mathbf{x}$ nor $\mathbf{y}$ dominates the other.

ii. If $\mathbf{z}$ is an imputation not in A, then there is at least one imputation $\mathbf{x}$ in A which dominates $\mathbf{z}$.

It must be emphasized again that the definition of solution in no way precludes the existence of imputations not in A which dominate one, or several, of the imputations which are in A. This occurred in the example we discussed. We shall return to this point, which is not without complications.

Some properties of the set F were used to suggest the definition of a solution, and it is easy to show that F does in fact meet the conditions of the definition, i.e., F is technically a solution.

We mentioned in section 8.3 that we want theories based on characteristic functions to be invariant under S-equivalence; i.e., two S-equivalent games should lead to results which are mapped into each other by the transformation of S-equivalence. This has been shown to be the case for the domination concept, and so it is true for solutions (von Neumann and Morgenstern, [1947]; McKinsey, [1950 b]).

▶ If I is the set of imputations of an n-person game, the notion of dominance imposes a relation over I. We symbolize this relation by $\succ$, where $\mathbf{x} \succ \mathbf{y}$ means $\mathbf{x}$ dominates $\mathbf{y}$. If I and I' are the sets of imputations of two different games, we say that there is a one-to-one correspondence between them if there can be found a singled valued function f from I onto I' with the property that if $\mathbf{x}'$ is in I' there is only one $\mathbf{x}$ in I such that $f(\mathbf{x}) = \mathbf{x}'$. The two sets of imputations are said to be isomorphic with respect to the relation of dominance if there is a one-to-one mapping between them which preserves the dominance relation, i.e., if $\mathbf{x}$ and $\mathbf{y}$ are in I,

$$\mathbf{x} \succ \mathbf{y} \quad \text{if and only if} \quad f(\mathbf{x}) \succ f(\mathbf{y}).$$

In effect, then, two imputation sets which are isomorphic have the same dominance structure.

Now, there is a possible converse to the theorem that dominance is preserved under S-equivalence, namely, that, if two games have imputation spaces which are isomorphic under domination, the two games have characteristic functions which are S-equivalent. This much more subtle result has been shown, by McKinsey [1950 b], to hold for zero-sum games.   ◀

9.2 SOME REMARKS ABOUT THE DEFINITION

Before discussing the mathematical results which have been obtained— and some which have not—concerning solutions, certain questions about the intuitive adequacy of the definition must be considered. The notion of "dominance with respect to a coalition T" is the conjunction of two quite distinct notions: The first condition, $v(T) \geqslant \sum_{i \text{ in } T} y_i$, formalizes the idea that "$\mathbf{y}$ is 'feasible' with respect to T," and the second one, $y_i > x_i$, for all i in T, states that "$\mathbf{y}$ is 'better' than $\mathbf{x}$ so far as the members of T are concerned." Of these two, there seems little reason to question the second but the first is open to some debate. The criticisms of feasibility are not so much an objection to the definition of a solution itself as to the representation of a normalized game by its characteristic function—a discussion begun in section 8.5. These objections evaporate completely in the special case where only the characteristic function is given, where the payoffs are assigned to coalitions, not to individuals.

For zero-sum games, it can be argued that if the coalition T forms it can never enforce more than $v(T)$ since the remaining players, who are assumed to be rational and unconstrained by any social limitations, will certainly form the coalition $-T$. In other words, for zero-sum games and unlimited collusion, an imputation $\mathbf{y}$ with $v(T) < \sum_{i \text{ in } T} y_i$ is certainly not feasible. If, however, we drop either of the conditions—either zero-sumness or unlimited collusion—then the condition of feasibility is subject to doubt. Furthermore, descriptively it is doubtful even in the zero-sum case, for the coalition may realistically count on the remaining players not to agree on a division of $v(-T)$ and so not to form the coalition $-T$.

Consider, then, the role the solution concept may play. A descriptive theory must be concerned with economic, military, and social conflicts of interest, and certainly not all of these reduce to the opposition of two coalitions. Yet if limitations on collusive arrangements prevent the opposing coalition $-T$ from forming, there is no *a priori* reason why the coalition T will be limited to receiving $v(T)$. Thus, it is questionable whether the theory of solutions can possibly be descriptive; of course, it may still play a normative role. But, as we pointed out, the feasibility condition is subject to doubt for non-zero-sum games. It is perfectly possible for the players in T to receive more than $v(T)$, even if $-T$ forms, since the cost to $-T$ of holding T down to $v(T)$ may be excessive. At the same time, most socially interesting conflicts of interest are not zero-sum. Thus, there is at least a doubt whether it is a suitable normative notion for those games which are of social interest. Finally, one may ask, in what sense is it normative? It certainly does not tell the players how to play the game, nor, indeed, does it specify what coalitions should form. Apparently a solution must be interpreted as a description of a set of possible payments, any of which might arise if the players choose strategies and form collusive arrangements as they "should."

It should be noted that solutions are only concerned with imputations and that they do not specify the coalition structures associated with each of the imputations, or even the set of all possible coalition structures associated with the imputations of the solution. One can argue that this, too, is a questionable feature of the definition, for an equilibrium state is presumably described not only by the payments received but also by the coalition structure. Certainly, in attempts to relate such a theory to data it is often easier to observe the coalition structure than the imputations.

We feel that the above *a priori* criticism is severe and not wholly idiosyncratic, since we find on p. 303 of McKinsey [1952 *a*] the statement: "Although a large part of von Neumann and Morgenstern's book (roughly 400 out of 600 pages) is devoted to games with more than two players, mathematicians generally seem to have been dissatisfied with the theory there developed." Nonetheless, the concept of solution is of great importance. To be sure, its importance is somewhat historical, but not entirely. The notion has afforded insights into problems which, prior to game theory, had received considerable analysis—analysis which, however, failed to bring out some of the subtleties involved (see, for example, section 9.4). Mathematicians, at least, have had warm admiration for the ingenuity of the definition, and it has received considerable study—to which we now turn.

9.3 SOME IMPLICATIONS OF THE DEFINITION

Our first main point is that a solution does not generally consist of a single imputation but, rather, of several. This is the case for the solution F of the three-person zero-sum game, discussed in section 9.1, and, indeed, it can be shown that a game which has a solution consisting of but one imputation is inessential (for definition, see end of section 8.2).

In addition to solutions having a finite number of imputations, such as the three in F, some solutions consist of an infinity—and not necessarily a countable infinity— of imputations.[1]   We shall give an example of such a solution below.

The second main point is that, aside from the multiplicity of imputations in a solution, most games that have been studied have many solutions. This possibility was suggested by our earlier observations that there may be imputations not contained in a solution which dominate imputations of that solution and that the relation of dominance is not transitive. Even in the three-person zero-sum game the solution F is not unique: If we choose c to be any fixed number in the range $0 \leqslant c < \frac{1}{2}$, and if we let x_1 and x_2 be chosen so that neither is negative and $x_1 + x_2 + c = 1$, then the set of imputations of the form

$$(x_1, x_2, c),$$

where x_1 and x_2 vary, forms a solution for each value of c.   We shall denote this solution by $F_3(c)$, where the numbers 3 and c indicate that the fixed amount c goes to player 3.   Equally well, the two sets of imputations obtained by moving c to player 1 and to player 2, i.e., the sets

$$(c, x_2, x_3) \quad \text{and} \quad (x_1, c, x_2),$$

are also solutions, which we shall call $F_1(c)$ and $F_2(c)$.   Since there are solutions for each possible value of c in the interval from 0 to $\frac{1}{2}$ (but not including $\frac{1}{2}$), there is a continuum of solutions, each of which contains a continuum of imputations.   Indeed, every possible imputation for the constant-sum three-person game is included in at least one solution! "Therefore in the case of the essential three-person game we have an embarrassing richness of solutions." [McKinsey, 1952 a.]

This abundance is not restricted to the three-person case.

The immediate question is how to interpret these solutions.   Von Neumann and Morgenstern divide the discussion into two parts.   First, they say that, of the several solutions, the one which is accepted depends upon "standards of behavior" which are moral or conventional rules imposed by society.   Thus, they say, if society accepts discrimination, one may find a solution of the type $F_i(c)$ where the position of c in the range 0 to $\frac{1}{2}$ is determined by the degree of discrimination tolerated by the society. Assuming c fixed, there is a question how the other two players will divide the amount $1 - c$, and this is a problem in bargaining which depends upon the relative bargaining abilities of the two players.   They do not

[1] A set contains a countable infinity of elements if we can enumerate them, i.e., if we can speak of a first, a second, and so on in such a way that each element has an integer associated to it.   Shapley [Kuhn, 1953 a] has conjectured that no solution ever contains a countable infinity of imputations.

say how it will be decided which player will be discriminated against or, in the case of the non-discriminatory solution F, which imputation will arise. Apparently this is a chance matter depending upon which coalition was first formed, or again, it may depend upon the relative bargaining abilities of the three players.   It is such discussion which gives this theory the *ad hoc* character mentioned earlier (see section 7.6).

Von Neumann and Morgenstern argue at some length that solutions are "stable."   They point out that, although an imputation not in a solution may dominate one in a solution and although it may be "preferable to an effective set of players, [it] will fail to attract them, because it is 'unsound' " [1947, p. 265].   And

· · · the attitude of the players must be imagined like this: If the solution [A] · · · is "accepted" by the players 1, · · · , n, then it must impress upon their minds the idea that only the imputations · · · [in A] are "sound" ways of distribution. [1947, p. 265.]

And

· · · the above considerations make it even more clear that only [A] in its entirety is a solution and possesses any kind of stability—but none of its elements individually.   The circular character · · · makes it plausible also that several solutions [A] may exist for the same game.   I.e., several stable standards of behavior may exist for the same factual situation.   Each of these would, of course, be stable and consistent in itself, but in conflict with all others.   [1947, p. 266.]

The full flavor of their argument is hard to recapture, and it can only be recommended that the reader turn to the discussions of solutions in their book.   That not all students of game theory have been completely persuaded by their arguments is indicated by the comment of McKinsey: "Some people have felt dissatisfied with the intuitive basis of this notion, however; and the question has been raised as to whether knowing a solution of a given n-person game would enable a person to play it with greater expectation of profit than if he were quite ignorant of this theory." [1952 *a*, p. 332.]   Of course, the force of this criticism depends upon one's view as to what is—or should be—the object of game theory.

9.4   THE SOLUTIONS OF A MARKET WITH ONE
SELLER AND TWO BUYERS

To a social scientist the ultimate value of any theory is the help it can give him in understanding and predicting the observed behavior of people in specific situations.   To show that the solution notion may serve this purpose, and because examples often crystalize an idea, we shall consider in detail an example of a market which consists of only three persons: a seller and two buyers.   This example is particularly suited to illustrative purposes because it is sufficiently simple to be analyzed by a "common sense" economic argument.   Its game theory solution has been presented by von Neumann and Morgenstern [1947, pp. 564–573], and it is also possible to analyze it in detail in terms of the other theories which have been offered (Chapters 10 and 11).   A comparison of the common aspects of these several analyses, and of their differences, is instructive.

Let the seller be called player 1 and the two buyers 2 and 3.   We shall suppose that the seller is in possession of a single indivisible commodity which he is willing to sell for a price.   Furthermore, we shall suppose that among the players there is an infinitely divisible and transferable commodity, which we shall call money, in terms of which the object is priced. Let the players 1, 2, and 3 value the object a, b, and c units of money respectively.   There is no loss of generality if we assume that $b \leqslant c$, and we may as well assume that $a < b$, for otherwise player 2 is not really a part of the market (since, if $a > b$, he values the object less than the person who already possesses it), and so there would be no point in treating it as a three-person game.   Because of the different valuations of the object, it is clear that this is a non-constant-sum game.   The question to be answered by any analysis we perform is what coalitions, if any, will form and what exchanges of money may be expected to take place.

Let us first determine the characteristic function of the game.   Since player 1 has the object and is not forced to sell, he can guarantee himself a value of a—the value he places on the object.   Since the other players are not forced to buy he cannot be certain of a value greater than a, so $v(\{1\}) = a$.   Equally well, since the other players do not possess the object and cannot be certain of getting it, they cannot be certain of a value in excess of 0; but since they are not forced to participate in the bargaining they can be certain of 0, so $v(\{2\}) = v(\{3\}) = 0$.   If the coalition $\{1, 2\}$ were to form, then the object would be in their possession and its worth to player 1 is a and to player 2 it is b.   The object cannot be removed from the coalition by a payment less than b, for, if player 1 were tempted to sell it for less, player 2 would pay him that much and continue to keep the object in the coalition.   On the other hand, the coalition cannot assure itself a value greater than b, so $v(\{1, 2\}) = b$.   Similar arguments show that the value of the game to $\{1, 3\}$ is c, to $\{2, 3\}$ it is 0, and to all three players it is c.   So, in summary, the characteristic function (not normalized) is:

$$v(\{1\}) = a, \qquad v(\{2\}) = v(\{3\}) = 0,$$

$$v(\{1, 2\}) = b, \qquad v(\{1, 3\}) = c, \qquad v(\{2, 3\}) = 0,$$

$$v(\{1, 2, 3\}) = c.$$

By definition, an imputation for this game is any triple (x_1, x_2, x_3) such that

$$x_1 + x_2 + x_3 = c, \quad \text{and} \quad x_1 \geqslant a, \quad x_2 \geqslant 0, \quad x_3 \geqslant 0.$$

The task now is to specify further which imputations may arise.

The first "theory" we shall present is the common sense economic analysis: Since there is only one unit of good under consideration, one buyer will be included in the transaction and the other excluded. Clearly, the stronger buyer, player 3, will be the one included, except when $b = c$, in which case either can participate. Since player 2 will pay up to b for the commodity player 1 can get at least b, and since player 3's limit is c he can get no more than that. If $c > b$, player 3 can always exclude player 2 from the bargaining by paying something more than b for the object. Thus, the only imputations which may arise, if this argument is valid, are those satisfying:

$$b \leqslant x_1 \leqslant c, \quad x_2 = 0, \quad x_3 = c - x_1. \tag{1}$$

As we shall see, these imputations are very important and common to all the theories; the several theories differ in what other imputations they include.

The second theory we shall consider is that of the core. It will be recalled (section 8.6) that when the argument used to limit n-tuples to imputations is carried to its logical conclusion (assuming no limitations on coalition formation) it results in the condition that

$$v(S) \leqslant \sum_{i \text{ in } S} x_i$$

for every subset S. The set of imputations satisfying this inequality was called the core. In terms of our new terminology, *the core consists of those imputations which are not dominated.* It was shown that for constant-sum games the core is empty, but for more general games this is not the case. If these inequalities are set up for the present example, it is easy to show they are equivalent to the three presented in eq. 1, i.e., the common sense argument leads to the core.

The third theory is that of solutions. It can be shown, for example, that the imputations described by eq. 1, plus those of the form

$$\left(x, \frac{c - x}{3}, \frac{2(c - x)}{3} \right),$$

where $a \leqslant x \leqslant b$, form one particular solution. There are others! This solution clearly contains many imputations not given by the common sense argument; presumably, the reason for the difference is that the definition of a solution takes into account the possibility of coalitions. These extra imputations presumably arise when players 2 and 3 form a coalition and agree not to pay more than b for the commodity. The exact location of the price between a and b depends upon their bargaining ability relative to player 1. Having bought the commodity, the problem of dividing the spoils remains—the one who keeps it must pay the other for his cooperation. The exact price, in terms of the selling price x, is given as one-third and two-thirds of $c - x$. Clearly, this function is highly special, and there must be other solutions with different schemes for splitting up the payment. Actually, it can be shown that, if f and g are any two monotonic decreasing functions of x such that

$$f(x) \geqslant 0, \qquad g(x) \geqslant 0, \qquad x + f(x) + g(x) = c,$$

for $a \leqslant x \leqslant b$, then the imputations of the core (eq. 1) plus those of the form

$$(x, f(x), g(x)),$$

where $a \leqslant x \leqslant b$, form a solution. Furthermore, all solutions are of this form. The core alone does not form a solution to this game because there exist imputations outside the core which are not dominated by any member of the core. However, the core must be in every solution, for if it were not then the imputations in the core, i.e., the undominated imputations, would fail to be dominated by an element of the solution, which is impossible. (Shapley [Kuhn, 1953 a] has conjectured that the intersection of all solutions of a game is the core, but this has not yet been proved. Gillies [1953 b] has given a necessary condition for the core to be a solution.) Now, according to von Neumann and Morgenstern, the choice of the pair of functions f and g, which determines the division of the spoils in terms of the selling price, and hence the choice of the solution, depends upon the "standards of behavior" of the society. Put more colloquially, it depends upon the "going" price for this type of cooperation.

One might be tempted to inquire why the interpretation of the solution involves mention of only the coalition $\{2, 3\}$. Although we could discuss this now, it will be more illuminating to reserve our comments until later (section 10.3).

9.5 FURTHER RESULTS ON SOLUTIONS

There exists a considerable literature on solutions that presents partial but detailed results about solutions for various classes of games. We do not propose to examine these theorems in great detail, for that would require too much space and too complex notation in comparison with the resulting gain in conceptual understanding of the theory. Instead, we shall only attempt to sketch out our impression of the state of the field. For the mathematically inclined reader we have supplemented this by an annotated set of references to the literature.

As we already know, for the three-person constant-sum game there are a plethora of solutions; this also appears to be the case for many of the four- and five-person constant-sum games. Yet, with this great richness of solutions for games with small n, it is still not known if every game possesses a solution. For example, it is not even known if every five-person game has a solution. From the first systematic presentation of n-person game theory to the present, the existence of a solution to every game has been considered the most important unresolved problem. The reader is referred to discussions in Kuhn [1953 a] and Wolfe [1955] for von Neumann's views on this problem and his suggestions as to research directions.

Previous results demonstrate that, at least for some games, there are vast numbers of solutions, but in these examples the structure of the solutions was comparatively easy to characterize. This would tend to suggest, and a mathematician would certainly hope, that solutions always possess a high degree of regularity which makes them comparatively simple to describe. Unfortunately, this is not the case, as Shapley [1952 c] has demonstrated. The exact nature of his result is not easily described (see below), but the gist of it has been neatly captured by Nash: Shapley's theorem shows that, if you sign your name in the geometrical space of the imputation set, then there is a game having a solution which contains, as an isolated part, your signature!

We have not cited all the results known about solutions, and, of those presented, we have eliminated so much detail that it may be difficult to appreciate fully their nature without consulting the original references. We hope nevertheless that enough of the "feel" of the situation has been conveyed so that the following summary observations will seem warranted. The variety and complexity of solutions in the games so far studied are overwhelming; their characterization and the corresponding proofs are involved and often subtle. It is doubtful that a mathematician could be found today holding any hope for a moderately simple and general characterization of solutions; the most optimistic goal is to classify a large portion of games into a number of subclasses such that the solutions of games in each class can be gradually characterized. This is surely an optimistic hope, for it has not yet been possible to characterize *completely* the solutions of games in any of the classes (except the three-person constant-sum game) so far isolated.

We may fairly conclude that, in addition to the conceptual difficulties mentioned in section 9.2, there are mathematical difficulties surrounding the solution notion, or, at least, the mathematical problem is difficult. At this stage it is not clear whether this will stimulate deeper insights into the concept or whether it will prove so discouraging that little more will be discovered about solutions.

Assuming that at least some people will be discouraged, there appear to be two other possibilities: (1) to single out some of the more regular solutions as more important than others and to study only these, and (2) to introduce new concepts more or less in competition with the solution notion. In section 9.6 we shall deal with an example of the first approach and in Chapters 10 and 11 with several examples of the second. But, before pursuing these topics, a more technical summary of the solution literature will be given for the mathematically oriented.

▶ In attempting to prove that solutions exist for all games, one possible procedure is to treat the domination relation over the space of imputations as an abstract relation over a set of points. The definition of solution can be given in this context without reference to the fact that, in game theory, the points will be imputations. The problem then becomes one of finding conditions which ensure the existence of a solution. In von Neumann and Morgenstern [1947] it was shown that, if a relation meets a condition, which we need not specify here, then a solution exists and it is unique; however, the condition is much too strong to be of general interest in the theory of games. Richardson [1946, 1953 a, 1953 b, 1955] has pursued this direction much further and has shown the existence of solutions under different and fairly weak assumptions, but to date he has not obtained a theorem which guarantees a solution to every game.

Earlier we mentioned very briefly the example produced by Shapley [1952 c], which shows how irregular solutions may be. The solution he presented is based, in part, on an arbitrary closed set C of an $(n - 3)$-dimensional subset of the space of imputations. This paper " · · · provides at one stroke a large fund of 'pathological' examples against which conjectures on the bahavior of · · · solutions can be tested." [1952 c, p. 1.] "The arbitrariness in the choice of C (for example, C may be a Cantor-type discontinuum) makes it easy to dispose of many conjectures concerning the regular behavior of · · · [solutions]." [1952 c, p. 2.]

The remaining papers on solutions are all concerned with the (partial) characterization of the solutions of specific classes of games.

It is known that every four-person constant-sum game has at least one solution, and the solutions of a few of these games have been studied in detail. Von Neumann and Morgenstern [1947] showed that these games can be put into one-to-one correspondence with the points of a cube, and they investigated some of these games in considerable detail; however, the only ones for which they gave all the solutions are those corresponding to the vertices of the cube. Later Mills [1954] determined all solutions corresponding to the edges of the cube, and he has some results concerning the faces. In the area of general-sum four-person and constant-sum five-person games, Nering [Kuhn, 1953 a] has found some solutions.

Another class of games for which detailed results are known are the *simple* ones, defined by the property that (in 0, 1 normalization) either $v(S) = 0$ or 1 for every coalition S.   A simple game is thus completely characterized by listing its so-called "winning" coalitions.   Von Neumann and Morgenstern studied various simple games for $n = 4$, 5, 6, and 7 and also certain more general cases.   They focused much of their attention on a type of solution known as the main solution.   Following the same tack, Gurk [Wolfe, 1955] has raised and answered a number of questions about the main solutions, and he has introduced another general and "natural" solution for simple games; in this connection, also see Isbell [1955]. Richardson has studied the main solutions of simple games which arise from finite projective space in a natural way.

Bott [1953] introduced a special class of simple games called the (n, k), or majority, games.   These are defined by the property

$$v(S) = \begin{cases} 0 & \text{if } |S| \leqslant k \\ 1 & \text{if } |S| > k, \end{cases}$$

where $k > n/2$ and $|S|$ denotes the number of players in S.   In this paper he examined the symmetric solutions.   In [1953 a], Gillies studied the non-symmetric or discriminatory solutions to such games.   In the words of Kuhn and Tucker, "Dropping symmetry, D. B. Gillies exhibits $\cdots$ a surprising variety of other solutions of (n, k)-games, all derived from Bott's symmetric solutions. Gillies' solutions are obtained by several methods which may carry over to a more general context: (1) by the addition of 'bargaining curves' (von Neumann and Morgenstern [1947], p. 501), (2) by inflation to larger games (ibid., p. 398), (3) by 'discrimination' (ibid., pp. 288–289) in which the non-discriminated players divide their take according to any solution to a smaller game, or (4) by partitioning the players into fixed subsets, assigning the spoils arbitrarily (i.e., in all admissible ways in one solution) among these subsets, and then dividing the spoils in any one subset according to the symmetric solution to a smaller game the players think they are playing."   [Kuhn and Tucker, 1953, p. 304.]

A third class of games which has received attention was defined by Shapley [1953 a].   A quota game is one for which it is possible to find n numbers ω_1, ω_2, $\cdots$, ω_n such that

$$v(I_n) = \omega_1 + \omega_2 + \cdots + \omega_n$$

and

$$v(\{i, j\}) = \omega_i + \omega_j, \quad \text{for all } i \text{ and } j, i \neq j.$$

"Shapley obtains families of solutions for the entire class of quota games, a class that contains some three-person games, all constant-sum four-person games, and a sizeable swath of all games with more than four players.   In a typical imputation in one of these solutions, all but two or three of the players receive their 'quotas' ω_i."   [Kuhn and Tucker, 1953, pp. 304–305.]   Kalisch [Kuhn, 1953 a] has generalized these results to what he calls m-quota games, i.e., games for which there exists an imputation $\boldsymbol{\omega} = (\omega_1, \omega_2, \cdots, \omega_n)$ such that $v(S) = \sum_{i \text{ in } S} \omega_i$ for all coalitions S with m members.

Several other references should be mentioned.   A game is symmetric if $v(S)$ depends only upon the number of elements in S.   Gelbaum [Kuhn, 1953 a] has presented some solutions to symmetric games.   Shapley [Wolfe, 1955] has characterized solutions for a type of game arising from a consideration of certain economic markets.   One interesting feature of this study is the existence of a phenomenon in the solution which can be interpreted as the market price.   The summaries of the two Conferences on Game Theory, held at Princeton in 1953 and 1955 [Kuhn, 1953 a; Wolfe, 1955], contain, in addition to reports of specific results, some discussion of the solution notion and the opinions of various workers about the future direction of the theory. Finally, many of the papers referred to above which either are unpublished or have only appeared as technical reports will in fact appear in print at about the same time this book is published; see Luce and Tucker [1958]. Our reason for not using this reference is that their table of contents was not settled at the time we were writing.   ◀

9.6 STRONG SOLUTIONS[2, 3]

The principal question to be discussed in this section is whether, aside from "standards of behavior," there are game-theoretic requirements which impose a greater stability on one solution than on another. This problem and the ideas here discussed were raised by Vickrey in some unpublished work [Kuhn, 1953 a; Vickrey, 1953].

The central idea is this: An imputation in a solution is likely to be adhered to if and only if any contemplated deviation to an imputation outside the solution will invite a corrective action, leading back into the solution, which results in a net loss to one of the players effecting the deviation. It will be recalled (section 9.1) that the solution

$$F = \{(\tfrac{1}{2}, \tfrac{1}{2}, 0), (\tfrac{1}{2}, 0, \tfrac{1}{2}), (0, \tfrac{1}{2}, \tfrac{1}{2})\}$$

of the three-person constant-sum game has the property that, although the imputation $(0, \tfrac{3}{4}, \tfrac{1}{4})$ dominates $(\tfrac{1}{2}, \tfrac{1}{2}, 0)$ of F, it is in turn dominated by $(\tfrac{1}{2}, 0, \tfrac{1}{2})$ of F, and the net result for player 2 is to pass from a payment of $\tfrac{1}{2}$ to one of 0.

With respect to a specific solution A, an imputation in A is called *conforming;* one not in A, *non-conforming.* Among the non-conforming imputations some dominate one or more conforming imputations; these are called *heretical imputations,* and an effective set for such a domination is called a *heretical set.* Vickrey discusses the example of the solution F as follows:

$\cdot\ \cdot\ \cdot$ in this case the movement to a non-conforming imputation $\cdot\ \cdot\ \cdot$ requires the cooperation of player 2, who though he may gain immediately, finds that although it may have been difficult to move from $(\tfrac{1}{2}, \tfrac{1}{2}, 0)$ to $(0, \tfrac{3}{4}, \tfrac{1}{4})$ it is now much easier for the couple $\{1, 3\}$ to organize a movement to the conforming imputation $(\tfrac{1}{2}, 0, \tfrac{1}{2})$ to the great discomfiture of 2. $\cdot\ \cdot\ \cdot$ If 2, finding himself

---

[2] Nothing in the remainder of this book depends upon the material in this section; however, it has not been starred because it is largely conceptual and the level of difficulty does not exceed that of other sections.

[3] Whenever we quote Vickrey in this section we shall change his symbols for imputations, coalitions, and solutions so that the notation is in conformity with the rest of the book.

now in the excluded position, attempts to negotiate with either 1 or 3 to move away from $(\frac{1}{2}, 0, \frac{1}{2})$, not only will 2 have to propose a heresy in which he gets less than the $\frac{1}{2}$ he started with in $(\frac{1}{2}, \frac{1}{2}, 0)$, but he will find that 1 and 3, having observed what happened to 2, will be very reluctant to join any such heretical coalition, and in fact may refuse to do so altogether. Either because the players all foresee this, or because after a short time they come to the conclusion as a result of experience that heresy is in the long run likely to lead to disaster for at least one of the heretics, they eventually will come to stick to the policy of staying at one of the approved imputations $\cdot \cdot \cdot$ [1953, p. 8.]

One might hope that all solutions would have this property, for it would serve as an added rationale for accepting them as a description of a stable social pattern, but they do not. To show this, we consider one of the discriminatory solutions of the three-person constant-sum game. Suppose we take $F_3(\frac{1}{4})$, which consists of all imputations of the form $(x_1, x_2, \frac{1}{4})$, where $x_1 + x_2 = \frac{3}{4}$ and $x_1 \geqslant 0$ and $x_2 \geqslant 0$. Suppose that the players are at the imputation $(\frac{7}{12}, \frac{2}{12}, \frac{1}{4})$ of this solution. The non-conforming imputation $(0, \frac{1}{2}, \frac{1}{2})$ is heretical since it dominates the given imputation of $F_3(\frac{1}{4})$, the effective set being $\{2, 3\}$. Of course, it is dominated by a conforming imputation, for example, by $(\frac{2}{12}, \frac{7}{12}, \frac{1}{4})$, the effective set being $\{1, 2\}$. But, observe, the net effect of the heresy has been a temporary gain for player 3 and a permanent gain of $\frac{5}{12}$ for player 2.

To be sure, there is no assurance that player 2 will always gain from a heresy. But

even if a return from [a heresy] to a conforming imputation $\cdot \cdot \cdot$ is made indirectly $\cdot \cdot \cdot$ so that it is possible for player 2 to be worse off [than originally], it is by no means certain from the characteristics of the game that player 2 will not be able to avoid such an eventual worsening of his position. And even if after one particular heretical excursion player 2 finds his position $\cdot \cdot \cdot$ worse $\cdot \cdot \cdot$, there is now nothing to prevent him from trying another heretical excursion, since player 3 whose cooperation he needs has nothing to lose by it in any event and stands to gain at least temporarily. In effect any player who is willing to engage in heretical excursions is at an advantage in bargaining for position among the approved imputations, over a player who eschews such tactics. It thus appears that in this case it will take a much stronger social sanction to compel adherence to the approved standard of behavior than where the standard of behavior conforms to the symmetrical solution $\cdot \cdot \cdot$ [1953, p. 9.]

Vickrey proposes the following two definitions. Let A be a solution, $\mathbf{x}$ an imputation of A, $\mathbf{y}$ a heretical imputation dominating $\mathbf{x}$ with the effective set T, and U the set of elements of A which dominate $\mathbf{y}$. A is said to be a *strong* solution if for every such $\mathbf{x}$, $\mathbf{y}$, and T and for every $\mathbf{z}$ in U there is at least one player i of T such that $z_i < x_i$. That is, a solution is strong if and only if every heresy necessarily results in a worsened position for at least one of the heretics once it is corrected by a dominating imputation of the solution.   It is important to recognize that the heretic who will be punished is not in general uniquely determined, which of course means that the argument against heresies occurring is much weaker than it would otherwise be.   On the other hand, A is said to be *weak* if for every x of A there exists at least one heretical y with effective set T such that for all z of A which dominate y, and all i in T, $z_i \geqslant x_i$.

For the constant-sum three-person game, as we have suggested, the symmetric solution F is strong, and all discriminatory solutions $F_i(c)$, $i = 1, 2, 3$, are weak.

For games with more than three players there are solutions which are neither strong nor weak, but rather there are intermediate notions of strength.   Primarily, however, one is interested in the strong solutions, for which every heresy is dangerous to some member of its heretical set.

At present, the only procedure available to determine the strong solutions is to determine all solutions and examine each one separately.   Thus, Vickrey has been restricted to studying such cases where all the solutions are known, i.e., to some four-person and some simple games.   In summary, he finds that

For constant-sum games, the concept of the strong solution has thus far appeared to be fairly effective in narrowing down the number of solutions that have to be accepted.   When it comes to the variable-sum games, unfortunately, it appears that much of the selectivity of insistence on strong solutions disappears.   For one and two person games, all solutions are already strong, while for three person games, it appears that insistence that · · · solutions be strong offers only a relatively small reduction in the range of possible imputations.   [1953, p. 32.]

No attempt has as yet been made to try out the effect of insisting on strong solutions for variable-sum games for more than three persons, so there is no way of telling whether the concept would prove more restrictive in such cases or not.   The complexities and variations possible between the extremes of strong and weak solutions already observed for the four-person constant-sum game indicate that the analysis of such games may prove to be extremely difficult.   On the basis of the experience with the three-person games, one is inclined to be not too sanguine.   The strong solution, that appears to be such a potent device for the simplification of the results of constant-sum games, may, it appears, be of relatively little value for the variable-sum games, although this tentative hypothesis is hardly more than a conjecture.   [1953, p. 35.]

*9.7 SOLUTIONS OVER DOMAINS DIFFERENT FROM IMPUTATIONS

When the concept of an imputation was first introduced (section 8.6) the question was raised whether it was not too restrictive; in particular the second condition, $v(I_n) = \sum_{i \text{ in } I_n} x_i$, was challenged as not following demonstrably from the principle of individual rational behavior. As Shapley has written,

The propriety of this restriction to the set of imputations may be challenged on several grounds. In the first place, it is not at all obvious that the notion of *group* rationality, as exemplified by the solution of an *n*-person game, must necessarily be a refinement of the principle of *individual* rationality, as embodied in the inequalities $[x_i \geqslant v(\{i\})]$. In the second place, it would seem methodologically more correct to study the consequences of the domination process separately from those of the blocking process.[4] One might even hope that the former, apparently the more powerful, might make the latter superfluous. (In that case, the restriction to [imputations] would be only a technical convenience, and would not prejudice the conceptual substructure of the theory.) Failing this, the restriction to [imputations] might better be applied (if it is desired to exclude "irrational" solutions) *after* stability under domination has been secured. [1952 *b*, p. 3.]

Shapley [1952 *b*] and Gillies [1953 *b*] have isolated the following four classes of *n*-tuples of payments:

$\bar{E}$ is the set of *n*-tuples **x** such that

$$\sum_{i \text{ in } I_n} x_i \leqslant v(I_n).$$

E is the set of *n*-tuples **x** such that

$$\sum_{i \text{ in } I_n} x_i = v(I_n).$$

$\bar{I}$ is the set of *n*-tuples in $\bar{E}$ such that

$$x_i \geqslant v(\{i\}) \qquad \text{for all players } i.$$

I is the set of *n*-tuples in *E* such that

$$x_i \geqslant v(\{i\}) \qquad \text{for all players } i.$$

We observe that *I* is the set of imputations, that $\bar{I}$ is obtained from *I* by dropping the group rationality condition, that *E* is obtained from *I* by dropping the individual rationality condition, and that $\bar{E}$ results from dropping both rationality conditions.

If the reader will turn back to section 9.1 he will see that neither the definition of domination nor that of solution directly employs the fact that we were dealing with imputations; they are concepts defined for any given set *C* of *n*-tuples, and at that time we specified $C = I$ (= the set of imputations). Shapley introduces the term *C*-stable for those sets *A* of *C* which satisfy the conditions of a solution, i.e.:

[4] By the "blocking process" Shapley means the refusal of player *i* to accept a payment of less than $v(\{i\})$.

i. No element in A dominates another element in A.

ii. Every element of C not in A is dominated by some element of A.

An I-stable set is therefore another way of speaking of a von Neumann-Morgenstern solution.

There are four theorems and several examples which establish the relations among the C-stable sets for the four classes of outcomes which have been defined.   First, a set A is $\bar{E}$-stable if and only if it is E-stable [Shapley, 1952 b].   That is to say, if one is concerned with stable sets, then it is immaterial whether one chooses $\bar{E}$ or E as the set of n-tuples, for no $\bar{E}$-stable set has outcomes lying outside the set E.   Thus, with respect to theories of stable sets where we do not choose to impose the condition of individual rationality, there is no loss of generality in imposing the condition of group rationality.   However, as we have made clear earlier, the condition of individual rationality appears to follow directly from the basic postulates of game theory, so our real concern is with the sets $\bar{I}$ and I.   Gillies [1953 b] has shown that a set A is $\bar{I}$-stable if and only if it is I-stable.   Thus, when the condition of individual rationality is imposed, there is no loss of generality so far as solution theory is concerned in the further restriction to imputations.   This theorem eliminates insofar as solution theory is concerned any *a priori* objections (such as those discussed in section 8.6) to restricting attention to imputations, but, of course, it does not obviate our remarks of section 8.6 that the solution notion is a means of bypassing the logical extension to all coalitions of the intuitive argument supporting group rationality.   Since the solution notion, or more generally that of a C-stable set, is not clearly acceptable, it may be desirable to create an entirely different theory—one in which the condition of group rationality may not be immaterial.

In the quotation given at the beginning of this section, Shapley points out that it would be interesting to know whether or not the notion of domination between outcomes implies individual rationality.   Given the above results, this reduces to an examination of the relation between I-stable and E-stable sets.   He has shown that a set A which is E-stable is also I-stable if and only if every outcome in A is an imputation, i.e., if A is a subset of I. In the reverse direction, he has shown that if A is a solution, i.e., an I-stable set, then A is E-stable if and only if for each player i it is possible to find an n-tuple $\mathbf{x}$ in A such that $x_i = v(\{i\})$.   These results strongly suggest that it is possible to have E-stable sets in which some, or indeed all, of the n-tuples are not imputations, i.e., in which individual rationality is violated.   They also suggest that there may be solutions which are not E-stable.   Shapley has exhibited examples of both possibilities (an E-stable set of any quota game with a weak player which includes no imputations, and the discriminatory solutions of the three-person constant-sum game which are not E-stable).

The picture, then, is clear insofar as solution theory is concerned. The condition of individual rationality must be imposed independently and as a direct consequence of the basic postulates of the model, because it does not follow from the notion of a stable set. The condition of group rationality, which seems *a priori* objectionable, may be assumed in solution theory because it results in no loss of generality.

9.8 SUMMARY

Solutions, the central equilibrium concept for games in characteristic function form, has been our topic throughout this chapter. An imputation **y** is said to *dominate* another imputation **x** *with respect to the coalition T* if it is both feasible, i.e.,

(i) $v(T) \geqslant \sum_{i \text{ in } T} y_i,$

and better than **x** for T, i.e.,

(ii) $y_i > x_i$, for every i in T.

y simply *dominates* **x** if there is some T such that it dominates **x** with respect to T. In terms of these concepts, a *solution* is defined to be a set A of imputations such that:

i. If **x** and **y** are in A, neither **x** nor **y** dominates the other.
ii. For any imputation **z** not in A, there is at least one **x** in A such that **x** dominates **z**.

A number of general qualitative facts about the concept were cited: most games have multiple solutions, and each of these consists of a (not necessarily finite) number of imputations; it has not been possible to characterize all the solutions of any very wide class of games; Shapley's example demonstrates that there are solutions with most irregular properties; and it is not known if every game has a solution. Quite a few papers have been published which characterize some of the solutions in many games, including all four-person constant-sum ones and some simple, quota, and symmetric games.

Conceptually, there seemed to be two difficulties. First, the feasibility condition of the domination concept seems none too plausible for non-constant-sum games or for situations where collusive changes are limited. Second, from among the many solutions, how is one to be selected, and, from that solution, how is an imputation isolated? Von Neumann and

Morgenstern argue that these are questions beyond the game framework, that the former depends upon "standards of behavior" in society and the latter upon the "bargaining abilities" of the players.

These difficulties and criticisms notwithstanding, the concept gives considerable insight into at least some situations in which coalition formation is allowed. This was illustrated by the simple example of a three-person market.

Vickrey's concept of a strong solution is the only attempt to narrow down the number of solutions to be considered. Roughly, a solution is called *strong* if the sequence—an imputation in the solution, a change to a non-conforming imputation, and a return to an imputation in the solution—always means that at least one of the players participating in the original deviation ultimately suffers a net loss. Thus, a strong solution has an inherent stability not possessed by other solutions, and so it might be expected to occur rather than one of the weaker solutions. There seem to be relatively few strong solutions in constant-sum games, but the restriction is not so effective for non-constant-sum ones.

Earlier we raised a question about restricting our attention to imputations, for there seems to be no valid argument based on individual rationality for imposing group rationality. In the final section we examined results concerned with this problem for solution theory. The central conclusion is this: so long as individual rationality is imposed, no loss of generality results in solution theory from the further restriction to imputations.

chapter 10

ψ-Stability

10.1 ψ-STABLE PAIRS

Aside from the solution concept and its ramifications, there are three other topics in characteristic function theory which have received attention. Although two of these (ψ-stability and reasonable outcomes) continue to be concerned with outcomes which can be argued to be reasonable, all three notions differ appreciably from the solution concept. For example, one of the salient differences of the definition we shall present in this chapter is that it does not deal with imputations or sets of imputations alone, but following the suggestion in section 7.6 it isolates pairs consisting of an imputation and a corresponding arrangement of the players into coalitions.

A second major difference between the present approach and the one previously examined will be the assumption that sheer profit is not always sufficient cause for a coalition to form. Intuitively, one senses that there are constraints in society limiting changes in coalition structures and that these are to a large degree non-rational. Nonetheless, one may conjecture that these constraints are a source of considerable social stability. Of course, one wonders what there is about society that introduces them, particularly if everyone is free to bribe or, to be less evaluative, to compensate others for their cooperation. It is doubtful that there is a single answer, but we would argue that one factor must be the expense and difficulty of complex communication which prevents players in a game from considering at one moment any but the simplest modifications of their alliance structures. Roughly speaking, society seems to exhibit a form of "friction." If such is the case, the problem then becomes one of formalizing these limitations on coalition changes so that the resulting model is both somewhat realistic and at the same time mathematically tractable. One proposal—the one we shall examine—was discussed in the latter part of section 7.6, namely, that these limitations be given by a rule ψ of admissible coalition changes which states for each partition of the players into coalitions those (admissible) sets of players which can take joint action. Since the following material rests heavily upon this notion, it may be advisable to reread section 7.6.

Assuming that such a function describing the limits of coalition change is given, Luce [1954, 1955 a, 1955 b] has attempted to construct a theory for games with side payments which parallels in spirit both the suggested approach to a cooperative theory when side payments are prohibited (section 7.9) and the definition of the core (section 8.6). In this approach one searches for equilibrium outcomes where, as we mentioned, one part of the outcome is the arrangement of the players into coalitions as given by a partition τ. Although one might like to use the strategies employed by the players as the other part of the outcome, these will not work here any more than in solution theory, for they do not tell what side payments were effected, and hence do not tell the total payment to each player. Thus, we are led again to use imputations rather than strategies as a description of the second aspect of the outcome. The task, therefore, becomes one of describing those pairs $[\mathbf{x}, \tau]$, where $\mathbf{x}$ is an imputation and τ is a coalition structure, which are in equilibrium when the game is described by its characteristic function v and when changes in collusive arrangements are limited according to a given function ψ.

Suppose that $[\mathbf{x}, \tau]$ is a candidate for an equilibrium pair, and let S be any of the coalitions in $\psi(\tau)$, i.e., S is any of the coalitions which might form if the players involved so desired. Suppose that the players who would constitute the coalition S add up the total amount they are receiving in the imputation $\mathbf{x}$ and find that this sum is less than the total amount the coalition S may expect to receive, i.e.,

$$\sum_{i \text{ in } S} x_i < v(S).$$

Then they have a distinct motive to change from their arrangements in τ, whatever they may have been, to form the coalition S; for by employing side payments—and this is their crucial role—each player of S can receive more than he is assigned in the imputation $\mathbf{x}$. Such a disruptive force is contrary to the idea of an equilibrium, so it would seem that a necessary condition for the pair $[\mathbf{x}, \tau]$ to be in equilibrium is that

$$v(S) \leqslant \sum_{i \text{ in } S} x_i.$$

We observe that this is exactly the same condition which seemed to hold for *all* coalitions if we accepted the logical consequences of the argument for group rationality needed to limit payments to imputations (section 8.6). The only difference is that here it is not imposed for all coalitions but only for those determined by the coalition structure τ and the given function ψ.

A second condition seems plausible if the pair $[\mathbf{x}, \tau]$ is to be in equilibrium. A player who participates in a non-trivial coalition (one having at least one other member than himself) must thereby benefit in the sense that he must receive more than he can assure himself under the most adverse circumstances, namely, opposition by a coalition of all the other players. This is, in a sense, an extension of the argument, based on individual rationality, which led to the condition $x_i \geqslant v(\{i\})$ for any equilibrium payments. It is now argued that not only must this condition be satisfied but, if i is in a non-trivial coalition, then $x_i > v(\{i\})$.

These two conditions are taken to characterize the equilibrium pairs, so in summary we make the following definition: A pair $[\mathbf{x}, \tau]$, where $\mathbf{x}$ is an imputation and τ is a coalition structure, is *ψ-stable* for the game with characteristic function v and given "boundary condition" ψ if the following two conditions are satisfied:

(i)   For every S in $\psi(\tau)$, $v(S) \leqslant \displaystyle\sum_{i \text{ in } S} x_i$;

and

(ii)   If $x_i = v(\{i\})$, then in the coalition structure τ player i is not in a coalition with any other players, i.e., $\{i\}$ is in τ.

In section 8.3 we pointed out that for an equilibrium definition to be satisfactory in characteristic function theory, it is necessary that it be invariant under S-equivalence. So, for this definition to be acceptable, two S-equivalent games should have corresponding ψ-stable pairs. They do.

A game having at least one ψ-stable pair is, itself, called ψ-stable; otherwise it is called ψ-unstable.

As was pointed out above, the core is, in effect, a special case of the ψ-stability concept. To be specific, if $\mathbf{x}$ is an imputation in the core, the pair $[\mathbf{x}, (\{1\}, \{2\}, \cdots, \{n\})]$ is ψ-stable for every ψ. Furthermore, if ψ'' denotes the function which admits every coalition as a possible change (see section 7.6) from every coalition structure, and if there is no ψ''-stable pair of the form $[\mathbf{x}, (\{1\}, \{2\}, \cdot \cdot \cdot, \{n\})]$, then the core is empty.

If the characteristic function v and the boundary condition ψ are left unspecified, as in the definition, little more can be said about the existence and nature of ψ-stable pairs. If, however, certain specific functions ψ are chosen, then it is reasonable to expect some theorems to hold for various special classes of characteristic functions. So far in the literature only one class of functions ψ has been examined; it can be roughly described as including some of those functions with the property: S is in $\psi(\tau)$ provided that there is a coalition T in τ which is not "too different" from S. For these functions certain general theorems do hold. At the beginning of the next section we shall give the precise definition of this class and state some of the theorems; the casual reader may, if he chooses, skip over this material, which is in small print, to the discussion of the faults and virtues of the ψ-stability concept.

10.2 CRITICISM

As background for a critical discussion of ψ-stability we shall present a summary of some of the theorems which have been proved:

▶A special class of boundary conditions must first be defined. Let k be a fixed integer lying in the range 1 to $n - 2$, where n is the number of players. We shall also denote the function we are about to define by k, i.e., for any τ, $k(\tau)$ will denote the set of admissible coalition changes from τ. The coalition S is included in $k(\tau)$ if and only if there is a coalition T in τ such that the sum of the number of players in S who are not in T and the number of players in T who are not in S does not exceed the integer k. Put in symbols, S is in $k(\tau)$ if and only if there exists a T in τ such that

$$|(S - T)\cup(T - S)| = |S - T| + |T - S| \leqslant k,$$

where $|S|$ denotes the number of players in S.[1] If we take the point of view that T is a basic coalition which is modifying "itself" by adding and subtracting players, we demand that the number of players expelled from T plus the number of new ones added in order to form S shall not total to more than k players. When we are using this function we shall, of course, speak of k-stability.

In all the following discussion we shall not admit the coalition of all players as a possible coalition structure. Under that restriction, the function $(n - 2)$ is the same as the function which admits every coalition as a possible change.

It is not difficult to see that increasing the size of k admits more coalitions as possible changes, and so, if $k' > k$, a pair which is k-stable may very well not be k'-stable; but a pair which is k'-stable is surely also k-stable.

[1] If S and T are sets, $S - T$ denotes the set of elements in S which are not in T, $S\cup T$ the set of elements in S or in T or in both, and $S\cap T$ the set of elements in both S and T.

[Many of the following theorems also hold for functions which differ from their $k(\tau)$ counterpart only by also including all coalitions which, themselves, have no more than $k + 1$ elements. This may be a useful modification for applications, e.g., if $k = 1$, then the modified function will always include the coalitions $\{i, j\}$, whereas they need not all be in $1(\tau)$.]

It can be shown that the three-person constant-sum game is 1-unstable. This is only a special case of the more general theorem that any constant-sum game with n players is $(n - 2)$-unstable. It is also a special case of a theorem for simple games which follows.

It will be recalled (section 9.5) that a game (not necessarily constant-sum) is called *simple* if, for every S, $v(S) = 0$ or 1. Those coalitions with $v(S) = 0$ are called *losing*; those with $v(S) = 1$, *winning*. Theorem: A simple game is k-stable if and only if either there is no winning coalition of $k + 1$ members or if there is at least one player who is a member of all such winning coalitions. Now consider the three-person constant-sum game: It is simple, and all two-element coalitions are winning; since no player is common to all of them, the game is 1-unstable.

More detailed results on the 1-stability of simple games are known. For instance, it is possible to describe the form of the 1-stable pairs, although that description is too complex to present here. More interesting and easier to describe are those simple games which are 1-unstable. To do so, we shall need the concept of a decomposable game, which is one consisting of two non-interacting games on complementary sets of players. To be precise, if there exists a set of players T such that for every coalition S

$$v(S) = v(S \cap T) + v(S - T),$$

then the game v is said to be *decomposable* into games on T and $-T$. Probably this notion is of little applied interest, but it must be considered, for there is nothing in the definition of a game which precludes the formal composition of two independent games into a third. Theorem: Any 1-unstable simple game with n players is decomposable into the three-person constant-sum game and the $(n - 3)$-person inessential game (for the definition of an inessential game see section 8.2). In other words, aside from the three-person constant-sum game and the trivial modifications of it to larger n by composing it with inessential games, all simple games are 1-stable.

The *symmetric* games form another class which has received some attention in the literature of solutions. They are characterized by the property that the value of a coalition depends only upon the number of elements in the coalition, i.e., we may write $v(S) = v(|S|)$. Theorem: If $v(i)$ is the characteristic function of a symmetric game in 0, 1 normalization, then the game is k-stable if and only if $v(i) \leqslant i/n$ for all integers i from 0 to and including $k + 1$.

A third class of games for which some stability results are known consists of Shapley's *quota* games, which, it will be recalled (section 9.5), are defined by the property that there exists an n-tuple $\omega = (\omega_1, \omega_2, \cdots, \omega_n)$ such that

$$v(I_n) = 1 = \omega_1 + \omega_2 + \cdots + \omega_n,$$

and

$$v(\{i, j\}) = \omega_i + \omega_j, \quad \text{for all } i \text{ and } j \text{ in } I_n, i \neq j.$$

Since $v(\{i, j\}) \geqslant 0$, it follows directly that there is at most one player i such that $\omega_i < 0$. Such a player, if he exists at all, is called *weak*. Theorem: A quota game is 1-stable if and only if it has no weak player. Some conditions for the k-stability of quota games are also known, but they are too fussy to present here. Of more interest are the following results on the nature of k-stable pairs. Theorem: If $[\mathbf{x}, \tau]$ is a k-stable pair of a quota game, then, either if n is odd or if n is even and $k \geqslant 2$, $\mathbf{x}$ must be the quota ω; in the case n is even and $k = 1$, then either $\mathbf{x} = \omega$ or each of the coalitions T in τ must have an even number of elements and $\mathbf{x}$ and ω must satisfy

$$v(T) = \sum_{i \text{ in } T} x_i = \sum_{i \text{ in } T} \omega_i.$$

The study of four-person constant-sum games, which are all quota games, becomes simply a matter of specializing the above results. We know by the constant-sum condition that they are all 2-unstable, but only those with a weak player are also 1-unstable. For those with no weak player, it is not very difficult to show that the only imputation which can arise in a 1-stable pair is the quota. The pair $[\omega, (\{1\}, \{2\}, \{3\}, \{4\})]$ is always 1-stable in such games, and, for most of them, it is the only 1-stable pair; however, there are certain games in which other coalition structures combined with ω are also 1-stable. ◀

Before criticizing the concept of ψ-stability, let us make certain summary comments concerning its virtues. Mathematically, the more restricted concept of k-stability is comparatively easy to work with, much easier, say, than the solutions of von Neumann and Morgenstern. Evidence of this is the fact that k-stability results are known for all constant-sum four-person games, for all simple games, for all symmetric games, and for all quota games; whereas, to date it has been possible to obtain information about solutions only for special cases of these classes of games. In cases where results are known from both theories, the k-stability results are generally the simpler to state but are still not intuitively obvious. Although the notion is mathematically plausible, for social science more is needed: the definition must have some intuitive merit and, possibly, some empirical usefulness. There are two points in its favor conceptually, and the first may also be valuable empirically. First, the state of equilibrium is described both in terms of an imputation and a coalition structure; conceptually, both seem to be relevent. Empirically, it is often easier to determine the coalition structure of an existing situation than to determine the payments made to the players (see, for example, the illustration discussed in section 12.2). Second, the spirit of both conditions of the definition is the same as that underlying the limitation of payments to imputations; in solution theory the defining properties depart somewhat from the considerations which led to imputations.

The discussion of the empirical merits and inadequacies will be left to Chapter 12, where a comparison with experimental data is made.

The criticisms of the notion fall naturally into six groupings:

i. *Restriction of payments to imputations.* When we first discussed n-tuples of payments to players we raised serious objections to the condition of

group rationality, i.e., $v(I_n) = \sum\limits_{i \text{ in } I_n} x_i$, and we pointed out that its
acceptance should force us to limit payments to the core, since, if the whole group of players could be rational, there was no reason to suppose that subgroups could not also be rational. The main idea of ψ-stability is to limit the subgroups to which this argument is applied, and, in general, one suspects these to be coalitions not very different from those already existing in the given coalition structure τ. By and large, these coalitions will not include the set of all players; yet the restriction of payments to imputations amounts to including the set of all players in ψ(τ). This is unacceptable. If, however, one examines the definition of ψ-stability, there is no reason why **x** cannot be any n-tuple satisfying $x_i \geqslant v(\{i\})$, for every i, and so the restriction to imputations did not actually create a basic flaw. Of course, it does not follow that any of the theorems stated for imputations are true in this more general context. By and large, the results are not greatly changed, but there is no point in going into the details.

ii. *Introduction of the functions* ψ. A very serious criticism of ψ-stability theory is the very introduction of the peculiar functions ψ, functions which are not explicit in most social situations. Where there are "standards of behavior" which are implicit, or at least vague, and which are not rigidly enforced, then, even if it is possible to estimate ψ, there is no assurance that the players will live up to one's estimates. (Some indication of how estimates of these boundary conditions can be made and used in a socially meaningful situation is given in section 12.2 where we discuss a game-theoretic analysis of certain congressional power distributions.) Assuming that in general such functions are not suitable, is there any hope of devising a more realistic theory which preserves the intuitive idea behind ψ-stability?

One possible approach would be to replace the rules of admissible changes by probabilistic statements. Presumably, one should assume that to each pair consisting of a coalition S and a coalition structure τ there is assigned a probability $p(S, \tau)$, which is interpreted as the probability that a change to S will be considered when the players are arranged in coalitions according to τ. We do not intend these to be interpreted as subjective probabilities existing in the player's minds; rather they are objective descriptions of the probability that a certain event will occur, namely: given that the arrangement τ exists, the probability that the members of the set S will in fact consider forming a coalition. These probabilities would, therefore, subsume a great number of facts about ease of communication, social limitations, intellectual limitations of the players, and so on. It is perfectly clear that it would be extremely difficult to get objective estimates of them, except, possibly, in situations which have recurred so often that frequencies can be observed. In any case, if they are assumed, the theory can presumably be constructed as before, except that assertions of the form "[x, τ] is ψ-stable" will have to be replaced by sentences such as "[x, τ] is stable with probability p." The successful development of such a theory appears to rest heavily upon the imposition of suitable regularity conditions on the probabilities—conditions which parallel the restriction of the general functions ψ to the particular class of functions k.

If, as may be the case, this whole direction seems too unreal, then it may be necessary to return to the foundations of game theory and to reconstruct them. The role of these functions is to introduce some sociological assumptions into the model. Such assumptions do not occur in its understructure, and evidently they are not necessary in two-person situations with strictly opposed interests, but we would suggest that both the difficulties of n-person theory and an intuitive consideration of conflict of interest indicate that sociological assumptions are necessary in general. It would be well to have them in the foundations.

iii. *The expectations upon which the equilibrium is based.* Basic to the notion of ψ-stability is the idea that the individuals who might potentially form the coalition S will compare with $v(S)$, i.e., with their total expected return when both S and $-S$ form coalitions, the sum total of their existing payments. This comparison might serve for a normative theory if the coalition structures of all ψ-stable pairs consisted of but two coalitions— actually this is rarely the case—but empirically it seems ridiculous. In society, we can observe the creation of coalitions which could not possibly benefit their members if all remaining participants in the conflict of interest were to band against them. Of course, such a unified opposition is not expected, and, experientially, this view is justified. Presumably, the participants in a coalition do have some expectation as to the reaction and reorganization of the other players to the new coalition, and it is on the basis of these estimates, which may be very long range, that the decision is reached whether to cooperate or not.

There are really two parts to this criticism. The first is, in large measure, a repetition of our earlier comments that in general the characteristic function is an inadequate representation of a conflict of interest. Nonetheless, whatever expectations the players have for the coalition S, it cannot be less than $v(S)$. Thus, $v(S)$ is a conservative estimate of their prospects. It follows that, whatever their estimates are, any pair which is ψ-stable using their estimates is also ψ-stable using the characteristic function, but the converse is not true. The set of ψ-stable pairs based on the characteristic function will include anything that is actually in equilibrium, plus, in general, some pairs which will not be in equilibrium in a more accurate model. Provided that the set of *ψ*-stable pairs is sufficiently restricted, as seems to be the case judging by the theorems proved, the idealization based on a characteristic function can still afford considerable information.

Another escape from this objection is to apply *ψ*-stability theory only to games which are actually given in characteristic function form: if S forms a coalition, it gets $v(S)$ regardless of what the other players do. In that case, we would probably want to add the condition that each coalition only divide up what it actually gets, i.e., we would only consider pairs $[\mathbf{x}, \tau]$ such that for every T in τ, $\sum_{i \text{ in } T} x_i = v(T)$.

The second criticism is much more profound. Once one admits that the players make predictions about the future in determining the worth of a change, one cannot argue that only immediately profitable coalitions will form. A coalition which results in an initial loss to its members may count on a reactionary realignment of the other players which is so unstable that, in turn, it will pass to another structure which allows substantial returns to the given coalition. Such a possibility seems meaningful only when there are limitations on coalition changes, but in that case it does seem realistic. We do not see how to cope with such possibilities within the present framework.

iv. *The unrealistically limited view of the players.* In contrast to the preceding criticism which suggested that there may be *ψ*-stable pairs which are not, in fact, in equilibrium, we shall now offer the criticism that there may also be *ψ*-unstable pairs which are actually in equilibrium. The basic point is substantially the same as that raised by Vickrey with regard to the solution concept (see section 9.6). A pair $[\mathbf{x}, \tau]$ is *ψ*-unstable if there exists an S in $\psi(\tau)$ such that $v(S)$ is larger than the sum of payments to players in S as given by $\mathbf{x}$, and the argument is that, since each can be made to profit, the change will be made. It is, however, quite conceivable that once the change is made, this will lead to other changes, and the net result may be an overall loss to one or more of the members participating in the change. Put another way, if the players have any foresight and predictive ability, they may reject certain immediately profitable changes on the grounds that ultimately they will be punished. If this is possible, it means that there may be pairs which are observably in equilibrium which are not predicted by the theory.

In section 10.3, where we shall discuss the *ψ*-stability analysis of the market of two buyers and one seller, we shall see that this is a real possibility. In that case it will be shown that an *a priori* analysis of the situation is sufficient to cause us to modify the *ψ* function to take into account simple long-range considerations. Whether this technique is generally useful to cope with such problems is not known.

v. *Existence.* Given a game v and a boundary condition ψ, there may not exist any ψ-stable pairs. This can even be true when the boundary conditions are extremely restrictive; for example, the three-person constant-sum game has no 1-stable pairs. As might be expected, relaxing the restrictions tends to reduce the number of stable pairs: when there are no restrictions, the only games with stable pairs are those with a nonempty core. Certain authors, notably von Neumann, have apparently felt it extremely important that, as for two-person zero-sum games, equilibrium states exist for all games, whatever the equilibrium concept may be. Von Neumann felt, for example, that the most important problem of n-person theory is a proof or counterexample to the theorem that every game possesses a solution [Kuhn 1953 a], and the implication is that a counter example would be very damaging to solution theory. We certainly agree that, mathematically, an existence theorem is very elegant to have. Furthermore, if we accept a solution as our idea of social equilibrium, then a proof would establish the existence of social equilibria under incredibly weak assumptions. Yet if the theorem does not hold for solutions— as it does not for ψ-stability—we should not be unduly disturbed, for it is perfectly plausible that there are social arrangements, just as there are physical ones, which are incapable of sustaining a state of equilibrium. This is not to say that these situations are uninteresting—an explosion hardly epitomizes dullness—but only that they are transient. To study transience a dynamic theory is essential, and, as we have repeatedly emphasized, game theory is not dynamic.

With the failure of general existence, the mathematical tasks are not eliminated nor, in general, is their difficulty reduced. Attempts must be made to characterize those games which do and those which do not possess equilibrium states and, for those with equilibrium states, to describe them.

vi. *Uniqueness and occurrence.* Most games which have a ψ-stable pair have more than one, just as most games have more than one solution. The uniqueness problem can be bypassed here in exactly the same way as in solution theory: by mentioning standards of behavior, bargaining ability, and the like. To us, this is no more acceptable now than it was before. What, then, can we say? It appears that one point of view is required if we treat the theory as normative, and another if we think of it as an attempt at description, however idealized. If it is normative, or if it is to be used in that way, the lack of uniqueness seems to be serious in much the same way as it was in Nash's equilibrium theory for non-cooperative games (section 7.8) and in solution theory. There will, in general, be a conflict of interest among the stable pairs with one player preferring one, another preferring a different one.   In no sense, then, does the theory tell a player how to behave, for if each attempts to arrive at his preferred pair without regard to the behavior of the others, a state of equilibrium will not usually be achieved.

If we look upon the theory as descriptive, then uniqueness does not seem to be a serious question.   Although it may still be charged that an adequate social theory should be able to predict, at least in probabilistic terms, which state will arise, the inability of an equilibrium theory to do so is not so much a criticism as a reminder that the evolution to an equilibrium point is a dynamic process for which there is no theory at present. The situation is reminiscent of certain physical problems in which there are several points of equilibria: the one which occurs depends, among other things, upon the initial point of the dynamic system, and to predict its occurrence requires a full dynamic theory, not just an equilibrium theory. The analogy seems close enough, and at least for the present, we shall assign the failure to predict the occurrence of equilibrium states to a lack of a full dynamic theory of coalition formation.

If this be true, then great care must be taken in assigning meaning to these equilibrium states, for, just as with physical systems, it appears that there can be equilibrium states which cannot be reached from any other state.   Consider, for example, a game having a ψ-stable pair $[\mathbf{x}, \tau]$ such that all the coalitions T of τ are losing, i.e., in 0, 1 normalization, $v(T) = 0$.[2] The question is whether such a pair could possibly have been reached from some other starting point through a series of intermediate steps using as the mechanism for change the one implicit in the ψ-stability concept. This mechanism is, of course, the following: when the payments are arranged according to the imputation $\mathbf{y}$, a change to a coalition T will occur only if $v(T) > \sum_{i \text{ in } T} y_i$.   But all the coalitions in τ are losing, so this would mean that $0 > \sum_{i \text{ in } T} y_i$, which is impossible for an imputation. Thus, such a pair cannot arise by a dynamic hunting process, and so we can only conclude that, although it is in equilibrium, its probability of occurrence is small.

Finally, among those equilibrium states which are likely to occur, there may be some which by our usual standards, and so presumably by the standards of those playing the game, would be described as socially undesirable.   Naturally, a static theory must be ignorant of such evaluations,

_____

[2] This does occur.   Any pair involving the coalition structure $(\{1\}, \{2\}, \cdots, \{n\})$ has this property.   Thus, for example, most four-person constant-sum games have only 1-stable pairs of this form.

but society is not. We should not, therefore, be surprised to find apparently stable states evaporating. One mechanism whereby this may be effected without changing the underlying game is to change the boundary conditions. We may expect to find a dynamic interaction between the equilibrium states given by a function ψ and the function itself—an interaction that first produces a modification of the boundary conditions, and the modified boundary condition, in turn, determines some new and, hopefully, more desirable equilibria. The boundary conditions may, first, be relaxed to render the undesirable equilibrium unstable, and later tightened to produce new equilibria. It is possible to think of the antitrust laws as a socially conscious attempt to increase the restrictions against the formation of new coalitions, thereby tending to preserve the status quo. The conscious manipulation of sociological restrictions to achieve predetermined states of equilibrium in conflict of interests is an important, if dangerous, social tool.

10.3 THE ψ-STABILITY ANALYSIS OF A MARKET WITH ONE SELLER AND TWO BUYERS

This section continues the example of a market consisting of a seller, who possesses a single indivisible commodity which he values at a units of money, and two buyers, who value it at b and c units respectively (section 9.4). The characteristic function was established to be:

$$v(\{1\}) = a, \qquad v(\{2\}) = v(\{3\}) = 0,$$

$$v(\{1, 2\}) = b, \qquad v(\{1, 3\}) = c, \qquad v(\{2, 3\}) = 0,$$

$$v(\{1, 2, 3\}) = c.$$

The imputations arrived at by a common sense argument were the core:

$$b \leqslant x_1 \leqslant c, \qquad x_2 = 0, \qquad x_3 = c - x_1. \tag{1}$$

All of these imputations were found to occur in every solution; however, the possibility of collusion as embodied in the solution concept led to the inclusion of other imputations.

Let us examine the results of ψ-stability theory for this example using the same assumption as in solution theory, namely: there are no limitations on coalition change. First, it is not difficult to show that any partition which contains either $\{1, 2\}$ or $\{2, 3\}$ as a coalition is unstable, so there are only two different τ's to consider:

i. A pair $[\mathbf{x}, (\{1\}, \{2\}, \{3\})]$ is stable if and only if $\mathbf{x}$ is in the core, i.e., it meets the conditions of eq. 1.

ii. A pair $[\mathbf{x}, (\{1, 3\}, \{2\})]$ is stable if and only if $\mathbf{x}$ meets the following slight modification[3] of eq. 1: $b \leqslant x_1 < c$, $x_2 = 0$, $x_3 = c - x_1$.

In other words, when there are no limitations on the possible realignments into coalitions, ψ-stability and the common sense argument are in substantial agreement, even though the stability notion takes coalition formation explicitly into account.   On the other hand, von Neumann and Morgenstern have accounted for the existence of imputations other than those in the core—in particular, those in which player 2 gets more than 0—as a consequence of coalition formation.   Where is the discrepancy?

In practice, at least, it is easy to see why the ψ-stability result is probably wrong.   Player 3 is in a position to threaten 2 in such a way that the coalition $\{1, 2\}$ is not really an admissible change—even though technically it is.   For 3 can point out to 2 that if they (2 and 3) form a coalition, then 2 will receive more than 0 (the amount he gets if the threatening and bargaining leads into the core); and if 2 tries to use a threat of defection from the coalition $\{2, 3\}$ in order to demand too much from 3, then 3 can threaten to force the outcome into the core.   Of course, if $\{2, 3\}$ forms and agrees not to let 1 get more than a, then 1 can come to 2 and offer him a little added profit in the coalition $\{1, 2\}$.   According to ψ-stability theory, this offer would be acceptable; however, at this point 3 would go to 1, offer him a little more to join 3 in a coalition, leaving 2 with nothing. Either 2 has the foresight to see this, or 3 can point it out to him, and in effect the coalition $\{1, 2\}$ is outlawed.   Equally well, player 3 in this process must agree not to accept overtures from 1, for this would just as surely put them into the core, with 1 getting more than they can hold him down to if they cooperate.   Apparently, therefore, if 2 and 3 agree to a coalition these long-range arguments lead to a self-imposed communication boundary condition of no changes.   The ψ-stability analysis of that case yields

$$a \leqslant x_1 < c, \qquad x_2 > 0, \qquad x_3 > 0, \qquad x_1 + x_2 + x_3 = c.$$

This set of imputations includes almost all the imputations contained in at least one solution.   The exceptions are when $x_1 = c$, or when $x_2 = 0$, or when $x_3 = 0$.   These imputations are permitted in the solutions because the initial feasibility of a point in the solution is never questioned; whereas the second part of the ψ-stability definition says, in essence, that some otherwise stable imputations are not really feasible.   The present analysis also includes some imputations not in the solutions, namely,

[3] The modification is introduced via the second condition in the definition of ψ-stability, namely, that a person will not participate on a non-trivial coalition unless he is made to do better than he could alone under the most adverse conditions.

those with

$$b \leqslant x_1 < c \quad \text{and} \quad x_2 > 0.$$

These extra cases seem to arise from the following possibility: Since 2 and 3 have agreed not to break up their coalition, the problem first amounts to a bargaining problem between 1 and $\{2, 3\}$, in which it is not obvious that 1 should get less than b, and after that is settled then 2 and 3 have a bargaining problem to divide the spoils. In practice, the latter bargain would probably be carried out first for all possible spoils, and then the former.

Although it is probably not reasonable to suppose that an arbitration scheme would be employed in this context, it is nonetheless amusing to see the results one obtains using Nash's scheme (section 6.5). Consider first the division of the spoils d between players 2 and 3. The status quo point for this bargain is $(0, c - b)$, since 2 can certainly get 0, and by threatening to form a coalition with 1 player 3 can keep 2 down to 0. Similarly, 3 can get at least $c - b$ in a coalition with 1, and by offering to buy at the price b player 2 can prevent 3 from getting more than $c - b$. The Nash solution to this bargain yields a return $(d - c + b)/2$ to player 2 and $(d + c - b)/2$ to player 3. In the bargain between player 1 and the coalition $\{2, 3\}$ the amount to be divided is c and the status quo point is clearly $(a, 0)$, so according to the Nash scheme player 1 receives $(c + a)/2$ and the coalition $(c - a)/2$. Substituting $d = (c - a)/2$, we find that player 2 gets $(2b - a - c)/4$ and player 3 gets $(3c - a - 2b)/4$.

Three points are worth noting: First, for such an arbitration to make sense it is necessary that $c + a \leqslant 2b$; otherwise player 2 receives less than 0. Second, the nearer b is to c, the more equal are the payments to 2 and to 3, as seems reasonable. Third, from the condition $c + a \leqslant 2b$ it follows that the payment to player 1 lies in the interval between a and b, and so this resolution of the conflict is in one of the solutions.

In summary, this example exhibits quite clearly one major fault of the ψ-stability notion, namely, that it does not grant the players any foresight, and so when there are no restrictions on communication it renders inadmissible the coalition structure $(\{1\}, \{2, 3\})$. Having a player react to immediate offers of gain without considering the long-range effects of the realignment sometimes yields conclusions at variance with one's intuition.

10.4 NON-TRANSFERABLE UTILITIES

The assumption in this and the preceeding two chapters that there exists a transferable utility in which side payments are effected is exceedingly restrictive—for many purposes it renders n-person theory next to useless.   Ideally, one wants a rich theory which abandons the transferability assumption while permitting side payments of the physical commodities resulting from a play of the game.   As we shall see, it is perfectly possible to extend such definitions of equilibrium behavior as solutions and ψ-stability to cover this more general case, but the theories are not at the moment rich; the resulting complexity is so great that nothing is known of their properties.

Abandoning transferability means that we shall no longer assume that there is a set of utility scales for the players and an infinitely divisible and desirable commodity such that transfers of this commodity result in increments and decrements of utility which sum to zero.   The transfer of physical goods will in no way be restricted except by their inherent indivisibility.   To be sure, such transfers will always cause transfers of utility, but we shall not assume that the utility sum is conserved for every side payment transaction.   Of course, once the transferability assumption is dropped, the characteristic function form of an n-person game no longer makes any sense, for a coalition will not have a unique joint utility for its commodity outcome—the joint utility will depend upon the distribution of goods among the players.   The task is to find an appropriate substitute for the characteristic function.

Instead of having a number attached to each coalition S, we shall suppose that there is a known lottery whose prizes are various bundles of goods, services, obligations, etc., which accrue to S as a whole.   To simplify the discussion, suppose the lottery is degenerate in the sense that there is but one prize.   Conceptually, there is no loss of generality in this restriction and the general case would be far more cumbersome to discuss. Let $C(S)$ denote the commodity bundle accruing to the coalition S and let $\mathfrak{I}(S)$ denote the set of all feasible physical distributions of $C(S)$ over the members of S.   If T represents one possible trade in $\mathfrak{I}(S)$ and if i is a member of S, then let $u_i(T)$ denote the utility of T for player i, where i chooses and fixes his unit and zero of measurement independently of the choices made by the other players.

A coalition S is defined to be *effective* for an n-tuple of utilities $\mathbf{x} = (x_1, x_2, \cdots, x_n)$ if there exists at least one T in $\mathfrak{I}(S)$ such that

$$u_i(T) \leqslant x_i, \qquad \text{for all } i \text{ in } S.$$

The n-tuple $\mathbf{y}$ is said to *dominate* the n-tuple $\mathbf{x}$ if there exists a (nonempty) coalition S such that:

(i) S is effective for $\mathbf{y}$,

and

(ii) $y_i > x_i$, for all i in S.

This binary relation of dominance on the set of ordered pairs of n-tuples allows us to extend the von Neumann and Morgenstern definition of a solution to the case of non-transferable utilities. Shapley and Shubik [1953], in an abstract in *Econometrica*, present a similar argument to show that transferability is not essential to n-person theory. They feel, as do we, that the conceptual validity of game theoretic applications does not depend upon the existence of a transferable utility.

The modification of ψ-stability is just as easy: A pair $[\mathbf{x}, \tau]$ is ψ-*stable* if and only if:

(i)   There do not exist S in $\psi(\tau)$ and T in $\mathfrak{Z}(S)$ such that $x_i < u_i(T)$ for all i in S,

and

(ii) $x_i \geqslant u_i[C(\{i\})]$, and the equality holds only if $\{i\}$ is a coalition of τ.

As we said, next to nothing is known about these definitions. Presumably, their properties depend upon particular assumptions one might make about $C(S)$, for it is clear that the mechanics of effecting side payments of physical commodities can be very complicated. For example, suppose that the joint payoff to the coalition $S = \{1, 2\}$ is $C(S) = \{A, B, C\}$, where A, B, and C are non-homogeneous non-divisible goods such as a house, a painting, and a car. Suppose that, if 3 were to join S to form $S' = \{1, 2, 3\}$, $C(S') = \{D, E\}$, where again D and E are non-homogeneous and non-divisible goods. The monetary equivalents for the commodities A, B, C, D, and E may differ widely from individual to individual, and it is not at all clear at this point how a coalition should decide whether it is profitable to add another player nor how it should go about dividing up the joint return. This is true even if they agree in advance to the principle of "equal" shares. In Chapter 14 we shall discuss some mechanisms for fair division in groups which conceivably may be relevant to such a theory, but no direct attack has been made using that or any other approach.

10.5 SUMMARY

There were two major points of contrast between this chapter and the preceeding one. First, the outcome of a game in characteristic function form was not taken to be simply an imputation, or even a set of them, but an imputation together with a coalition structure. Second, equilibrium behavior was assumed to arise not only from the strategic possibilities inherent in the game but also from communication limitations, i.e., a sort of social "friction." We took this to mean an idealized boundary condition of the type described in section 7.6, and the equilibrium theory presented amounted to requiring that the restrictions of the core shall be met by just the admissible coalition changes.   Put another way, a pair $[\mathbf{x}, \tau]$, where $\mathbf{x}$ is an imputation and τ a coalition structure, is unstable if one of the admissible changes is profitable.   Formally, in a game with characteristic function v and boundary condition ψ, the pair $[\mathbf{x}, \tau]$ is called ψ-stable if:

(i) $v(S) \leqslant \sum_{i \text{ in } S} x_i$, for every S in $\psi(\tau)$,

and

(ii) $x_i = v(\{i\})$ implies $\{i\}$ is one of the coalitions of τ.

Some properties of the definition were given for a restricted family of functions ψ and several broad classes of games, including the four-person constant-sum, simple, quota, and symmetric games.   These results seem to have the virtues of being both non-obvious and compact.

A number of criticisms were examined, of which three seem serious. First, the function ψ is not generally explicit in social situations, and no theory has been offered as to how it can be determined.   Nonetheless, as we shall see in Chapter 12, it is possible to make plausible choices for some situations.   Second, ψ-stable pairs are not generally unique, so, as in solution theory, the problem remains how to select just one.   We had no better answer here than in the preceeding chapter.   Third, the players are assumed to have far too limited views of the future.   Specifically, a pair $[\mathbf{x}, \tau]$ is unstable whenever there is an admissible and immediately profitable change, no matter what that change may precipitate.   Judging by our analysis of the three-person market, this fault is serious.

Clearly, the utility assumptions underlying characteristic function theory are highly idealized and cannot often be expected to be met in life; hence, one wonders whether the theories developed and the concepts introduced have any applicability to more realistic situations.   In the final section we outlined how it is possible to extend both the idea of a solution and of a ψ-stable pair to the case where goods are transferable, but where the utilities do not meet the restrictive assumptions previously made.   Thus, characteristic function theory, although idealized, does not appear to be a totally special case; this is important since the more general theory has not been worked on at all.

chapter 11

REASONABLE OUTCOMES

AND VALUE

11.1 REASONABLE OUTCOMES: THE CLASS B

Another author who has considered possible limitation on the outcomes of a game is Milnor [1952]. He has suggested three different systems of "reasonable" conditions, each of which isolates a subset of the set $\bar{E}$ of outcomes, i.e., of the set of n-tuples $\mathbf{x}$ such that

$$\sum_{i \text{ in } I_n} x_i \leqslant v(I_n).$$

In doing so, he has taken " · · · the point of view that it is better to have the set too large rather than too small. Thus it is not asserted that all points within one of our sets are plausible as outcomes; but only that points outside these sets are implausible." [1952, p. 2.]

The status of these three notions, the first of which will be described in this section, tends to be a bit obscure because very little work has been done on them and so only a few of their mathematical properties are known. In the literature their only use has been in the analysis of an experiment which we shall describe in section 12.3. We may, therefore, be charged with devoting too much space—three sections—to their discussion; however, we feel that each one has some conceptual interest and that their relationships to other concepts should be explored.

Our pattern of presentation will be to devote a section to each of Milnor's concepts, giving his defense of it, some criticisms of it, some of its mathematical properties, variations on the concept, its relation to other concepts, and its analysis of the market example we have previously studied (see section 9.4). Because some of the arguments we shall use require consideration of potential changes of coalition alignments, these sections may tend to seem a little involved; we can only ask the reader to bear with us.

The first set, B, consists of all those outcomes **x** *of* $\bar{E}$ *such that for every player* i, $x_i \leqslant b(i)$, *where* $b(i)$ *is defined to be the largest incremental contribution player* i *makes to any coalition, i.e.,*

$$b(i) = \max_S [v(S) - v(S - \{i\})].$$

The argument offered by Milnor in support of this condition is extremely simple: "In any play of the game, player i will wind up in some coalition S. The players $S - \{i\}$ would be foolish to keep i in their coalition if he tries to get so much that they could do better without him." [1952, p. 3.] To us, this supporting argument seems quite weak and perhaps less reasonable than the formal definition of B itself, for there exist games having the following property: There are non-overlapping coalitions S and T, and players i and j in S such that i is not tempted to move out of S provided that j stays in S, but if j moves to T then i can profit by moving from $S - \{j\}$ to $T \cup \{j\}$. In such a case, if i is important to S, it may behoove the coalition to pay j more than his incremental contribution in order to keep both i and j.[1]

From a normative point of view, we are inclined to agree with Milnor that it seems unreasonable to pay any player more than his maximal incremental contribution to *any* coalition, for that seems to be the strongest

---

[1] A specific example of the situation described is any simple game (see section 10.2 for definition) where S contains at least three players, including i and j, T is the complement of S, and

(i) $S, S - \{j\}$, and $T \cup \{i, j\}$ are winning,

and

(ii) $S - \{i\}$, T, $T \cup \{i\}$, and $T \cup \{j\}$ are losing.

It seems reasonable for S to pay j more than $v(S) - v(S - \{j\}) = 1 - 1 = 0$ in order to keep both i and j in the coalition, for, should j bolt to T, then $T \cup \{j\}$ might attempt to attract i to form the winning coalition $T \cup \{i, j\}$. Were they successful, the remaining players of S, $S - \{i, j\}$, would lose everything.

threat that he can employ against a particular coalition. Descriptively, however, this may not be correct. Consider, for example, that S has formed, where S is one of the coalitions in which player i makes his largest incremental contribution, i.e., $v(S) - v(S - \{i\}) = b(i)$. Player i might attempt to demand more than $b(i)$ on the grounds that if he does not get more he will defect from S, that this in turn will cause several other players to defect from S, and that in the ensuing chaos and realignment into coalitions the members of $S - \{i\}$ would probably be worse off than if they had paid him. Of course, the other members of S might counter with the equally plausible argument that i too will be worse off, since no other coalition will allow him more than $b(i)$. At this point arguments involving interpersonal comparisons of utilities could enter. It is difficult to see how to give this formal meaning. However, from a normative position—e.g., were we asked to arbitrate such a game—we feel that no more than $b(i)$ should be paid to player i. Moreover, judging by the data reported in section 12.3, we would suspect that "most" people in a characteristic function game would agree to limit imputations to the set B.

The following properties of the set B are known: For the three-person constant-sum game, B contains the set I of imputations. For four-person constant-sum games, B does not contain all of I, but, in at least one of these games B does include a sizeable portion of I. In general, it can be shown that B includes both the Shapley value (to be discussed in section 11.4) and all the von Neumann-Morgenstern solutions.[2] It is not difficult to show that the imputations of the k-stable pairs of any simple game and of any four-person constant-sum game are in B, but this is not generally the case.[3] The reason for the incompatibility of these two concepts is that Milnor considers all potential coalition structures and allows only one player defections, whereas ψ-stability considers only restricted changes from the coalition structure τ in the pair. (Incidentally, from the fact that the imputation of a ψ-stable pair need not be in B, coupled with Milnor's result that every imputation of a von Neumann-Morgenstern solution must be in B, it follows immediately that the imputations of stability theory are not necessarily included in any of the solutions.)

---

[2] Gillies [1953 b], as well as Milnor, has obtained this result.

[3] An example is the game with characteristic function

$$v(S) = \begin{cases} 0, & \text{if the number of players in } S \text{ is not more than 2,} \\ |S|/n, & \text{if the number of players in } S, |S|, \text{ is more than 2.} \end{cases}$$

The pair $[(0, \cdots, 0, 1), (\{1\}, \{2\}, \cdots, \{n\})]$ is 1-stable since $v(\{i, j\}) = 0$. It is easily seen that for this game $b(i) = 3/n$, so for $n \geqslant 4$,

$$x_n = 1 > 3/n = b(i).$$

Thus, the imputation $(0, 0, \cdots, 0, 1)$, which is in a 1-stable pair, is not in B.

▶Let us apply this concept to the market example described in section 9.4. It is not difficult to show:

$$b(1) = c, \qquad b(2) = b - a, \qquad \text{and} \qquad b(3) = c - a.$$

As can be seen from our previous discussion of this example, these are not strong limitations. As a special case of the result mentioned above, B includes all imputations in von Neumann-Morgenstern solutions; it also includes all the imputations in any of the ψ-stable pairs. ◀

11.2 REASONABLE OUTCOMES: THE CLASS L

Milnor's second class of reasonable outcomes is, in some ways, conceptually allied to ψ-stability, and so to the core. If we make the following suppositions about coalition formation, the condition arises naturally:

i. The bargaining in a game leads to the formation of a coalition which is opposed by its complement.

ii. There are no limitations on preplay communication, so any coalition deemed profitable may form.

iii. In order for a coalition S to form, it must distribute its total payoff in such a manner that each subset S' of S is given at least $v(S')$.

If $\mathbf{x}$ is any imputation and S a coalition, let $x(S)$ denote the sum of payments to the players in S, i.e., $x(S) = \sum_{i \text{ in } S} x_i$. Suppose that $\mathbf{x}$ is an imputation which arises and is in equilibrium when the players are arranged according to the coalition structure $\tau = (T, -T)$, and suppose that (iii) is satisfied. Then, for any set of players S, condition (iii) implies that

$$x(S') \geqslant v(S'),$$

where $S' = S \cap T$ (the part common to S and T), and

$$x(S - S') \geqslant v(S - S').$$

Adding these two inequalities, we get as a necessary condition

$$x(S) \geqslant v(S') + v(S - S').$$

These coalitions S and S' depend upon the particular coalition structure τ, whereas condition (ii) implies that any such structure may form as needed; thus, we wish to suppress any particular reference to τ. So as a weak lower bound on the imputation $\mathbf{x}$ it is reasonable to take the minimum of the right-hand expression over all possible coalitions T, which amounts to finding the minimum over all subsets S' of the set S:

$$x(S) \geqslant \min_{\substack{S' \\ S' \text{ a subset of } S}} [v(S') + v(S - S')].$$

The expression on the right is denoted by $l(S)$ and the set of imputations satisfying this inequality for all S is called the set L. Summarizing: L consists of all those outcomes $\mathbf{x}$ of $\bar{E}$ such that, for each S,

$$x(S) \geqslant l(s) = \min_{\substack{S' \\ S' \text{ a subset of } S}} [v(S') + v(S - S')]. \cdot$$

▶Given the above motivation for the class L, the following more refined bound may be suggested. If both the coalition structure $(T, -T)$ and the outcome $\mathbf{x}$ result, then, by the argument used for ψ-stability, it is untenable that either

$$x(T) < v(T) \qquad \text{or} \qquad x(-T) < v(-T).$$

Thus, we should restrict our attention to those cases where the opposite inequalities hold, i.e., for each S, $\mathbf{x}$ should satisfy

$$x(S) \geqslant \min \{v[S \cap T] + v[S \cap (-T)]\},$$

where the minimum runs over all those coalitions T such that

$$x(T) \geqslant v(T) \qquad \text{and} \qquad x(-T) \geqslant v(-T). \qquad \blacktriangleleft$$

The following results concerning L are known. For the three- and four-person constant-sum games, L is exactly the intersection of the set B with the set I of imputations. This cannot be generally true, however, for we know that the intersection of B and I includes the von Neumann-Morgenstern solutions and the Shapley value, and an example can be given of a game with a solution not wholly in L and with its value not in L. It is not known if L is always non-empty, but at least for many classes of games it is not the empty set. Indeed, for many games the condition is far too weak to be of much interest (see the market example below and the experiment in section 12.3).

▶For the market of two buyers and one seller, a simple computation shows that

$$l(\{1\}) = a, \qquad l(\{2\}) = l(\{3\}) = 0, \qquad l(\{1, 2\}) = a, \qquad l(\{1, 3\}) = a,$$
$$l(\{2, 3\}) = 0, \qquad \text{and} \qquad l(\{1, 2, 3\}) = a.$$

Thus, the set of outcomes L consists of all $\mathbf{x}$ such that $x_1 \geqslant a$, $x_2 \geqslant 0$, and $x_3 \geqslant 0$, which means, for example, that all imputations are included in L. In this case, then, the condition imposes no real limitations. $\qquad \blacktriangleleft$

There are possible variants on the class L which stem from the same or weaker assumptions. Let us drop the first assumption that a coalition is invariably opposed by its complement, but let us retain (ii) and (iii). If the coalition T forms we may suppose that its members will decide how to share the spoils accruing to it, provided it remains intact. Such decisions, however, will not generally be disclosed to the other players in the game. If a subset T' of $-T$ enters into negotiation with a subset S' of T to form a new coalition, it behooves the members of S' to mislead the others as to their actual joint return in T as a means of trying to get more from the new coalition. It thus can happen that $x(T'\cup S') < v(T'\cup S')$, where $\mathbf{x}$ is an outcome (not an imputation), without the participants knowing it. Imperfect information as to the imputation under consideration can, therefore, prevent certain profitable changes from occurring even though there is completely free preplay communication. On the other hand, if a subset of a coalition in a coalition structure τ is to receive less than it can command in the most adverse situation, it will know it and either demand more from its "parent" coalition or defect. Thus, it is reasonable to require as a necessary condition for a pair $[\mathbf{x}, \tau]$, where $\mathbf{x}$ is an outcome, to be in equilibrium that $x(S) \geqslant v(S)$, for every S which is a subset of a coalition T in τ. This is simply a case of ψ-stability theory where $\psi(\tau)$ consists of all subsets of the coalitions in τ. Note that when we introduced the idea of the boundary condition ψ we described it in terms of lack of communication, whereas here we have assumed completely free (but not quite honest) communication.

We may go on to add a second condition which is in a somewhat different spirit. Suppose T_i and T_j are two coalitions in the structure τ, and suppose $v(T_i \cup T_j) > v(T_i) + v(T_j)$. It should then be clear to the members of $T_i \cup T_j$ that they can come to a mutually profitable arrangement, and so the coalition structure will be disrupted. Thus, we conclude that $[\mathbf{x}, \tau]$ will be in equilibrium only if:

(i)   For every subset S of a coalition in τ, $x(S) \geqslant v(S)$,

and

(ii)   For every subcollection of coalitions in τ, say $T_{i_1}, T_{i_2}, \cdots, T_{i_s}$,
$$v(T_{i_1} \cup T_{i_2} \cup \cdots \cup T_{i_s}) = v(T_{i_1}) + v(T_{i_2}) + \cdots + v(T_{i_s}).$$

If τ consists of but the single coalition of all players, then $[\mathbf{x}, \tau]$ is in equilibrium if and only if $\mathbf{x}$ is in the core.

This definition is closely related to, but by no means identical to, the ψ-stability notion when $\psi(\tau)$ consists of all subsets of coalitions in τ and the union of any set of coalitions in τ. A boundary condition of this general sort is used in section 12.3 to analyze some experimental data.

11.3 REASONABLE OUTCOMES: THE CLASS D

Milnor's final concept of a reasonable outcome is also somewhat similar in spirit to the ψ-stability definition. If we denote by δ a possible total payment to a coalition S, then the question to be raised is the conditions under which it is reasonable for S to demand δ. We attack this by examining when δ is unreasonable. If the complement of S, $-S$, can *effectively* keep S from receiving δ, then it seems reasonable to say that δ is unreasonable. This $-S$ can do, it is argued, if there exists an outcome **x** with these two properties:

(i)   It is feasible for $-S$ in the sense that it can get its part of **x**,

and

(ii) $x(S) < \delta$ and, so long as S demands as much as δ, S is unable to disrupt the coalition $-S$ by causing defections to occur.

The outcome **x** is feasible for $-S$ if

(i') $x(-S) \leqslant v(-S)$.

In a similar way, let us try to give formal meaning to the second property. Suppose that a subset T of $-S$ were to defect to S, then, if S demands and gets an amount δ as its share of the total income, there remains an amount $v(S \cup T) - \delta$ to be distributed among the players in T. To make the defection attractive to T, it is necessary that the amount that they can receive as a result of the defection exceed the amount they were assured of by the outcome **x**. Thus, S will be unable to disrupt $-S$, and demand δ for itself at the same time, if

(ii') $x(T) > v(S \cup T) - \delta$, for every subset T of $-S$.

By setting $T = -S$ in (ii') and noting that $v(I_n) \geqslant x(I_n)$ we obtain

$$\delta > v(I_n) - x(-S) \geqslant x(I_n) - x(-S) = x(S).$$

So the following formal definition is set up: a payment δ for a coalition S is an *unreasonable demand* if there is an outcome **x** such that conditions (i') and (ii') are both met. We observe that a necessary (but not sufficient) condition for δ to be unreasonable is that $\delta > v(S)$; this follows from (ii') when T is the empty set.

Conversely, δ is called *reasonable* if it is not unreasonable, i.e., if and only if, *whenever* **x** *is feasible for* $-S$ $[x(-S) \leqslant v(-S)]$, S *can lure a subset* T *to defect from* $-S$ [by giving T more than $x(T)$] *and still have at least* δ *left for itself* [i.e., $v(S \cup T) - x(T) > \delta$].

The set D is defined to consist of all outcomes such that the sum of payments to *each* subset of players is reasonable.

▶A more formal definition of D can be given. Consider any outcome **x**, then, if S is to get some set T to defect from $-S$, it cannot expect to get any more than $v(S \cup T) - x(T)$. Presumably, S would try to attract that T which allows it the maximum return:

$$\max_{\substack{T \\ T \text{ a subset of } -S}} [v(S \cup T) - x(T)].$$

If we now look at this from $-S$'s point of view, $\mathbf{x}$ will be so chosen that it both is feasible and minimizes the above quantity. Thus δ is unreasonable if and only if

$$\delta > d(S) = \min_{\substack{\mathbf{x} \\ x(-S) = v(-S)}} \left\{ \max_{\substack{T \\ T \text{ a subset of } -S}} [v(S \cup T) - x(T)] \right\}.$$

So the set D of reasonable outcomes consists of those outcomes $\mathbf{x}$ such that for each subset S, $x(S) \leqslant d(S)$. In other words, $d(S)$ is the most that S can enforce if it considers joining forces with various subsets of $-S$, assuming that $-S$ does its best to hold S down. ◄

Relatively little is known about the set D, but an example can be given (see the market example below) where neither the Shapley value nor a von Neumann-Morgenstern solution are included in D.   On the other hand, for the three-person constant-sum game, the intersection of D and of E (see section 9.7) is very closely related to the symmetric solution F.   It is the set of payoffs spanned by the three imputations $(\frac{1}{2}, \frac{1}{2}, 0)$, $(\frac{1}{2}, 0, \frac{1}{2})$, and $(0, \frac{1}{2}, \frac{1}{2})$ of F—i.e., the set of imputations of the form

$$x_1 \left(\frac{1}{2}, \frac{1}{2}, 0 \right) + x_2 \left(\frac{1}{2}, 0, \frac{1}{2} \right) + x_3 \left(0, \frac{1}{2}, \frac{1}{2} \right) = \left(\frac{x_1 + x_2}{2}, \frac{x_1 + x_3}{2}, \frac{x_2 + x_3}{2} \right),$$

where x_1, x_2, x_3 are non-negative and sum to 1.

▶ For the market example, $d(\{1\}) = c$, $d(\{2\}) = 0$, $d(\{3\}) = c - b$, $d(\{1, 2\}) = c$, $d(\{1, 3\}) = c$, and $d(\{2, 3\}) = c - a$. Thus, for $\mathbf{x}$ to be in D it is necessary that $x_2 = 0$, so D neither includes the von Neumann-Morgenstern solutions nor all the imputations from the ψ-stable pairs.   It does, however, include the core, and, if we restrict our attention to *imputations* in D, that subset of D is identical to the core. ◄

One can raise conceptual objections to the class D, and the simplest way is to consider an example.   Suppose we consider the three-person game with characteristic function (in 0, 1 normalization):

$$v(\{1, 2\}) = \frac{1}{2}, \qquad v(\{1, 3\}) = 1, \qquad v(\{2, 3\}) = 0.$$

(This is the market example with $a = 0$, $b = \frac{1}{2}$, $c = 1$.)   Consider the imputation $\mathbf{y} = (\frac{1}{16}, \frac{1}{16}, \frac{14}{16})$.   Player 2's return of $\frac{1}{16}$ is an unreasonable demand because $\{1, 3\}$ can enforce $(\frac{1}{2}, 0, \frac{1}{2})$ and 2 cannot lure either 1 or 3 into a coalition.   But, if $\mathbf{y}$ is the initial point of the argument, *why should player 3 be instrumental in reducing 2 from $\frac{1}{16}$ to 0 at the expense of reducing himself from $\frac{14}{16}$ to $\frac{1}{2}$?*   This suggests that, at least in this case, the discussion leading to the notion of a "reasonable demand" collapses, unless it can be argued that $\mathbf{y}$ is not a feasible initial point to begin a hypothetical preplay process leading to equilibrium.   To forestall this escape, let us establish that it is a plausible starting point.   Clearly, the coalition $\{2, 3\}$ can form initially and, as a unit, bargain with player 1.   As we know from Chapter 6, anything can happen in the bargaining situation, and in particular 1 might get $\frac{1}{16}$ and $\{2, 3\}$ might get $1\frac{5}{16}$. Again, 2 and 3 must bargain for the quantity $1\frac{5}{16}$. Player 3 can insist on giving 2 only $\frac{1}{16}$ and if 2 objects and combines with 1, 3 can easily lure 1 away and reduce 2's return to zero. Of course, player 2 can also threaten, and what will happen depends upon psychological variables not included in the model. This is not our problem here; all we wish to do is to point out that the imputation **y** is not a ludicrous starting point. In summary, one may say that the stability of the coalition $\{2, 3\}$ with the imputation **y** arises from the fact that the demands of both the players are unreasonable, and any attempt by one to hurt the other will in fact hurt both.

The following summary comments on Milnor's three subsets of outcomes seem to be justified. As we shall see in section 12.3, in one experiment the outcomes did for the most part lie in the sets B and L but not in D. Furthermore, conceptually it seems reasonable that they should lie in both B and L, for the conditions defining B and L are quite weak and they do not attempt to take into account the interlocking threat relations among the several coalitions. The set D is more difficult to comment on, for its rationalization is somewhat complicated. We like the spirit of the idea in that it brings into play the threat power possessed by subsets which will defect from a coalition if there is an assured profit to that action, but it does not follow the argument through to see what reactions and counterreactions will probably ensue. The above example suggests that it fails to capture certain salient aspects of the threat situation. In addition, when there are a large number of players the following doubts about D as a descriptive idea seem relevant: its defense is based upon the supposition that a coalition is always opposed by its complementary coalition; the condition is required to hold for all subsets which, we have previously argued, may not be reasonable for a descriptive theory or for certain types of normative theories; and there is absolutely no indication of the coalition structure which one should find associated with an outcome lying in D.

11.4 VALUE

In contrast to the two preceding chapters and the first sections of this chapter, we shall no longer be concerned with possible equilibrium outcomes for games, but rather with the notion of an *a priori* evaluation of the entire game by each of its players. Shapley writes,

In attempting to apply the theory [of games] to any field, one would normally expect to be permitted to include, in the class of "prospects," the prospect of having to play a game. The possibility of evaluating games is, therefore, of critical importance. So long as the theory is unable to assign values to the games typically found in application, only relatively simple situations—where games do not depend on other games—will be susceptible to analysis and solution. [1953 b, p. 307.]

For two-person zero-sum games we have seen (Chapter 4) that the maximin value yields a sensible and unique evaluation of the game for each of the players. But certainly the value $v(\{i\})$ arrived at using two-person theory is not a suitable evaluation of the worth of an n-person game to player i, since the whole point of joining coalitions in essential games is to do better than $v(\{i\})$.

Suppose that there were a perfectly acceptable equilibrium theory for n-person games and that from it one could show, for a particular game, that the imputations $\mathbf{x}^{(1)}, \mathbf{x}^{(2)}, \cdots, \mathbf{x}^{(m)}$ were in equilibrium. Then player i could expect to get one of the amounts $x_i^{(1)}, x_i^{(2)}, \cdots, x_i^{(m)}$, depending upon which equilibrium imputation obtained. In order, therefore, to find his *a priori* expectation it is necessary to know the probabilities of occurrence of each of these various equilibrium states, and that presumably requires a dynamic theory. This, then, is a blind alley, and some other approach must be found. Not all is wasted, however, for viewing the problem of *a priori* evaluations in this way makes at least one thing clear. There is no reason to expect the evaluation to be one of the equilibrium outcomes. Suppose the imputation $\mathbf{x}^{(j)}$ occurs with probability p_j, then the *a priori* expected return to player i is

$$y_i = p_1 x_i^{(1)} + p_2 x_i^{(2)} + \cdots + p_m x_i^{(m)},$$

and, in general, $\mathbf{y} = (y_1, y_2, \cdots, y_n)$ will be different from any of the $\mathbf{x}^{(j)}$'s. Thus, however we arrive at an evaluation, there is no reason to expect or to desire that it fall within one of the classes of outcomes that have been isolated so far: the core, a solution, those of ψ-stability theory, or the sets B, L, or D.

Since an approach based upon equilibrium imputations appears likely to fail, we must backtrack to the notion of the characteristic function which underlies the present equilibrium concepts and find an evaluation which depends upon the set of values of $v(T)$ for all coalitions T. Just what function of the characteristic function would be reasonable to select is not, on the face of it, obvious, and certainly any arbitrary definition would be questioned and countered by other suggestions. An alternative procedure—one of extreme power and persuasiveness, and one which is too little known in the social sciences—is to specify the function by certain properties one feels it must have. These should be chosen so that all, or most, readers would agree that they are intuitively acceptable in a function whose purpose is an *a priori* evaluation of the game. Once they are set down formally, the mathematical task is to determine whether there are any functions meeting the given requirements, and, if so, whether there is only one or more than one. It is also useful, but sometimes difficult, to give a formula for the function(s), or a systematic procedure whereby it can be determined for specific cases. Such was the approach taken by Shapley [1953 *b*]; he listed three *apparently* weak conditions, and then, surprisingly, he was able to show that these uniquely determine an evaluation function—that there can be only one function satisfying the three conditions, and that there is one. He has called the function so determined for each player the *value* of the game for that player.

We start out first with the idea that a player's evaluation of a game is a real number, so we may symbolize it as $\phi_i(v)$, where i denotes the player and v the characteristic function of the game. Since the numbering of the players is arbitrary, we may always renumber them in any way we like by a permutation of the original system. This will cause the characteristic function to look different even though it represents the same underlying game, but, since these are only notational differences, players who correspond under the relabeling should have the same value. So Shapley's first condition is:

i. Value shall be a property of the abstract game, i.e., if the players are permuted, then the value to player i in the original game shall be the same as the value to the permutation of player i in the permuted game. Stated a bit differently, the value to a player should not depend upon the labeling used to abstract the game.

If we consider a fixed game with characteristic function v, then, although the n-tuple of values $[\phi_1(v), \phi_2(v), \cdots, \phi_n(v)]$ may not occur as an equilibrium outcome, it would be strange if it could not be an outcome. For example, it would be unacceptable if the sum of the individual values were to total to more than could possibly be obtained from the game situation. Surely then, one of the players would be overevaluating the worth of the game to himself. Consequently, one is tempted to impose the requirement that the n-tuple of values be an imputation; it is actually sufficient to require a little less.

ii. The individual values of the game shall form an additive partition of the value of the whole game, i.e.,

$$\phi_1(v) + \phi_2(v) + \cdots + \phi_n(v) = v(I_n).$$

On the other hand, if value to a player is to be interpreted as an *a priori* expectation, it seems that the equality sign in condition (ii) is *much too strong*—indeed, it represents an n-fold combination of wishful thinking.

Next, consider a player i who is participating in two different games with characteristic functions v and w, say. He has an evaluation for each of these games: $\phi_i(v)$ and $\phi_i(w)$. Now if we could think of these two games as being a single game, let us call it u, then he would have an evaluation $\phi_i(u)$, but, since we assume that u is but a renaming of the two given games, we should have

$$\phi_i(u) = \phi_i(v) + \phi_i(w).$$

The next thing to consider is whether we can treat the two games as a single one. Let us suppose that v is a game over the set of players R and that w is a game over the set S. Although in our preceeding discussion we assumed that R and S overlapped, at least to the extent of player i, we shall now be more general and suppose that they may or may not overlap. It is a trivial matter to extend both v and w to the set of all players, $R \cup S$. If T is a subset of $R \cup S$, we define

$$v(T) = v(R \cap T) \qquad \text{and} \qquad w(T) = w(S \cap T).$$

This is to say, in the game v, a coalition T has exactly the strength given by those members of T who are actually in the game, i.e., those of T who are also in R; the members from S who are not in R contribute nothing. · Now, the two games are defined over the same set of players. Consider what may be called the sum[4] of these two games, denoted by $u = v + w$, and defined by the condition that, if T is a subset of $R \cup S$,

$$u(T) = v(T) + w(T).$$

It is easy to see that u is a characteristic function, and so it will serve as the single game representing the two given ones. Thus, the third condition can be written as:

iii. If v and w are two games and if $v + w$ is defined as above, then

$$\phi_i(v + w) = \phi_i(v) + \phi_i(w).$$

The last condition is not nearly so innocent as the other two. For, although $v + w$ is a game composed from v and w, we cannot in general expect it to be played as if it were the two separate games. It will have its own structure which will determine a set of equilibrium outcomes which may be very different from those for v and for w. Therefore, one might very well argue that its *a priori* value should not necessarily be the sum of the values of the two component games. This strikes us as a flaw in the concept of value, but we have no alternative to suggest.

If these three conditions are accepted, then Shapley has shown that one need not—dare not—demand more of a value, for they are sufficient to

---

[4] This notion includes, as a special case, the concept of a decomposable game defined in section 10.2.

determine ϕ_i uniquely, and, indeed, one can obtain an explicit formula for it, namely:

$$\phi_i(v) = \sum_{S \text{ a subset of } I_n} \gamma_n(s)[v(S) - v(S - \{i\})],$$

where s is the number of elements in S and $\gamma_n(s) = (s - 1)!(n - s)!/n!$. The symbol $k!$ stands for $k(k - 1) \cdots 3 \cdot 2 \cdot 1$, when k is a positive integer, and $0! = 1$. Let us examine this formula in more detail. It is a summation over all subsets of the set of players, with a typical term consisting of a coefficient—which we shall discuss presently—multiplying $[v(S) - v(S - \{i\})]$. If i is not a member of S, then $S - \{i\} = S$, so the term becomes zero; thus the formula depends only upon those coalitions involving i. It amounts, therefore, to a weighted sum of the incremental additions made by i to all the coalitions of which he is a member. It may be useful to carry out the calculation in a few simple cases. First, consider the general two-person game:

$$\phi_1 = \frac{1!0!}{2!} [v(\{1, 2\}) - v(\{2\})] + \frac{0!1!}{2!} [v(\{1\}) - v(\phi)]$$
$$= \tfrac{1}{2}[v(\{1, 2\}) + v(\{1\}) - v(\{2\})].$$

If the two-person game is zero-sum, i.e., $v(\{1, 2\}) = 0$ and $v(\{2\}) = -v(\{1\})$, then $\phi_1 = v(\{1\})$, which establishes that the Shapley value is in fact a generalization of the minimax value.

For the general three-person game:

$$\phi_1 = \frac{2!0!}{3!} [v(\{1, 2, 3\}) - v(\{2, 3\})] + \frac{1!1!}{3!} [v(\{1, 2\}) - v(\{2\})]$$
$$+ \frac{1!1!}{3!} [v(\{1, 3\}) - v(\{3\})] + \frac{0!2!}{3!} [v(\{1\}) - v(\phi)].$$

If we substitute into this formula the values of the characteristic function of the three-person constant-sum game in 0, 1 normalization we find $\phi_1 = \tfrac{1}{3}$, which, considering the perfect symmetry of that game, is the desired answer.

▶The Shapley value for the market situation previously discussed in sections 9.4, 10.3, 11.1, 11.2, and 11.3 is:

$$\phi_1 = a/3 + b/6 + c/2$$
$$\phi_2 = -a/6 + b/6$$
$$\phi_3 = -a/6 - b/3 + c/2.$$

It is easy to show that, if $a < b < c$, then $\phi_1 > \phi_3 > \phi_2$. This ordering conforms well to one's intuition about the situation, except that, possibly, the previous results would suggest that player 3 is not invariably worse off than 1. Such must not be the case, otherwise we could have deduced only $\phi_1 \geqslant \phi_3$. ◀

To return to the coefficients, any one who has dealt at all with simple probability models will recognize them as very familiar. Indeed, Shapley's

> . . . result can be interpreted by imagining the random formation of a coalition of all the players, starting with a single member and adding one player at a time. Each player is then assigned the advantage accruing to the coalition at the time of his admission. In this process of computing the expected value for an individual player all coalition formations are considered as equally likely. [Kuhn and Tucker, 1953, p. 303.]

The above theorem, which has been stated here (and was first presented) only for characteristic functions, is actually a special case of a similar theorem proved by Shapley [1953 c] to hold in a much more general context.

11.5 VALUE AS AN ARBITRATION SCHEME

In addition to the questions we have raised about the axioms for an *a priori* value, one can also question why we should ever be concerned with such an evaluation of the entire game by each of its players. What operational use will be made of it? So long as this is uncertain, it is difficult to criticize the axioms in a fully convincing manner. What we propose to do in this section is to consider an alternative interpretation of the axioms and to criticize them from that point of view. We shall look on the value in the same spirit as in our previous discussion of arbitration schemes for two-person games—as an arbitration scheme for *n*-person games in characteristic function form. In fairness to Shapley, we must point out he has never given this interpretation, and so the fact that the value possesses difficulties as an arbitration scheme should not be laid at his doorstep.

Suppose that an *n*-person game arises directly in characteristic function form; this assumption permits us to bypass the question whether or not the reduction from normal form has retained the full strategic flavor of the game. It is certainly possible that the dynamics of preplay bargaining and threatening may actually result in an outcome which is not Pareto optimal, i.e., a different and achievable outcome may be preferred by all the players. Suppose that the players are aware of this possibility in advance and wish to avoid it; then they might turn to an impartial outsider to suggest an outcome which is "fair" to each of them in terms of the strategic aspects of the game. In other words, the game will be resolved by an arbiter, whose role was discussed at length in Chapter 6. The question is whether we can accept Shapley's conditions as desiderata for an arbitration scheme.

The first condition merely requires that the arbiter shall be guided by the strategic role of each player and not by his label. The second condition demands that he restrict his attention to Pareto optimal outcomes, i.e., the players should not be able to point out to him outcomes which they would all prefer. The third condition is harder to rationalize. It says, in effect, that, if a game can be decomposed into two games, the value assigned to a player shall be the sum of the values assigned him in each of the component games *treated in isolation.*

Although we took exception to the second condition when value is interpreted as a reasonable *a priori* expectation, there is certainly no objection to it as an arbitration condition. The first seems equally acceptable. It is only the third that seems to give trouble—serious trouble. This we may best illustrate by examples. Suppose v and w are two characteristic functions in 0, 1 normalization on three players, where:

$$v(\{1, 2\}) = v(\{1, 3\}) = 0, \qquad\qquad v(\{2, 3\}) = 1$$
$$w(\{1, 2\}) = 1, \qquad\qquad w(\{1, 3\}) = w(\{2, 3\}) = 0.$$

If $u = v + w$, then the characteristic values of the one-person coalitions are still zero, and

$$u(\{1, 2\}) = u(\{2, 3\}) = 1,$$
$$u(\{1, 3\}) = 0,$$
$$u(\{1, 2, 3\}) = 2.$$

For the games v and w, the imputations $(0, \frac{1}{2}, \frac{1}{2})$ and $(\frac{1}{2}, \frac{1}{2}, 0)$ respectively seem to be reasonable arbitrated outcomes, and they are in fact the Shapley values. Thus, the Shapley value for the game u is $(\frac{1}{2}, 1, \frac{1}{2})$.

Is this a reasonable arbitrated outcome for u? To us it seems questionable. Certainly, if either the game v or w is *treated in isolation,* then the coalition $\{1, 3\}$ makes no sense at all. But in the game u, or what is the same thing if v and w are being *jointly considered,* it does make good sense. It remains true that, as a coalition, they command nothing, but they also can hold 2 to nothing. So a bargain exists between $\{1, 3\}$ and $\{2\}$ with two units for them to share. There are two possible outcomes, each of which seems plausible. Either:

i. Each coalition $\{1, 3\}$ and $\{2\}$ receives an equal share of the total incremental gain, 2 units, and the outcome is the Shapley value, **or**

ii. Each player receives an equal share of the incremental gain of 2 units, giving rise to the arbitrated outcome $(\frac{2}{3}, \frac{2}{3}, \frac{2}{3})$, which is different from the Shapley value.

To anyone accepting case (ii) as the fair arbitration of this game, the Shapley value is discredited for that purpose; thus, we need only give our attention to those who find the reasoning leading to case (i) agreeable.

What we shall do is this: modify the characteristic functions slightly so that the argument which led to case (i) fails to yield the Shapley value, whereas the argument leading to case (ii) does yield the value.

To this end, let v be modified into v' by changing the payment to $\{2, 3\}$ so $v'(\{2, 3\}) = 0$. In this case the Shapley value is $(\frac{1}{3}, \frac{1}{3}, \frac{1}{3})$. If we let $u' = v' + w$, then

$$u'(\{1, 2\}) = 1,$$
$$u'(\{1, 3\}) = u'(\{2, 3\}) = 0,$$
$$u'(\{1, 2, 3\}) = 2.$$

The Shapley value is, of course, $(\frac{5}{6}, \frac{5}{6}, \frac{1}{3})$. In the game u' one would expect the coalition $\{1, 2\}$ to form and to bargain with player 3 about his potential contribution of 1 unit to $\{1, 2\}$. Again, two possible divisions seem plausible:

i'. $\{1, 2\}$ and $\{3\}$ each get one-half of the marginal unit, which gives rise to the outcome $(\frac{3}{4}, \frac{3}{4}, \frac{1}{2})$, which is not the Shapley value, or

ii'. Each player gets an equal share of the marginal unit, giving rise to $(\frac{5}{6}, \frac{5}{6}, \frac{1}{3})$, the Shapley value.

Thus, whichever of these plausible arguments you prefer for arbitration, there is an example where it fails to yield the Shapley value and the alternative argument does arrive at the value. The basic trouble, as we see it, with Shapley's third condition is that it is unreasonable to demand that players involved in two games play each in isolation of the other. Nonetheless, *if* we were called upon to arbitrate an n-person game in characteristic function form, we would use the value for lack of any explicit alternative.

Since we have proposed this interpretation of value as an arbitration scheme, it would be well to close by examining how it relates to arbitration in the two-person case (Chapter 6). Let the characteristic function be denoted

$$v(\{1\}) = v_1, \qquad v(\{2\}) = v_2, \qquad \text{and} \qquad v(\{1, 2\}) = c.$$

In this form, players 1 and 2 are engaged in a bargain for a total of c units where (v_1, v_2) is the status quo point. The Nash bargaining solution for this case is easily shown to be identical to the Shapley value. On the other hand, it will be recalled that for two-person cooperative games in normal form we found fault with the Shapley value. However, the trouble in that case stemmed from the fact that value depends only upon the characteristic function, and in many cases this does not adequately reflect the threat powers of the players in the normalized game (see the example of section 8.5).

chapter 12

APPLICATIONS OF
n-PERSON THEORY[1]

12.1 THE A PRIORI POWER DISTRIBUTIONS OF VOTING SCHEMES

Possibly the most interesting published application of n-person game theory (as of late 1956) to a social science problem is an attempt to estimate the *a priori* power distributions inherent in various legislative voting procedures. In a very readable article, Shapley and Shubik [1954] have suggested that the notion of value discussed in the preceding chapter is suited to this purpose. Indeed, if one accepts the three conditions Shapley stated as necessary for an *a priori* evaluation of a game, it is the only function which is suitable.

Consider the passage of bills at the federal level in the United States. In most cases, there are only two ways a bill can be passed: either by a simple majority in each house of Congress plus the president's signature, or by a two-thirds' majority in each house overriding the presidential veto. All other combinations fail to pass a bill. Let us treat the president, the senators, and the representatives as the players in a simple game, i.e., a game whose characteristic function assumes only the values 0 or 1. A "coalition" is defined to be winning and to have the value 1 if it can pass

---

[1] We have not attempted to summarize the complex of applications to economic theory discussed by Shubik in *Competition, Oligopoly and the Theory of Games* [1957].

a bill in one of these two ways; if not, it is called losing and has the value 0. In effect, this postulates that the power of all coalitions which can pass bills is equal and that the power of all which cannot pass them is equal— more of this later.   It is easy to show that for any reasonable legislative scheme, where by reasonable we mean that two non-overlapping coalitions cannot both be winning, this definition results in a characteristic function.

It is now possible to compute the Shapley value for each of the persons in this simple game using the formula given in section 11.4.   The interpretation of that formula reduces to the following: the probability that a given individual will be pivotal in transforming a losing coalition into a winning one when the final winning coalition is built up by random selections from among the players.   It is true, of course, that in an existing legislative body the formation of coalitions is not random; however, when considering a voting scheme in advance one cannot know what deviations from randomness will occur, so it may be argued that the value gives a suitable *a priori* estimation of relative power positions.

Whether we are willing to accept the value as such an estimate depends largely upon our willingness to accept Shapley's third, and controversial, condition.   For this situation, it says that, if a player participates in two such legislative schemes, his *a priori* evaluation of the two together is simply the sum of the values he would assign to the two schemes independently.   Thus, for example, were we to consider not only Congress as a voting body but also congressional committees as a separate scheme, then Shapley's third condition requires that we assume the value of the game consisting of Congress together with its committees to be the sum of the values of the two component games.   In effect, then, the condition stipulates that, insofar as an *a priori* evaluation is concerned, we must suppose that there is no power interaction between Congress and its committees; this strikes us as unrealistic.

Be that as it may, it continues to be interesting to see what the value yields in specific instances.   Considering Congress, the indices for a single representative, a single senator, and the president are in the proportions 2:9:350.   In terms of the House of Representatives, the Senate, and the presidency the *a priori* power proportions are 5:5:2.   Had Congress not been permitted to override the presidential veto, the power indices would be approximately 1:1:2, with the House having slightly less power than the Senate.   This result is not at all obvious and entails considerable calculation.

"The effect of a revision [of legislative procedure] usually cannot be gauged in advance except in the roughest terms; it can easily happen that the mathematical structure of a voting system conceals a bias in power

distribution unsuspected and unintended by the authors of the revision."
[1954, p. 787.]   Without committing ourselves as to the naivete or intentions of its authors, it is amusing to compute the *a priori* power distribution
of the United Nations Security Council and to speculate on the propa-
ganda value these figures might once have had.   It will be recalled that
the Council consists of eleven members of whom five—the "Big Five"—
have vetoes.   To pass a substantive resolution there must be no vetoes and
seven affirmative votes.   Shapley and Shubik report that 98.7% of the
power lies in the hands of the "Big Five" and only a total of 1.3% resides
with the other six members.   "Individually, the members of the 'Big
Five' enjoy a better than 90 to 1 advantage over the others."   [1954,
p. 791.]

There is little point in reporting the calculations of other special cases;
if the reader is interested in the *a priori* power distribution of a particular
legislative scheme he will find that the computation is straightforward,
though it may be laborious.

It is important to emphasize the nature of the measure used in these calculations; Shapley and Shubik make the point forcefully:

In a multicameral system such as · · · [Congress], it is obviously easier to
defeat a measure than to pass it.   A coalition of senators, sufficiently numerous,
can block passage of any bill.   But they cannot push through a bill of their own
without help from the other chamber.   This suggests that our analysis so far has
been incomplete—that we need an index of "blocking power" to supplement the
index already defined.   To this end, we could set up a formal scheme similar to
the previous one, namely: arrange the individuals in all possible orders and
imagine them casting *negative* votes.   In each arrangement determine the person
whose vote finally defeats the measure and give him credit for the block.   Then
the "blocking power" index for each person would be the relative number of times
he was the "blocker."

Now it is a remarkable fact that the new index is exactly equal to the index of
our original definition.   We can even make a stronger assertion: *any scheme for
imputing power among the members of a committee system either yields the power index defined
above or leads to a logical inconsistency.*   [1954, p. 789.]

The precise formulation of the last statement, which we would regard as
misleading as it stands, is the theorem stated in the preceding section.
Shapley and Shubik's assertion is true if we accept the three conditions as
necessary to a "scheme for imputing power," and not otherwise.

12.2 POWER DISTRIBUTIONS IN AN IDEALIZED LEGISLATURE

An *a priori* evaluation of power distributions based upon a random selection of participants to form a winning coalition may be suitable for discussing proposed legislative schemes, but it will hardly satisfy a political

scientist faced with the analysis of an existing legislature, whose behavior depends upon certain more or less limiting social and political factors, in addition to the formal prescription of winning and losing coalitions.   In Congress, for example, the members of both chambers are divided into two parties, and party affiliations create a basic coalition structure which is reflected, to some degree, in almost every vote.   In other words, certain coalitions have a much greater than chance expectancy of occurring and others a much lower expectancy.   An analysis of congressional power structures must take, at least, this empirical fact into account.

In the remainder of this section we shall describe such an analysis using ψ-stability theory applied to an idealized model of Congress.   This example should not be taken as an attempt at an accurate analysis of congressional power distributions—it is much too simplified for that—but rather as an effort to demonstrate that some of the techniques of game theory may be suited to such a study.

As in the Shapley-Shubik model, Luce and Rogow [1956] suppose that the legislative scheme is described by the characteristic function of a simple game.   The result of passing a bill is certain "power" rewards which are distributed among the members of Congress; in our previous language, such a power distribution is taken to be an imputation.   One idealization occurs here: it is supposed that power is a divisible and transferable commodity, and in some measure it is but certainly not in the neat ways of game theory.   The general problem is to find which power distributions coupled with which coalition structures are stable.   A simpler problem, the one Luce and Rogow attacked, is to choose a particular coalition structure such as the two-party system and ask what power distributions render it stable.   This particular structure is of interest because it has existed for so long.

Obviously, the problem as formulated is vague, since the word stable has not been defined.   If we take it to mean ψ-stable for some function ψ, the heart of the problem is to decide what to take for ψ.   Note that, by restricting their attention to one coalition structure, they need not know ψ except for that one structure!   This makes life very much simpler. They choose ψ to represent limitations on defections from the party structure.   Without going into full details, they divide each party in Congress into two non-overlapping subsets: the potential defectors and the diehards, i.e., those who are unwilling to defect from their party (for the bill under consideration).   In the case that they have examined, the potential defectors may defect to the other party or not, but the formation of a coalition by the defectors from both parties is excluded.   It can be included, if one chooses, at the expense of some more labor in the calculations.   The remaining specification of the functions ψ depends upon whether or not the president can defect from his party to the other or not. Several different cases of the function arise depending upon the sizes of the resulting coalitions: whether they form a two-thirds majority, a simple majority, or a minority (it is assumed, for simplicity, that similar majorities exist in both houses of Congress).

Considering each of these special cases along with the several assumptions one can make about the two-party coalition structure—the majority party has simple majority only or a two-thirds majority, and the president is or is not a member of the majority party—there result a total of 36 different cases. Each case is examined separately, using the definition of ψ-stability, to determine which imputations with the two-party coalition structure are stable. Actually, Luce and Rogow did not use the definition as it is given in section 10.1; rather, they waived the second condition which states that a person is not a member of a non-trivial coalition unless such membership is profitable. Their rationale for this change was that passing a bill is not a one-shot affair but part of an ongoing process, and so coalition membership might well be sustained, at least for a period, even if it does not produce immediate rewards. This difficulty, and its arbitrary resolution, raises the question whether it is often feasible to isolate a game from its more general social context and to suppose that it can be played without regard to these outside factors. In principle there is no problem, for either the game can be enlarged to include these factors or the utility functions of the players can be chosen to take them into account. In practice, however, neither alternative is particularly useful, and so certain *ad hoc* tricks have to be employed.

Note well how oversimplified this model is. It fails to take into account many facts of known importance: the interaction among bills, the disparate returns of power and prestige depending upon which coalition passes the bill, the whole role of the important congressional committees, the possibility of filibuster in the Senate, etc. On the other hand, such limitation on party defection as pressure from constituents, party discipline, pressure from lobbies, etc., are built into the description. To be sure, it would be very difficult to specify exactly what function ψ holds at the time of a particular vote, but this does not prevent us from examining all possible functions of the given general type to determine what general conclusions hold.

The calculation of the stable distributions for a particular function ψ is very easy, and the reader is referred to the Luce-Rogow paper for an example. It is worth noting that for the given assumptions it amounts to little more than a formalization of the ordinary arguments one uses when discussing the location of power. It may be charged that considerable machinery has been employed to find out what is nearly obvious by common sense arguments, and that, therefore, it is all a fraud.   So it is if no more sophisticated studies are undertaken; however, with the formalism illustrated by a simple case, it is comparatively easy to see how to generalize the elementary analysis to more complex cases—to both more complicated characteristic functions representing the rewards accruing to coalitions and to more complex boundary conditions representing the sociopolitical limitations on coalition change.

From the calculations of the 36 special cases, six qualitative conclusions were drawn, which we shall repeat.   Whether or not they accord with reality, they are in a form which is meaningful to a political scientist and they can be evaluated by him in the light of current theory and data. They are:

1. In all cases the arrangement of Congress into two opposed party coalitions is stable provided the power is distributed as indicated.   In very many cases, however, it is necessary to form coalitions other than along party lines in order to produce a winning coalition, i.e., to pass a bill.   In only one case are the limitations so stringent that no working majority can form: this is when the president is of the minority party and will not defect to the majority, the majority has only a simple majority even with the defectors from the minority, and the minority does not have a simple majority even with the defectors from the majority. What is interesting is that in only one case of the 36 can such an impasse result.

2. In all circumstances, the president is weak when the majority party— whether he is a member of it or not—has a two-thirds majority.   If this model has any relation to reality, we must conclude that a president should fear a real congressional landslide for *either* party.

3. The president possesses power (from voting considerations) only when neither party can muster more than a simple majority even with the help of the defectors from the other party.

4. The only circumstances when the minority party is the holder of any power is when the president is in the minority party and he is unwilling to defect to the majority.

5. Under all conditions, if the defectors from [the majority] party added to [the minority] party fail to form a majority, then the diehards of [the majority] party possess power.   The only other case in which they possess power is when the president is a member of [the majority] party, he is unwilling to defect, and [the minority] party plus the defectors from [the majority] party form only a simple majority · · · · .

6. The only case when the [minority] party diehards possess any power is when the president is a member of their party, he is unwilling to defect, [the majority] party has only a simple majority, and [the minority] party plus the defectors from [the majority] party form either a majority or a two-thirds majority · · · · .   [1956, p. 91.]

There can be no question that the model described is too idealized to be of much interest itself; its principal merit is in illustrating how one equilibrium theory of *n*-person games might be used to study Congress. With refinements, which will complicate both the characteristic function and the boundary conditions and so create hundreds—instead of tens—of cases to examine, qualitative conclusions of a similar, but much more subtle, nature should result.

12.3 AN EXPERIMENT

Notably lacking in all of our discussion have been data, or even the mention of data. In part this may be attributed to the realization that game theory is inadequate as a descriptive theory; human beings simply do not have the perception, the memory, or the logical facility assumed by any of the theories. But two other reasons are actually more important. Assuming that we wish to carry out empirical studies of a coalition theory, it is necessary to know the characteristic function, and this would seem to entail knowing the normal form. We have already pointed out the great difficulty of determining the normal form of most existing game situations and, even assuming that known, the extensive calculations required to solve the two-person games on which the characteristic function is based. Yet without the characteristic function, we cannot know what any of the theories predict. A second difficulty exists even if the characteristic function is assumed to be known: what does the principal theory—von Neumann and Morgenstern's solutions—predict? In discussing the outcome of an experiment performed at RAND, Kalisch, Milnor, Nash, and Nering remark,

> It is extremely difficult to tell whether or not the observed results corroborate the von Neumann-Morgenstern theory. This is partly so because it is not quite clear what the theory asserts. According to one interpretation a "solution" represents a stable social structure of the players. In order to test this theory adequately, it would probably be necessary to keep repeating a game, with a fixed set of players, until there seemed to be some stability in the set of outcomes which occurred. One could then see to what extent the outcomes of this final set dominate each other and to what extent other possible imputations are not dominated by them. [1954, p. 313.]

Of these two difficulties, the second seems less important, for even if it is not possible to interpret solution theory there do exist other theories which are easily given empirical meaning. The problem of determining the characteristic function is more profound, and it seems to us that the most important development for empirical verification will be a practical method to calculate the characteristic function of actual situations. Probably the most significant contribution social scientists can make in this area is a feasible method for the approximate determination of characteristic functions. In section 12.4 we shall describe one such proposal, but we might as well admit now that it appears to raise as many problems as it solves.

In the laboratory these problems can be bypassed, at least in part, by presenting the game to the subjects directly in terms of the characteristic function. This is exactly what Kalisch, Milnor, Nash, and Nering did [1952, 1954]. We shall report only the portion of their experiment which was concerned with two four-person constant-sum games. To some of the subjects these games were presented in what amounted to 0, 1 normalization, and to others in an S-equivalent form. The subjects were told what each coalition would receive if it formed, and they were given 10 minutes to form coalitions and to agree upon payments, the agreements being reported to an umpire. He announced the agreements to the group, and if there was no dissension he held the players rigidly to their formal agreements at the end of the bargaining. The authors point out that there were, in addition, numerous informal agreements which were not processed through the umpire but which were kept in good faith.

We feel that the general qualitative impressions of the authors are of sufficient importance to be quoted at length:

> There was a tendency for members of a coalition to split evenly, particularly among the first members of a coalition. Once a nucleus of a coalition had formed, it felt some security and tried to exact a larger share from subsequent members of a coalition. The tendency for an even split among the first members of a coalition appeared to be due, in part, to a feeling that it was more urgent to get a coalition formed than to argue much about the exact terms.
>
> Another feature of the bargaining was a tendency to look upon the coalitions with large positive values as the only ones worth considering, often overlooking the fact that some players could gain in a coalition with a negative value to their mutual benefit · · · .
>
> Coalitions of more than two persons seldom formed except by being built up from smaller coalitions. Further coalition forming was usually also a matter of bargaining between two groups rather than more.
>
> A result of these tendencies was that the coalition most likely to form was the two-person coalition with the largest value, even though this coalition did not always represent the greatest net advantage for the participants; and this coalition usually split evenly. Thus it frequently happened that the player with apparently the second highest initial advantage got the most of the bargaining. The player with the apparently highest initial advantage was most likely to get into a coalition, but he usually did not get the larger share of the proceeds of the coalition.
>
> Initially the players were more inclined to bargain and wait or invite competing offers. This remained true to some extent in those games where the situation did not appear to be symmetric. However, later and in those games which were obviously symmetric, the basic motive seemed to be a desire to avoid being left out of a coalition. Hence there was little bargaining, and the tendency was to try to speak as quickly as possible after the umpire said "go," and to conclude some sort of deal immediately. Even in a game which was strategically equivalent to a symmetric game, the players did not feel so rushed. A possible reason

might be that some players felt they were better off than the others whether or not they got into coalitions, while others felt that they were worse off whether or not they got into coalitions. They seemed to pay little attention to the fact that the net gain of the coalition was the same to all · · · .

Personality differences between the players were everywhere in evidence. The tendency of a player to get into coalitions seemed to have a high correlation with talkativeness. Frequently, when a coalition formed, its most aggressive member took charge of future bargaining for the coalition. In many cases, aggressiveness played a role even in the first formation of a coalition; and who yelled first and loudest after the umpire said "go" made a difference in the outcome.

In the four-person games, it seemed that the geometrical arrangement of the players around the table had no effect on the result; but in the five-person game, and especially in the seven-person game, it became quite important. Thus in the five-person game, two players facing each other across the table were quite likely to form a coalition; and in the seven-person game, all coalitions were between adjacent players or groups of players. In general as the number of players increased, the atmosphere became more confused, more hectic, more competitive, and less pleasant to the subjects. The plays of the seven-person game were simply explosions of coalition formations.

Despite the exhortation contained in the general instructions to instill a completely selfish and competitive attitude in the players, they frequently took a fairly cooperative attitude. Of course, this was quite functional in that it heightened their chances of getting into coalitions. Informal agreements were always honored. Thus it was frequently understood that two players would stick together even though no explicit commitment was made. The two-person commitments which were made were nearly always agreements to form a coalition with a specified split of the profits, unless a third player could be attracted, in which case the payoff was not specified. This left open the possibility of argument after a third party was attracted, but such argument never developed. In fact, the split-the-difference principle was always applied in such cases. [1954, pp. 306–308.]

We have quoted at such length for three reasons. First, it is important when evaluating the results that the reader have some flavor of the procedure and of the performance. Second, it is interesting that the coalition changes were effected, in the early stages, one person at a time and, in the later stages, by one small coalition joining with another. Third, certain aspects of the experimental procedure seem undesirable and could easily be eliminated. The geometrical effects, though possibly interesting in some applications, are not desirable in a study of human response to characteristic functions. To eliminate these one might employ telephone communication or a variant on the Bavelas partitioned table for small group studies [Christie, 1956]. The latter wou'd require the use of written messages, which incidentally would give a permanent record of the bargaining. It would have the slowing effect that any written communication has, but it is not clear that this would be any disadvantage in this case. Further, in the small group work it was observed that a high

TABLE 12.1.   Results From RAND Experiment for the Four-Person Constant-Sum Game: $v(\{1, 2\}) = \frac{3}{4}$, $v(\{1, 3\}) = \frac{1}{2}$, $v(\{1, 4\}) = \frac{1}{4}$

			Imputation				Coalition Structure
			Players				
			1	2	3	4	
Game 1		1	.00	.40	.30	.30	$\{1\}, \{2, 3, 4\}$
		2	.00	.43	.43	.15	$\{1\}, \{2, 3, 4\}$
		3	.13	.38	.38	.13	$\{1, 4\}, \{2, 3\}$
	Runs	4	.13	.44	.44	.00	$\{1, 2, 3\}, \{4\}$
		5	.25	.50	.13	.13	$\{1, 2\}, \{3, 4\}$
		6	.43	.43	.00	.15	$\{1, 2, 4\}, \{3\}$
		7	.19	.44	.38	.00	$\{1, 2, 3\}, \{4\}$
		8	.44	.44	.00	.11	$\{1, 2, 4\}, \{3\}$
	Average		.20	.43	.25	.12	
Game 4		1	.38	.00	.25	.38	$\{1, 3, 4\}, \{2\}$
		2	.00	.42	.42	.17	$\{1\}, \{2, 3, 4\}$
		3	.29	.00	.46	.25	$\{1, 3, 4\}, \{2\}$
	Runs	4	.38	.54	.00	.08	$\{1, 2, 4\}, \{3\}$
		5	.37	.53	.00	.10	$\{1, 2, 4\}, \{3\}$
		6	.13	.38	.38	.13	$\{1, 4\}, \{2, 3\}$
		7	.38	.54	.00	.08	$\{1, 2, 4\}, \{3\}$
		8	.29	.00	.42	.29	$\{1, 3, 4\}, \{2\}$
	Average		.28	.30	.24	.18	
	Value		.25	.33	.25	.17	
	Quota		.25	.50	.25	.00	

These data have been adapted from Kalisch *et al.* [1954] by transforming them into 0, 1 normalization.   Game 1 was presented to the subjects in what amounted to normalized form; game 4 was in an *S*-equivalent form.

degree of anonymity was preserved, and this might allow more ruthless competition than was obtained at RAND.   One may also question the decision to have a 10-minute time limit; it probably created a sense of urgency which did not permit the players to reflect about their decisions. It is doubtful that a sophisticated response to the situation can be expected in 10 minutes.

Each of the four-person games was played eight times, a total of eight subjects being employed.   Changes in the grouping of players were made for each play of the game to prevent the formation of permanent coalitions. Since the games were constant-sum, the characteristic functions in 0, 1 normalization can be described by their values for three two-person coalitions.   The first game is described by:

$$v(\{1, 2\}) = \tfrac{3}{4}, \quad v(\{1, 3\}) = \tfrac{1}{2}, \quad \text{and} \quad v(\{1, 4\}) = \tfrac{1}{4};$$

TABLE 12.2. **Results from RAND Experiment for the Symmetric Four-Person Constant-Sum Game:** $v(\{1, 2\}) = v(\{1, 3\}) = v(\{1, 4\})$ $= \frac{1}{2}$

			Imputation				Coalition Structure
			Players				
			1	2	3	4	
Game 2		1	.45	.13	.38	.05	{1, 4}, {2, 3}
		2	.48	.20	.33	.00	{1, 2, 3}, {4}
		3	.19	.19	.31	.31	{1, 4}, {2, 3}
	Runs	4	.25	.31	.44	.00	{1, 2, 3}, {4}
		5	.21	.19	.31	.29	{1, 4}, {2, 3}
		6	.28	.19	.31	.23	{1, 4}, {2, 3}
		7	.00	.40	.51	.09	{1}, {2, 3, 4}
		8	.00	.30	.43	.28	{1}, {2, 3, 4}
	Average		.23	.23	.38	.16	
Game 3		1	.25	.25	.25	.25	{1, 2}, {3, 4}
		2	.00	.26	.36	.38	{1}, {2, 3, 4}
		3	.34	.33	.34	.00	{1, 2, 3}, {4}
	Runs	4	.38	.36	.00	.26	{1, 2, 4}, {3}
		5	.25	.25	.25	.25	{1, 3}, {2, 4}
		6	.00	.36	.28	.36	{1}, {2, 3, 4}
		7	.25	.25	.25	.25	{1, 2}, {3, 4}
		8	.25	.25	.25	.25	{1, 2, 3, 4}
	Average		.21	.29	.25	.25	
	Value		.25	.25	.25	.25	
	Quota		.25	.25	.25	.25	

These data have been adapted from *Kalisch et al* [1954] by transforming them into 0, 1 normalization. Game 3 was presented to the subjects in what amounted to normalized form; game 2 was in an *S*-equivalent form which concealed the symmetry.

the second game by:

$$v(\{1, 2\}) = v(\{1, 3\}) = v(\{1, 4\}) = \frac{1}{2}.$$

We shall refer to these respectively as the non-symmetric and symmetric games. To see the exact form of the characteristic functions given to the subjects, consult p. 305 of Kalisch *et al.* [1954].

A summary of the imputations (normalized) and coalition structures which arose in these experiments is given in Tables 12.1 and 12.2. Probably the most striking feature about these data is the apparent difference between the behavior in the *S*-equivalent games. Whether or not there is a real difference is, however, difficult to say. It is by no means adequate to look at the two average *n*-tuples and to state that these exhibit differences which are beyond experimental error; for it is not clear what the average means. If we possessed what we were certain was a correct equilibrium theory, then we could expect any of the imputations which occurred to be predicted by that theory; however, there would be no reason to expect the average imputation to be one of those predicted by that theory. Nonetheless, intuitively one senses that there is a difference in the response of the subjects to S-equivalent games, and one might be tempted to conclude that the subjects did not always get to the base of the matter. Some of the analysis given below suggests that the results must be given a somewhat more subtle interpretation than this.

Let us consider the relation between data and theory for the several theories which have been offered.

Core. Since these are constant-sum games, the core is empty.

Solutions. As we suggested by the quotation at the beginning of this section, the experimenters did not know what the von Neumann-Morgenstern theory asserts for the experiment, and so no comparison was made by them.

ψ-stability. The prediction of stability theory depends, of course, upon the choice of the boundary condition ψ. For example, were we to choose the function defined in section 10.2, i.e., any coalition may consider adding a player not in the coalition or it may consider expelling any one from it, then the only stable imputation is Shapley's quota. In the symmetric case (Table 12.2, games 2 and 3) the only stable coalition structure is $(\{1\}, \{2\}, \{3\}, \{4\})$; in the non-symmetric case (Table 12.1, games 1 and 4) both that structure and $(\{1, 2, 3\}, \{4\})$ are stable. We see that almost without exception these predictions are not confirmed; however, so far as the imputations are concerned the predictions tend to be in the right direction for the games in 0, 1 normalization, as is argued in Luce [1955 a]. All this is none too convincing, however.

Let us, therefore, reconsider the experimenters' comments. There is a suggestion that a different function ψ should be used, namely: a coalition may consider expelling a single member, as before, but it can only consider adding a member who is not already in a coalition. This is to say, for example, if $(\{1, 2\}, \{3, 4\})$ exists, no changes can occur except the breakup of a coalition; whereas, if $(\{1, 2\}, \{3\}, \{4\})$ exists, coalition $\{1, 2\}$ may consider adding either player 3 or 4 as well as breaking up, and players 3 and 4 may consider forming a coalition.

Using this boundary condition, Luce [1955 c] has obtained the predictions shown in Tables 12.3 and 12.4. For clarity, the data have been regrouped according to the equilibrium coalition, and a comparison between the data and the predictions is given. We have required that the data be within one percentage point of the predicted value, for this is the magnitude of the round-off error introduced in the reduction to 0, 1 normalization.   It should be noted that, with this boundary condition, the limitations on the imputations are less restrictive than with the original suggestion.   For example, when there are two two-person coalitions, the predictions are essentially trivial, and so we shall not consider those cases further.   In the other cases non-trivial predictions are obtained.   For the symmetric game, the data are compatible with the predictions in 8 out of the 9 non-trivial cases.   Run 8 of game 3 probably should not be interpreted as a failure for the following reason: the theory says that the pair consisting of the quota and ($\{1\}$, $\{2\}$, $\{3\}$, $\{4\}$) is stable, but by the nature of the experiment this structure could never achieve an imputation, since the total payment to the players would be 0.   Thus, in practice, the only way to achieve an imputation with this coalition structure is for the four players to call themselves a coalition, and divide the proceeds according to the quota; this is how they were, in fact, divided.   For the non-symmetric game there are 13 non-trivial cases (i.e., cases where there are no two person coalitions), of which 10 confirm the theory.   In one of the failures (run 1 of game 1) the error is five percentage points in a prediction of 75.   In the other two (runs 4 and 7 of game 1), the observed coalition structure is stable only if the imputation is the quota, and the data differ considerably from that.

Thus, if we ignore the questionable case of the coalition of four players, there were a total of 21 non-trivial cases, of which 18 yield data in agreement with the theory.   Of the three failures, one disagrees by only a small amount.   In all three cases of failure, the runs involve the S-equivalent form of the non-symmetric game, not the 0, 1 normalization.   If we are willing to accept this theory as accurate, then these results certainly reconfirm the belief that the subjects did not fully grasp the logic of the (non-symmetric) games presented in S-equivalent form.   Furthermore, if Tables 12.3 and 12.4 are examined, there appears to be a tendency for a particular coalition structure to appear either in the 0, 1 normalized game or in the S-equivalent game, but not in both.   This certainly suggests that the mode of presentation affected the dynamics of arriving at a stable pair, and therefore the chance of it arising, but that it had less effect—though still some—on the group decision whether a pair was stable or not once it was reached.

Reasonable outcomes.   Only once in the four-person games did a player get as much as or more than the bound $b(i)$.   In no cases did the outcomes of these games lie outside the set L; however, it is quite a weak restriction for these two games.   In both cases $l(S) = 0$ for S having 0, 1, or 2 players and $l(I_4) = 1$.   In the symmetric case, $l(S) = \frac{1}{2}$ for S having 3 players.   In the non-symmetric case $l(\{1, 2, 3\}) = \frac{1}{2}$, and for all other

TABLE 12.3.   Comparison of Results from the Non-Symmetric Four-Person Constant-Sum Game (RAND Experiment) with ψ-Stability Predictions

Coalition structure and Corresponding ψ-Stable Imputations	Game No.	Run No.	Observed Imputation				Incompatibilities between ψ-Stability Theory and Data
			1	2	3	4	
$(\{1\}, \{2, 3, 4\})$	1	1	.00	.40	.30	.30	$x_2 + x_3 = 0.70 < 0.75$
$x_2 + x_3 \geq 0.75$	1	2	.00	.43	.43	.15	None
$x_2 + x_4 \geq 0.50$	4	2	.00	.42	.42	.17	None
$x_3 + x_4 \geq 0.25$							
$x_1 = .00$							
$(\{2\}, \{1, 3, 4\})$	4	1	.38	.00	.25	.38	None
$x_1 + x_3 \geq 0.50$	4	3	.29	.00	.46	.25	None
$x_1 + x_4 \geq 0.25$	4	8	.29	.00	.42	.29	None
$x_3 + x_4 \geq 0.25$							
$x_2 = 0.00$							
$(\{3\}, \{1, 2, 4\})$	1	6	.43	.43	.00	.15	None
$x_1 + x_2 \geq 0.75$	1	8	.44	.44	.00	.11	None
$x_1 + x_4 \geq 0.25$	4	4	.38	.54	.00	.08	None
$x_2 + x_4 \geq 0.50$	4	5	.37	.53	.00	.10	None
$x_3 = 0.00$	4	7	.38	.54	.00	.08	None
$(\{1, 2, 3\}, \{4\})$	1	4	.13	.44	.44	.00	Incompatible
$x_1 = 0.25, x_2 = 0.50$	1	7	.19	.44	.38	.00	Incompatible
$x_3 = 0.25, x_4 = 0.00$							
$(\{1, 4\}, \{2, 3\})$	1	3	.13	.38	.38	.13	None
$x_1 + x_4 = 0.25$	4	6	.13	.38	.38	.13	None
$x_2 + x_3 = 0.75$							
$(\{1, 2\}, \{3, 4\})$	1	5	.25	.50	.13	.13	None
$x_1 + x_2 = 0.75$							
$x_3 + x_4 = 0.25$							

The function ψ is described in text.   In all predictions, the condition $x_i > 0$ for members of non trivial coalitions is omitted; it was always confirmed.   The comparison is required to hold only to the nearest hundredth, the round-off error of the computations.

three-person coalitions $l(S) = \frac{1}{4}$.   For these games, the variation on the set L described at the end of section 11.2 yields exactly the same conditions as the ψ-stability analysis—provided, of course, the ψ suggested above is selected.   For treating data of this type, this variation of L appears to have the important advantage over ψ-stability that no arbitrary decision, i.e., choice of ψ, is required.   Furthermore, it says something about

TABLE 12.4.   Comparison of Results from the Symmetric Four-Person Constant-Sum Game (RAND Experiment) with ψ-Stability Predictions

Coalition Structure and Corresponding ψ-Stable Imputations	Game No.	Run No.	Observed Imputation 1 2 3 4	Incompatibilities between ψ-Stability Theory and Data
$(\{1\}, \{2, 3, 4\})$ $x_2 + x_3 \geq .50$ $x_2 + x_4 \geq .50$ $x_3 + x_4 \geq .50$ $x_1 = .00$	2 2 3 3	7 8 2 6	.00 .40 .51 .09 .00 .30 .43 .28 .00 .26 .36 .38 .00 .36 .28 .36	None None None None
$(\{3\}, \{1, 2, 4\})$ $x_1 + x_2 \geq .50$ $x_1 + x_4 \geq .50$ $x_2 + x_4 \geq .50$ $x_3 = .00$	3	4	.38 .36 .00 .26	None
$(\{1, 2, 3\}, \{4\})$ $x_1 + x_2 \geq .50$ $x_1 + x_3 \geq .50$ $x_2 + x_3 \geq .50$ $x_4 = .00$	2 2 3	2 4 3	.48 .20 .33 .00 .25 .31 .44 .00 .34 .33 .34 .00	None None None
$(\{1, 4\}, \{2, 3\})$ $x_1 + x_4 = .50$ $x_2 + x_3 = .50$	2 2 2 2	1 3 5 6	.45 .13 .38 .05 .19 .19 .31 .31 .21 .19 .31 .29 .28 .19 .31 .23	None None None None
$(\{1, 2\}, \{3, 4\})$ $x_1 + x_2 = .50$ $x_3 + x_4 = .50$	3 3	1 7	.25 .25 .25 .25 .25 .25 .25 .25	None None
$(\{1, 3\}, \{2, 4\})$ $x_1 + x_3 = .50$ $x_2 + x_4 = .50$	3	5	.25 .25 .25 .25	None
$(\{1, 2, 3, 4\})$ None	3	8	.25 .25 .25 .25	Incompatible (see text)

The function ψ is described in text.   In all predictions, the condition $x_i > 0$ for members of non-trivial coalitions is omitted; it was always confirmed.   The comparison is required to hold only to the nearest hundredth, the round-off error of the computations.

restrictions on τ independent of any consideration of imputations [see condition (ii) on p. 242], and these may be far from trivial in more complicated games. In their original presentation of this material [1952], Kalisch *et al.* also compared the data with the upper bound $d(S)$, and they found that in most of the experimental runs at least one set S received more than $d(S)$. They concluded that " $\cdots$ the function $d(S)$ seems to have no relation with the way the game was actually played." [1952, p. 27.]

Value. Although there is no particular reason to expect any specific equilibrium outcome to be the Shapley value, one might argue, much as we did when introducing that concept, that it should predict the average equilibrium outcome. If so, then it makes sense to compare the average imputation with the value. In the symmetric game the value and the quota are identical. It will be seen from Table 12.2 that the average imputation for the game presented in 0, 1 normalization (game 3) is quite close to the value, whereas the average for the S-equivalent game is not. For the non-symmetric game, the reverse pattern seems to be true. Actually, the value is not a bad indicator of a player's expectation when we simply average over the 16 cases of each game, without regard to the differences in presentation of the characteristic function. However, if our comments above about the form of presentation of a game affecting the dynamics of coalition formation but not affecting the existence of equilibrium outcomes are substantially correct, we cannot generally expect the value to predict the average imputation. For, by varying the mode of presentation, we should be able to change the probability of various equilibrium imputations occurring, and thus change the average imputation.

What are we to conclude from this experiment? This is difficult to say, for, although it is clear that the results do not coincide exactly with any present theory, it is a question how much the outcome was influenced by the experimental technique. One senses from the author's comments that the time pressure was heavily felt, and that seems to be an ideal way to prevent the players from being all knowing—a basic assumption of the model. Furthermore, the geometrical obstacles to coalition formation are certainly not a phenomena encompassed by the theory, except possibly in stability theory by means of an appropriate choice of the boundary function ψ; however, it is doubtful that these obstacles played an important role in the four-person games. As far as confirmation of theories is concerned, one would conclude from these data that an equilibrium imputation does lie in the sets B and L. Also, with a few exceptions which may be the result of experimental technique, there does exist a boundary function ψ such that the outcomes are ψ-stable pairs. Of course, there may exist other functions which would do better. At least for this experiment, the value seems to be an adequate predictor of the average imputation, but we doubt that this is a general proposition. The stability analysis certainly suggests that it is reasonable to treat the outcome as a pair consisting of an imputation and a coalition structure, since the data become quite coherent when grouped according to coalition structures.

Possibly the most significant fact suggested by this experiment, and one we expect to be generally true, is that subjects not only respond to the strategic aspects of the characteristic function but they are also influenced by its mode of presentation. This is almost certainly true for the dynamics of coalition formation, and there is some indication that it may be true for the equilibrium behavior of players.

12.4 ARE "REAL" GAMES EVER "ABSTRACT" GAMES?

It is trivial to create an experimental situation which satisfies the *rules* of an extensive game (see section 3.3), but this does not ensure that it will be a game in extensive form, as we took pains to point out in section 3.5. Three conditions beyond the rules were added which must be met before it is a game. These were interpreted as describing the players: each player has a utility function over the set of lotteries generated from the outcomes, each attempts to maximize his own expected utility, and each is assumed to know the extensive game in full—in particular, to know *all* of the utility functions. If the game is taken in normal form, these assumptions remain the same except that the players are each assumed to know the structure of the normal form in full, i.e., each knows the strategy sets and the payoff functions of *all* the players.

Since we interpreted the maximization principle in such a way that it is tautologically true, the first two assumptions are both verified simply by showing that a utility function exists for each player. Certainly, at present it cannot be claimed that this has been shown to be true (even approximately) in a wide variety of situations, but there is some slight evidence from simple experiments on the utility of money in gambling situations suggesting that it may not be totally unrealistic (see section 2.8). At least there is the possibility that such functions exist, which leaves the third—the knowledge—assumption to be considered. Possibly it is met in certain extremely simple situations, but in any experiment of significant complexity or in any situation occurring in life we seriously doubt that this assumption is tenable, even as a first approximation. If that be so, then we are forced to admit that the answer to our section heading is No—that most interesting cases of conflict of interest are not in fact games either in extensive or normal form.

Having admitted this, the question arises as to what they are. There seem to be two conceptually different suggestions. One—probably the more realistic—says that each player is to some degree uncertain as to the utility functions of the others, and that he is forced to treat his problem as one of decision making under uncertainty. This point of view will be explored to some extent in the next chapter, where we enter into the general problem of individual decision making under uncertainty. The other suggestion attempts to extend the game theory framework slightly in such a way as to weaken the knowledge assumptions, but, at the same time, to continue to utilize some of the formalism of game theory. Possibly the most important feature of this generalization is the technique it suggests for overcoming some of the difficulties in finding characteristic functions. However, the generalization is full of weaknesses. In addition to those of its own, it has most of the shortcomings of game theory: there are no suitable sociological assumptions in the underlying structure, and it supposes the existence of a transferable utility (see sections 7.7 and 10.4). Be that as it may, let us examine the idea briefly; for a fuller statement see Luce and Adams [1956].

Although each player may not correctly perceive another player's payoff function, it is still conceivable that he will behave as if he postulates utility functions for each of the other players which he "believes" they are trying to maximize. This we shall assume. Thus, to each player i there will be associated n payoff functions $M_i{}^j, j = 1, 2, \cdots, n$. $M_i{}^i$ denotes player i's true payoff function, $M_i{}^j$ the payoff function he believes player j is attempting to maximize (when in fact it is $M_j{}^j$), and $M_k{}^i$ the payoff function k believes i is attempting to maximize. Except for this change, the model remains the same: each player has a set S_i of pure strategies, and the others know this, and each attempts to maximize his own payoff function $M_i{}^i$. The only difference is that each player thinks he is participating in a different game, e.g., player i thinks he is in the game with the payoff functions $M_i{}^1, M_i{}^2, \cdots, M_i{}^i, \cdots, M_i{}^n, j$ in the game with payoffs $M_j{}^1, M_j{}^2, \cdots, M_j{}^j, \cdots, M_j{}^n$, etc. Such a structure is called a *game with misperceptions*, or an *m*-game for short. It reduces to an ordinary game when there are no "misperceptions," i.e., when $M_i{}^j = M_j{}^j$, for all i and j.

Obviously, this model is far from the most general possible, and it is not at all clear that existing and experimental game-like situations can be realistically abstracted as *m*-games. Another generalization which avoids dealing with uncertainty directly is to suppose that the players do not correctly perceive each other's strategy sets. Such must be the case, for example, in much technological competition, where research may change the strategies available to a particular producer. By keeping such developments secret he deludes the other players as to his strategy set. Still another form of erroneous perception seems common. Player j may have a perception of i's perception of k's utility function which is, in fact, different from i's perception of k's utility function. Of course, such misperceptions of misperceptions can be carried as many steps as one chooses, but with little likelihood of profit. Indeed, whether it is valuable to go from a game to an m-game is debatable; certainly it has yet to be conclusively shown.

But to continue with the idea, since each player i believes he is in a game with payoffs $M_i{}^j$, a characteristic function v_i can be computed; this is called player i's *subjective characteristic function.* From the objective game $M_1{}^1$, $M_2{}^2$, $\cdots$, $M_n{}^n$ an objective characteristic function v can also be computed, but it appears to be of less interest. Clearly, if the m-game is in fact a game, $v_i = v$ for all i. There are now two questions to be considered, one theoretical and the other practical: What sort of theoretical superstructure can be raised on n subjective characteristic functions, and in what way is it possible to determine these subjective functions?

The question of a theory is far from adequately handled by Luce and Adams, and there seems little point is reproducing their discussion here except to say that they attempt to reduce the structure once again to a single set function. As they recognize, their attempt is unsatisfactory because it rests on an *ad hoc* interpersonal comparison of utilities.

Of more interest is their idea for dealing with the practical problem, which is, of course, of some magnitude. Determining a characteristic function was a serious problem when we had only one in game theory, and n of them surely does not make it easier. In principle, the solution exists: from each person find not only his own preference pattern, but also his beliefs as to the patterns of the others. From these construct the subjective payoff functions and then solve the necessary two-person zero-sum games to get the subjective characteristic functions. But this is simply not feasible.

Within the context of m-games it may make sense to try to determine these characteristic functions directly without passing through the normal form. The idea is almost trivially simple. The subject is to report his preferences in paired comparisons between coalitions and lotteries involving coalitions (including those coalitions he would not actually be in). If his preferences meet certain consistency requirements, then a characteristic function can be constructed which, for reasons that will be given, is plausibly interpreted as his subjective characteristic function. We could let the word "preference" be an undefined primitive, as in utility theory, which must be given a suitable realization by the experimenter. However, since coalitions are fairly specific alternatives, let us try to spell out what we want the subject to have in mind when he says that he prefers one coalition to another. We would instruct him to forget, for the moment, which player he is. Rather he is to approach the whole conflict structure as an outsider who, on the basis of his choices, would be assigned to a player role in the situation. When deciding between two coalitions, he is to imagine that he would be placed randomly in one of the player roles of the coalition he chooses. Thus, he is to decide whether he would like to be an "average" member of one coalition or the other. For a lottery, a chance device with known probabilities will decide which coalition he is to imagine he is in, and, as before, his role (and therefore payoff) within that coalition will be randomly decided. A mathematical result is given below which argues for this particular interpretation of preference.

In whatever way we attempt to realize the primitive idea of preference, we shall suppose that it satisfies the several axioms of Chapter 2 which lead to a linear utility function. In addition we shall impose another axiom, one which makes a certain amount of sense for coalitions, though in general it is not meaningful. Consider two disjoint coalitions R and S with $|R|$ and $|S|$ members, respectively. We shall assume that $R \cup S$ is preferred or indifferent to the lottery in which R arises with probability $|R|/(|R| + |S|)$ and S with probability $|S|/(|R| + |S|)$. We may argue for this condition as follows: The probability of taking a particular player role in the two alternatives is exactly the same, since in the coalition $R \cup S$ it is $1/(|R| + |S|)$ and in the gamble the probability of being any member of R is $|R|/(|R| + |S|) \cdot 1/|R| = 1/(|R| + |S|)$ and of being any member of S is $|S|/(|R| + |S|) \cdot 1/|S| = 1/(|R| + |S|)$. However, just as in game theory, the strategic possibilities of $R \cup S$ are never inferior to those of the separate subcoalitions R and S, so, given that the probabilities of being a particular player are the same in both cases, he should never prefer the gamble to $R \cup S$.

From the assumption that preference meets the utility axioms, we know that it can be represented by a linear utility function u which is unique up to a positive linear transformation. From this function we generate a whole class of functions defined over the coalitions of the game. A typical one is v, where for all coalitions S

$$v(S) = c|S|[u(S) - u(\phi)] + \sum_{i \text{ in } S} a_i,$$

and where c is a positive constant, and $a_1, a_2, \cdots, a_n$ are arbitrary constants. The class is generated as we let $c, a_1, a_2, \cdots,$ and a_n vary over their possible domains. This class of functions has these pertinent properties:

i. Every member of the class is a characteristic function. (This follows from the extra axiom that was imposed.)

ii. Any two members in the class are S-equivalent, and any characteristic function S-equivalent to a member of the class is also in the class. In other words, a given utility function u generates a whole equivalence class of characteristic functions.

iii. Any positive linear transformation of u, $au + b$, where $a > 0$, generates exactly the same equivalence class of characteristic functions as does u.

Summarizing these three points: if a subject's preferences among lotteries of coalitions satisfy the axioms we have assumed they do, then an S-equivalence class of characteristic functions is naturally associated with his preference relation. The suggestion is that these functions be interpreted as the class of characteristic functions S-equivalent to his subjective characteristic function. One strong argument for doing so is the following.

Suppose that a person actually does have a subjective characteristic function v for the given situation; this could be the objective characteristic function of the game, or it could be his calculation of his subjective characteristic function, or it could be arbitrarily given. It does not matter so long as he knows it numerically. Suppose that he is placed in the experiment described above and that he proposes to use v as best he can to arrive at his decisions. If his choice is between two coalitions R and S, and if he is to be randomly placed in the role of one of the players, a plausible index for comparing R and S is

$$\frac{v(R)}{|R|} \quad \text{versus} \quad \frac{v(S)}{|S|}.$$

For a lottery in which R occurs with probability p and S with probability $1 - p$, a plausible index is the expected value of the index for each coalition separately, i.e.,

$$p\,\frac{v(R)}{|R|} + (1 - p)\,\frac{v(S)}{|S|}.$$

Should he actually use this index to determine his answers, then it can be shown that the resulting preference relation must satisfy all the axioms we have assumed and that the S-equivalence class of characteristic functions generated by the above scheme includes the given characteristic function v.

One apparent objection to this last result is the observation that, had our subject not used v but rather some S-equivalent v', in general the preference relation would be different. Indeed it would, but one can easily show, nonetheless, that the two different patterns must generate the same S-equivalence class of characteristic functions.

It is not known at present whether this technique is experimentally realizable and, if it is, whether the axioms are met. The alternatives involved here are, it is true, special cases of the abstract alternatives of Chapter 2, but they also possess certain peculiar features which lead one to doubt whether the utility axioms hold. These alternatives are very complicated: a person must evaluate dispassionately both the coalitions he is in and those he is not in, he must imagine what it is like to be an "average" member of each of these coalitions, and he must consider lotteries having coalitions as prizes. This seems so taxing to the imagination that one can fairly doubt that he will be consistent in the sense of the utility axioms. Furthermore, the added axiom, which ensures superadditivity, is also suspect. To be sure, the coalition $R\cup S$ should be preferred to the lottery when all of the players are thought to be rational, for it has the greater strategic potentialities; but, when the evaluations are obtained as suggested, it may well happen that a person will prefer the lottery on the grounds that effective cooperation in a larger group is more difficult to achieve than in a smaller one.

Assuming, however, that the technique is feasible, it would certainly be interesting to know what pattern of coalition preferences the subjects would have in an experiment similar to the one described in the last section, and in particular whether they would faithfully reproduce the monetary worths of the coalitions. There is reason to suppose that utility for money is not linear with money, but in all likelihood this would be a small effect. Possibly a more striking effect would arise from the failure of the subjects to respond to the true relative advantages of the coalitions when the characteristic function is not presented in normalized form. It will be recalled that the experimenters had the impression that their subjects were much impressed by large monetary values and paid little regard to the actual incremental effects. If this is generally so, we might expect to determine subjective characteristic functions somewhat at variance with the objective ones—at least for non-normalized games—which in turn would mean the S-equivalent objective payoffs are not truly S-equivalent as games, if they are games at all.

chapter 13

Individual Decision Making Under Uncertainty

13.1 INTRODUCTION AND STATEMENT OF PROBLEM

Possibly the best way to begin this chapter is to reread section 2.1, where we discussed the classification of decision making according to whether it is by a group or an individual and according to whether it is being carried out under conditions of certainty, risk, or uncertainty. For the ten intervening chapters we have been concerned with individual decision making in a very particular context of uncertainty known as a game. In a game the uncertainty is due entirely to the unknown decisions of the other players, and, in the model, the degree of uncertainty is reduced through the assumption that each player knows the desires of the other players and the assumption that they will each take whatever actions appear to gain their ends. Traditionally, the game model is not called decision making under uncertainty; that title is reserved for another special class of problems which lie in the domain of uncertainty. These problems, which we shall discuss presently, have for the most part grown up and been examined in the statistical literature, for they are very much involved in an understanding of experimental evidence and in drawing appropriate inferences from data.

The gist of the problem is simple to state.   A choice must be made from among a set of acts A_1, A_2, $\cdots$, A_m, but the relative desirability of each act depends upon which "state of nature" prevails, either s_1, s_2, $\cdots$, s_n. The term "state of nature" will be more fully explicated later, but we hope the idea is intuitively clear.   As the decision maker, we are aware that one of several possible things is true; which one it is is relevant to our choice, but we do not even know the relative probabilities of their truth— or, indeed, if it is even meaningful to talk about probabilities—let alone which one obtains.   A simple example will illustrate the dilemma; this one is due to Savage [1954]:

> Your wife has just broken five good eggs into a bowl when you come in and volunteer to finish making the omelet.   A sixth egg, which for some reason must be either used for the omelet or wasted altogether, lies unbroken beside the bowl. You must decide what to do with this unbroken egg.   Perhaps it is not too great an oversimplification to say that you must decide among three acts only, namely, to break it into the bowl containing the other five, to break it into a saucer for inspection, or to throw it away without inspection.   Depending on the state of the egg, each of those three acts will have some consequence of concern to you, say that indicated by Table 13.1.

TABLE 13.1

State

Act	Good	Rotten
Break into bowl	Six-egg omelet	No omelet, and five good eggs destroyed
Break into saucer	Six-egg omelet and a saucer to wash	Five-egg omelet and a saucer to wash
Throw away	Five-egg omelet, and one good egg destroyed	Five-egg omelet

In general, to each pair (A_i, s_j), consisting of an act and a state, there will be a consequence or outcome.   We assume that our subject's preferences among these outcomes, and among hypothetical lotteries with these outcomes as prizes, are consistent in the sense that they may be summarized by means of a utility function (see Chapter 2).   If we arbitrarily choose some specific utility function, in other words, choose the origin and a unit of measurement, then we can summarize the decision problem under uncertainty (d. p. u. u.) as in Table 13.2.   Here u_{ij} is the utility associated to the consequence of the pair (A_i, s_j).   So the problem reduces to: Given an m by n array of numbers u_{ij}, to choose a row (act) which is optimal in some sense—or, more generally, to rank the rows (acts) according to some optimality criterion.

Somewhat more must be said about the states of nature.   With respect to any decision problem, the set of "states of nature" is assumed to form a mutually exclusive and exhaustive listing of those aspects of nature which are relevant to this particular choice problem and about which the decision maker is uncertain. Although this characterization is quite vague, often there is a natural enumeration of the possible, pertinent, states of the world in particular contexts. We assume that there is a "true" state of the world which is unknown to the decision maker at the time of choice.

TABLE 13.2

Acts	s_1	s_2	$\cdots$	s_j	$\cdots$	s_n
A_1	u_{11}	u_{12}	$\cdots$	u_{1j}	$\cdots$	u_{1n}
A_2	u_{21}	u_{22}	$\cdots$	u_{2j}	$\cdots$	u_{2n}
.	.	.		.		.
.	.	.		.		.
A_i	u_{i1}	u_{i2}	$\cdots$	u_{ij}	$\cdots$	u_{in}
.	.	.		.		.
.	.	.		.		.
A_m	u_{m1}	u_{m2}	$\cdots$	u_{mj}	$\cdots$	u_{mn}

States

One extreme possibility we know how to treat—namely, risk. In that case a probability distribution over the set of states is known—or, better yet, the decision maker deems it suitable to act as if it were known. For example, suppose in the omelet problem described above, the husband—a scientifically minded farmer—"knows" that in a random sample of six eggs the conditional probability of the sixth egg's being rotten when the other five are good is 0.008. Thus, he may view breaking the sixth egg into the bowl as the lottery: 0.992 probability of the six-egg-omelet prize and 0.008 probability of the no-omelet-and-five-good-eggs-destroyed prize. In other words, an *a priori* probability distribution over the states "good" and "rotten" allows one to structure the problem as one of decision making under risk—as a choice among lotteries.

In general, if an *a priori* probability distribution over the states of nature exists, or is assumed as meaningful by the decision maker, then the problem can be transformed into the domain of decision making under risk. In particular, if the probabilities of states $s_1, s_2, \cdots, s_n$ are $p_1, p_2, \cdots, p_n$, respectively, $\left(\text{where } \sum_{j=1}^{n} p_j = 1, p_j \geq 0\right)$, then the utility index for act A_i is its expected utility, i.e., $u_{i1}p_1 + u_{i2}p_2 + \cdots + u_{in}p_n$. The act having the maximum utility index is chosen, and we say that this act is "best against the given *a priori* probability distribution." (Equivalently, we can think of the decision problem as a game: the decision maker is player

1 who has strategies A_1, A_2, $\cdots$, A_m; "nature" is player 2 who has strategies s_1, s_2, $\cdots$, s_n; the payoff to 1 for the strategy pair (A_i, s_j) is u_{ij}; and, if 1 knows that 2 is employing the mixed strategy $(p_1 s_1, p_2 s_2, \cdots, p_n s_n)$, 1 should adopt a strategy (act) which is best against this mixed strategy, i.e., against the given *a priori* probability distribution.)

Thus, one extreme assumption leads us to a problem we have already examined in detail. Let us, therefore, turn to the other extreme in which we assume that the decision maker is "completely ignorant" as to which state of nature prevails. This phrase "completely ignorant" is vague, we know, and it has led to much philosophical controversy. The vagueness will be considerably diminished when later we attempt to cope axiomatically with decision making under uncertainty; however, perhaps it can now be reduced some by an illustration. Let us again examine the omelet problem, but with the cast changed. Instead of a scientific farmer, suppose the omelet is completed by a city boy unaccustomed to the ways of eggs. Furthermore, assume that the five eggs already broken were white, whereas the sixth is speckled brown and (to the city boy!) of unusual size. He doesn't have the faintest idea what to expect, having had no previous experience in matters of this kind. Nonetheless, he must make a decision, which leads to the question of criteria for decision making when the states are completely uncertain.

13.2 SOME DECISION CRITERIA

We shall now list, but only partially discuss, certain criteria which have been offered to resolve the decision problem under uncertainty, which we shall abbreviate as d. p. u. u. A criterion is well-defined if and only if it prescribes a precise algorithm which, for any d. p. u. u., unambiguously selects the act(s) which is (are) tautologically termed "optimal according to the criterion."

In each of the following criteria we shall suppose that we are given a d. p. u. u. having acts A_1, A_2, $\cdots$, A_m, states s_1, s_2, $\cdots$, s_n, and utility payoffs u_{ij}, $i = 1$, $\cdots$, m and $j = 1$, $\cdots$, n.

The maximin criterion. To each act assign its security level as an index. Thus, the index for A_i is the minimum of the numbers u_{i1}, u_{i2}, $\cdots$, u_{in}. Choose that act whose associated index is maximum—i.e., choose the act which maximizes the minimum payoff. Thus, each act is appraised by looking at the worst state for that act, and the "optimal choice" is the one with the best worst state.

We have seen in the theory of games that the optimal security level often can be raised by allowing randomizations over acts. Consider, for example:

$$
\begin{array}{c c}
 & \begin{array}{c c} s_1 & s_2 \end{array} \\
\begin{array}{c} A_1 \\ A_2 \end{array} & \left[\begin{array}{c c} 0 & 1 \\ 1 & 0 \end{array}\right].
\end{array}
$$

In this case, the security level for each act is 0, but if we permit randomization between A_1 and A_2 the security level can be raised to $\frac{1}{2}$ by using $(\frac{1}{2}A_1, \frac{1}{2}A_2)$. This is the hedging principle discussed in section 4.7. It is suggested that the reader review section 4.10, which dealt with the appropriateness and interpretation of a randomized strategy (act).

The maximin principle can be given another interpretation which, although often misleading in our opinion, is sufficiently prevalent to warrant some comment. According to this view the decision problem is a two-person zero-sum game where the decision maker plays against a diabolical Miss Nature.[1] The maximin strategy is then a best retort against nature's minimax strategy, i.e., against the "least favorable" *a priori* distribution nature can employ. We recall that in a two-person zero-sum game the maximin strategy makes good sense from various points of view: it maximizes 1's security level; and it is good against player 2's minimax strategy, which there is reason to suspect 2 will employ since it optimizes his security level and, in turn, it is good against 1's maximin strategy. In a game against nature, however, such a cyclical reinforcing effect is completely lacking.

Nonetheless, just because a close conceptual parallelism between a d. p. u. u. and a zero-sum game is lacking, it does not follow that the maximin procedure is not a wise criterion to adopt. It has the merit that it is extremely conservative in a context where conservatism *might* make good sense. We will have more to say about this later.

(It is customary in the literature to consider negative utility, disutility, or loss, as an index appraising consequences. With that orientation the decision maker, therefore, attempts to minimize the maximum loss he runs from adopting an act—i.e., he "minimaxes" instead of "maximining." Consequently, the principle described above is usually called the *minimax principle*.)

The following simple example exhibits a possible objection to the maximin principle:

$$
\begin{array}{c c}
 & \begin{array}{c c} s_1 & s_2 \end{array} \\
\begin{array}{c} A_1 \\ A_2 \end{array} & \left[\begin{array}{c c} 0 & 100 \\ 1 & 1 \end{array}\right].
\end{array}
$$

[1] In a recent lecture to statisticians one of the authors spoke of "diabolical Mr. Nature." The audience reaction was so antagonistic that we have elected the path of least resistance.

Since A_1 and A_2 have security levels of 0 and 1 respectively, A_2 is preferred to A_1 relative to the maximin criterion. This remains true even if randomized acts are considered. Some consider this unreasonable, and to emphasize their objection they point out that this criterion would still select A_2 even if the 1 were reduced to 0.00001 and the 100 increased to 10^6. These critics agree that act A_2 is reasonable *if* player 2 is a conscious adversary of 1, for then 2 should choose s_1, and A_2 is best against s_1; but, they emphasize, nature does not behave in that way, and if we are completely ignorant about the true state of nature, then they claim A_1 is manifestly better.

The minimax risk criterion (suggested by Savage [1951] as an improvement over the maximin (utility) criterion). This criterion can be suggested by continuing the analysis of the above d. p. u. u. If s_1 is the true state, then we have no "risk" or "regret" if we choose A_2, but some "risk" if we choose A_1; if s_2 is the true state, then we have no risk if we choose A_1 and a good deal of risk if we choose A_2. Schematically:

$$
\text{Utility Payoffs} \qquad \text{``Risk'' Payoffs}
$$

$$
\begin{array}{c}
 \\
A_1 \\
A_2
\end{array}
\begin{array}{cc}
s_1 & s_2 \\
\left[\begin{array}{cc} 0 & 100 \\ 1 & 1 \end{array}\right]
\end{array}
\rightarrow
\begin{array}{c}
 \\
A_1 \\
A_2
\end{array}
\begin{array}{cc}
s_1 & s_2 \\
\left[\begin{array}{cc} 1 & 0 \\ 0 & 99 \end{array}\right].
\end{array}
$$

In terms of "risk" payoffs, A_1 has a possible maximum risk of 1, whereas A_2 has a possible maximum risk of 99. Consequently, A_1 minimizes the maximum risk. However, if randomization is permitted, neither A_1 nor A_2 is optimal.

The general procedure goes as follows:

i. To a d. p. u. u. with utility entries u_{ij}, associate a new table with risk payoffs r_{ij}, where r_{ij} is defined as the amount that has to be added to u_{ij} to equal the maximum utility payoff in the jth column.

ii. Choose that act which minimizes the maximum risk index for each act.

To illustrate the "reasonableness" of a criterion based upon risk payoffs rather than utility payoffs, consider some d. p. u. u. with money payoffs and a decision maker whose utility function is linear with money. Now suppose this d. p. u. u. is modified by giving a $10 bonus to the decision maker, regardless of his choice, provided a particular state, say s_3, turns out to be the true state. This bonus, so it is argued, cannot alter the strategic aspects of the decision problem, hence the preference pattern among acts should be identical for both the original and the modified problem. This amounts to saying that adding a constant to any column of the payoff array should not change the preference ordering of acts. In particular, then, the arrays

$$\begin{bmatrix} 0 & 100 \\ 1 & 1 \end{bmatrix} \quad \text{and} \quad \begin{bmatrix} 0 + a & 100 + b \\ 1 + a & 1 + b \end{bmatrix}$$

should be strategically equivalent for any a and b. By setting a equal to -1 and b equal to -100, we get

$$\begin{bmatrix} -1 & 0 \\ 0 & -99 \end{bmatrix},$$

which is the negative of the risk payoff array. Therefore, the maximin criterion for this payoff array is the same as the minimax criterion for the risk array.

In criticism of this proposal, we quote from Chernoff [1954]:

Unfortunately, the minimax regret [risk] criterion has several drawbacks. First, it has never been clearly demonstrated that differences in utility do in fact measure what one may call regret [risk]. In other words, it is not clear that the "regret" of going from a state of utility 5 to a state of utility 3 is equivalent in some sense to that of going from a state of utility 11 to one of utility 9. Secondly, one may construct examples where an arbitrarily small advantage in one state of nature outweighs a considerable advantage in another state. Such examples tend to produce the same feelings of uneasiness which led many to object to the [maximin utility] criterion.

A third objection which the author considers very serious is the following. In some examples the minimax regret criterion may select a strategy [act] A_3 among the available strategies[2] A_1, A_2, A_3, and A_4. On the other hand, if for some reason A_4 is made unavailable, the minimax regret criterion will select A_2 among A_1, A_2, and A_3. The author feels that for a reasonable criterion the presence of an undesirable strategy A_4 should not have an influence on the choice among the remaining strategies.

Chernoff's third objection to the minimax risk principle is a variation on our old theme of the "independence of irrelevant alternatives." There is an obvious modification of the minimax risk principle which copes with the problem of non-independence of irrelevant alternatives—but, unfortunately, it has its own, more serious fault. Roughly, the idea is: instead of comparing an act with all others to ascertain the risk, which introduces the difficulties when new acts are added, simply make paired comparisons between acts. *Relative to the universe of any two acts,* and for each state, determine the risk of taking each act. Of the two acts, choose the one whose maximum risk is least. An optimal act is then defined as one which is preferred or indifferent, when compared in this way, to every other act. This procedure is unsatisfactory because there are d. p. u. u.'s in which

---

[2] Chernoff uses letters d_1, d_2, d_3, and d_4.

intransitivities occur, and so for these cases it fails to lead to an unambiguous optimal act.   An example is the d. p. u. u.

$$
\begin{array}{c}
 & \begin{array}{ccc} s_1 & s_2 & s_3 \end{array} \\
\begin{array}{c} A_1 \\ A_2 \\ A_3 \end{array}
\left[\begin{array}{ccc}
10 & 5 & 1 \\
0 & 10 & 4 \\
5 & 2 & 10
\end{array}\right]
\end{array}
\qquad \text{(payoff in utility units).}
$$

The procedure outlined yields the following:

(i) A_1 over A_2 for: A_1 has a maximum risk of 5 (from s_2) whereas A_2 has a maximum risk of 10 (from s_1).

(ii) A_2 over A_3 for: A_2 has a maximum risk of 6 (from s_3) whereas A_3 has a maximum risk of 8 (from s_2).

(iii) A_3 over A_1 for: A_3 has a maximum risk of 5 (from s_1) whereas A_1 has a maximum risk of 9 (from s_3).

Consequently, none of the three acts can be optimal since each is less preferred (in a paired comparison) than one of the others.

This same example also illustrates Chernoff's third objection to the minimax risk criterion.   Restricting ourselves to acts A_2 and A_3, that criterion selects A_2 as optimal and A_3 *as non-optimal*.   When A_1 is added, the risk matrix is

$$
\begin{array}{c}
 & \begin{array}{ccc} s_1 & s_2 & s_3 \end{array} \\
\begin{array}{c} A_1 \\ A_2 \\ A_3 \end{array}
\left[\begin{array}{ccc}
0 & 5 & 9 \\
10 & 0 & 6 \\
5 & 8 & 0
\end{array}\right]
\end{array}
\qquad \text{(payoff in risk units)}
$$

and A_3 *is then optimal* since its maximum risk is a minimum among the maximum risks.

The pessimism—optimism index criterion of Hurwicz.   The maximin utility and the minimax risk criteria are each ultraconservative (or pessimistic) in that, relative to each act, they concentrate upon the state having the worst consequence.   Why not look at the best state, or at a weighted combination of the best and worst?   This, in essence, is the Hurwicz [1951 a] criterion.

For act A_i, let m_i be the minimum and M_i the maximum of the utility numbers $u_{i1}, u_{i2}, \cdots, u_{in}$.   Let a fixed number α between 0 and 1, called the pessimism-optimism index, be given.   To each A_i associate the index $\alpha m_i + (1 - \alpha)M_i$, which we shall term the α-index of A_i.   Of two acts, the one with higher α-index is preferred.

Note that, if $\alpha = 1$, the above procedure is the maximin (utility) criterion, whereas if $\alpha = 0$, it is the maximax (utility) criterion.   If neither of these are satisfactory, then how does one decide what α to use?   One way is to see what seems reasonable in certain simple classes of d. p. u. u.'s, for example, in the class:

$$
\begin{array}{c} \\ A_1 \\ A_2 \end{array}
\begin{array}{c} s_1 \quad s_2 \\ \left[\begin{array}{cc} 0 & 1 \\ x & x \end{array} \right] \end{array}
\quad \text{(utility payoff).}
$$

The α-indices of A_1 and A_2 are $1 - \alpha$ and x respectively. Consequently, if one can choose an x such that A_1 and A_2 are indifferent, then one can impute an α-level to oneself. For example, if A_1 and A_2 are indifferent for $x = \frac{3}{8}$, then α must be $\frac{5}{8}$. Thus, by resolving a simple decision problem an α-level can be chosen empirically, which, in turn, can be employed in more complicated decisions.

But there are also objections to this criterion; one may be illustrated by the following example:

$$
\begin{array}{c} \\ A_1 \\ A_2 \\ (\frac{1}{2}A_1, \frac{1}{2}A_2) \end{array}
\begin{array}{c} s_1 \quad s_2 \quad s_3 \\ \left[\begin{array}{ccc} 0 & 1 & 0 \\ 1. & 0 & 0 \\ \frac{1}{2} & \frac{1}{2} & 0 \end{array} \right] \end{array}
\quad \text{(utility payoff).}
$$

Suppose the α-level of $\frac{1}{4}$ is chosen. The α-indices of A_1 and A_2 are each $\frac{1}{4} \cdot 0 + (1 - \frac{1}{4}) \cdot 1 = \frac{3}{4}$, whereas the index of $(\frac{1}{2}A_1, \frac{1}{2}A_2)$ is $\frac{1}{4} \cdot 0 + (1 - \frac{1}{4}) \cdot \frac{1}{2} = \frac{3}{8}$. Consequently, although A_1 and A_2 are each optimal, the procedure of tossing a fair coin and taking A_1 if heads and A_2 if tails is not optimal. Critics of the Hurwicz criterion claim that any randomization over optimal acts (according to a particular criterion) should itself also be optimal according to that criterion. Remember that a randomization which uses only optimal acts will ultimately cause the decision maker to adopt one of these optimal acts!

A second possible criticism of the Hurwicz criterion is that it resolves the following d. p. u. u. counter to one's best intuitive judgment:

$$
\begin{array}{c} \\ A_1 \\ A_2 \end{array}
\begin{array}{c} s_1 \quad s_2 \quad s_3 \quad \cdots \quad s_i \quad \cdots \quad s_{100} \\ \left[\begin{array}{ccccccc} 0 & 1 & 1 & \cdots & 1 & \cdots & 1 \\ 1 & 0 & 0 & \cdots & 0 & \cdots & 0 \end{array} \right]. \end{array}
$$

According to any α-level Hurwicz criterion, both acts A_1 and A_2 have an α-index of $1 - \alpha$, and so they are considered indifferent; however, if one is "completely ignorant" concerning which is the true state, then, the critics argue, A_1 is manifestly better than A_2. But, in defense of Hurwicz, *is* A_1 clearly better than A_2? What seems to be implied here is that the "true" state is "more likely" to be one of the states s_2 to s_{100} than s_1. This, however, is not what Hurwicz intuits about the notion of "complete ignorance," for he would assert that "complete ignorance" implies the above d. p. u. u. is strategically equivalent to

$$
\begin{array}{c}
\begin{array}{cc} s_1' & s_2' \end{array} \\
\begin{array}{c} A_1 \\ A_2 \end{array}
\begin{bmatrix} 0 & 1 \\ 1 & 0 \end{bmatrix}.
\end{array}
$$

A complete characterization of what he means by the term "complete ignorance" can best be given in axiomatic form (see section 13.4).

The criterion based on the "principle of insufficient reason." The criterion of insufficient reason asserts that, if one is "completely ignorant" as to which state among s_1, s_2, $\cdot$ $\cdot$ $\cdot$, s_n obtains, then one should behave as if they are equally likely. Thus, one is to treat the problem as one of risk with the uniform *a priori* probability distribution over states, and to each act A_i assign its expected utility index,

$$
\frac{u_{i1} + u_{i2} + \cdot \cdot \cdot + u_{in}}{n},
$$

and choose the act with the largest index.

At this juncture, it would be apropos to digress into the philosophical foundations of probability and to review the special role of the principle of insufficient reason in relation to these foundations. But we shall resist this temptation, for to do the topic justice would require a sizable digression, and there are already excellent expository accounts of this material. (See, for instance, Arrow [1951 *b*], Nagel [1939], and Savage [1954]; each of these references, in turn, gives a relatively complete bibliography.) We will confine ourselves to a few simple remarks.

The principle of insufficient reason, first formulated by Jacob Bernoulli (1654–1705), states in boldest terms that, if there is no evidence leading one to believe that one event from an exhaustive set of mutually exclusive events is more likely to occur than another, then the events should be judged equally probable. This principle is extremely vague, and its indiscriminate use has led to many nonsensical results. Writers since Bernoulli's time have attempted to add qualifications to the principle and to specify limited interpretations so as to avoid some of the more blatant contradictions.

From an empirical point of view, one difficulty with the principle is this: Suppose we are confronted with a real problem in decision making under uncertainty, then our first task is to give a mutually exclusive and exhaustive listing of the possible states of nature. The rub is that many such listings are possible, and in general these different abstractions of the same problem will, when resolved by the principle of insufficient reason, yield different real solutions. For instance, in one listing of the states we might have: s_1, the organism remains fixed; s_2, the organism moves.   In another equally good listing we might have: s_1, the organism remains fixed; s_2, the organism moves to the left; s_3, the organism moves to the right.   We can further complicate our description of the possible states of nature by noting which leg first moves, whether the animal raises its head or not, etc.

There is a counterargument to this objection.   Although it may be true that there are various acceptable interpretations as to what constitutes a state in a given real problem, it is not true that we will feel that the states are "equally likely" in each interpretation.   In other words, care must be exerted in the choice of states if one wishes to use this principle.   As it stands, this defense is weak in that there is a crying need for an empirical clarification of the term "equally likely."   Eventually, we shall examine two suggested clarifications.   The first, an axiomatic treatment due to Chernoff [1954], characterizes his notion of "complete ignorance" in such a manner as to justify logically the principle of insufficient reason. This will be described in section 13.4.   In the second, the equally likely assignment gains empirical meaning through the "practical" suggestions for probability assignments offered by the personalistic school of probability (see section 13.5).

Incidentally, the arguments against the principle of insufficient reason become even more cogent when there are an infinite set of pertinent states of nature, for then it is difficult to single out a natural parametrization, or enumeration, of the states for which a suitable generalization of the "equally likely" criterion is appropriate.

Before we turn to the axiomatic studies of decision criteria, what of the poor decision maker who is now totally confused by the pros and cons of the above criteria?   Can he, in desperation, compromise by adopting some sort of arbitrary composite of the criteria?   Subsequently, we will suggest some plausible composites; however, for the present, the following example must be included as a note of caution, for some *apparently* acceptable compromises may not be so acceptable after all.

Take the case of a decision maker who cannot crystallize his preferences among the maximin criterion, the Hurwicz criterion with $\alpha = \frac{3}{4}$, and the principle of insufficient reason.   He thus decides to define one act as preferable to another if and only if a majority of these three criteria register this preference.   The following d. p. u. u. establishes that this compromise procedure is not well defined:

$$
\begin{array}{c}
 & \begin{array}{ccc} s_1 & s_2 & s_3 \end{array} \\
\begin{array}{c} A_1 \\ A_2 \\ A_3 \end{array} & \left[\begin{array}{ccc} 2 & 12 & -3 \\ 5 & 5 & -1 \\ 0 & 10 & -2 \end{array} \right]
\end{array} \qquad \text{(utility payoff)}.
$$

Preferences according to:

Maximin criterion	A_2 over A_3 over A_1
Hurwicz criterion[3] ($\alpha = \frac{3}{4}$)	A_3 over A_1 over A_2
Principle of insufficient reason	A_1 over A_2 over A_3

A majority of the criteria select A_1 over A_2, A_2 over A_3, and A_3 over A_1—an intransitivity. The majority decision principle applied in social welfare contexts (Chapter 14) leads to the same embarrassing intransitivities of preference. The reasons are analogous.

13.3 AXIOMATIC TREATMENT: THE AXIOMS NOT REFERRING TO "COMPLETE IGNORANCE"

Instead of applying specific proposed decision criteria to carefully selected decision problems, thereby determining whether or not each criterion complies with our intuitive criteria (which we deem to be reasonable), let us, as so often before, invert the procedure. Let us cull from our intuitions certain reasonable desiderata for decision criteria to fulfill, which we can then investigate both as to compatibility with one another and as to their logical implications. Our axiomatic presentation mainly follows Chernoff [1954], but it is also a curious mixture of the works of Milnor [1954], Hurwicz [1951 a], Savage [1954], Arrow [1953], and unpublished comments by Rubin.

There are two distinct types of axiomatic approaches in the literature. In one the criterion must establish for each d. p. u. u. a complete ordering of the available acts. As in the four criteria we have previously mentioned, this is usually effected by attaching a numerical index to each act. In the other approach, a criterion isolates an "optimal" subset of acts, but it does not attempt to rank non-optimal ones. Of course, this can be thought of as a complete ordering of all acts—but into just two categories: optimal and non-optimal! We will follow the latter procedure, for it is closer to the natural demands of the problem area.

Let A' and A'' be two arbitrary but specific acts in a decision problem. We define the following preliminary notions.

i. $A' \sim A''$: means that the acts are *equivalent* in the sense that they yield the same utilities for each state of nature.

ii. $A' \succ A''$: means that A' *strongly dominates* A'' in the sense that A' is preferred to A'' for each state of nature.

---

[3] The α-indices of A_3, A_1, and A_2 are $\frac{1}{4}(10) + (\frac{3}{4})(-2) = 1$, $\frac{1}{4}(12) + (\frac{3}{4})(-3) = \frac{3}{4}$, and $(\frac{1}{4})(5) + (\frac{3}{4})(-1) = \frac{2}{4}$, respectively.

iii. $A' \succsim A''$: means that A' *weakly dominates* A'' in the sense that A' is preferred to A'' for at least one state and is preferred or indifferent to A'' for all other states.

Since any d. p. u. u. is characterized by a class of acts $\mathcal{G}$, a set of states of nature S, and a utility function u, we may symbolically identify the d. p. u. u. with the triple $(\mathcal{G}, S, u)$. A decision criterion associates to each d. p. u. u., i.e., to each $(\mathcal{G}, S, u)$, a subset $\hat{\mathcal{G}}$ of $\mathcal{G}$; the acts in $\hat{\mathcal{G}}$ are called *optimal* for $(\mathcal{G}, S, u)$ relative to the given criterion. $\hat{\mathcal{G}}$ is called the *choice* or *optimal* set.

Desiderata for criteria

Axiom 1. *For any d. p. u. u. $(\mathcal{G}, S, u)$, the set $\hat{\mathcal{G}}$ is non-empty, i.e., every problem can be resolved.*

Axiom 2. *The choice set for d. p. u. u. does not depend upon the choice of origin and unit of the utility scale used to abstract the problem.*

Axiom 3. *The choice set is invariant under the labeling of acts, i.e., the real acts singled out as optimal should not depend upon the arbitrary labeling of acts used to abstract the problem.*

Axiom 4. *If A' belongs to $\hat{\mathcal{G}}$ and $A'' \succsim A'$ or $A'' \sim A'$, then A'' belongs to $\hat{\mathcal{G}}$.*

Axioms 1 through 4 are quite innocuous in the sense that, if a person takes serious issue with them, then we would contend that he is not really attuned to the problem we have in mind.

An act A' is said to be *admissible* if there is no act A in $\mathcal{G}$ such that $A \succsim A'$, i.e., A' is admissible if A' is not weakly dominated by any other act.

Axiom 5. *If A' belongs to $\hat{\mathcal{G}}$, then A' is admissible.*

Axiom 5 is equivalent to:

Given A', if there exists an A such that $A \succsim A'$ (that is, if A' is not admissible), then A' does not belong to $\hat{\mathcal{G}}$.

It should be noted that as they were originally stated neither the maximin principle nor the Hurwicz α-criteria satisfy axiom 5; however, both can be appropriately modified in a trivial manner. To see the problem, consider the following d. p. u. u.:

$$
\begin{array}{c}
 \begin{array}{ccc} s_1 & s_2 & s_3 \end{array} \\
\begin{array}{c} A_1 \\ A_2 \end{array} \left[\begin{array}{ccc} 0 & 1 & 3/4 \\ 0 & 1 & 1/2 \end{array} \right].
\end{array}
$$

The strategy A_2 is not admissible, since $A_1 \succsim A_2$; however, A_1 and A_2, and all randomizations between them, have the same security level, 0, and the same Hurwicz α-index. Consequently, any randomized act is optimal according to these criteria. We can modify them to meet axiom 5 either by deleting all acts which are not admissible, or by deleting from the class of optimal acts those which are not admissible. This point suggests the next axiom.

Axiom 6. *Adding new acts to a d. p. u. u., each of which is weakly dominated by or is equivalent to some old act, has no effect on the optimality or non-optimality of an old act.*

Example. A gentleman wandering in a strange city at dinner time chances upon a modest restaurant which he enters uncertainly. The waiter informs him that there is no menu, but that this evening he may have either broiled salmon at \$2.50 or steak at \$4.00. In a first-rate restaurant his choice would have been steak, but considering his unknown surroundings and the different prices he elects the salmon. Soon after the waiter returns from the kitchen, apologizes profusely, blaming the uncommunicative chef for omitting to tell him that fried snails and frog's legs are also on the bill of fare at \$4.50 each. It so happens that our hero detests them both and would always select salmon in preference to either, yet his response is "Splendid, I'll change my order to steak." Clearly, this violates the seemingly plausible axiom 6. Yet can we really argue that he is acting unreasonably? He, like most of us, has concluded from previous experience that only "good" restaurants are likely to serve snails and frog's legs, and so the risk of a bad steak is lessened in his eyes.

This illustrates the important assumption implicit in axiom 6, namely, that adding new acts to a d. p. u. u. *does not alter one's a priori information as to which is the true state of nature.* In what follows, we shall suppose that this proviso is satisfied. In practice this means that, if a problem is first formulated so that the availability of certain acts influences the plausibility of certain states of nature, then it must be reformulated by redefining the states of nature so that the interaction is eliminated.

Axiom 6 can be strengthened to the following form of the principle of the independence of irrelevant alternatives:

Axiom 7. *If an act is non-optimal for a d. p. u. u., it cannot be made optimal by adding new acts to the problem.*

A typical violation of axiom 7 is this incongruous exchange.

DOCTOR: Well, Nurse, that's the evidence. Since I must decide whether or not he is tubercular, I'll diagnose tubercular.

NURSE: But, Doctor, you do not have to decide one way or the other, you can say you are undecided.

DOCTOR: That's true, isn't it?   In that case, mark him not tubercular.
NURSE: Please repeat that!

The example given at the end of the discussion of the minimax risk criterion shows that axiom 7 rules out the minimax risk principle.

Note that axiom 7 does not prevent an optimal act from being changed into a non-optimal one by adding new acts; this is true even if none of the new acts is optimal.   Therefore, one might wish to strengthen axiom 7 to:

Axiom 7′.   *The addition of new acts does not transform an old, originally non-optimal act into an optimal one, and it can change an old, originally optimal act into a non-optimal one only if at least one of the new acts is optimal.*

A further strengthening of axiom 7 is:

Axiom 7″.   *The addition of new acts to a d. p. u. u. never changes old, originally non-optimal acts into optimal ones and, in addition, either*

(i)   *All the old, originally optimal acts remain optimal,*

or

(ii)   *None of the old, originally optimal acts remain optimal.*

The all-or-none feature of axiom 7″ may seem a bit too stringent, but one can offer this rationalization for it.   Suppose that the merit of each act can be summarized by a single numerical index which is independent of the other acts available.   Then the optimal set of the original problem is composed of all the acts with the highest index.   Now, among the new acts either there is one with a higher index, which therefore annihilates all the old optimal acts, or there is not and the original optimal set is left intact.   A severe criticism of axiom 7″ is that it yields unreasonable results when it is coupled with either of the more palatable axioms 5 and 6.   Take, for example, the following d. p. u. u.:

$$
\begin{array}{c}
 & s_1 \; s_2 \; s_3 \; s_4 \\
A_1 & \left[\begin{array}{cccc} 0 & 4 & 2 & 2 \\ 4 & 0 & 0 & 4 \end{array}\right]. \\
A_2 &
\end{array}
$$

It is reasonable that some criterion should allow both A_1 and A_2 in the optimal set.   Now add an A_3 whose utilities are

$$A_3 \; [4 \quad 0 \quad 0.1 \quad 4]$$

Since A_2 is weakly dominated by A_3, axiom 5 implies that act A_2 cannot remain optimal.   But one may very well want also to keep A_1 as optimal, in violation of 7″.   The rationalization of axiom 7″ (namely, that each act can be fully appraised by a single index) is apparently not suitable.

This is suggested by the fact that acts A_2 and A_3 have the same indices according to the maximin (utility), minimax risk or regret, and Hurwicz (for any α-index) criteria. The criterion based on the principle of insufficient reason, however, does satisfy axiom 7″.

▶ There is still another variation on the theme of the independence of irrelevant alternatives, which is especially suited to finding the logical consequences of some combinations of these axioms.

Axiom 7‴. *An act A' is optimal only if it is optimal in the paired comparisons between A' and A, for all A in $\mathfrak{A}$.*

This axiom enables us to transform the decision problem into a series of paired comparisons between acts and to eliminate those acts which are not optimal in any one of these comparisons. We will not, however, use this condition. ◀

Axiom 7 and its different versions are somewhat controversial. Each of these rules out the minimax risk or regret principle. We are most sympathetic to axioms 7 and 7‴. The others, 7′ and 7″, are slightly harder to see through (i.e., they are a little less intuitive), so let us suspend judgment until some of their consequences are stated.

The next axiom is due to Rubin. To suggest it, suppose a decision maker is given two decision problems having the same sets of available acts and states but differing in payoffs. Suppose the second problem is trivial in the sense that the payoff depends only upon the state and not upon the act adopted. In other words, in the array representing problem 2, all entries in the same column are the same. If the decision maker knows only that he is playing problem 1 with probability p and problem 2 with probability $1 - p$ when he has to adopt an act, then he should adopt an act which is optimal for problem 1, since problem 2, which enters with probability $1 - p$, is irrelevant as far as his choice is concerned. It is straightforward to formalize this requirement into an axiom, but we will be content merely with the following suggestive formulation.

Axiom 8. *Consider a probability mixture of two d. p. u. u.'s with the same sets of actions and states. If the second d. p. u. u. has payoffs which do not depend upon the act chosen, then the optimal set of the mixture problem should be the same as the optimal set of the first d. p. u. u.*

Axiom 8 can be shown to imply that *adding a constant to each entry of a column of a d. p. u. u. does not alter the optimal set.* Instead of Rubin's axiom, perhaps it would have been simpler to take the italicized consequence as the axiom; however, we feel, as do Rubin and Chernoff, that this property is not as intuitively compelling as the axiom given.

Axiom 8 goes a long way towards selecting a criterion. For example, it rules out the maximin criterion and all the Hurwicz α-criteria. Therefore, we should be careful before we accept or reject it. First, to argue against the axiom, these points may be raised:

i. As stated, the axiom is not intuitive enough to be given the status of a basic desideratum.

ii. Consider the following problems:

$$
\begin{array}{cc}
\text{Problem 1} & \\
 & \begin{array}{cc} s_1 & s_2 \end{array} \\
\begin{array}{c} A_1 \\ A_2 \end{array} & \left[\begin{array}{cc} 0 & -9 \\ -10 & 0 \end{array}\right],
\end{array}
\qquad
\begin{array}{cc}
\text{Problem 2} & \\
 & \begin{array}{cc} s_1 & s_2 \end{array} \\
\begin{array}{c} A_1 \\ A_2 \end{array} & \left[\begin{array}{cc} 1000 & 0 \\ 1000 & 0 \end{array}\right],
\end{array}
\qquad
\begin{array}{cc}
\text{Problem 3} & \\
 & \begin{array}{cc} s_1 & s_2 \end{array} \\
\begin{array}{c} A_1 \\ A_2 \end{array} & \left[\begin{array}{cc} 500 & -9/2 \\ 495 & 0 \end{array}\right],
\end{array}
$$

where, it will be noted, problem 3 is a mixture of the other two in which each is played with probability $1/2$. Intuitively, a plausible method for analyzing these d. p. u. u.'s is to be somewhat pessimistic and to behave as if the less desirable state is somewhat more likely to arise. The extreme example of this rule is the maximiner who focuses entirely on the undesirable state, but our point holds equally well for one who emphasizes the undesirable state only slightly. In problem 1, s_1 is less desirable, and so one is led to choose A_1. In problem 3, s_2 is less desirable, and so one might be led to choose A_2. But if one subscribes to axiom 8, the same alternative must be chosen in both cases, and so we are led to doubt the axiom.

iii. Axiom 8, when added to axiom 3 (i.e., the choice set is invariant under labeling of acts) and to axiom 7 (i.e., the addition of acts cannot make a non-optimal act optimal), both of which are extremely reasonable, yields the following result: *If an optimal act of a given d. p. u. u. is equivalent to a probability mixture of two other acts, then each of these acts is also optimal.*[4] For example, in the d. p. u. u.

$$
\begin{array}{cc}
 & \begin{array}{cc} s_1 & s_2 \end{array} \\
\begin{array}{c} A_1 \\ A_2 \\ A_3 \end{array} & \left[\begin{array}{cc} 0 & 2 \\ 1 & 0 \\ 1/2 & 1 \end{array}\right],
\end{array}
$$

if A_3 is optimal, so are A_1 and A_2, since A_3 is equivalent to $(1/2 A_1, 1/2 A_2)$. This also implies the result that one need *never resort to randomized acts* in this type of decision problem. Since, it is contended, this consequence is absurd, one should discard the weakest link in the argument leading to it. Therefore, axiom 8 should go.

Now, to argue against these arguments point by point:

i. Rubin's axiom is not only intuitively meaningful but it seems perfectly reasonable. This is a matter of taste!

---

[4] This proposition is referred to as the anticonvexity property of the optimal set.

ii. The very compelling *a priori* quality of Rubin's axiom argues against the analysis which led us to choose A_1 in problem 1 and A_2 in problem 3. Certainly, the intuitive analysis cannot be used without restriction, for it would also lead us to choose A_2 again in

$$
\begin{array}{c}
\begin{array}{cc} s_1 & s_2 \end{array} \\
\begin{array}{c} A_1 \\ A_2 \end{array}
\left[\begin{array}{cc} 500 & -0.01 \\ 100 & 0 \end{array} \right],
\end{array}
$$

and that seems counterintuitive. We suspect that most people who are unaware of axiom 8 would find it difficult to resolve problem 3 above and that they could easily be persuaded to choose either A_1 or A_2; however, once they become aware of the axiom they will find it acceptable and will use it to decide upon A_1 in that problem.

iii. Is the assertion that one need never resort to randomized strategies in a d. p. u. u. so absurd? Maybe not, for one can cite many "reasonable" criteria which lead to an optimal non-randomized act for any d. p. u. u. Furthermore, there are arguments against randomization; for example, part of the discussion found in section 4.10, where we examined the operational interpretation of randomized strategies and cast some doubt upon their applicability, can be taken over almost verbatim. Finally, Chernoff [1954, p. 438] argues as follows "It would seem that the need for randomization depends on the statistician's need to oversimplify the statement of his problem because with limited computational ability he cannot take full advantage of the actual relationships involved. Generally, the simplification has the effect of *combining states of nature which are equivalent when random samples are insisted upon.*" [Italics ours.] This discussion leads naturally to the next axiom.

Axiom 9. *If A' and A'' are both optimal for a d. p. u. u., a probability mixture of A' and A'' is also optimal, i.e., the optimal set is convex.*

Remember that a probability mixture using A' or A'' will in fact choose either A' or A'', and, if they are both optimal, certainly any mixture should be. This seems very palatable; however, it rules out all Hurwicz's criteria with $\alpha < 1$. Put in another fashion, if we are committed to using some one of the criteria of the Hurwicz family, and if we impose axiom 9, then we must choose $\alpha = 1$, that is, the maximin (utility) criterion.

Hurwicz would argue, facetiously perhaps, that it does not grieve him too much to be forced into the $\alpha = 1$ camp, for that is where he started from in the first place. He only invented the pessimism-optimism index as a modification of the maximin criterion in order to appease those souls who were unwilling to endorse its pessimistic approach.   However, he would continue, axiom 9 is not as innocuous as it seems.   If axiom 9 were a consequence of some other more basic axioms, he would not object too much, but it does not seem to him to warrant the status of an axiom. Suppose A_1 and A_2 are both optimal acts.   It is true that a mixture such as ($\frac{1}{2}A_1$, $\frac{1}{2}A_2$) will, operationally, result in a selection of one of the two optimal acts.   Nonetheless, the mixture may evoke a psychological response in its own right, and, before it is known which optimal act is adopted, there is no compelling reason why the anticipation of the mixture must be as good as either A_1 or A_2.   For example, an optimist might like both A_1 and A_2 because in each case he can look forward to very desirable returns if certain states obtain; however, with the randomization all expected returns will be mitigated, and so the anticipation is not nearly so pleasant.   Of course, the counterargument is that the apparent reasonableness of the axiom simply demonstrates the irrationality of the optimist's wishful thinking.   So the battle is joined.   The present authors are very partial to the axiom and believe the argument against it is rather weak.

So far we have not tried to characterize the notion of "complete ignorance."   Our purpose in postponing this discussion is obvious: Axioms 1 through 9 are pertinent to decision making where one is not "completely ignorant" of the true state.   It is interesting that, even without committing ourselves on the notion of "complete ignorance," acceptance of axioms 1 through 9 serves to eliminate the maximin criterion (eliminated by axiom 8), the minimax risk or regret (eliminated by axiom 7 or any of its variations), and the Hurwicz α-criteria (eliminated by axiom 8 and, for $\alpha < 1$, by axiom 9).   Nonetheless, axioms 1 through 9 are compatible: the criterion based on the principle of insufficient reason, for example, satisfies all of them.

The following theorem is basic:

To each criterion which resolves all d. p. u. u.'s in such a manner as to satisfy axioms 1, 3, 4, 5, 7', 8, and 9, there is an appropriate a priori distribution over the states of nature which is independent of any new acts which might be added, such that an act is optimal (according to the criterion) only if it is best against this a priori distribution.

Note, this theorem does *not* say that if an act is best against this *a priori* distribution then it is optimal according to the criterion.   It only says the converse.   The theorem indicates that, if we are committed to axioms 1, 3, 4, 5, 7', 8, and 9, our first step should be to search for a suitable *a priori* distribution.   What distribution is chosen will, naturally, depend upon the information we possess concerning the true state of nature.

13.4   AXIOMATIC TREATMENT: THE AXIOMS REFERRING TO "COMPLETE IGNORANCE"

Now we turn to the question of "complete ignorance."   Consider the following:

Axiom 10.   *For any d. p. u. u., the optimal set should not depend upon the labeling of the states of nature.*

Obviously, if we have reason to suspect that a given state of nature is quite likely the true state whereas another state is quite likely not the true state, then in any abstraction of the problem we wish to distinguish between these two states.   Or, if we number the states of nature in a given problem in such a manner that the lower the number the more likely we feel that it is the "true" state, then certainly we want to keep the labeling of the states in mind and axiom 10 would not be at all appropriate.   Loosely speaking, whenever axiom 10 is not appropriate, we are not in the realm of "complete ignorance."

There is a tendency to read too much into this axiom.   Some hold that adopting axiom 10 is essentially equivalent to assuming that each state is equally likely.   Although this is true when a suitable collection of the other axioms is added to 10 (see below), it is not true for 10 alone, or for 10 and certain of the other axioms.   For example, if axiom 7''' is accepted (i.e., A' is optimal only if it is optimal in each paired comparison), then axiom 10 has the following interpretation: If A' is optimal and if the utilities for A'', $[u(A'', s_1), u(A'', s_2), \cdots, u(A'', s_n)]$, are a permutation of those for A', $[u(A', s_1), u(A', s_2), \cdots, u(A', s_n)]$, then A'' is also optimal.   This does not require that the states of nature be equally likely, since the maximin criterion, for example, satisfies this requirement.

It is very easy to see the role that axiom 10 plays when appended to axioms 1, 3, 4, 5, 7', 8, and 9.   As a consequence of these other axioms, almost everything hinges on an *a priori* probability distribution over the states of nature.   Yet, if we must be indifferent to the labeling of the states, it can be shown that the only possible *a priori* distribution must make each state equally likely, i.e., it must be the one which assigns the probability $1/n$ to each state if there are n states in all.

Thus, by coupling axiom 10 with the theorem we stated for these seven axioms, we know that an act is optimal only if it yields the highest average utility (the average being taken over all n utilities associated with the act and where each utility number is given weight $1/n$).   But with axiom 10 added it can be shown that the "only if" assertion can be strengthened to "if and only if," i.e., *if an act has the highest average utility, then it is indeed optimal.* To round out the picture, the same result holds if for axiom 7' one substitutes axioms 6 and 7. (Note that 7' implies 7 directly, and when it is bolstered by 4 and 5 it also implies 6.)

In summary, then, axioms 1 through 10 (actually 2 is not needed) characterize the criterion based on the principle of insufficient reason, i.e., it is the unique criterion which satisfies them. This result is due to Chernoff [1954].

The maximiners and minimaxers, however, argue that, although axiom 10 is all right, it does not go far enough in characterizing the notion of "complete ignorance." For example, consider the two d. p. u. u.'s

$$\textbf{D. P. U. U. 1} \qquad\qquad \textbf{D. P. U. U. 2}$$

$$\begin{array}{c}\\ A_1 \\ A_2 \end{array}\begin{array}{cccc} s_1 & s_2 & s_3 & s_4 \\ \left[\begin{array}{cccc} 6 & 2 & 2 & 2 \\ 0 & 5 & 5 & 5 \end{array}\right] \end{array} \quad \text{and} \quad \begin{array}{c}\\ A_1 \\ A_2 \end{array}\begin{array}{cc} s_1 & s_2 \\ \left[\begin{array}{cc} 6 & 2 \\ 0 & 5 \end{array}\right]. \end{array}$$

According to the criterion based on the principle of insufficient reason, A_2 is optimal for d. p. u. u. 1 and A_1 for d. p. u. u. 2. But *if one is truly completely ignorant* about the true state in each problem aren't these problems identical? In d. p. u. u. 1, s_2, s_3, and s_4 can be strategically lumped into one state—call it s*. True, s* is "not less likely" to be true than either s_2, s_3, or s_4, but if we are completely ignorant we cannot say anything about s_1 versus s*. The principle of insufficient reason interprets complete ignorance as "each state being equally likely," so s* must be treated as if it were "three times as likely" as s_1, and, therefore, this criterion chooses A_2. But, in considering s* as more likely than s_1, one admits that he is *not completely ignorant.* According to some, the very essence of complete ignorance is to treat d. p. u. u.'s 1 and 2 as equivalent. They would add that one is almost never in a state of *complete* ignorance, but they would insist that, if one wants to list reasonable desiderata for criteria which purport to handle this case, the following axiom is indispensable.

Axiom 11. *If a d. p. u. u. is modified by deleting a repetitious column (i.e., collapsing two states which yield identical payoffs for all acts into one), then the optimal set is not altered.*

Axiom 11 can be strengthened to:

Axiom 11'. *If a d. p. u. u. is modified by deleting a column which is equivalent to a probability mixture of other columns, then the optimal set is not altered.*

If one feels strongly about the criterion based on the principle of insufficient reason and also wants to endorse axiom 11, the two can be combined into this criterion: In any d. p. u. u. delete all repetitious columns, and in this modified d. p. u. u. choose those acts having the highest average payoff (equal weights).   This criterion fails to satisfy axiom 7 or any of its variations.   For example, consider the following d. p. u. u.:

$$
\begin{array}{c c c c}
 & s_1 & s_2 & s_3 \\
A_1 & \begin{bmatrix} 11 & 0 & 0 \\ A_2 & 0 & 10 & 10 \\ A_3 & 9 & 9 & 0 \end{bmatrix}
\end{array}
$$

If the choice is confined to A_1 or A_2, A_1 is optimal (since by axiom 11 s_3 is deleted).   If A_3 is added to A_1 and A_2, then s_3 cannot be deleted, and according to this criterion A_2 is changed from non-optimal to optimal whereas A_1 is changed from optimal to non-optimal.   Thus, any variant of axiom 7 is contradicted.

Axioms 10 and 11 together are said to characterize "complete ignorance."   Although axioms 10 and 11 are compatible, and axioms 1 to 9 are compatible, all eleven obviously are not.   Something will have to be deleted, and one possible candidate is Rubin's axiom 8—which amounts to saying that the addition of a constant to a column has no effect on the optimal set.   The Hurwicz α-criteria, modified to the extent of deleting all weakly dominated acts before applying the criteria, satisfy axioms 1 through 6, plus any version of 7, plus 10 and 11.   The maximin (utility) criterion, modified in the same way, satisfies these and axiom 9 in addition.

Arrow [1953], modifying a result due to Hurwicz [1951 a], has proved the following result: If a criterion satisfies axioms 1, 3, 4, 7″, 10, 11, then it takes into account only the minimum and maximum utility associated with each act.   However, the particular way these maxima and minima are to be used to select a specific act as best is left unresolved by the group of axioms.   For example, all the Hurwicz α-criteria are compatible with this axiom set.   Another compatible criterion is: An act is optimal if and only if either its minimum is larger than the minimum of any other act or, when there are ties for the largest minimum, it has the largest maximum among those acts with the largest minimum.

Suppose that we let m denote the minimum utility associated with an act and M the maximum, then if we accept this axiom set (1, 3, 4, 7″, 10, 11) the crux of the problem is to decide upon an ordering between pairs (m', M') and (m'', M'').   If we also demand that axiom 2 be met, the criterion must yield the same ordering when we change the utilities by a linear transformation.   Thus, if the criterion selects (m', M') over $(m''$,

M'') then it must also select $(am' + b, aM' + b)$ over $(am'' + b, aM'' + b)$, where $a > 0$. In this connection, the following can be shown: If, for the d. p. u. u.

$$\begin{array}{c} & \begin{array}{cc} s_1 & s_2 \end{array} \\ \begin{array}{c} A_1 \\ A_2 \end{array} & \left[\begin{array}{cc} 0 & 1 \\ x & x \end{array} \right], \end{array}$$

there exists a number α such that we would say A_1 is optimal for all $x \leqslant 1 - \alpha$, and A_2 is optimal for all $x \geqslant 1 - \alpha$, and if we demand that a criterion yielding this decision also satisfy axioms 1, 2, 3, 4, 7$''$, 10, and 11, then it must be Hurwicz's with index α.

The approach just used, which will be employed again in the next section, warrants a comment. We first commit ourselves to a class of axioms, thereby restricting the class of potential criteria. Second, we consider a simple class of d. p. u. u.'s for which we feel able to make subjective commitments as to the optimal sets. If our choice of axioms and special cases is clever, then by using the axioms we can logically extend the consistent decisions given for a simple class of d. p. u. u.'s to a precise formula which resolves all d. p. u. u.'s.

▶ Milnor [1954] states a set of requirements for reasonable decision criteria, where the criteria do not select an optimal set of acts but yield a complete (transitive) ordering for all acts. The analysis is much simpler in these terms. We outline his work here with a minimum of comments. In parenthesis after each axiom we give the nearest corresponding statement in terms of optimal sets.

1. **Ordering.** *All acts must be completely ordered.* (1.)
2. **Symmetry.** *The ordering is independent of labeling of rows and columns.* (3 and 10.)
3. **Strong domination.** *Act A' is preferred to A'' if A' strongly dominates A''.* (4 and 5.)
4. **Continuity.** *If A' is preferred to A'' in a sequence of d. p. u. u.'s, then A'' is not preferred to A' in the limit d. p. u. u.* [A sequence of d. p. u. u.'s converge to a limiting d. p. u. u. if the utility numbers for each (act, state) pair converge to the utility number of the (act, state) pair of the limit d. p. u. u.] (No correlate.)
5. **Linearity.** *The ordering is not changed by linear utility transformations.* (2.)
6. **Row adjunction.** *The ordering between old rows is not changed by adding a new row.* (7, 7$'$, 7$''$, 7$'''$.)
7. **Column linearity.** *The ordering is not changed by adding a constant to a column.* (8.)
8. **Column duplication.** *Adding an identical column does not change the ordering.* (11.)
9. **Convexity.** *If A' and A'' are indifferent in the ordering, then neither A' nor A'' is preferred to $(\frac{1}{2}A', \frac{1}{2}A'')$.* (9.)
10. **Special row adjunction.** *Adding a weakly dominated act does not change the ordering of old acts.* (6.)

Milnor summarizes his results in the table.

Axiom	Laplace	Wald	Hurwicz	Savage
1. Ordering	⊗	⊗	⊗	⊗
2. Symmetry	⊗	⊗	⊗	⊗
3. Str. dom.	⊗	⊗	⊗	⊗
4. Continuity	x	⊗	⊗	⊗
5. Linearity	x	x	⊗	x
6. Row adj.	⊗	⊗	⊗	$\cdots$
7. Col. lin.	⊗	$\cdots$	$\cdots$	⊗
8. Col. dup.	$\cdots$	⊗	⊗	⊗
9. Convexity	x	⊗	$\cdots$	⊗
10. Sp. row adj.	x	x	x	⊗

In this tabulation Laplace refers to the criterion based on the principle of insufficient reason, Wald to the maximin utility criterion, Hurwicz to the α optimism-pessimism criteria, and Savage to the minimax risk or regret criterion. An x means the criterion and the axiom are compatible. Each criterion is characterized by the axioms marked ⊗.

Note that, unlike Chernoff's characterization of the Laplace criterion, Milnor's does not require the convexity axiom. This discrepancy seems strange until it is recalled that Milnor's axioms 1 and 6 are stronger then their correlates in Chernoff's system. Milnor demands a complete ordering, not just an optimal set, and his sixth axiom corresponds to axiom 7″ (cf. p. 289) which is stronger than axiom 7 used by Chernoff.

Another point of discrepancy is Milnor's use of strong domination and continuity. All four of the criteria satisfy these conditions, but they would not if weak domination (i.e., axiom 5, p. 287) were employed instead of strong domination. To see this, consider the d. p. u. u.

$$\begin{array}{c} \\ A_1 \\ A_2 \end{array}\begin{array}{cc} s_1 & s_2 \\ \left[\begin{array}{cc} 0 & 4 \\ 1/n & 3 \end{array}\right]. \end{array}$$

By the maximin utility criterion, A_2 is preferred to A_1 for all n, but in the limit as n increases we obtain

$$\begin{array}{c} \\ A_1 \\ A_2 \end{array}\begin{array}{cc} s_1 & s_2 \\ \left[\begin{array}{cc} 0 & 4 \\ 0 & 3 \end{array}\right], \end{array}$$

so by weak domination A_1 is preferred to A_2. Thus, that criterion cannot satisfy both weak domination and continuity.   ◀

13.5 THE CASE OF "PARTIAL IGNORANCE"

A common criticism of such criteria as the maximin utility, minimax regret, Hurwicz α, and that based on the principle of insufficient reason is that they are rationalized on some notion of *complete* ignorance. In practice, however, the decision maker usually has some vague partial information concerning the true state. No matter how vague it is, he may not wish to endorse any characterization of complete ignorance (e.g., axiom 10 or 11), and so the heart is cut out of criteria based on this notion. The present section is devoted to suggestions for coping with this hiatus between complete ignorance and risk.

As background for this discussion, consider a contestant on the famous $64,000 quiz show who has just answered the $32,000 question correctly. His problem is whether to choose act A_1, to try for $64,000, or to choose act A_2, to stop at $32,000. His d. p. u. u. takes the form:

The $64,000 question is one that the contestant

	s_1 = could answer	s_2 = could not answer
A_1 = try for $64,000	Obtain $64,000 (taxable) plus prestige, publicity, etc.	Obtain a consolation prize of a Cadillac, plus knowledge that $32,000 (taxable) was lost
A_2 = stop	Obtain $32,000 (taxable), get less prestige and publicity than for the (A_1, s_1) pair	Same as (A_2, s_1) pair

We assume that in utility terms the problem reduces to the form:

$$\begin{array}{cc} & \begin{array}{cc} s_1 & s_2 \end{array} \\ \begin{array}{c} A_1 \\ A_2 \end{array} & \begin{bmatrix} 1 & 0 \\ x & x \end{bmatrix}. \end{array}$$

Let us suppose, further that no other contestant has ever tried for the $64,000 question. For all our contestant knows, the difficulty of the question can run the gamut from the impossible to "What was the color of Washington's white horse?" Everything hinges on his appraisal of the relative possibilities of s_1 and s_2. He might take the point of view that he is completely ignorant of the true state, but it is much more likely that he would take into consideration such intangibles as: (a) the public reaction against the sponsor if the question were too difficult; (*b*) the bad precedent that would be set if the question were too easy; and (*c*) the trend in question difficulty in going from $4000 to $8000, from $8000 to $16,000, and from $16,000 to $32,000. Although the problem surely is not in the realm of complete ignorance, it is not obvious how this vague information can be systematically processed.

Suppose, after due deliberation, the contestant chooses A_1. We can then assert that he behaved as if it were meaningful to assign an *a priori* probability to s_1 of x or greater.[5] Conversely, one is tempted to say that, if the "subjective probability" of s_1 is x or greater, then A_1 should be chosen. It is this net of ideas which will be partially formulated now.

We shall first report on the school led by Savage [1954], which holds the view that by processing one's partial information (as evidenced by one's responses to a series of simple hypothetical questions of the Yes-No variety) one can generate an *a priori* probability distribution over the states of nature which is appropriate for making decisions. This reduces the decision problem from one of uncertainty to one of risk. The *a priori* distribution obtained in this manner is called a *subjective* probability distribution.

Savage, in his *The Foundations of Statistics*, "develops, explains, and defends a certain abstract theory of behavior of a highly idealized person faced with uncertainty." The theory is based on a synthesis of the works of Bruno de Finetti on a personalistic view of probability and of the modern theory of utility due to von Neumann and Morgenstern. Since Savage expounds his position with vigor and clarity, we shall merely attempt to capture what, to our minds, is the most salient contribution of his school. Furthermore, we shall not follow Savage's development of the subject; rather we shall graft the new concepts onto the development given in the two previous sections.

Let $s_1, s_2, \cdots, s_n$ be a labeling of the possible states of nature for some concrete decision problem. Each of these labels refers to specific real world phenomena and we (in the role of a decision maker) might feel that some states are more plausible than others. Suppose, furthermore, that after reflection we are convinced that we want to be consistent when facing problems of this type—consistent in the sense that our adopted decision criterion should satisfy axioms 1, 3, 4, 5, 7', 8, and 9. Since these axioms do not in any way refer to our state of ignorance concerning the true state of nature, we are free to commit ourselves to them independent of any information we possess or subjective feelings we have as to the relative plausibility of the different states. Now, as we previously noted, any

---

[5] An equally valid interpretation of this single choice is that the subject applied a Hurwicz criterion with index $\alpha \leqslant 1 - x$.

criterion which satisfies these axioms must select as optimal a subset of the acts which are best against some specific *a priori* distribution. Furthermore, this *a priori* distribution is independent of the particular acts available in a given problem (as long as the states $s_1, s_2, \cdots, s_n$ are involved) since adding new acts does not change non-optimal acts into optimal ones. Thus, it is reasonable to assert that if there exists an "appropriate" *a priori* probability distribution over the states, then this distribution depends solely upon our state of information concerning $s_1, s_2, \cdots, s_n$. The strategy now is to consider a series of simple hypothetical d. p. u. u.'s with these states of nature, to resolve them according to our best intuitive judgement, and then to use these commitments to infer a plausible *a priori* distribution.

Let us illustrate the procedure by a case which involves three specific states $s_1, s_2,$ and s_3. In order to generate an "appropriate" *a priori* distribution over these states let us introduce two hypothetical acts, A_1 and A_2, such that their consequences for the various states have the following monetary equivalences:

$$
\begin{array}{c}
\quad\quad s_1 \quad\ s_2 \quad\ s_3 \\
\begin{array}{c} A_1 \\ A_2 \end{array}
\left[\begin{array}{ccc}
\$0 & \$0 & \$100 \\
\$y & \$y & \$y
\end{array} \right].
\end{array}
$$

Adjust act A_2, i.e., y, until we are indifferent between A_1 and A_2. Suppose the point of indifference (which is assumed to exist) is at \$65. Suppose, further, that we are indifferent between obtaining \$65 for certain and getting \$100 with an objective probability of 0.8 and \$0 with an objective probability of 0.2. Hence the utilities of \$0, \$65, and \$100 can be taken as 0, 0.8, and 1. In utility payoffs we have

$$
\begin{array}{c}
\quad\quad s_1 \quad\ s_2 \quad\ s_3 \\
\begin{array}{c} A_1 \\ A_2 \end{array}
\left[\begin{array}{ccc}
0 & 0 & 1 \\
0.8 & 0.8 & 0.8
\end{array} \right].
\end{array}
$$

Now, indifference between A_1 and A_2 is compatible with an *a priori* distribution only if the *a priori* probability of s_3 is 0.8. If we have no preferences about the states themselves then, as a check and possible short cut, we could ask ourselves: "If we were given the alternative (*a*) of obtaining a prize of x dollars if s_3 turns out to be true and nothing if s_1 or s_2 were true, versus the alternative (*b*) of obtaining a prize of x dollars with objective probability p and nothing with objective probability $1 - p$, for what p would we be indifferent?" To check, we would require that indifference come at $p = 0.8$ independent of the value of x, so long as it is positive! In a similar manner, we could force ourselves to accept a probability assignment for s_2 and for s_1. In practice, however, one's choices for a series of problems—no matter how simple—usually are not consistent. For example, the *a priori* probability assignments for s_1, s_2, s_3 may not add up to 1. Once confronted with such inconsistencies, one should, so the argument goes, modify one's initial decisions in such a manner as to be, consistent. Let us assume that this jockeying—making snap judgments, checking on their consistency, modifying them, again checking on consistency, etc.—leads ultimately to a bona fide *a priori* distribution. Now, if we wish our decision criterion both to satisfy the axioms stated above and to yield results that agree with our by now consistent set of preferences for simple hypothetical problems, then we are committed to a criterion which selects as optimal only acts which are best against this *a priori* distribution.

▶ To describe precisely what Savage means by a consistent set of preferences, we must outline briefly his postulates for a personalistic theory of decision. The assumed ingredients of the decision problem are:

    i. The set of *states* of the world—a set S with (an infinite number of) elements s, s', $\cdots$ and with subsets E, E', $\cdots$ called *events*.

    ii. The set of *consequences*—a set C with elements c, c', $\cdots$.

    iii. The set of *acts*—a set $\mathcal{A}$ with elements A, A', $\cdots$.

    iv. An assignment to each act-state pair (A, s) of a consequence from C which is denoted by $A(s)$.

    v. A binary relation $\succsim$ between pairs of acts which is interpreted to mean "is preferred or indifferent to."

Savage then postulates and defines the following:

Postulate 1.   *The relation $\succsim$ is a weak ordering of the acts, i.e., every pair of acts is comparable and the relation is transitive.*

Definition.   The expression "$A \succsim A'$ given E" means that, if acts A and A' are modified so that their consequences are the same for every state not included in the event E, but if they are not changed for the states in E, then the modification of A is preferred or indifferent to the modification of A'.

This definition is not well defined unless the preference relation between modified acts is required not to depend upon the particular agreement selected for states not in E. The next postulate makes this assumption indirectly.

Postulate 2.   *Conditional preference, as defined above, is well defined.*

Definition.   If $A(s) = c$ and $A'(s) = c'$ for every s in S, then we define $c \succsim c'$ if and only if $A \succsim A'$.

The given A and A' of this definition are called "constant" acts since their consequences are independent of which state holds. The relation $\succsim$ is extended to the set of consequences by identifying each consequence with the constant act which yields it for each state.

Definition.   An event ϕ is called *null* if every pair of acts are indifferent given ϕ, i.e., for every A and A', $A \succsim A'$ given ϕ and $A' \succsim A$ given ϕ.

Postulate 3.   *If E is a non-null event and $A(s) = c$ and $A'(s) = c'$ for all s in E, then $A \succsim A'$ given E if and only if $c \succsim c'$.*

This asserts that conditional preferences do not affect consequence preferences.

Definition. The event E is said to be *not more probable than* the event E' if, whenever

(i) c and c' are any two consequences such that $c > c'$,

(ii) $A(s) = c$ for s in E and $A(s) = c'$ for s not in E,

and

(iii) $A'(s) = c$ for s in E' and $A'(s) = c'$ for s not in E',

then $A' \succsim A$.

Postulate 4. *Probabilitywise, any two events are comparable.*

Postulate 5. *There is at least one pair of acts which are not indifferent.*

Postulate 6. *Suppose $A > A'$. For each consequence c, no matter how desirable or undesirable it may be, there exists a sufficiently fine partitioning of S into a finite number of events such that if either A or A' is modified to yield c for any single event of the partition the preference for A over A' is not changed.*

Postulate 7. *Let A' be an act and let A_s' be the constant act which agrees with A' for the state s. Then,*

(i) $A \succsim A_s'$ *given E for all s in E implies $A \succsim A'$ given E,*

and

(ii) $A_s' \succsim A$ *given E for all s in E implies $A' \succsim A$ given E.*

From these seven postulates Savage is able to show (among other things) the following two theorems.

Theorem. *There exists a unique real-valued function P defined for the set of events (subsets of S) such that*

(i) $P(E) \geqslant 0$ *for all E,*

(ii) $P(S) = 1$,

(iii)   *If E and E' are disjoint, then $P(E \cup E') = P(E) + P(E')$,*

and

(iv)   *E is not more probable than E' if and only if $P(E) \leqslant P(E')$.*

P is called the *personalistic probability* measure reflecting the individual's reported feelings as to which of a pair of events is more likely to occur.

Theorem. *There exists a real-valued function u defined over the set of consequences having the following property: If E_i, where $i = 1, 2, \cdots, n$, is a partition of S and A is an act with consequence c_i on E_i, and if E_i', where $i = 1, 2, \cdots, m$, is another partition of S and A' is an act with consequence c_i' on E_i', then $A \succsim A'$ if and only if*

$$\sum_{i=1}^{n} u(c_i)P(E_i) \geqslant \sum_{i=1}^{m} u(c_i')P(E_i').$$

The function u is called a *utility* function. As in the von Neumann-Morgenstern theory, it is unique up to a positive linear transformation.   ◄

A primary, and elegant, feature of Savage's theory is that no concept of objective probability is assumed; rather a subjective probability measure arises as a consequence of his axioms. This in turn is used to calibrate utilities, and it is established that it can be done in such a way that expected utilities correctly reflect preferences. Thus, Savage's contribution—a major one in the foundations of decision making—is a synthesis of the von Neumann-Morgenstern utility approach to decision making and de Finetti's calculus of subjective probability.

To transform vague information concerning the states of nature into an explicit *a priori* probability distribution, the decision maker has had to register consistent choices in a series of simple hypothetical problems involving these states. No one claims that this is an easy task, but some go so far as to assert that in some contexts even these preliminary choices are too difficult to make with any confidence. They hold, further, that, if consistent responses are forced, the results are not very reliable and to build upon them is a mistake. They feel, introspectively, that, if one could instantaneously wipe out the memory of one's past choices and if the process for obtaining a subjective *a priori* distribution were immediately repeated, the new *a priori* distribution could easily be quite different from the old one.

There are two suggestions in the literature, Hurwicz [1951 b] and Hodges and Lehman [1952], designed to cope partially with this problem. Let A be a generic act in α (the decision maker's strategy set); let $\mathbf{x}$ denote the generic randomized act in X (the set of all randomized acts); let s be a generic state of nature in S (nature's state set); let $\mathbf{y}$ denote an *a priori* probability distribution over S; and let Y be the set of all *a priori* probability distributions. As we have seen, Savage suggests that partial knowledge can be utilized to find a unique *a priori* distribution $\mathbf{y}^{(0)}$, and the decision maker is to choose an A which is best against $\mathbf{y}^{(0)}$. Hurwicz goes in the other direction: he suggests that *partial ignorance* over S can be effectively processed to yield *complete ignorance* over some subset $Y^{(0)}$ of Y. That is, although our knowledge may be insufficient to choose a specific *a priori* distribution in Y, it may be adequate to eliminate certain *a priori* distributions—let the remaining class be $Y^{(0)}$. Hurwicz proposes that the *a priori* distributions in $Y^{(0)}$ should be treated as new states of nature about which one is totally ignorant, and that a criterion based on complete ignorance over these states should be utilized. For example, let $M(\mathbf{x}, \mathbf{y})$ be the utility payoff when the decision maker chooses the randomized act $\mathbf{x}$ and when $\mathbf{y}$ is the *a priori* distribution. To apply the Hurwicz α-criterion, associate to each act **x** the α-index,

$$\alpha m_{\mathbf{x}} + (1 - \alpha)M_{\mathbf{x}},$$

where $m_{\mathbf{x}}$ and $M_{\mathbf{x}}$ are, respectively, the minimum and maximum payoffs[6] which result from **x** as the *a priori* distribution **y** runs over its domain $Y^{(0)}$. Choose an act which yields the highest α-index.

The spirit of Hurwicz's proposal is quite clear, and there are contexts[7] where we feel his specific proposal can be employed. In general, however, we feel that his suggestions are too vague to resolve the problem. Operationally, how does one characterize the elements of $Y^{(0)}$? Even if all "reasonable" **y** are included in $Y^{(0)}$, can't some **y**'s be "more reasonable" than others? Maybe one could capture this differential plausibility for **y**'s in $Y^{(0)}$ by an *a priori* distribution on $Y^{(0)}$. But why stop there? There is a next level, and a next, etc. Of course, expedient compromises can be made, and Hurwicz's original hope still has merit: that from a lot of special decisions about $Y^{(0)}$, one will come closer to extracting faithfully one's partial information about the states than by a forced choice of an *a priori* distribution.

Independently of Hurwicz, Good [1950] has offered much the same suggestion for processing information; however, he subsquently used the maximin criterion rather the α-criteria.

Hodges and Lehmann [1952] also take the position that, in practice, information about states of nature often lies somewhere between complete ignorance and a precise specification of an *a priori* distribution. For example, an *a priori* distribution $\mathbf{y}^{(0)}$ might seem likely and yet not be sufficiently reliable to base decisions on. An act which is best against $\mathbf{y}^{(0)}$ might involve a large risk if some state actually turns out to be true. (Note that Hodges and Lehmann, like most statisticians, phrase their results in terms of risk payoffs rather than in utility payoffs.) So they propose that: (*a*) An act (maybe randomized) be found which minimizes the maximum risk; let its maximum risk be C. (*b*) On the basis of the quantity C and the context of the problem, choose a quantity C_0, greater than C, to serve as the maximum tolerable risk. (*c*) Choose an act **x** which is best against $\mathbf{y}^{(0)}$ *subject to the condition that the act has a maximum*

---

[6] There is some question here of the existence of the minimum and maximum; however, from a mathematical point of view, this can be taken care of easily.

[7] Let the two states of nature be whether a subject does or does not have tuberculosis, and suppose that from medical statistics the proportion of people having T.B. is known to be π. Because the subject is self-selected, we may be unwilling to say that the *a priori* probability of T.B. is π; but we may find it acceptable to say that it is anything greater than or equal to π, and, conceivably, we might behave as if we were completely ignorant as to which value it has in this interval.

risk not greater than C_0.[8]   Naturally, the choice of C_0 will depend upon how much confidence we have in $\mathbf{y}^{(0)}$.

13.6 GAMES AS DECISION MAKING UNDER UNCERTAINTY

The problem of individual decision making under uncertainty can be considered as a one-person game against a neutral nature.   Some of these ideas can be applied indirectly to individual decision making under conflict, i.e., where the adversary is not neutral but a true adversary.   In a two-person non-zero-sum, non-cooperative game, let us refer to player 1 as "the decision maker" and to 2 as "the adversary."   The decision maker wishes to choose an "optimal" set from the set of possible strategies (acts) available to him.   One *modus operandi* for the decision maker is to generate an *a priori* probability distribution over the states (pure strategies) of his adversary by taking into account both the strategic aspects of the game and what "psychological" information is known about his adversary, and to choose an act which is best against this *a priori* distribution.   To determine such a subjective *a priori* distribution, the decision maker might imagine a series of simple hypothetical side bets whose payoffs depend upon the strategy his adversary employs.   This is easier said than done, however, since the decision maker cannot ignore the possibility that his adversary will attempt to hypothecate such a procedure for him and will adjust his choice of strategy accordingly.   In other words, the decision maker's very selection of an *a priori* distribution for his adversary sets up indirect forces to alter this initial choice.   If such is the case, one can argue that the decision maker should keep on modifying the *a priori* distribution until this alleged indirect feedback no longer produces any change—until there exists an equilibrium in the decision maker's mind. We suspect that, roughly, this is the way games of strategy are actually played.   If in a given situation the theory is clear cut and if a decision maker knows that his adversary will comply with the theory, then, in a sense, the theory defines the decision maker's choice of an *a priori* distribution for his adversary.

Of course, a decision maker may not feel very confident in his subjective appraisals of his adversary, and so he might want to compromise in some way.   For example, he might use the compromise suggested by Hodges and Lehmann [1952] and discussed in the preceding section.   They state:

---

[8] Once $\mathbf{y}^{(0)}$ is chosen, the payoff for $\mathbf{x}$ is a linear function.   Mathematically, then, the problem is one of minimizing a linear function subject to linear inequalities—i.e., a linear-programing problem (see section 2.3 and Appendix 5).   Because of the equivalence between linear programing and two-person zero-sum game theory, it is reasonable that game theory should be pertinent in proving theorems in this area.   It is!

"The formulations given here may be applicable also to games played against an opponent rather than against Nature. This would be the case (in the two-person zero-sum game) if one believed from past experience that the opponent is likely to make certain mistakes. One could then take advantage of these and still protect oneself in case the opponent has improved." Or, the decision maker might use the Hurwicz proposal and maximin, or he might use an α-index over some suitably chosen restricted class of *a priori* distributions.

▶ It is interesting to reconsider the appropriateness of the axioms for a reasonable criterion when nature is replaced by an intelligent adversary. Axioms 1 through 5 seem equally acceptable in this interpretation. Axioms 6 and all versions of 7—the independence of irrelevant alternatives—are open to the obvious criticism that adding a new act for the decision maker can affect the strategic position of the adversary and therefore the decision maker should reappraise the relative merits of the old acts. The minimax risk criterion of Savage, which was mainly criticized on the basis of its non-independence of irrelevant alternatives, should therefore be re-evaluated. Rubin's axiom 8 must be modified slightly in order for it to make sense in this context. Recall that game 1 is played with probability p and game 2 is played with probability $1 - p$. Assume that in game 2 the payoffs to player 1 are constant within any column and that the payoffs to player 2 are constant within any row (remember the game is non-zero-sum). In this case, the modified axiom asserts that the decision maker should behave in the same way both in the mixture of the two games and in game 1. This modified axiom seems just as reasonable in this context as the original did in its context. Axiom 9 (convexity of the optimal set) is just as reasonable as before. Axiom 10 (the optimal set for the decision maker should not depend upon the labeling of states for his adversary) seems more universally applicable in the conflict context than it did in the original context. As originally proposed, this axiom was designed to capture the notion of complete ignorance, but no such interpretation need be implied by its use in the present context. Axiom 11 needs to be slightly modified: If two columns have identical payoffs for both the decision maker and his adversary, then one column can be deleted without changing the decision maker's optimal set. Axiom 11' is modified in an analogous manner. The modified axioms 11 and 11' seem quite reasonable.

Certain weak implications follow from these axioms. For example, the modified axiom 8 rules out the Hodges-Lehmann proposal, the maximin (utility) criterion (even modified for admissibility), and the Hurwicz α-criterion (even when applied to restricted subsets of *a priori* distributions for the adversary). We cannot conclude from any subset of these axioms, however, that *all* optimal acts for a specific game must be best against some specific *a priori* distribution for the adversary.

For a two-person game-like situation where each player knows his own payoff, but nothing about his opponent's payoff, the axioms designed for decision making under uncertainty can be interpreted directly: they are all meaningful—but not necessarily reasonable. In essence, a player can treat his opponent's pure strategies as states, and his opponent's choice as the "true state." Certain two-person non-zero-sum, non-cooperative games—especially with imperfections of knowledge—are close to this ideal type.   ◀

Relatively little is known about n-person games ($n \geqslant 2$) against nature. By this we mean the following: Let α_1, α_2, $\cdot \ \cdot \ \cdot$, α_m be the pure strategies of player 1; let β_1, β_2, $\cdot \ \cdot \ \cdot$, β_n be the pure strategies of player 2; let the states of nature be denoted by s_1, s_2, $\cdot \ \cdot \ \cdot$, s_t. Corresponding to a triple (α_i, β_j, s_k) there are payoffs $a_{ij}^{(k)}$ and $b_{ij}^{(k)}$ to 1 and 2 respectively. If the players have "no information" about the "true state," what should they do? An example of a two-person game against nature—the prisoner's dilemma game repeated an indefinite number of times—was described in Chapter 5. Another two-person game against nature, which Robbins [1950] calls the "competing estimation problem," has been considered. Two statisticians, with the same experimental evidence, have to estimate an unknown parameter (e.g., the mean of a normal distribution) and the payoff is $+1$ to the statistician whose guess comes closest to the true parameter and -1 to the other statistician. Robbins, however, considers only the case where an *a priori* distribution over the states (i.e., the set of parameter values) is given, or where such a distribution can be partially inferred from past problems.

▶ The two-person game against nature may be given an alternative "realistic" interpretation. As before, let $a_{ij}^{(k)}$, $b_{ij}^{(k)}$ be the payoffs to players 1 and 2 when state s_k is true, but let us suppose that s_k is not under the control of a "neutral" nature, but of a benevolent third party, which we can conveniently think of as the government or as a planner. The motivation is this: In Chapter 5 we established that in games without preplay communication, equilibria can exist which are far from Pareto optimal. A planning agency, instead of dictating the actions of its subjects, could attempt a form of planned decentralization which would exploit the selfish aims of the players of the "game." The problem for the planner is to so tinker with the rules of the game that the members of the society in pursuing their own ends will be forced into an equilibrium which is Pareto optimal, or nearly so. In effect, then the planner controls s_k in a manner such that the payoffs $a_{ij}^{(k)}$, $b_{ij}^{(k)}$ are jointly desirable from the planner's viewpoint. He can thus consider himself a third player (or in the general case, an $(n + 1)$st player), with, as a payoff, a composite index which takes into account both the "social desirability" of the payoffs to the other players and the penalty (psychological, political, and financial) the planner pays because of his involvement. One type of strategy a planner can profitably use in some situations is to be unpredictable, and so to play artificially the role of nature insofar as the other players are concerned.   ◀

Milnor [1951] defines a pair of mixed strategies $(\mathbf{x}^{(0)}, \mathbf{y}^{(0)})$ to be in equilibrium if each strategy of the pair is optimal in the game against nature when the other strategy of the pair is used by the other player. Existence of equilibrium pairs depends upon the optimality criterion employed in the games against nature; Milnor gives certain requirements on optimality criteria to ensure existence. The conceptual generalization from two-person to n-person non-cooperative games against nature is straightforward, and Milnor's results also apply to this case.

Many conflict situations, not bona fide games because the player's knowledge is limited (e.g., with respect to strategy domains, utility payoffs, etc.), can be considered formally to be games against nature. These games, however, are quite difficult to formulate realistically, since it is necessary to specify each player's *a priori* information about the states of nature.

13.7 STATISTICAL DECISION MAKING—FIXED EXPERIMENTATION

Classical statistical inference is usually compartmentalized into two categories: (*a*) the theory of testing hypotheses, and (*b*) the theory of estimation. The theory of confidence estimation is then introduced as a conceptual generalization of the theory of (point) estimation, but in technical detail it is more intimately connected with the theory of testing hypotheses. For our purposes it will be easier to categorize inference problems according to: (*a*) the number of states of nature, (e.g., exactly two states, a finite number of states, a continuum of states), (*b*) the number of pure terminal acts available, and (*c*) the type of experimental evidence which is available or can be obtained.

In each case, our strategy will be to reduce the statistical decision problem to one of decision making under uncertainty. We will adopt the formulation of the statistical decision problem due to Wald [1950 *a*], and we will show where the more classical formulations fit into the overall picture.

Illustration. An example of a two-state, two-act problem where the type of experimentation is fixed.

The diagnostic problem of deciding whether or not a particular patient is tubercular can be systematized as follows:

	State of Nature	
	s_1 = Patient is Tubercular	s_2 = Patient is not Tubercular
A_1 = Assert Patient is Tubercular	Classify tubercular correctly	Misclassify a non-tubercular
Act A_2 = Assert Patient is not Tubercular	Misclassify a tubercular	Classify non-tubercular correctly

Often, to help decide which act to choose, an experiment $\mathcal{E}$ is performed on the subject (e.g., an X-ray, a sputum test, a guinea pig test, or some combination of these). Let $\mathcal{O}_1, \mathcal{O}_2, \cdots, \mathcal{O}_r$ be the set of possible outcomes[9] of experiment $\mathcal{E}$. A *decision rule* (or strategy) is an overall pre-

---

[9] The set of outcomes $\{\mathcal{O}_1, \mathcal{O}_2, \cdots, \mathcal{O}_r\}$ is called the *sample space* of $\mathcal{E}$.

scription which associates[10] to each outcome a precise terminal act. Since there are two acts possible for each outcome, and r possible outcomes, there are 2^r possible decision rules. Let us list these as D_1, D_2, $\cdots$, $D_i \cdots$, D_{2^r}. Obviously we would like to adopt a rule which associates A_1 to the outcomes which are "most likely" to occur when s_1 is true and A_2 to outcomes which are "most likely" to occur when s_2 is true.

To formalize this, suppose that as part of the givens of the problem we are told the probability of each outcome when s_1 is true and when s_2 is true. To evaluate the decision rule D_i, we first compute its performance under s_1 and then under s_2.

Let

$P_1(A_2 \mid D_i)$ = the probability that, when s_1 is true, experiment $\mathcal{E}$ results in an outcome for which D_i associates act A_2 (i.e., the probability of D_i resulting in a misclassification of an s_1 patient).

$P_2(A_1 \mid D_i)$ = the probability that, when s_2 is true, experiment $\mathcal{E}$ results in an outcome for which D_i associates act A_1 (i.e., the probability of D_i resulting in a misclassification of an s_2 patient).

In statistical lingo, $P_1(A_2 \mid D_i)$ and $P_2(A_1 \mid D_i)$ are the probabilities of errors of types 1 and 2, respectively. Naturally, one wishes to choose D_i to make these two probabilities small. The rub is that, if one of these probabilities is decreased by a judicious choice of a decision rule, the other is invariably increased, and so a tug of war exists.[11]

The consequence of a (D_i, s_1) pair is a lottery with "prizes" (A_1, s_1), the correct classification of a tubercular, and (A_2, s_1), the incorrect classification of a tubercular, with probabilities $1 - P_1(A_2 \mid D_i)$ and $P_1(A_2 \mid D_i)$, respectively. Similarly, the consequence of a (D_i, s_2) pair is a lottery with "prizes" (A_1, s_2), the incorrect classification of a non-tubercular, and (A_2, s_2), the correct classification of a non-tubercular, with probabilities $P_2(A_1 \mid D_i)$ and $1 - P_2(A_1 \mid D_i)$ respectively. Assume that the decision maker's preferences for lotteries involving the consequences of the pairs (A_1, s_1), (A_1, s_2), (A_2, s_1) and (A_2, s_2) may be faithfully summarized

[10] More formally, a decision rule is a *function* whose domain is the sample space of $\mathcal{E}$ and whose range is the set of terminal acts $\{A_1, A_2\}$.

[11] This problem is solved in the classical Neyman-Pearson sense by the following *ad hoc* rule: Select a value α_0 (like 0.05 or 0.01) and find the rule D which minimizes $P_2(A_1 \mid D)$ subject to the restriction that $P_1(A_2 \mid D) \leqslant \alpha_0$. How α_0 is chosen is usually not made explicit, but it obviously should depend upon the relative seriousness of different types of errors (i.e., on utility preferences) and on our *a priori* partial knowledge concerning the relative likelihoods of s_1 and s_2.

by a (linear) utility function, and let the utilities of the four basic consequences be

$$
\begin{array}{cc}
 & s_1 \quad s_2 \\
\begin{array}{c} A_1 \\ A_2 \end{array}
\begin{bmatrix} u_{11} & u_{12} \\ u_{21} & u_{22} \end{bmatrix},
\end{array}
$$

where an arbitrary but fixed choice of origin and unit has been made. Thus, the utility for the consequence

i. (D_i, s_1) is $u_{11}[1 - P_1(A_2 \mid D_i)] + u_{21}P_1(A_2 \mid D_i) \equiv u(D_i, s_1)$, say.
ii. (D_i, s_2) is $u_{12}P_2(A_1 \mid D_i) + u_{22}[1 - P_2(A_1 \mid D_i)] \equiv u(D_i, s_2)$, say.

With these assumptions and notations, our original problem shapes up as follows: To choose among the acts $D_1, D_2, \cdots, D_{2^r}$ (these acts in this modified problem are really decision rules for the statistical decision problem), given the two states of nature s_1 and s_2 and the utility payoff array:

$$
\begin{array}{c}
 \\
D_1 \\
D_2 \\
\cdot \\
\cdot \\
\cdot \\
D_{2^r}
\end{array}
\begin{array}{cc}
s_1 \qquad\qquad s_2 \\
\begin{bmatrix}
u(D_1, s_1) & u(D_1, s_2) \\
u(D_2, s_1) & u(D_2, s_2) \\
\cdot & \cdot \\
\cdot & \cdot \\
\cdot & \cdot \\
u(D_{2^r}, s_1) & u(D_{2^r}, s_2)
\end{bmatrix}
\end{array}
\qquad \text{(utility payoff).}
$$

In this formulation, the problem is nothing but a decision problem under uncertainty (d. p. u. u.), and our previous discussion is directly applicable.

Now, let us return to the general case where there are n states of nature $s_1, s_2, \cdots, s_n$ and m acts $A_1, A_2, \cdots, A_m$. The analysis of the example can be extended in the obvious way, as we shall see. As part of the data of the problem we are given:

1. The utility payoffs u_{ij}, $i = 1, 2, \cdots, m$, $j = 1, 2, \cdots, n$, where u_{ij} is the utility of the consequence associated with the act-state pair (A_i, s_j).

2. An experiment $\mathcal{E}$, and a probability distribution over the set of possible outcomes of $\mathcal{E}$ for each state of nature.

A decision rule D assigns to each possible outcome of $\mathcal{E}$ a unique act.[12] Consider the consequence of a decision rule D when s_j is true, i.e., a lottery whose prizes are the consequences of the act-state pairs (A_1, s_j), $(A_2, s_j), \cdots, (A_m, s_j)$. The probability of the prize (A_i, s_j) is the probability that, if s_j is true, an outcome of $\mathcal{E}$ will occur such that D prescribes A_i. Let this probability be denoted by $P_j(A_i \mid D)$. Hence,

---

[12] If there are r possible outcomes, then there will be m^r possible decision rules.

the consequence of a (D, s_j) pair is a lottery whose utility is

$$P_j(A_1 \mid D)u_{1j} + P_j(A_2 \mid D)u_{2j} + \cdots + P_j(A_m \mid D)u_{mj},$$

which we call $u(D, s_j)$. The appraisal of D is then given by

$$\begin{array}{ccccc} & s_1 & s_2 & \cdots & s_n \\ D: [& u(D, s_1) & u(D, s_2) & \cdots & u(D, s_n)]. \end{array}$$

We have thus succeeded in transforming the problem into a choice among decision rules, where the payoff for each decison rule depends upon which state of nature is true.   This is the typical form of a d. p. u. u., and so our previous discussion applies directly.

Suppose that an *a priori* distribution is given[13] over the states of nature, and let the probability of the states be $P(s_1)$, $P(s_2)$, $\cdots$, $P(s_n)$. If the experiment $\mathcal{E}$ were not performed, the choice problem would be one of risk rather than uncertainty, and the utility of act A_i would be

$$P(s_1)u_{i1} + P(s_2)u_{i2} + \cdots + P(s_n)u_{in}.$$

In this case, what purpose does performing the experiment $\mathcal{E}$ serve? Since the likelihood of an outcome of $\mathcal{E}$ depends upon the true state, it seems reasonable that, once the outcome of $\mathcal{E}$ is known, the probability assignment over the states should be altered.   Let $\mathcal{O}$ be the outcome of $\mathcal{E}$, and let the conditional probability of s_j given $\mathcal{O}$ be denoted[14] by $P(s_j \mid \mathcal{O})$. But now that $\mathcal{E}$ has been conducted and $\mathcal{O}$ observed, we are back to the original problem of the optimal selection of an act when the probabilities of the states of nature are known—however, these probabilities are now $P(s_1 \mid \mathcal{O})$, $P(s_2 \mid \mathcal{O})$, $\cdots$, $P(s_n \mid \mathcal{O})$.   The utility of act A_i, when $\mathcal{O}$ is observed, is

$$P(s_1 \mid \mathcal{O})u_{i1} + P(s_2 \mid \mathcal{O})u_{i2} + \cdots + P(s_n \mid \mathcal{O})u_{in},$$

and the act which has the highest utility is optimal.   Thus, to each outcome $\mathcal{O}$ we can associate the act which has the highest weighted average of utility payoffs—the weights depending upon outcome $\mathcal{O}$.   This prescription, which associates to each outcome $\mathcal{O}$ one of the particular acts

---

[13] Recall that, in the Savage subjectivist school, an *a priori* distribution is always meaningful and essentially given.

[14] It can be shown that

$$P(s_j \mid \mathcal{O}) = \frac{P(\mathcal{O} \mid s_j)P(s_j)}{P(\mathcal{O} \mid s_1)P(s_1) + P(\mathcal{O} \mid s_2)P(s_2) + \cdots + P(\mathcal{O} \mid s_n)P(s_n)},$$

where $P(\mathcal{O} \mid s_j)$ is the probability (likelihood) of $\mathcal{O}$, given that s_j is true. This expression is known as *Bayes' formula*.

described above, is known as the *Bayes decision rule* against the *a priori* distribution $\{P(s_1), P(s_2), \cdots, P(s_n)\}$.

The following point can be easily verified: Among all decision rules D, the Bayes rule maximizes the index

$$P(s_1)u(D, s_1) + P(s_2)u(D, s_2) + \cdots + P(s_n)u(D, s_n),$$

which is associated to each decision rule D. This result nicely tidies up the loose ends of decision making under *risk* in light of additional experimental evidence. In short, the initial probability distribution over the states is changed to the conditional one given by the outcome of the experiment, and then one proceeds as in the case of no experimental evidence.

▶The final topic of this section can be given the elliptic heading "On the equivalence of two methods of randomization." Suppose that $D^{(1)}, D^{(2)}, \cdots, D^{(r)}$ are different (non-randomized) decision rules. If experiment $\mathcal{E}$ has the outcome $\mathcal{O}$, let $D^{(i)}(\mathcal{O})$ denote the act specified by rule $D^{(i)}$. A probability mixture over decision rules, $(p_1D^{(1)}, p_2D^{(2)}, \cdots, p_rD^{(r)})$, where the p_i are non-negative and sum to 1, is analogous to a mixed strategy. Operationally, if such a probability mixture is chosen and the experiment has the outcome $\mathcal{O}$, then act $D^{(i)}(\mathcal{O})$ is adopted with probability p_i. We observe that, although p_i does not depend upon $\mathcal{O}$, $D^{(i)}(\mathcal{O})$ of course does.

Instead of taking mixtures over decision rules, a more general scheme is to define for each possible outcome of $\mathcal{E}$ a probability mixture over the acts. The number of acts used and the probabilities with which they are employed can depend upon the outcome. For example, if $\mathcal{O}'$ occurs we might adopt $(\frac{1}{4}A_3, \frac{1}{2}A_5, \frac{1}{4}A_9)$, whereas if $\mathcal{O}''$ occurs we might adopt $(\frac{1}{5}A_2, \frac{1}{5}A_6, \frac{2}{5}A_9, \frac{1}{5}A_{13})$. Any rule of this type which assigns a mixture of acts to each outcome is called a *randomized decision rule*. The problem is this: Given a randomized decision rule, does there always exist an appropriate probability mixture of non-randomized rules (i.e., rules that prescribe a definite act to each outcome) which will yield the same results? Put another way, are we unduly restricting ourselves by first considering non-randomized rules and then allowing probability mixtures over these, instead of allowing for randomized rules initially? The answer to the first question is Yes; to the second, No. We can exactly match any randomized rule with a probability mixture of non-randomized ones provided the set of outcomes of $\mathcal{E}$ is finite, and, even if the outcome set of $\mathcal{E}$ is infinite, very modest assumptions on the probability measures involved are sufficient to show that for each randomized rule there is an equivalent probability mixture of non-randomized rules— equivalent in the sense that they yield the same utility payoffs for each state of nature.   ◀

13.8 STATISTICAL DECISION MAKING— EXPERIMENTATION NOT FIXED

We consider now the same type of problem as in the preceding section, except that the experiment is not necessarily prescribed in advance.

Again we assume that we have acts $A_1, \cdots, A_m$ and states $s_1, s_2, \cdots, s_n$ and that preferences for consequences of act-state pairs are tautologically mirrored by a utility function. As to experimentation, we might have, for example, a set of possible experiments $\mathcal{E}^{(1)}, \mathcal{E}^{(2)}, \cdots, \mathcal{E}^{(n)}, \cdots$, where $\mathcal{E}^{(1)}$ is an experiment which makes a single observation, $\mathcal{E}^{(2)}$ makes two observations by repeating $\mathcal{E}^{(1)}$ twice, $\cdots$, $\mathcal{E}^{(n)}$ repeats $\mathcal{E}^{(1)}$ n times, etc. We might be interested in the number of observations we should take before coming to a terminal decision. Or, to take a more complicated case, we might wish to employ a sequential plan of experimentation where the decision on taking another observation is made to depend upon the previous observations. Or fancier still, we might wish to make the decision as to the type of observation to be taken at a given stage dependent upon the previous history of experimentation. In short, in the present framework we want to tolerate all sorts of sequential or non-sequential designs of experiments, questionnaires, sampling procedures, etc. We only require that any decision rule (strategy) which the decision maker adopts for experimentation and for eventual terminal action should be explicit in the sense that it must assert unequivocally, prior to any experimentation, exactly what is to be done at each stage as a function of the information available at that stage. Thus for each rule (strategy) one can list, at least conceptually, *all* the possible outcomes of experimentation and of terminal action. The problem in all its complexity reduces simply to a choice among decision rules (strategies).

Let us first evaluate D's performance when s_j is true. It is assumed that each possible outcome that is compatible with D can be given a utility index. This utility index will be a composite of two types of considerations: (*a*) the cost of obtaining the particular outcome (including the cost of time, labor, materials, etc.) and (*b*) the losses due to wrong terminal decisions. But, conditional upon the knowledge that s_j is true, we can again (conceptually) compute the likelihood of each outcome which is compatible with strategy D. Thus, when s_j is true, to each D we have associated a massive lottery: the prizes are the consequences associated to (outcome, s_j) pairs weighted according to probabilities which are computed on the basis of s_j's validity. Let $u(D, s_j)$ be the utility of this lottery, then D is appraised by

$$
\begin{array}{cccc}
s_1 & s_2 & & s_n \\
D: [u(D, s_1), & u(D, s_2), & \cdots, & u(D, s_n)].
\end{array}
$$

Once again we have reduced the given problem to the typical form of a d. p. u. u., and our discussion of this case applies directly.

The following example, due to Radner and Marshak [1954] will serve to illustrate some of the points raised above and to suggest others as well:

We, the decision maker, are given a coin whose probability of landing heads or tails is unknown (to us). The coil is to be tossed by a specific mechanism, and the outcome—heads or tails— will be noted by a reputable outsider but not told to us. We have the choice of guessing heads or tails and our payoff is:

$$
\begin{array}{c}
\phantom{A_1 = \text{Guess Heads}} s_1 = \text{Heads} \quad s_2 = \text{Tails} \\
\begin{array}{c}
A_1 = \text{Guess Heads} \\
A_2 = \text{Guess Tails}
\end{array}
\left[
\begin{array}{cc}
\text{Win \$10} & \text{Lose \$10} \\
\text{Lose \$10} & \text{Win \$10}
\end{array}
\right].
\end{array}
$$

As stated so far, the problem is an ordinary d. p. u. u. Now, let experimentation be introduced. Prior to making our guess, we are given the opportunity to observe this particular mechanism toss the given coin any odd number of times at a flat rate of c dollars per toss. Assume we must state in advance the number of tosses to be made. We will confine ourselves to decision rules that can be summarized by a pair of numbers (n, m), where n refers to the (odd) number of observations to be taken and where m has the following interpretation: if the number of heads is less than m, guess tails; if greater than or equal to m, guess heads. Intuitively, for any n, the most reasonable m is $n/2$, but let us not prejudge the problem.

Since there is obviously an upper bound for n, the choice problem involves a finite number of decision rules.

Let us make the assumption that repeated tosses are independent and that in the long run the ratio of the number of heads to tosses will "stabilize" to some number p, which will be interpreted as the objective probability of the specific coin turning up heads when tossed by the given mechanism. The number p can take on all values from 0 to 1 inclusive, and each value of p will be identified with a possible state of nature. In other words, we have a continuum of states.

▶ Let us evaluate the decision rule (n, m) under the assumption that p is true. Assume that the utility of a dollars is a units (i.e., the utility of money is linear in money). Furthermore, let $B(m, n, p)$ denote probability of getting at least m heads in n tosses when the probability of a head at each toss is p. The evaluation of decision rule (n, m) if p is true is:

Original Toss	No. of Heads in n Trials	Utility of Outcome	Probability
H	Less than m	$-10 - cn$	$p[1 - B(m, n, p)]$
H	At least m	$10 - cn$	$pB(m, n, p)$
T	Less than m	$10 - cn$	$(1 - p)[1 - B(m, n, p)]$
T	At least m	$-10 - cn$	$(1 - p)B(m, n, p).$

To find the utility of (n, m) when p is true (i.e., $u[(n, m), p]$) one must sum the utility of each outcome times its probability over all possible outcomes. When this is done and the expression is simplified, we get:

$$u[(n, m), p] = -cn + 10(1 - 2p) + B(m, n, p)[20(2p - 1)]. \quad \blacktriangleleft$$

The optimal decision rule according to the maximin utility criterion, is *not to take any observations whatsoever regardless of the cost of c,* even if this were as low as 100 observations per penny! Essentially, the reason is that, regardless of the outcome of experimentation, there always remains the possibility that $p = \frac{1}{2}$—and in that case knowing p will not help us. If we are completely pessimistic in outlook why spend any money whatsoever sampling? Just take heads with probability $\frac{1}{2}$. Now suppose one is completely optimistic. Then the best p is 0 or 1, and, if we take exactly 1 observation, we are sure to determine which it is. The Hurwicz α-criterion asserts that at most one observation is ever necessary, and it should be taken only if α, the optimism-pessimism index, is greater than 20 c. In other words if c is a penny, then one should take 1 observation only if $\alpha > 0.2$.

These solutions both go counter to intuition, and therefore the reasonableness of the maximin and the Hurwicz α-criteria is further cast into doubt. The example illustrates a major criticism of these criteria, namely: *They focus so strongly on the best and worst states of nature that often they do not permit one to gather negative information about the plausibility of such states, no matter how slight the cost.*

The minimax risk criterion, on the other hand, does not turn up its nose so easily at cheap experimental evidence. Recall, however, that a major criticism leveled at the minimax risk criterion is that it does not satisfy the independence of irrelevant alternatives axiom. Radner and Marschak illustrate this in terms of the above example as follows: Suppose c is a penny. If we are confined to rules of the form $(n, n/2)$ (i.e., we choose heads if and only if the majority of the n tosses are heads), then the minimax risk criterion yields as the optimal rule: 19 observations with probability 0.9 and 21 observations with probability 0.1. If, however, we may choose from all rules of the form (n, m), where m is not restricted to $n/2$, then the minimax risk criterion suggests taking 37 observations with probability 0.2 and 39 observations with probability 0.8 and (this is the interesting point) heads should be adopted *if and only if heads appears on a majority of the observed tosses.* Thus, in this case when we add a richer variety of rules from which to choose, the criterion selects as optimal a rule which, although originally available, was then non-optimal!

13.9 COMPLETE CLASSES OF DECISION RULES

In a given statistical decision problem, let $\mathfrak{D}$ denote the set of all randomized or non-randomized decision rules (acts) and let D, D', D'', etc., denote specific elements of $\mathfrak{D}$. We have assumed that it is possible in principle to associate a utility payoff $u(D, s)$ to each D in $\mathfrak{D}$ and s in S

(set of states of nature). This may be conceptually plausible, but in many practical examples it is not mathematically feasible with present techniques; however, it is not uncommon that the statistician can prove for specific inference problems

$$u(D', s) < u(D'', s), \qquad \text{for all } s \text{ in } S,$$

(i.e., D'' strongly dominates D') without ever explicitly computing the values of $u(D', s)$, $u(D'', s)$ for any s.

This observation suggests the following definitions which are employed by statisticians:

i. *A complete class of decision rules* is a subset $\mathfrak{D}_0$ of $\mathfrak{D}$ such that for every D in $\mathfrak{D}$ but not in $\mathfrak{D}_0$ there exists a D' in $\mathfrak{D}_0$ which weakly dominates D. [That is,

$$u(D', s) \geqslant u(D, s), \qquad \text{all } s \text{ in } S,$$

and $>$ holds for some s]. A statistician has nothing to lose if he confines his attention to a complete class.

ii. *A minimal complete class of decision rules* is a complete class such that no proper subset of it is also complete.

Recall that a decision rule, D', was said to be *admissible* if D' was not weakly dominated by any other rule D in $\mathfrak{D}$. In decision problems where the sets of terminal acts, states of nature, and outcomes of experimentation are all finite, it can be shown that the set of *all* admissible rules forms a minimal complete class of decision rules; however, in more complex cases this need not be so. Consider the trivial counterexample where there is exactly one state of nature, no experiment, and a countable infinity of terminal acts $A_1, A_2, \cdots$; let the utility of A_i be $1 - 1/i$. Hence A_1 is weakly dominated by A_2, which is weakly dominated by A_3, etc. In this example, every act is weakly dominated by some other act, and so no admissible act exists. Obviously, there exist complete classes. For example, $\{A_j, A_{j+1}, \cdots\}$ is a complete class, but so is $\{A_{j+1}, A_{j+2}, \cdots\}$, and we easily see that there is *no* minimal complete class for this problem.

Statistical analysis of a decision problem is usually broken up into two parts.

Part 1. To ascertain the existence of a minimal complete class, and to characterize it; or, if no minimal complete class exists, to characterize a "reasonably small" complete class.

Part 2. To select an "optimal" decision rule from a complete class.

Although our discussion of various decision criteria has been directed mainly towards the problems of part 2, the bulk of current research publications in statistical decision theory are devoted to topics in part 1.

Although these topics are not conceptually difficult, the mathematical techniques employed are often quite profound. It is commonly thought that the results of two-person zero-sum game theory are important for statistical decision theory solely because of its relation to the maximin (utility) and minimax risk criteria. *However, the most sophisticated and important uses of the two-person theory are in the existence questions of part* 1. Roughly, one finds theorems of this type: such and such a class of decision rules is complete if and only if each game in such and such a family of infinite two-person zero-sum games has a value and player 1 has a maximin strategy. In other words, existence questions in complete class theory are intimately related to existence questions for induced two-person games with an infinite number of pure strategies (which are briefly discussed in Appendix 7). Sizeable portions of Wald's *Statistical Decision Functions* and Blackwell and Girshick's *Introduction to the Theory of Games and Statistical Decisions* are devoted to (1) the existence theory for two-person games with an infinite number of strategies, (2) the relation of complete class theory to game theory, and, (3) the applications of 2 to classical statistical inference problems.

Recent contributions to the existence theory for games with an infinite number of pure strategies have had the peculiar effect of minimizing the importance of games in statistical decision theory. The mathematical techniques employed in these game theory papers can be applied directly to existence questions in statistical decision problems, and no explicit mention of game theory or the minimax theorem need be made. Consequently, future mathematical books on statistical decision theory probably will de-emphasize the importance of game theory.

13.10 CLASSICAL STATISTICAL INFERENCE VERSUS MODERN STATISTICAL DECISION THEORY: SOME VERY BRIEF COMMENTS

Since most social scientists are quite familiar with the conventional topics of statistical inference—testing hypotheses, point estimation, and confidence interval estimation—it is appropriate to trace the relation between these matters and the problem of decision making under uncertainty in the light of experimental evidence. To do this, let us consider a specific policy problem.[15] A new vaccine is developed for immunization against a certain disease, whose effects the Public Health Department

[15] Simpler problems than this one could have been chosen to illustrate our points; however, this one does have the decided advantage in that it shows that the analysis is applicable to cases where there is more than one parameter and where inferences are to be made about a quantity which is a function of the several parameters.

wishes to investigate statistically prior to making any recommendation. Let us suppose they are willing to employ the following model: For those who have and have not been vaccinated, let $p_1^{(0)}$ and $p_2^{(0)}$ denote, respectively, the "true" probabilities that an individual chosen at random from the population will contract the disease within a fixed period of time. Let $\lambda^{(0)} = p_1^{(0)}/p_2^{(0)}$. In practice it is quite possible that $p_1^{(0)}$ and $p_2^{(0)}$ will vary from time period to time period, since there may be epidemics of the disease, etc., but for the purpose of this analysis let us suppose that their ratio $\lambda^{(0)}$ remains invariant.

The following three problems are traditionally considered:

i. *To test the hypothesis* that the true value $\lambda^{(0)}$ is greater than some pre-assigned quantity λ^* versus the alternative that it is smaller than or equal to λ^*.

ii. *To point estimate* the value of $\lambda^{(0)}$ in the sense of guessing, on the basis of a sample, a number which is "close" to the true value $\lambda^{(0)}$.

iii. *To interval estimate* the value of $\lambda^{(0)}$ in the sense of guessing, again on the basis of a sample, an interval which has a "good chance" of containing the true value $\lambda^{(0)}$.

The solution to any of these problems is not usually an end in itself, but rather serves to influence a policy decision. Although it is true that the real world terminal actions which can be employed, the losses due to wrong terminal actions, and the costs of experimentation are not explicitly introduced into any of the problems as formulated, such considerations will certainly influence some arbitrary procedural commitments which must be made to resolve such problems. This will soon be evident.

In what follows we let E be a generic symbol for the strategy of experimentation, and we let D denote a typical decision rule which associates to each outcome of the experiment an appropriate guess for the particular problem at hand. We consider the three problems separately.

First, the testing of an hypothesis: Suppose a strategy pair (E, D) is given, then whether we guess that $\lambda^{(0)} \geqslant \lambda^*$ or not depends upon the chance outcome of an experiment, and the probability of that outcome depends, in turn, upon the true states $p_1^{(0)}$ and $p_2^{(0)}$. Of course, we do not know $(p_1^{(0)}, p_2^{(0)})$; nonetheless, by a probabilistic analysis, we can in principle determine the probability that (E, D) leads to the guess that $\lambda^{(0)} \leqslant \lambda^*$, given the assumption that $(p_1^{(0)}, p_2^{(0)})$ is a specific pair of numbers (p_1, p_2). Symbolically, we can denote this probability as

$$P_{(p_1, p_2)}[\text{guess that } \lambda^{(0)} \leqslant \lambda^*, \text{ given } (E, D)].$$

Ideally, when $p_1/p_2 \leqslant \lambda^*$ this probability should be (close to) 1, and when $p_1/p_2 > \lambda^*$ it should be (close to) 0. By complicating E in one way or another, e.g., by taking a very large sample, we could presumably adjust D to come close to this ideal. Furthermore, if we hold E fixed, then D can be so chosen that for some of the (p_1, p_2) pairs the resulting probabilities are close to the ideal—but only at the expense of not being near the ideal for other pairs. In practice, such a conflict is resolved by an analysis (outside of the formal model) of such factors as alternative policy decisions, losses resulting from incorrect decisions, the cost of experimentation, *a priori* subjective information about the true values, etc. The very choice of the value λ^* depends upon such an unformalized analysis.

Possibly the most prevalent procedure is to select in advance a number α_0 called a significance level. Often $\alpha_0 = 0.05$ or 0.01 are used. One then demands of an (E, D) pair that

$$P_{(p_1, p_2)}[\text{guess that } \lambda^{(0)} \leqslant \lambda^*, \text{ given } (E, D)] \leqslant \alpha_0,$$

for all (p_1, p_2) such that $p_1/p_2 \geqslant \lambda^*$, and that for certain specific pairs (p_1, p_2) with $p_1/p_2 \leqslant \lambda^*$ the resulting probabilities be "reasonably" large. The more sophisticated problems of inference center about the choice of E, rather than about the choice of D for a given E.

When the testing-of-an-hypothesis procedure is used, presumably two possible policy actions are contemplated—one for the guess that $\lambda^{(0)} \leqslant \lambda^*$ and the other for $\lambda^{(0)} > \lambda^*$. If, however, there are many more feasible actions, then it seems more appropriate to base the terminal action upon an estimation of the value $\lambda^{(0)}$; this observation leads one to the problem of point estimation.

In this problem a decision rule D associates to each outcome $\mathcal{O}$ of an experiment E a guess $D(\mathcal{O})$ of the value $\lambda^{(0)}$. Such a rule is called an *estimator* in this context. Naturally, for those $\mathcal{O}$'s which are "likely" to occur when $(p_1^{(0)}, p_2^{(0)})$ is true we want $D(\mathcal{O})$ to be near to $\lambda^{(0)}$. Since, again, we do not know the true values $(p_1^{(0)}, p_2^{(0)})$, we must examine the estimate made by D for arbitrary values (p_1, p_2). At least conceptually, we can perform a probabilistic analysis of the situation to determine the probability that the guess of the parameter $\lambda^{(0)}$ falls in some specific interval when the values are assumed to be (p_1, p_2) and when the strategy pair (E, D) is employed. Pictorially, the result for each (p_1, p_2), E, and D will be a function of the type shown in Fig. 1. The area under the curve in any given interval represents the probability that, when (p_1, p_2) is true, E will result in an outcome such that D yields an estimate for $\lambda^{(0)}$ which falls in the given interval. In statistical parlance, this curve is known as the probability density function, or p. d. f., of the estimator D when experiment E is employed and (p_1, p_2) is the true parameter. Our aim is to choose (E, D) such that, when the pair (p_1, p_2) is true, the resulting p. d. f. is concentrated about the ratio p_1/p_2 for all such pairs. Then, no matter what the true values $(p_1^{(0)}, p_2^{(0)})$ are, D will be a good estimator of $\lambda^{(0)}$.

In classical estimation the following tack is taken. For the triple $[(p_1, p_2), E, D]$, let $m[(p_1, p_2), E, D]$ denote the mean (or average) of the p. d. f. associated with it, and let $\sigma[(p_1, p_2), E, D]$ be its standard deviation. Our hope, in this case, is to choose E and D so that

$$\{m[(p_1, p_2), E, D] - p_1/p_2\} \qquad \text{and} \qquad \sigma[(p_1, p_2), E, D]$$

are both small. In case (E, D) is such that

$$m[(p_1, p_2), E, D] = p_1/p_2, \qquad \text{for all } p_1, p_2,$$

the estimation procedure is said to be *unbiased*. For unbiased estimators, a reasonable index of the performance of (E, D) is the standard deviation.

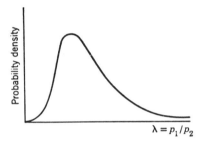

FIG. 1

More generally, a weighted average of $\{m[(p_1, p_2), E, D] - p_1/p_2\}$ and $\sigma[(p_1, p_2), E, D]$ is used as a measure of performance. These classical procedures do not explicitly incorporate into the formal model those initial conditions which state the losses due to incorrect terminal decisions; one must try to take such considerations into account by the interpretation given to the measures being "small." This gets confusing in some situations; for example, in some cases it may be much worse to overestimate $\lambda^{(0)}$ than to underestimate it by the same amount.

More modern work attempts to take the consequence of erroneous terminal decisions into account. It is assumed that as part of the initial data one is given a function L, where $L(p_1, p_2; \lambda)$ represents the loss (disutility) of guessing λ for $\lambda^{(0)}$ when $\lambda^{(0)}$ really is p_1/p_2. Thus, if E leads to an outcome $\mathcal{O}$ to which the estimator D associates the guess $D(\mathcal{O})$, then the loss is $L[p_1, p_2; D(\mathcal{O})]$ when (p_1, p_2) is actually true. To appraise (E, D) with regard to wrong guesses, but not other factors such as the cost of experimentation, we can use the expected value of the loss, i.e., $L[p_1, p_2; D(\mathcal{O})]$ is multiplied by the probability (likelihood) that E leads to $\mathcal{O}$ when (p_1, p_2)

is true, and these quantities are summed (integrated) over all possible outcomes Θ. Since this expected value depends upon (p_1, p_2), E, and D we may denote it by $L[(p_1, p_2), (E, D)]$. The problem again is to choose (E, D) so that this quantity is "small" for all (p_1, p_2). As before, this can be done by the choice of E, e.g., by taking a very large sample, and with E held constant we can vary D so as to make the loss index small for some (p_1, p_2) values at the expense of allowing it to be large for others. To resolve the latter conflict, certain arbitrary decision criteria must be employed.

For many policy purposes, point estimation seems to be a dangerous tool, for what in a given instance is the "best guess" of a parameter may, indeed, be a "poor guess" in actuality. A more sophisticated analysis should yield a probability statement concerning the deviations between the guess and the true value. From this knowledge, one can compute the amount of "confidence" to hold in the estimate. For example, suppose we know that, regardless of the true value $(p_1^{(0)}, p_2^{(0)})$, a strategy (E, D) will lead with a probability of 0.99 or higher to a guess which does not deviate from $\lambda^{(0)}$ by more than 0.05; then this information can and should be exploited in policy decisions. It is such observations which led to the theory of confidence estimation.

Let the rule D associate to each outcome Θ of experiment E an interval of values instead of a unique guess for $\lambda^{(0)}$; this interval we may denote by $D(\Theta)$ (note the change in meaning for this symbol). To be useful, such an interval should both be small and include $\lambda^{(0)}$. In much the same way as before, we can ascertain for each (p_1, p_2) pair the probability that E will result in an outcome to which D assigns an interval containing p_1/p_2. Suppose that strategy (E, D) has the property that *for each* (p_1, p_2) the probability is 0.99 or greater that D associates to the outcome of E an interval containing p_1/p_2. This means that if a given experiment E leads to Θ, then we can assert that "$D(\Theta)$ covers $\lambda^{(0)}$ with confidence 0.99", that is, our batting average over a large number of identical situations will approach 0.99.

But why use 0.99; why not 0.95 or 0.50? One answer is that the confidence level chosen, or the type of interval used—this is often a choice since D is not necessarily unique—depends upon the particular action problem to which the method is applied. This does not seem entirely satisfactory, for, if the end product is to be the terminal action problem, why break it artificially into two stages: first an interval estimate and then its conversion into an action? This is by no means an easy question to answer, and the best defense we know of is entirely pragmatic: the two-stage procedure is often more efficient than dealing with the action problem directly. For example, the set of possible terminal actions, the losses due to wrong action, etc., may not have been thought through at the time of experimentation, so, in lieu of a solution to the unformulated problem, the confidence interval statement constitutes a preliminary report on the outcome of the experiment. Furthermore, to introduce losses into a problem is not usually easy, so the fewer actions that have to be considered the better; sometimes knowing a confidence interval helps to eliminate certain of the feasible terminal actions, thereby reducing some of the delicate and time consuming appraisals of losses.

To a growing number of statisticians, this defense of confidence intervals as a preliminary to the second stage of analysis seems weak and spurious. If a confidence interval is merely to be used as a tool to summarize experimental observation, many would claim that there are more informative summaries of statistical data which, when used by an expert, can be efficiently converted into action choices. Others hold, however, that, although often there are technically more informative summaries than confidence interval estimations, the latter are especially easy to understand, to assimilate, and to internalize. This point is usually shrugged off by the remark, "It is but a matter of training."

In summary, the essential difference between classical inference and modern decision theory is this: only in the latter model is a formal attempt made to incorporate the actual terminal acts and the specific economic and psychological losses attributable to wrong terminal decisions. The modern work attempts, in a sense, to come closer to the real world problem by introducing into the framework of the model more of the initial conditions. The classical work, on the other hand, left many of these considerations outside the formal model, only to incorporate them indirectly and informally via such concepts as significance levels, confidence levels, and lengths of confidence intervals.

Thus, it is held, the modern theory is handicapped as an inferential tool in scientific research. How can a scientist realistically appraise the losses from falsely rejecting or accepting a research hypothesis? Or how can he evaluate the losses in estimating a parameter when this estimate may be used for a variety of purposes—some of which may be unknown or irrelevent to him? Since these evaluations do not seem possible, perhaps one should not assume them known. Furthermore, in scientific reporting the problem often is to select an experiment whose results are most likely to be maximally informative in some sense, not to arrive at an explicit terminal action. In that event, it is argued, a classical approach is much more sensible.

Decision theorists make several rejoinders. Even though the classical theory does not explicitly introduce losses, eventually they appear implicitly, for otherwise how can one decide upon significance levels, confidence levels, sample sizes, etc.?   Second, statistical results which are obtained for a given loss structure—a structure which is typically introduced more for reasons of mathematical tractibility than for realism in mirroring economic and psychological considerations—often remain "qualitatively reasonable" for classes of loss structures which are "qualitatively similar" to the original one.   Finally, if information is what is desired, then this requirement should be formalized and attempts should be made to introduce the appropriate information measures as a part of the loss structure.   This hardly ends the controversy, however, for decision theorists are only too aware that such a program is easier suggested than executed!

13.11 SUMMARY

An individual decision-making problem under uncertainty (d. p. u. u.) where the act and state spaces are finite, was formulated as follows: given an m by n matrix $[u_{ij}]$, where u_{ij} is the person's utility for the consequence associated with act A_i when nature is in state s_j, find the subset of acts which are in some sense "optimal."   Of course, the intriguing problem is: what constitute reasonable criteria for optimality?

If one can say that the individual has an *a priori* distribution over the states of nature, the problem becomes one of decision making under risk. This is simply dealt with by calculating the expected utility of the acts and choosing those with the largest value.   It remained, therefore, to deal in the chapter with the case where there is no apparent *a priori* probability assignment over the states.

If having an *a priori* probability assignment over the states is one extreme, the other should be "complete ignorance" about the "true state."   We avoided the temptation to discuss that concept initially, for we surely would have floundered in semantic debate over the meaning of "complete ignorance"; rather, we left the notion purposely vague while going on to discuss certain decision criteria.   By analyzing some of the pros and cons (mostly cons!) of criteria such as maximin utility or minimax loss, minimax regret, principle of insufficient reason, and the pessimism-optimism index, we hoped that the reader would glean some idea as to what several authors have meant by the phrase "complete ignorance." We then altered our tack.   Instead of analyzing and criticizing specific proposed criteria, we culled from our intuition a number of reasonable desiderata which a respectable decision criterion ought to fulfill.   Naturally, we hoped as a result to be better able to discriminate among suggested candidates for a definition of optimality and, perhaps, to be able to investigate whether or not our intuitions of desirability are themselves consistent.

In section 13.3 we discussed requirements (axioms) which do not in any way refer to the individual's knowledge of the likelihood of the states of nature. Axioms 1 through 5, and to a lesser extent 6, seem quite innocuous and, so far as we are aware, all serious proposals for criteria satisfy them. These axioms are: existence (1); invariance under utility transformations (2); invariance under labeling of acts (3); weak dominance property (4); admissibility (5); and irrelevance of weakly dominated acts (6). Axioms 7, 7', 7'', and 7''' were variations on the theme of the independence of irrelevant alternatives. Axiom 7—a non-optimal act shall not become optimal through the addition of new acts—suffices to rule out the minimax regret principle. Axiom 8, which implies that a constant added to a column shall not alter the optimal set and, together with 3 and 7, implies that the optimal set shall be anticonvex, rules out the maximin utility (minimax loss) criterion and all the Hurwicz α-criteria. For the Hurwicz α-criteria, Axiom 9—convexity of the optimal set—implies $\alpha = 1$, i.e., the maximin utility criterion. Axioms 1, 3, 4, 5, 7', 8, and 9 imply that there exists an *a priori* distribution which is independent of any new acts that might be added and which has the property that any act which is optimal for the problem must be best against this distribution.

The notion of complete ignorance, which was bypassed earlier, was captured in section 13.4 by two axioms: invariance of the optimal set under labelings of the states of nature (10) and under deletions of repetitious columns (11). Chernoff has shown that the criterion based on the principle of insufficient reason is characterized by axioms 1 through 10. The Hurwicz α-criteria, modified by adding an admissibility requirement, satisfy all the axioms save 8 and 9, and they are the only criteria satisfying axioms 1 through 4, 7', 10, and 11 plus a continuity requirement for simple choices.

If we demand that a criterion completely order all of the acts instead of selecting an optimal set, then the analysis is much simpler. We presented Milnor's axiomatic analysis, which pithily characterizes such criteria (see chart on p. 298).

What is known of the no man's land between complete knowledge and complete ignorance was the topic of section 13.5. If one endorses axioms 1 through 6, 7', 8, and 9, then, as we have said, all optimal acts are optimal against a particular *a priori* probability distribution over the states of nature. Following Savage, we indicated how this distribution can be generated from consistent answers to a hypothetical list of simple questions. We also mentioned suggestions of Hurwicz and of Hodges and Lehmann for coping with partial ignorance.

The appropriateness of these axioms when nature is replaced by an intelligent adversary was explored in section 13.6. Some were not directly meaningful, and they had to be modified slightly.   A number of these considerations seemed pertinent as guides in non-zero-sum non-cooperative games and, especially, in games where the adversary's payoff is imperfectly known.

Next, we turned to statistical decision making.   Again there is a set of acts, a set of states, and a utility payoff for (the consequence of) each act-state pair.   The new wrinkle added is this: An experiment can be performed in which the likelihood of any of its potential outcomes depends upon which is the "true" state of nature.   Thus, additional partial knowledge about the states of nature can be gleaned by experimentation.   How should this partial knowledge be processed?

If an *a priori* probability distribution over the states is already known, then experimental evidence merely alters it (by means of Bayes Theorem); after the experiment, any act is chosen which is best against the new—the *a posteriori*—distribution over the states.   If an *a priori* distribution is not known or assumed, then the analysis is a bit more involved.   Instead of choosing among terminal acts, we now choose among decision rules, i.e., among rules which associate to each possible experimental outcome a specific act.   The consequences of any decision rule–state pair is a well-defined lottery, and so it is evaluated by its expected utility.   This processing of the experimental possibilities reduces the statistical decision problem to the previously considered problem of decision making under uncertainty.

If several experiments are available, the selection among them, as well as the choice among actions, can be incorporated into the decision rule.   Of course, the consequence now of any decision rule-state pair is a more complicated lottery, for not only must we evaluate the consequences of different actions for each state of nature but also the experimental costs incurred.

In the final section we stated the nature of some of the classical problems of statistical inference (testing hypotheses, point estimation, and confidence interval estimation), and the essential differences between the classical and more modern approaches were indicated.   Wherever possible, the modern decision model formally incorporates the actual terminal acts which can be taken and the specific economic and psychological losses attributable to wrong terminal decisions.   The classical approach left many of these considerations outside the formal model, only to introduce them indirectly and non-formally via such concepts as significance levels, confidence levels, and lengths of confidence intervals.

chapter 14

GROUP DECISION MAKING

14.1 INTRODUCTION

Democratic theorists, economic as well as political, have long wrestled with the intriguing ethical question of how "best" to aggregate individual choices into social preferences and choices. In this chapter we shall survey some recent formal contributions to this topic.

The next three sections of the chapter are devoted to describing Arrow's basic work in formulating *a* problem of social choice and to stating his central (negative) result. This theorem—the voting paradox—is examined at some length in section 14.5, and various schemes for avoiding it are described. One class of schemes—those which utilize individual strengths of preferences—is of such importance and so thoroughly interlocked with the unresolved question of interpersonal comparisons of utility that we have treated it separately in section 14.6. The next two sections deal with two aspects of majority rule: conditions restricting the possible individual preferences in such a way that majority rule is a plausible voting procedure and its game-theoretic features. The penultimate section is devoted to games of "fair division," and the final one is a summary.

14.2 SOCIAL CHOICE AND INDIVIDUAL VALUES: PRELIMINARY STATEMENT

Our section heading, "Social Choice and Individual Values," is the title of Arrow's book [1951 a], a work which initiated much of the research we shall examine. As it is a basic reference for this and the following three sections and since we shall concentrate only on the formal aspects of Arrow's work, we urge the reader to go to his volume for historical background material and for guidance through the literature. We have taken the liberty both of altering his notation slightly and of making minor modifications in his formulation.

In its most general terms, the problem is to define "fair" methods for amalgamating individual choices to yield a social decision. As interpreted by Arrow, this becomes a question of "combining" individual preference patterns over various states of affairs to generate a single preference pattern for the society composed of these individuals. Some of the more common methods or procedures for passing from the preferences of individuals among social alternatives to a preference for the society are: convention, custom, religious code, authority, dictatorial decree, voting, economic market institutions, etc. Not all of these are usually considered fair and representative schemes, so part of our task will be to decide what we might mean by a procedure which takes into account the welfare of the members of society, i.e., by a "welfare function." We may best indicate those aspects of the problem we wish to abstract by first considering some very simple examples; this will lead into the general formulation of the next section.

Consider a society of two individuals, 1 and 2, each of whom have preferences for the two possible alternatives x and y. An individual can either (a) prefer x to y, (b) prefer y to x, or (c) be indifferent between x and y. These three cases are designated by R^1, R^2, and R^3, respectively. (The reason for this notation will appear later.) Schematically, we have:

R^1	R^2	R^3
x	y	$x - y$
y	x	

where the more preferred alternative is written above the less preferred. Let $\Re = \{R^1, R^2, R^3\}$ be the set of possible preference relations over the alternatives x and y. If individual 1 prefers x to y and 2 is indifferent between x and y, then the individual patterns of preference for this society can be summarized by the ordered pair (R^1, R^3). Conversely, to every ordered pair of elements from $\Re$ there corresponds a given pattern of order preferences for the members of this society.   The set of ordered pairs of elements of $\Re$ will be denoted by $\Re \times \Re$.

TABLE 14.1

1	2	F_1	F_2	F_3	F_4
R^1	R^1	R^1	R^1	R^1	R^2
R^1	R^2	R^3	R^1	R^1	R^3
R^1	R^3	R^1	R^1	R^1	R^2
R^2	R^1	R^3	R^1	R^2	R^3
R^2	R^2	R^2	R^1	R^2	R^1
R^2	R^3	R^2	R^1	R^2	R^2
R^3	R^1	R^1	R^1	R^3	R^2
R^3	R^2	R^2	R^1	R^3	R^3
R^3	R^3	R^3	R^1	R^3	R^1

The first two columns of Table 14.1 list all of the elements of $\Re \times \Re$, i.e., all the possible ordered patterns of preference between two alternatives for a society of two individuals.   The third column, labeled F_1, associates to each element of $\Re \times \Re$ an element of $\Re$.   Thus, the third column exhibits one method of amalgamating the individuals' choices into a social preference pattern.   For example, procedure F_1 sends (R^2, R^3) into R^2, and the interpretation is this: if 1 prefers y to x and 2 is indifferent between x and y, then rule F_1 "combines" these choices into a social preference for y over x.   Columns F_2, F_3, F_4 represent other possible procedures for passing from tastes of individuals to the choice for the society.   Procedure F_2 can be characterized as an *imposed* procedure since the choice of the society does not depend upon the choice of the individuals.   In procedure F_3 the choice of the society depends only upon individual 1's choice and not upon 2's, so we may term it *dictatorial*. F_4 does not seem a reasonable procedure; nevertheless, it is a method of amalgamation.

In each case, the F_i represents a function with all of $\Re \times \Re$ as its domain of definition and with $\Re$ as its range.   Naturally, there are other functions from $\Re \times \Re$ to $\Re$; indeed, there are $3^9 = 19,683$ such functions.   Most of these functions, like F_4, are more appropriately referred to as "illfare" rather than "welfare" functions.   This raises the problem: what can one intuitively mean by a "welfare" function?

Consider, for example, the following requirement. If all individuals have the same preference, then society should have this common preference. For our two-person two-alternative situation, the requirement amounts to: $(R^1, R^1) \rightarrow R^1$, (where the $\rightarrow$ is read: "is sent into"), $(R^2, R^2) \rightarrow R^2$, $(R^3, R^3) \rightarrow R^3$. It cuts down the number of possible functions mapping $\mathfrak{R} \times \mathfrak{R}$ into $\mathfrak{R}$ from 3^9 to 3^6. However, if we accept this restriction, then we can also argue that it is reasonable to insist that (R^1, R^3) not be mapped into R^2. First, note that (R^1, R^3) results from (R^3, R^3) by 1 changing his mind in favor of alternative x. If society was originally indifferent between x and y (which follows from the first requirement), it seems perverse for society to change in favor of y when 2's choice is held fixed and 1 changes in favor of x. It is by "reasonable" conditions such as these that we propose to eliminate many of the initially possible functions from the category of "welfare" functions. In section 14.4 the full development will be given.

Now that we have some idea of what we shall be doing, let us point out several aspects of the general problem that we shall *not* discuss. Suppose that F_4 is the method of social choice used in the society consisting of 1, 2. If 1 prefers x to y he will certainly not state this, i.e., R^1, for that ensures R^1 is not chosen by the society. He will register either R^2 or R^3. If he thinks that 2 will say R^1 or R^2, he will choose R^2; if he thinks that 2 will say R^3, he will choose R^3. Such strategy aspects of the problem will not be discussed. We will assume that the choices of the individuals are part of the data of the problem. Alternatively, we might try to impose a condition on the function to the effect that it never benefits an individual to misrepresent his actual tastes; however, this idea seems to be very difficult to formalize, and no attempt will be made to use it.

A possible social rule is: a given coin is tossed; individual 1 dictates the choice of society if it turns up heads; otherwise 2's choice is the social choice. This rule is certainly well prescribed, but it does not constitute a function from $\mathfrak{R} \times \mathfrak{R}$ into $\mathfrak{R}$. For the pair (R^1, R^2) this rule prescribes R^1 with probability p, where p is the probability of heads, and R^2 with probability $1 - p$, i.e., (R^1, R^2) is mapped into the mixture $[pR^1, (1 - p)R^2]$. The difficulty with such procedures is that the social choice for fixed individual preferences can differ with repeated trials. Of course, this may be a desirable feature; nonetheless, we shall not discuss rules of this kind. We demand that the rule map $\mathfrak{R} \times \mathfrak{R}$ into $\mathfrak{R}$, not into probability distributions on $\mathfrak{R}$.

Finally, we assume that the given data is in the form of simple preferences, e.g., 1 prefers x to y and 2 prefers y to x. It is not of the form: 1 *strongly* prefers x to y and 2 *weakly* prefers y to x. Such considerations lead into problems of subjective utility and interpersonal comparisons of utilities, and we prefer first to analyze the problem of social choice without these complications. Later we shall consider how the model changes when such features are introduced.

Returning to our problem, one can gain a sense of its full complexity by examining three alternatives and a society of three people. Let α consist of three alternatives labeled $x, y,$ and z, i.e., $\alpha = \{x, y, z\}$. All the possible preference-or-indifference relations on α are:

TABLE 14.2

R^1	R^2	R^3	R^4	R^5	R^6	R^7	R^8	R^9	R^{10}	R^{11}	R^{12}	R^{13}
x	x	y	y	z	z	x		y		z	$x - y$	$x - z$	$y - z$	$x - y - z$
y	z	x	z	x	y	$y - z$	$x - z$	$x - y$	z	y	x
z	y	z	x	y	x					

Thus in R^8, for example, y is preferred to both x and z, and x and z are indifferent. Note that, for every pair of distinct elements u, v belonging to α (i.e., u is either $x, y,$ or z, and v is either $x, y,$ or z), either u is preferred to v or v is preferred to u or u and v are indifferent. Further, preferences are never allowed to be intransitive. Of course, in actual practice an individual may have intransitive preferences (e.g., x preferred to y, y preferred to z, z preferred to x), in which event this model does not apply. One way to avoid intransitive responses is to insist that the individual choose an element from the set

$$\mathcal{R} = \{R^1, R^2, R^3, \cdots, R^{13}\},$$

i.e., rank-order the alternatives.

As before, to each ordered triple of elements of $\mathcal{R}$ there is a corresponding pattern of choices for the individuals, and conversely. Thus to the triple (R^3, R^8, R^2) we have: individual 1 prefers y to x to z; 2 prefers y to x, y to z, and is indifferent between x and z; and 3 prefers x to z to y. Thus, a function F from $\mathcal{R} \times \mathcal{R} \times \mathcal{R}$ into $\mathcal{R}$ can be interpreted as a procedure for passing from the individuals' preferences to the social preference.

14.3 GENERAL FORMULATION OF PROBLEM

As the number of alternatives and people are increased beyond three it becomes more and more cumbersome to exhibit all the possible preference orders on the alternatives. Therefore, a simple, compact terminology applicable to all cases is needed. The following is suitable:

i. *Alternatives.* Let $\alpha = \{x, y, \cdots, z\}$ be a set of alternatives.

ii. *Individuals.* Let the individuals of the society be denoted by 1, 2, $\cdots, i, \cdots, n$.

iii. *Preferences.* For each individual i and any alternatives u and v, one and only one of the following holds:

(*a*) "*i* prefers *u* to *v*," which is written as uP_iv,
(*b*) "*i* prefers *v* to *u*," which is written as vP_iu,
(*c*) "*i* is indifferent between *u* and *v*," which is written as uI_iv.

If "*i* does not prefer *v* to *u*" we write $v\bar{P}_iu$. This is clearly equivalent to: either uP_iv or uI_iv holds. We also demand that each individual be consistent in his preferences in the sense that P_i, I_i, and $\bar{P}_i$ are each assumed to be transitive.

We have already seen that for two alternatives there are three possible preference orderings and for three alternatives there are thirteen possible preference orderings. In general, let $\mathfrak{R} = \{R^1, R^2, \cdots, R^m\}$ be the possible preference orderings of the alternatives, where *m* depends upon and increases rapidly with the number of alternatives. By a *profile of preference orderings* for the individuals of the society, we shall mean an *n*-tuple of orderings, $(R_1, R_2, \cdots, R_n)$, where R_i is the preference ordering for the *i*th individual.[1] The set of all possible profiles of preference orderings will be denoted by $\mathfrak{R}^{(n)} = \mathfrak{R} \times \mathfrak{R} \times \cdots \times \mathfrak{R}$.

iv. By a *social welfare function* (or "constitution," or "arbitration scheme," or "conciliation policy," or "amalgamation method," or "voting procedure," etc.), we shall simply mean a rule which associates to each profile of preference orderings (i.e., to each element of $\mathfrak{R}^{(n)}$) a preference ordering for the society itself. If *F* denotes such a rule we shall symbolically write:

$$(R_1, R_2, \cdots, R_n) \xrightarrow{F} R,$$

which means that rule *F* "combines" the profile of orderings $(R_1, R_2, \cdots, R_n)$ to yield the ordering *R* for the society.

Obviously, there are many conceivable functions from $\mathfrak{R}^{(n)}$ to $\mathfrak{R}$; however, if we interpret the mathematics as a social problem, there are several requirements a function should satisfy before we would consider it "ethically" acceptable. The requirements of "ethical acceptability" are purely subjective value judgments not subject to mathematical derivation; however, in a mathematical treatment of their implications such desiderata have to be phrased as explicit mathematical statements.

We certainly do not wish to imply that there does or should exist a universal ethic, or even that an individual's "ethical standard" should be invariant over different social choice problems. When we assert that a set of conditions or restrictions on *F* is "reasonable," all that is implied is that we *think* there are some situations for which enough individuals will subscribe to the conditions to warrant investigating their implications.

---

[1] Note the different role of subscripts and superscripts. R^i means the *i*th preference relation in the listing $\mathfrak{R}$ of all of them; whereas R_i means the preference relation for person *i* in a particular profile of individual preferences.

Imposing conditions on F of course restricts the class of permissible functions. Arrow has shown that some seemingly innocuous requirements narrow F down to the point where it is either *imposed* or *dictatorial*. However, he has also shown that, if the domain of definition of F is suitably restricted (single-peakedness condition; see section 14.7), then socially desirable welfare functions do exist—socially desirable in the tautological sense of satisfying the stated prerequisites.

Summarizing, the procedure to be adopted is as follows: There are many specific social welfare functions which are well defined; however, particular functions are often adversely criticized because they fail to satisfy some "socially desirable" criteria. Hence, instead of considering specific functions, we shall attempt to capture our intuition of "desirable" by explicitly stating properties that any social welfare function should satisfy. We shall examine a set of conditions, each of which individually has merit as well as a long and illustrious history, but which collectively are inconsistent, i.e., *no* social welfare function exists which fulfills all these conditions.

As such a conclusion may seem odd, the following example should serve to suggest why it is not so trivial to find "reasonable" welfare functions

Simple majority rule. Corresponding to each profile of preferences, let society prefer the alternative u to the alternative v if and only if a majority of the individuals prefer u to v. The following profile of choices for alternatives x, y, and z illustrates the difficulty with this rule:

R^1	R^4	R^5
x	y	z
y	z	x
z	x	y

Since a majority of individuals prefer x to y, y to z, and z to x, society selects the intransitive relation: xPy, yPz, zPx. Consequently, the rule does not tell us what element of $\mathfrak{R}$ to associate to (R^1, R^4, R^5). To be sure, we could alter the simple majority rule to let (R^1, R^4, R^5) map into R^{13}, i.e., $(xIyIz)$, but then we would have a different rule—one which has other faults.

14.4 CONDITIONS ON THE SOCIAL WELFARE FUNCTION AND ARROW'S IMPOSSIBILITY THEOREM

The cases where the number of alternatives is one or two can be easily handled separately. Indeed, for one alternative no analysis is necessary! Hence we shall confine our attention to situations having three or more alternatives. The number three plays an important role, since intransitivities can only occur on three or more elements.

It is true that simple majority rule led to an intransitive result for the profile (R^1, R^4, R^5), but this would be of no import in a particular society of three members if we had reason to believe that this much disagreement would never occur.   For then we need not demand that the function be defined on all of $\mathcal{R}^{(3)}$, but only on a certain restricted part of $\mathcal{R}^{(3)}$.   The smaller the domain, the easier it is to construct "reasonable" welfare functions.   In the extreme when only complete agreement is allowed, the task is trivial.   Some groups, possibly because of self-selection or because of a common ethic, do not often exhibit a wide divergence of opinions, and for such societies majority rule probably never will be embarrassed by an intransitive set of orders.   For more discussion of this point, see section 14.7.

We choose, however, to confine our attention to the mathematically more interesting case where the domain of the welfare function is rich enough to make our task formidable.   For ease of presentation we shall assume that the domain of the welfare function is all of $\mathcal{R}^{(n)}$; i.e., we require the function F to be defined for *all conceivable profiles of individual orderings of the alternatives*.   Actually, the other conditions that will be imposed on F continue to be incompatible even when the domain is restricted to certain proper subsets of $\mathcal{R}^{(n)}$.   However, certain of the restricted domains given in the literature (Arrow [1951 a] and Inada [1954, 1955], do not allow one to establish the contradiction asserted, as has been pointed out by Blau [1957].

We summarize the above discussion as:

Condition 1.   (a) *The number of elements (alernatives) in $\mathcal{A}$ is greater than or equal to three.*

(b) *The social welfare function F is defined for all possible profiles of individual orderings.*

(c) *There are at least two individuals.*

As background for the second condition we consider a special example. Let $\mathcal{A} = \{x, y, z\}$, and let the society have exactly three members.   Consider the pattern (R^1, R^{11}, R^{13}), namely:

R^1	R^{11}	R^{13}
x	$x - z$	$x - y - z$
y	y	
z		

and suppose that F_0 is a welfare function such that society prefers y to z for the profile (R^1, R^{11}, R^{13}).   Now consider the profile (R^{10}, R^{11}, R^8), namely:

$$R^{10} \qquad R^{11} \qquad R^8$$

$$
\begin{array}{ccc}
x - y & x - z & y \\
z & y & x - z
\end{array}
$$

If the first profile, (R^1, R^{11}, R^{13}), is modified by the first and third members' pushing y up while keeping x and z fixed, then the second profile, (R^{10}, R^{11}, R^8), results. It thus seems "reasonable" that, since F_0 selects y in preference to z for the former profile, it should also do so in the latter one. Stated alternatively, a social welfare function is not "reasonable" if, when the members choose (R^1, R^{11}, R^{13}), society prefers y to z, and when the members choose (R^{10}, R^{11}, R^8), society does not prefer y to z.

Condition 2 (positive association of social and individual values). *If the welfare function asserts that x is preferred to y for a given profile of individual preferences, it shall assert the same when the profile is modified as follows:*

(a) *The individual paired comparisons between alternatives other than x are not changed,*

and

(b) *Each individual paired comparison between x and any other alternative either remains unchanged or it is modified in x's favor.*

To arrive at the next condition, consider the case of four alternatives w, x, y, z and suppose that for some profile of individual preference a specific welfare function F_0 states that society prefers alternative x to alternatives w, y, and z. That is, given the particular choices for the individuals, the rule F_0 says that x is the "best" (most preferred) alternative from the set $\{w, x, y, z\}$.

Now suppose we restrict their consideration to the alternatives $\{x, y, z\}$. Presumably, the individuals might change their preferences among x, y, z; but suppose they do not! If all paired comparisons made by the individuals between elements x, y, z do not change, then isn't it "reasonable" to expect that, since x is socially best in $\{w, x, y, z\}$, it is also socially best in $\{x, y, z\}$? To be very concrete, let us quote Arrow [1951 a, p. 27] on the rank-order method of voting used in clubs:

With a finite number of candidates, let each individual rank all the candidates, i.e., designate his first choice candidate, second choice candidate, etc. Let preassigned weights be given to the first, second, etc., choices, the higher weight to the higher choice, and then let the candidate with the highest weighted sum of votes be elected. In particular, suppose that there are three voters and four candidates, x, y, z, and w. Let the weights for the first, second, third, and fourth choices be 4, 3, 2, and 1, respectively. Suppose that individuals 1 and 2 rank the candidates in the order x, y, z, and w, while individual 3 ranks them in the order z, w, x, and y. Under the given electoral system, x is chosen. Then, certainly, if y is deleted from the ranks of the candidates, the system applied to the remaining candidates should yield the same result, especially since, in this case, y is inferior to x according to the tastes of every individual; but, if y is in fact deleted, the indicated electoral system would yield a tie between x and z.

As another example, let $\alpha = \{x, y, z\}$. For the profiles

$$R^{10} \qquad R^1 \qquad R^4$$

$$
\begin{array}{ccc}
x - y & x & y \\
z & y & z \\
 & z & x
\end{array}
$$

and

$$R^9 \qquad R^2 \qquad R^8$$

$$
\begin{array}{ccc}
z & x & y \\
x - y & z & x - z \\
 & y &
\end{array}
$$

the preference relations between x and y are identical for each individual, so the argument is that for both profiles society should reach the same choice between x and y. The counterargument, however, notes that when 2 changes from

$$
\begin{array}{ccc}
x & & x \\
y & \text{to} & z \\
z & & y
\end{array}
$$

he seems to be indicating that he prefers x to y "more" in the latter than in the former pattern. That is, alternative z is not *irrelevant* in appraising the "strength" of preferences for x versus y, and so it is not "unreasonable" for society to prefer y to x in the first profile and x to y in the second. In Arrow's voting example, the claim is that when w drops out of the race the information regarding the preferences of the individuals is *changed*, even though the paired comparison judgments remain constant.

The following example, taken from Goodman and Markowitz [1951], further illustrates this point. A host has two dinner guests to whom he is willing to serve either coffee or tea but not both. Instead of asking which each prefers, coffee or tea, this subtle host gets them to rank a whole class of drinks. Two possible profiles of responses are:

	1's Preferences	2's Preferences
	Coffee	Tea
	Postum	Coffee
	Milk	Postum
Profile 1	Lemonade	Milk
	Hot chocolate	Lemonade
	Coca-Cola	Coca-Cola
	Tea	Hot chocolate

and

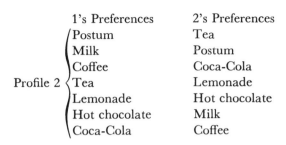

For profile 1, the host deems it fair to serve coffee and for profile 2 to serve tea. He reasons that the other drinks are not irrelevant, and their introduction permits him to appraise the relative intensities of preferences for coffee versus tea.

Yet, is such reasoning really plausible! One can argue that this procedure introduces a notion of interpersonal comparison of utility, and, *if* it is desirable to do that, then this is surely a naive way of doing it. Indeed, *if* such a statement as "1 prefers coffee to tea more than 2 prefers tea to coffee (in profile 1)" has any meaning at all, then an alternative rationalization of profile 1 can be given to show that the host should have served tea, namely:

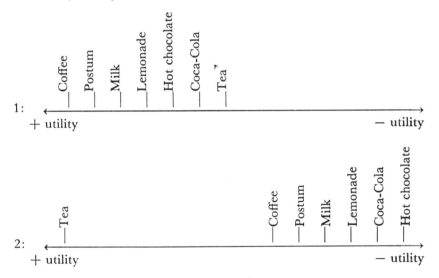

Having argued both sides of the question of the independence of irrelevant alternatives, we would claim that it is a sufficiently "reasonable" condition to warrant an investigation of its implications.

Condition 3 (independence of irrelevant alternatives). *Let* $\mathcal{Q}_1$ *be any subset of alternatives in* $\mathcal{Q}$. *If a profile of orderings is modified in such a manner that each individual's paired comparisons among the alternatives of* $\mathcal{Q}_1$ *are left invariant, the social orderings resulting from the original and modified profiles of individual orderings should be identical for the alternatives in* $\mathcal{Q}_1$.

If $\mathcal{Q}_1$ is simply any two-element set, then condition 3 reduces to:

If two profiles are such that each individual's paired comparisons between two alternatives x and y, say, are identical in both profiles, the society's ordering of x versus y should be identical for both profiles.

This condition is extremely powerful. For example, suppose that we treat the individuals of a two-person society symmetrically in the sense that, if one prefers y to z and the other z to y, society shall be indifferent between y and z. Let us try to arrive at the social ordering for the profile:

1	2
x	z
y	x
z	y

Since both individuals prefer x to y, it is plausible that society should prefer x to y. On the other hand, for x and z, y is irrelevant by condition 3, so the symmetry assumption requires that society be indifferent between x and z; similarly, x is irrelevant when comparing z and y, so symmetry leads to social indifference between z and y. But indifference is a transitive relation, so society is indifferent between x and y, contrary to our conclusion above that society prefers x to y.

Condition 4 (citizen's sovereignty). *For each pair of alternatives x and y, there is some profile of individual orderings such that society prefers x to y.*

Consider any welfare function for which condition 4 is not satisfied; then there exists a pair of alternatives x, y such that x is not preferred to y regardless of the preference orderings of the individuals. In other words, the citizens of the society do not exercise any sovereignty with respect to the pair x versus y, and so we could say that society's ordering "x is not preferred to y" is *imposed*.

Condition 5 (non-dictatorship). *There is no individual with the property that whenever he prefers x to y (for any x and y) society does likewise, regardless of the preferences of other individuals.*

If a welfare function does not satisfy condition 5, then this distinguished individual is a dictator in the sense that if he prefers one alternative to another so does the society; the rest of the community has an effect on society's choice between two alternatives only if he is indifferent between them.

Arrow's impossibility theorem states that conditions 1, 2, 3, 4, and 5 are *inconsistent*. That is, there does *not* exist any welfare function which possesses the properties demanded by these conditions. Stated alternately, if a welfare function satisfies conditions 1, 2, and 3, then it is either imposed or dictatorial.

▶The central steps of the proof can be outlined without too much rigorous argument. With respect to a given social welfare function a subset V of individuals is said to be *decisive* for the ordered pair (x, y) if whenever the members of V each prefer x to y society does likewise—regardless of what members not in V have to say about x versus y. In other words, if V is decisive for (x, y), then the coalition V can always enforce x over y by having each member express preference for x over y.

By condition 2, a set V is decisive for (x, y) if and only if society prefers x to y when all of the members of V prefer x to y and all other individuals prefer y to x.

From conditions 1, 2, 3, and 4 it is easy to prove

Pareto optimality: The set of all individuals is *decisive* for every ordered pair (x, y), i.e., if each individual prefers x to y, so does society.

Arrow felt that conditions 2 and 4 were more basic than Pareto optimality, so he chose not to introduce it as a basic axiom. Later, Inada [1955] attempted to show that Arrow's conditions 1 and 3, Pareto optimality, and a slightly modified but equally innocuous version of 5 also lead to an inconsistency, but, as Blau [1957] points out, his formulation possesses the same slight flaw as Arrow's. His result is, however, true with our version of condition 1 substituted for Arrow's, and it is also true with our condition 1 somewhat relaxed, but not to the extent done in these papers. Another formulation, based directly on decisive sets, is given by Weldon [1952].[2]

Arrow's fifth condition can be rephrased as: No individual is decisive for every ordered pair of alternatives.

A proof of Arrow's impossibility theorem is as follows:

i. Suppose V is a minimal decisive set, i.e., it is decisive for some x against some y, whereas no proper subset is decisive for any ordered pair of alternatives. Such a set must exist since, as we mentioned above, the set of all individuals is decisive and the individuals can be removed one at a time until the remaining set is no longer decisive for any pair. This proves the existence of a minimal decisive set, which cannot be the empty set since otherwise its complement, the set of all individuals, would not be decisive for some pair.

ii. Let j be a specific individual in V, W the remaining individuals in V, and U the set of all individuals not in V. Since the society has at least two individuals, U and W may not both be empty. Let z be any third alternative, and consider the following profile of orderings:

---

[2] Care must be taken with Weldon's paper, for it appears as if Arrow's theorem is proved without using the condition of the independence of irrelevant alternatives, when in fact it is used in the proof.

$\{j\}$	W	U
x	z	y
y	x	z
z	y	x

iii. Since x is preferred to y for all i in $V = \{j\}\cup W$, and since V is decisive for x against y, society prefers x to y, i.e., xPy.

iv. If society preferred z to y, this would mean that W is decisive for z against y since all members of W prefer z to y whereas all others prefer y to z, which is contrary to the choice of V as a minimal decisive set. Therefore, society does not prefer z to y, i.e., $z\bar{P}y$.

v. By transitivity, xPy and $z\bar{P}y$ imply xPz.

vi. But j is the only individual who prefers x to z, so $\{j\}$ is decisive for x against z. Thus, $\{j\}$ cannot be a proper subset of V, so $\{j\} = V$, and by hypothesis $\{j\}$ is decisive for x against y.

vii. Since we now know that $\{j\}$ is decisive for x against any z different from x, it is sufficient to show that it is decisive for any w, different from x, against z and for w against x. Suppose $w \neq x$, and consider the profile

$\{j\}$	U
w	z
x	w
z	x

By Pareto optimality, wPx, and, since $\{j\}$ is decisive for x against z, xPz. Thus, by transitivity, wPz, so $\{j\}$ is decisive for w against z. Next, consider the profile

$\{j\}$	U
w	z
z	x
x	w

Since $\{j\}$ is decisive for w against z, wPz, and, by Pareto optimality, zPx. Therefore, by transitivity, wPx, so $\{j\}$ is decisive for w against x, as was to be shown. We have therefore established that $\{j\}$ is decisive for all pairs and so $\{j\}$ is a dictator. But this is impossible, so no function exists which meets the five conditions. ◀

14.5 DISCUSSION OF THE ARROW PARADOX

What are the ramifications of Arrow's Impossibility Theorem? Social choice problems still arise and must be resolved, and so the impasse must be side-stepped. As we see it, there are two distinct possibilities: (i) to keep Arrow's formulation of the problem and to reject one of his conditions as being too restrictive; or (ii) to alter the formulation of the problem itself by changing the givens and/or the demands on society's final product.

With respect to the first possibility, we feel that the only possible bone of contention can be the independence of irrelevant alternatives. To deny the other conditions is tantamount, in our view, to changing the problem and will be better discussed under (ii). In the next section we will consider several proposals which fail to meet the independence of irrelevant alternatives condition, but which do address themselves to the same problem Arrow considered.

To illustrate once again a possible objection to the independence axiom, consider the case of two individuals in roughly the "same economic positions" and let their profile for two alternatives A_1 and A_2 be:

Individual 1: A_1 preferred to A_2
Individual 2: A_2 preferred to A_1,

and suppose the welfare rule says A_1 and A_2 are indifferent for society. Now let us add the following alternatives: $1000 given to each, $1 given to each, $-$1 given to each, $-$1000 given to each. Suppose with these new alternatives included, the preference profile takes the form:

Individual 1: ($1000 each) preferred to ($1 each) preferred to A_1 preferred to A_2 preferred to ($-$1 each) preferred to ($-$1000 each).
Individual 2: A_2 preferred to ($1000 each) preferred to ($1 each) preferred to ($-$1 each) preferred to ($-$1000 each) preferred to A_1.

In a situation of this kind one might be inclined to say that these new alternatives are not irrelevant, and *if* we could believe 1 and 2's orderings, then a "welfare" rule should choose A_2 over A_1 for this society. If we reject the axiom of the independence of irrelevant alternatives, we can inject judiciously chosen hypothetical alternatives into the picture to serve as a base line for evaluating comparative strengths of preference. However, if the members of society realize that these extraneous alternatives are not really feasible, it will behoove them to lie about their true tastes and to play a game of strategy.

As to reforming the problem itself, there are several ways to proceed.

i. *Restrict the domain of the welfare function.* Arrow required a social solution for every profile of individual preferences over three or more alternatives, no matter how chaotic it might be; possibly we should accept a less universal rule. For example, we might assume that there is some underlying structure to preferences which prohibits extreme divergences of opinion. This hedge, in essence, weakens Arrow's condition 1, but in so doing it really changes the problem. We will return to this topic later (section 14.7).

ii. *Weaken the demands on the range of welfare functions.* Arrow demands that both the individuals and society completely order all the alternatives. Wouldn't it suffice to have society choose only one or a few alternatives as "optimal?" (Recall that in the previous chapter we concentrated on selecting optimal sets, not on finding a complete ordering of acts.) Arrow, we understand, did at first phrase his impossibility theorem in terms of choice sets for society and rankings for individuals; however, he elected to publish his work for complete social rankings, because the exposition and proofs are simpler for that case. Furthermore, that assumption is not without interest, since, when individual rankings that are obtained at one time period are used to reach social decisions at a later time period, certain alternatives may turn out not to be feasible, e.g., a candidate may die. If, however, a complete social ordering is determined, then the choice set is easily found for any subset of the alternatives. In any event, since Arrow's axioms are still inconsistent when translated into choice set terminology, we cannot avoid the impasse merely by this expedient.

iii. *Obtain more data on individual values.* Had we demanded from each individual only his optimal choice and not his whole ranking, there would have been no difficulty in amalgamating these into a social choice. At the other extreme, instead of asking each individual only to order the alternatives, we could ask him to order all of the orderings of the alternatives. For example, with three alternatives there are 13 orderings, and each individual could be asked to order this set of 13 orderings. Such additional information enables one to extract some data about the strengths of an individual's preferences among the alternatives. (Incidentally, if individuals ordered orderings, then a socially optimal set amounts to a social ordering of the alternatives.) Or, looked at another way, if society's terminal action is an ordering of alternatives, then it seems reasonable to ask each individual to rank society's terminal actions (i.e., each individual should give an ordering of orderings). But even if this complication is introduced, Arrow's axioms can be given a direct interpretation by treating an ordering of alternatives as a basic "alternative," and his impossibility result still obtains. One should not misinterpret this result: In no way does it say that we should not solicit more information about each individual's preferences; it only says that this particular information does not allow those of us who are committed to Arrow's axioms to bypass the logical inconsistency.

Before going on to the next level of individual information, it is interesting to draw an analogy between the social choice problem as formulated by Arrow and the decision problem under uncertainty discussed in the previous chapter. When this identification is made and the conditions imposed by Arrow are suitably translated into slightly different terminology, a known result from the theory of decision making under uncertainty yields Arrow's impossibility theorem as a corollary. Furthermore, the analogy will be suggestive of several new tacks for the social choice problem. Finally, the format to be presented is quite flexible, and, in particular, it is well suited to introducing more refined information about individual values.

Let $A_1, A_2, \ldots, A_m$ be "alternatives" (these play the role of the "acts" of Chapter 13), and let $s_1, s_2, \cdots, s_n$ be "individuals" or "subjects" (these play the role of the "states of nature" of Chapter 13). Consider the payoff array

$$
\begin{array}{c}
\begin{array}{ccccc} s_1 & \cdots & s_j & \cdots & s_n \end{array} \\
\begin{array}{c} A_1 \\ \cdot \\ \cdot \\ \cdot \\ A_i \\ \cdot \\ \cdot \\ \cdot \\ A_m \end{array}
\left[
\begin{array}{ccccc}
 & & \cdot & & \\
 & & \cdot & & \\
 & & \cdot & & \\
\cdots & u(A_i, s_j) & \cdots & & \\
 & & \cdot & & \\
 & & \cdot & & \\
 & & \cdot & & \\
 & & & &
\end{array}
\right],
\end{array}
$$

where $u(A_i, s_j)$ is a number which, in comparison with other numbers in the same column, reflects something about s_j's preference for A_i. In attempting to determine how much significance to give to the numbers of the payoff array, it is helpful to decide when two arrays are strategically equivalent. The m numbers $u(A_1, s_j), \cdots, u(A_m, s_j)$ of column j reflect s_j's preference for alternatives in the sense that the higher the number, the more preferred the alternative; ties reflect his indifferences. Hence, each payoff array induces a profile of individual orderings. Now, if we wish to abstract away completely the notion of preference strengths, as Arrow does, then we should treat as equivalent two payoff arrays which induce the same profiles of individual order preferences—two such arrays we will call *order equivalent*. Any strictly monotonic transformation of the numbers in any column yields an *order equivalent* array.

For payoff arrays, Arrow's conditions can be paraphrased as follows:

Condition 1. *A social ordering shall be associated to every possible array;* $m \geqslant 3, n \geqslant 2$.

Condition 2 (positive association of individual values). *If a given array is modified by adding positive quantities to some entries in the ith row, and if society originally preferred A_i to A_k, it shall do likewise after the modification.*

Condition 3 (independence of irrelevant alternatives). *Adding new rows shall not change society's ordering for old rows.*

There is a clear analogy between the axiomatic development of decision making given in Chapter 13 and the present discussion of welfare axiomatics. The (act, state of nature) pair of the decision problem under uncertainty is replaced by the (alternative, subject) pair of the welfare problem. (It is interesting to return to the axioms discussed in the last chapter and to interpret these in the welfare context.) Note that in Arrow's formulation of the problem it is necessary to impose the condition that *order equivalent arrays shall yield the same social ordering of alternatives.* This condition implies, among other things, that the social ordering for alternatives should be invariant under both positive linear transformations (utility transformations) of the payoff entries and the addition of a constant to any column. Since each row is identified with an n-tuple of numbers, ordering rows is equivalent to finding a complete ordering for the set of n-tuples of numbers. But now there are so many constraints on the ordering of these n-tuples (constraints dictated by conditions 1 and 3, Pareto optimality, and the requirement that order equivalent arrays induce identical social orderings) that only a lexicographical ordering is possible.[3] This means that social preference is determined by a serial dictatorship in which the ordering dutifully follows the preference dictates of one subject, however, when he is indifferent, another specific subject takes over, etc. The dictator always has his way; the dictator's wife has her way when her lord is indifferent; the dictator's mistress exerts her influence only when the dictator is indifferent and does not have to appease his wife, etc.

iv. *Introduce risky alternatives, lotteries, and utilities.* To gain more information about strength of preference among alternatives, lotteries (probability mixtures) of the basic alternatives $A_1, \cdot\,\cdot\,\cdot, A_m$ can be introduced. If we are interested only in society's ranking of $A_1, \cdot\,\cdot\,\cdot, A_m$, then Arrow's irrelevancy axiom requires that individual preferences recorded for lotteries (or for any other alternatives for that matter) shall not have any influence whatsoever. But let us leave this objection to one side and continue along with the lottery idea for a while. The next objection to using lotteries is that individual preferences among lotteries reflect not only strengths of preferences but also attitudes toward gambling itself, and these attitudes are completely irrelevant in the welfare model which, in essence, is group decision making under certainty, not under risk. It can be argued in rebuttal that if $A_1, \cdot\,\cdot\,\cdot, A_m$ are available to society so are mixtures of them, and we should not prejudge the problem by ignoring them. If bona fide mixtures are undesirable for society, this should become apparent from the analysis.

Whenever the set of basic alternatives is inflated by considering lotteries, it is customary to restrict the individual complete orderings of the lotteries to those which can be summarized by a linear utility indicator. That is, for individual s_j there must exist m numbers $u(A_1, s_j), \cdot\,\cdot\,\cdot, u(A_m, s_j)$, which reflect s_j's preferences for $A_1, \cdot\,\cdot\,\cdot, A_m$ in such a way that any

[3] This result is a simple modification of a theorem due to Blackwell and Girshick [1954, p. 118].

lottery $L' = (p_1'A_1, \cdots, p_m'A_m)$ is preferred to any $L'' = (p_1''A_1, \cdots, p_m''A_m)$ if and only if

$$\sum_{i=1}^{m} p_i'u(A_i, s_j) > \sum_{i=1}^{m} p_i''u(A_i, s_j).$$

With these assumptions, we can summarize the profile of individual preferences for lotteries in terms of an m by n array, where the entry of the ith row and the jth column is $u(A_i, s_j)$. But since each subject's m utility numbers are defined only up to a positive linear transformation (i.e., the origin and unit of measurement are not determined), two arrays should be treated as equivalent, utilitywise, if each column of one can be obtained from the corresponding column of the other by such a transformation. We can now see one tremendous difference between the social choice problem and the decision-making problem under uncertainty. In the latter problem it is meaningful to compare the utility for the (act, state) pair (A_2, s_5) with the utility for the (act, state) pair (A_6, s_4). However, in the former problem, comparing $u(A_2, s_5)$ with $u(A_6, s_4)$ is meaningless (as it now stands!) since this involves an *interpersonal* comparison between subject 5 and subject 4.

It can be shown (simply by modifying the Blackwell and Girshick theorem mentioned in footnote 3) that conditions 1 and 3, Pareto optimality, and invariance under individual utility transformations, again lead to *serial dictatorship*, so it is clear that, if we are to exploit utility preferences of individuals, some inner bond must be hypothesized between the utility scales of different subjects. But how can this be meaningfully accomplished? We turn in the next section to some proposals.

*14.6 SOCIAL CHOICE PROCEDURES BASED ON INDIVIDUAL STRENGTHS OF PREFERENCES

Social choice procedures which attempt to reflect individual *strengths* of preference (not merely rankings) must eventually contend with the interpersonal comparison problem. Either a commensurating unit, or a base of reference, or both have been suggested. Goodman and Markowitz [1952] suggest an operational procedure for making interpersonal comparisons of differences of individual strengths of preferences. Nash's work on the bargaining problem can be generalized directly to yield a social choice procedure for groups of arbitrary size. In this generalization no interpersonal comparison of utility is made, but the bargain situation has a natural base of reference—the status quo point. Hildreth [1953] also utilizes linear utility scales to capture strengths of preferences; but, unlike

Nash's work, his scheme in essence establishes a commensurating unit and a common origin of measurement by singling out a pair of social states which serve as a base of common reference.   We will comment on these procedures in turn.

The commensurating unit used by Goodman and Markowitz is a variation of the just-noticeable-difference (j. n. d.) notion used by sensory psychologists.   Very roughly and not in the words of the authors, individual 1 prefers alternative A_1 to A_2 more than 2 prefers alternative A_3 to A_4 if 1 "establishes" more discernible potential distinct indifference levels[4] between A_1 and A_2 than 2 does between A_3 and A_4.   The authors assume that "each individual has only a finite number of indifference levels or 'levels of discretion.'   . . . A change from one level to the next represents the minimum difference which is discernible to an individual."

Let A_1, A_2, $\cdots$, A_m be m alternatives and s_1, s_2, $\cdots$, s_n be n individuals.   The profile of individual orderings and strengths of preferences can be summarized in an m by n array where the entry in the (i, j)th cell is $u(A_i, s_j)$.   The numerical quantities are confined to be *integers*, $\cdots$ -2, $-1, 0, 1, 2, \cdots$, having the following interpretation: If $u(A_i, s_j) = 10$ and $u(A_k, s_j) = 6$, say, then s_j prefers A_i to A_k, and s_j can discern exactly four indifference levels between A_i and A_k.   From this interpretation it is easy to see that two (integral-valued) arrays are strategically equivalent if one can be obtained from the other by adding or subtracting integers from each column.   The authors require: (i) that to each integral-valued array there be associated a complete ordering of alternatives; (ii) that adding an integer to a column shall not change the social ordering; (iii) Pareto optimality; (iv) that adding new alternatives shall not change the social ordering of the old alternatives; and (v) that the social ordering of alternatives shall not depend upon the labeling of alternatives or individuals.

We have already commented on (i) and (ii).   Pareto optimality needs no discussion at this stage of the game.   Requirement (iv) is an irrelevancy axiom.   Note that, in obtaining the payoff matrix, hypothetical, unattainable, and *really* irrelevant alternatives are considered by each individual in order to ascertain his discernible number of distinct potential indifference levels between any two available alternatives under consideration.   But once the payoffs are determined, then deleting an originally available alternative or adding an available alternative shall not change the existing payoffs and shall not alter society's choice for existing paired comparisons.

[4] They are, of course, making the extremely plausible assumption that indifference is not a transitive relation—that, for example, one can be indifferent between a and b, between b and c, and between c and d but still prefer a to d.   Think, if you please, of adding pepper to food one grain at a time.

Conditions (i) and (iv) together state that the social ordering is fully determined by society's choices for paired comparisons of alternatives. In condition (v), the part about the labeling of alternatives is perfectly innocuous, but the part about invariance under labeling of individuals is a major assumption. This is an egalitarian, symmetry, or democracy assumption which may or may not be applicable depending upon the context of the problem.

From these five requirements it can be shown that society *must* order alternatives strictly according to their average payoffs—or, what operationally amounts to the same thing, society's "utility" for an alternative is the sum of the individual "utilities" for that alternative. Note that the Goodman-Markowitz requirements are mathematically identical to Milnor's axioms characterizing the Laplace criterion (i.e., the criterion based on the principle of insufficient reason) in the decision-making problem under uncertainty.

Goodman and Markowitz do not spell out in sufficient detail just how an individual can determine experimentally the number of distinct indifference levels between two specific alternatives. One simple way is to employ, at the outset, a "sufficiently rich" set α of alternatives which includes, among others, all the alternatives that the individuals and society will be asked to rank. Each individual can then assign integers to the alternatives of this canonical set as follows: the numerical assignment to A_i is exactly one unit higher than that to A_k if and only if both A_i is preferred to A_k and there is no alternative in α which is more preferred than A_k and less preferred than A_i. Any specific social choice problem will then involve a small subset of available alternatives from α, but each of these will have a specific index for each individual.

It is difficult for us to see how the set α can be initially defined. However, even if we assume, for the time being, that operational numerical assignments to alternatives can be effected, there still remains the real question: Does the number of "discernible discretion levels" between a pair of alternatives really measure *strength of preference* for a single individual, and should this be a basis for interpersonal comparisons in the context of social choice? The next example is none too encouraging.

Consider two individuals s_1 and s_2 who have to select one of two candidates A_1 or A_2; candidate A_1 is preferred by s_1, and A_2 by s_2. To resolve the strength of preference problem, these voters are also asked to rank some non-available candidates, A_3, A_4, $\cdots$, A_{100}. Voter s_2 is very discerning, and he ranks the candidates A_2 over A_3 over $A_4 \cdots$ over A_{99} over A_{100} over A_1. On the other hand, voter s_1 is dedicated to a single issue and he divides all candidates into two camps, namely, those who are "for" it and those who are "against" it; he is indifferent among the "for's" and indifferent among the "agin's." In this case s_2 will have his way since he can discern 99 indifference levels between A_2 and A_1, whereas s_1 can only discern one indifference level between them. But who is to say that s_2 feels more strongly than s_1?

Arrow [1951 d] points out that "the use of the *minimum sensible* as a commensurating unit for utility functions goes back to F. Y. Edgeworth [1881]." More recently, it has been used by W. E. Armstrong [1939, 1951]. Armstrong takes the point of view that, although individual preference is a transitive relation, indifference is not. For example, A_1 might be preferred to A_3 and at the same time A_2 might be indifferent to both A_1 and A_3!

▶ Recently, Luce [1956 a] has given a general formulation of preference relations where strict preference is transitive and indifference is not, which is substantially equivalent to the idea of minimum sensible utility differences. These relations, which are called semiorders, induce a natural weak order on the set of alternatives; and, if this weak order can be represented by a numerical utility function, the semiorder can be represented by that function plus two just-noticeable difference functions—one giving the just-perceived increases, and the other the decreases. Furthermore, the converse holds: if three functions over a set of alternatives meet certain weak and natural conditions, then a semiorder is induced on the set. This generalizes the non-probabilistic discrimination model used in sensory psychology to arbitrary sets, and it does not suppose at the outset that just-noticeable differences are equal to each other. These questions arise: when, if ever, in the domain of preferences, is it reasonable to assume j. n. d.'s are equal; and, supposing they are equal for each individual, when it is reasonable to equate the j. n. d.'s between people?

In later work, unpublished at the time this book was written, this model has been generalized to a probabilistic theory of utility (see Appendix 1 and Luce [1956 b]). It is supposed that the underlying set of alternatives consist of all probability mixtures of a given set of basic alternatives. (*Note:* In the above example of two candidates, which was devised to cast doubt on equating j. n. d. units between people, we probably could not have drawn the same conclusion had we taken all mixtures of the two candidates as the basic set of alternatives). Under certain sufficient conditions (see Appendix 1), it has been shown that a linear utility function must have constant j. n. d. functions, so the j. n. d. may be taken to be the unit. As to whether it is at all plausible to equate these between people, one can say at least this: it makes no sense to equate them unless a certain theoretical psychophysical parameter is the same for all people. It is an empirical question, which has not yet been answered, whether this parameter exists, and whether it varies from person to person or is constant. Every indication, however, is that it varies, and, if so, one can show that the interpersonal comparison resulting from equating j. n. d.'s depends upon a purely artifactual parameter, namely, the probability cut off one uses to decide whether alternatives are indifferent or not. Therefore, equating j. n. d.'s appears not to be an acceptable solution to the interpersonal comparison problem; and, even if the parameter does not vary, it remains an open question whether this unit would be a satisfactory solution to the problem.   ◀

Next, we shall briefly reconsider Nash's work [1950 *b*] on the bargaining problem and discuss its relation to the problem of social choice. An *n*-person bargain is a simple generalization of the two-person case. There are *n* individuals 1, 2, $\cdots$, *n* and individual *i* comes to the "market" with a bundle of goods $T_i{}^*$, so the status quo point is the *n*-tuple $(T_1{}^*,$ $T_2{}^*, \cdots , T_n{}^*)$. Let $\mathfrak{I}$ be the set of all possible trades or reapportionments of the whole set of goods, and let $\mathbf{T} = (T_1, T_2, \cdots , T_n)$ be a generic trade in $\mathfrak{I}$, where T_i represents *i*'s share of the goods in trade **T**. Each of the *n* individuals have preference orderings over the trades in $\mathfrak{I}$, and the problem is to amalgamate a profile of individual preferences over the set of trades to obtain an optimal trade (or trades) for the *n*-person society. There is one basic difference between the *n*-person $(n > 2)$ bargain and the two-person bargain: coalitions are possible, and so the full complexities of *n*-person game theory arise. If we wish the "optimal" choice for society to reflect the strategic aspects of coalition structures, a simple modification of the Nash two-person bargain analysis is not appropriate. However, in some contexts, coalitions (other than a coalition of all players) are explicitly prohibited by the rules of the game; in other contexts, even though coalitions might not be prohibited, the players may seek arbitration which explicitly avoids the strategic aspects of coalition formation. Since we have already considered at length topics in *n*-person theory which, in some sense, can be thought of as social choice procedures which reflect coalition strengths (e.g., Shapley's value and Milnor's reasonable sets), we now turn to the case where coalitions are either explicitly or implicitly *ignored*.

For completeness, we will state briefly an *n*-person modification of Nash's axioms for the two-person bargain.

Axiom 0. *Each individual's preferences for trades and lotteries over trades can be summarized by a linear utility indicator.*

If **T** is a trade in $\mathfrak{I}$, let $u_i(\mathbf{T})$ denote player *i*'s utility for **T** and let $\mathbf{u}(\mathbf{T})$ be the *n*-tuple $[u_1(\mathbf{T}), u_2(\mathbf{T}), \cdots , u_n(\mathbf{T})]$ representing the utilities of **T** for the *n* players. For brevity, we write $\mathbf{u}^* = (u_1{}^*, u_2{}^*, \cdots , u_n{}^*)$ for $\mathbf{u}(\mathbf{T}^*)$. The set of all *n*-tuples **u** obtained by letting **T** vary over its domain $\mathfrak{I}$ will be denoted by *R*. Thus, a version of a bargain is the pair $(R, \mathbf{u}^*)$.

A social choice function associates to each bargain $(R, \mathbf{u}^*)$ a single point $\mathbf{u}' = (u_1{}', \cdots , u_n{}')$, which belongs to *R*. The element $\mathbf{u}'$ is said to be *the solution* of the version $(R, \mathbf{u}^*)$, and the trade or lottery whose utility appraisal is $\mathbf{u}'$ will be termed *the solution* of the real bargain (relative to the specific choice function employed.) We shall assume for mathematical convenience that there exists at least one point $\mathbf{u} = (u_1, u_2, \cdots , u_n)$

in R such that $u_i > u_i^*$ for $i = 1, 2, \cdots, n$. Nash then demands of a social choice function:

Axiom 1. *The solution of the real bargain shall not depend upon the utility scales (origins and units of measurement) used to abstract the problem.*

Axiom 2 (Pareto optimality). *If $\mathbf{u}'$ is a solution of $(R, \mathbf{u}^*)$ then*

(i)   *$\mathbf{u}'$ belongs to R,*

(ii)   *There does not exist a $\mathbf{u}''$ in R distinct from $\mathbf{u}'$ for which $u_i'' \geqslant u_i'$, $i = 1, 2, \cdots, n$.*

Axiom 3 (independence of irrelevant alternatives). *Adding new (trading) alternatives, with the status quo point kept fixed, shall not change an old alternative from a non-solution to a solution.*

Axiom 4 (symmetry). *Let a version of a bargain have the properties that*

(i) $u_i^* = u_j^*$, i *and* $j = 1, 2, \cdots, n,$

and

(ii)   *If $\mathbf{u} = (u_1, \cdots, u_n)$ belongs to R, any permutation of the components of $\mathbf{u}$ also yields an element of R,*

then $u_i' = u_j'$, i *and* $j = 1, 2, \cdots n,$

where $\mathbf{u}' = (u_1', u_2', \cdots, u_n')$ *is the solution of $(R, \mathbf{u}^*)$.*

Nash's proof for $n = 2$ can be modified directly to yield: $\mathbf{u}'$ *is a solution of $(R, \mathbf{u}^*)$ if and only if $\mathbf{u}'$ belongs to R, $u_i' > u_i^*$ for all i, and*

$$(u_1' - u_1^*)(u_2' - u_2^*) \cdots (u_n' - u_n^*)$$
$$\geqslant (u_1 - u_1^*)(u_2 - u_2^*) \cdots (u_n - u_n^*)$$

for all $\mathbf{u} = (u_1, u_2, \cdots, u_n)$ in R such that $u_i > u_i^*$ for all i.

Hildreth [1953] gives a set of conditions for social orderings which bears some relationship to this modification of the two-person bargain. Each social state X in Hildreth's work can be identified with an n-tuple $(X_1, X_2, \cdots, X_n)$, where X_i represents "a specification of amounts of commodities to be received and furnished by the [ith] individual over a given period of time or a probability combination of such specifications." Hildreth also assumes that each individual ordering of social states can be represented by a utility indicator. (Actually Hildreth's axioms on individual preferences are not quite strong enough to yield an individual linear utility scale, but his intent certainly is to establish such scales.)

Both Hildreth and Nash demand individual linear utility scales, Pareto optimality, and some form of symmetry condition. But here the similarity ends. Nash only requires the selection of a best element for society (i.e., a unique solution); Hildreth requires a complete ordering of alternatives by society. Nash explicitly demands that his solution be invariant up to individual utility transformations, whereas Hildreth makes no such demand—indeed, his suggested schemes establish both a common utility origin and unit of utility measurement. We will describe Hildreth's technique for accomplishing this below. Nash, like Arrow and Goodman and Markowitz, invokes an irrelevancy axiom; Hildreth does not demand this.

To summarize, Hildreth requires: (i) individual utility scales, (ii) complete order for society, (iii) the usual Pareto optimality, and (iv) strong symmetry. The strong symmetry condition consists of two parts.

(a) *Invariance with respect to labeling of individuals.* This is like the Goodman-Markowitz symmetry requirement, and it implies Nash's symmetry condition.

(b) *Similar treatment for similar individuals.* This will require some explanation.

Let $X^{(ij)}$ mean the social state which results if the ith and the jth elements of X are interchanged. In other words, i and j get in $X^{(ij)}$ what i and i get, respectively, in X; the returns to the other individuals are kept fixed. Individuals i and j are said to be *similar* if and only if (1) i reacts to any paired comparison X versus Y the way j reacts to $X^{(ij)}$ versus $Y^{(ij)}$, and (2) other individuals treat i and j indifferently in the sense that they are each indifferent between X and $X^{(ij)}$ for all X. Let S be any class of alternatives such that (1) S contains all lotteries with elements of itself as prizes, and (2) if X is in S, so is $X^{(ij)}$. By *similar treatment for similar individuals*, i and j, is meant that, if X^* is optimal for society for the class of alternatives S, i and j are both indifferent between X^* and $X^{*(ij)}$.

Hildreth next adds an assumption which is innocuous enough as it stands, but, as he applies it, it is clearly the weakest link in his argument. He assumes:

"*There exist two states, say X^*, Y^*, for which the following hold*

(a) $X_i^* = X_j^*$, $Y_i^* = Y_j^*$ (all i, j),
(b) $X^*P_i\, Y^*$ (all i)."

We can find no fault in saying two such states X^* and Y^* exist—our objection comes from the fact that they might be used as reference points for interpersonal comparisons. If that is the intent, then some logical underpinning is sorely needed. How does one choose an X^* and Y^* from the set of potential candidates? The author does not say anything about this, but implications can be drawn from the use he makes of X^* and Y^*.

Hildreth shows that the above conditions (plus two operational conditions which "are not motivated by ethical considerations but by the desire for [technical] power and convenience of application [mathematical tractability]") are consistent in the sense that there exist procedures which satisfy all his desiderata.   To this end he proceeds as follows:

i. Choose utility scales for each individual such that

$$u_i(X^*) = 1, \; u_i(Y^*) = 0, \qquad \text{for } i = 1, 2, \cdots, n,$$

where $u_i(X)$ represents i's utility for alternative X.

ii. Let g be any continuous, monotonic increasing, and strictly concave function.   Associate to each X a social utility index $U_g(X)$ where

$$U_g(X) = g[u_1(X)] + g[u_2(X)] + \cdots + g[u_n(X)].$$

A procedure which orders alternatives according to the magnitude of the social utility index U_g satisfies all of Hildredth's requirements; this establishes consistency of the requirements.

▶Still another way to attempt to deal with group decision making is suggested by the following example.   Three industrialists decide to form a corporation with themselves as board of directors.   They anticipate that future disagreements will arise, and, to forestall wrangling, they invite a consultant to draw up a constitution giving the rules for resolving any potential division of opinion.   The three men are sincere in their efforts to cooperate, and there is no question about their falsifying their individual values.   Essentially, the constitution must provide a technique for amalgamating their individual values in any situation to obtain a choice for the corporation.   The consultant ascertains

(i) That the strengths of individual preferences are important.
(ii) That the individuals are to be treated asymmetrically since their initial capital investments, etc., differ.

The consultant might proceed as follows.   Using a set of alternatives that the industrialists conceivably might be called upon to rank, he determines each individual's utility function over them.   Presumably, these utilities reflect strengths of preferences.   The consultant next takes a big jump: he assumes that it is meaningful to postulate a common utility scale—even though he does not know how to find it.   That is, if u_1, u_2, u_3 are the utility indicators of the three individuals, he assumes there are (unknown) constants a_1, b_1, a_2, b_2, a_3, b_3 such that the utility indicators

$$a_1 u_1 + b_1, \qquad a_2 u_2 + b_2, \qquad a_3 u_3 + b_3$$

are expressed in a common interpersonal unit.   In terms of these utility scales, each alternative A can be given an index, $u(A)$, for the corporation as a whole, namely:

$$u(A) = w_1[a_1 u_1(A) + b_1] + w_2[a_2 u_2(A) + b_2] + w_3[a_3 u_3(A) + b_3],$$

where w_1, w_2, w_3 are weights which reflect the relative influence or importance of the industrialists.   The a's, b's, and w's are unknown at this point of analysis.

By simple algebra

$$u(A) = c_0 + c_1 u_1(A) + c_2 u_2(A) + c_3 u_3(A),$$

where $c_0 = w_1 b_1 + w_2 b_2 + w_3 b_3$, $c_1 = w_1 a_1$, $c_2 = w_2 a_2$, $c_3 = w_3 a_3$.

The consultant now asks the industrialists as a *group* to rank-order some of the alternatives as best as they can. From this series of group responses, the quantities c_1, c_2, c_3 are estimated. These estimates are then used to get group indices on the basis of which group action is taken in other more complex situations.

This method is similar to Savage's procedure (discussed in the last chapter) for getting *a priori* probability weights for the states of nature. There a series of hypothetical problems were first posed, and then *a priori* weights were derived from an analysis of responses. The problem of estimating c_1, c_2, c_3 may present difficulties since a group may not be consistent with the model, in which case either an error theory has to be introduced or the group itself can be asked which triplet of c's comes closest to predicting the group responses for the alternatives they have considered.

Note that no attempt has been made to recover either the individual a and b parameters or the w parameters. Indeed, any attempt would be futile since they are clearly non-identifiable. This analysis does *not* enable one to make either interpersonal comparisons of utility or an analysis of the distribution of influence or importance. ◀

▶Let us draw another analogy, this time between social choice and individual choice. Consider, for example, an individual who must order a set of alternatives, or stimuli, each of which is complex in the sense that it evokes reactions with respect to several attributes or "psychological dimensions." The following analysis might be plausible: each attribute orders the alternatives, and the individual amalgamates this profile of preferences over the relevant attributes when giving his individual ordering. Interpersonal utility comparisons for the social choice problem are replaced in this model of individual choice by interattribute utility comparisons. We imagine that most people feel it is easier to make operational sense of interattribute comparisons within an individual's mind than to make sense of interpersonal comparisons. Anyhow, we do. But why? Primarily because the individual himself can compare his psychological reaction to one (alternative, attribute) pair versus another pair. In unpublished work, Abramson [1956] has given a mathematical formulation of this model. Analogously, in the social choice problem, an arbiter might conceivably ask himself whether he would prefer being in s_5's shoes if A_2 were adopted or in s_4's shoes if A_6 were adopted. Although such considerations are undoubtedly employed by professional conciliators, it seems a most unhappy way of resolving interpersonal comparisons. It imposes the tastes of an outsider and, according to our way of thinking, oversteps the natural prerogatives of an arbiter. ◀

14.7 MAJORITY RULE AND RESTRICTED PROFILES

We have seen that, if very dissimilar individual orderings are permitted, simple majority rule can lead to an intransitive set of social preferences. This leaves open whether there are any reasonable restrictions on the profiles of individual orderings—reasonable in the sense that they do not rule out all practical, non-trivial cases—such that majority rule never leads to an inconsistent social ordering. If so, this is one way of side-stepping Arrow's impossibility theorem.   The answer is that, if we do not demand too much versatility of a social choice function, functions (including majority rule) do exist which satisfy conditions 2, 3, 4, and 5 of section 14.4. One "reasonable" restriction on the set of possible profiles of individual orderings, which has been studied independently by Coombs [1950, 1952, 1954] and Black [1948 a, b, c], is our present topic.

Let us, for simplicity, exclude all indifferences or ties in rankings throughout this section.

In attitude measurement theory, Coombs noticed that often when a group of individuals rank a set of stimuli many of the potential rankings are conspicuous by their absence.   For example, suppose that a large group of individuals ranking the stimuli (or alternatives) $\{A_1, A_2, A_3, A_4\}$ yield only the following 7 of the 24 possible rankings (recall that indifferences are excluded):

1	2	3	4	5	6	7
A_1	A_2	A_2	A_3	A_3	A_3	A_4
A_2	A_1	A_3	A_2	A_2	A_4	A_3
A_3	A_3	A_1	A_1	A_4	A_2	A_2
A_4	A_4	A_4	A_4	A_1	A_1	A_1

An explanation, or rationalization, for these 7 and not the other 17 being present is the existence of a "unidimensional underlying continuum." Suppose the alternatives A_1, A_2, A_3, A_4 are situated on some attribute continuum (e.g., a liberalism-conservatism scale when the alternatives are political candidates) as shown:

Assume each individual s_i has an ideal on this continuum and that he ranks the alternatives according to their "distances" from this ideal.   Thus, in the above diagram, s_i would register the pattern of decreasing preferences: A_3, A_2, A_1, A_4.   It can be verified that, if A_1 to A_4 are ordered as shown and if the distance from A_1 to A_2 is less than that from A_3 to A_4, each of the seven rankings (and only these!) can be generated by varying the location of the ideal points.   These seven rankings are said to be compatible with an underlying joint *quantitative* scale.

Obviously, not every profile of individual rankings is compatible with stimuli and individual ideals placed jointly on the same continuum and with rankings determined according to the distances from stimuli to ideal points.   Furthermore, if a profile of individual orderings is compatible with an underlying joint quantitative scale, then the positions of the stimuli and individual ideals are not uniquely located; however, certain constraints on the ordering of stimuli and individual ideals, and on the relative sizes of distances between stimuli, are dictated by the manifest data (i.e., the profile of individual rankings).

For the time being, let us assume that we are omniscient to the extent of knowing the *exact* position of the alternatives and the individuals on a joint continuum. Given this underlying structure for a profile of individual rankings, how can one pass from it to a ranking for society? Coombs [1954] suggests that a *reasonable ranking for society is that given by the median individual along the continuum* (assume for convenience that there are an odd number of individuals so that the median is well defined). Goodman [1954] has pointed out that in this model *the ranking given by the median individual yields the same ordering as that generated by simple majority rule*. This is simple to see. If the median individual prefers A_i to A_j (i.e., if his ideal is closer to A_i than to A_j), then a majority of individual ideals are closer to A_i than to A_j, so a majority of individuals prefer A_i to A_j. As a trivial by-product of this result, we note that, if a profile is compatible with an underlying Coombsian joint quantitative scale, simple majority rule yields a transitive ordering for society.

In practice, however, Coombs has noted that most profiles of individual rankings cannot be rationalized by a joint quantitative scale, so he has suggested a simple modification. A profile of individual rankings is said to be compatible with an underlying *qualitative* (in distinction to *quantitative*) joint scale if there exists some overall complete ordering of both alternatives and individual ideals such that

i. If, in terms of the overall ordering, s_i is not between the two alternatives of a paired comparison, the alternative closer to s_i in the ordering is preferred.

ii. If s_i is between the two alternatives, then s_i may make either choice.

For example, suppose a profile of individual orderings contain the seven rank orderings given on p. 354 plus the eighth ranking: A_2 over A_3 over A_4 over A_1. These eight rankings *cannot* be generated by any joint *quantitative* scale; this we can see as follows. The original seven rankings require that the distance between A_1 and A_2 be less than the distance between A_3 and A_4. The ideal point of an individual registering the eight ranking must be to the left of the midpoint between A_2 and A_3 (since A_2 is preferred over A_3), and, therefore, A_1 must be preferred to A_4, contrary to assumption. However, these eight rankings are compatible with an underlying *qualitative* joint scale. For example, an individual whose ideal is between A_2 and A_3 (when the alternatives are ranked A_1, A_2, A_3, A_4) can register the ranking A_2, A_3, A_1, A_4. Indeed, all that is required of such an individual is that he rank A_2 over A_1 and A_3 over A_4.

Before we introduce qualitative joint scales into the social choice problem, let us relate them to Black's work. Black defines a profile of individual rankings to satisfy the *single-peakedness condition* if there is some basic ordering of the alternatives such that, in passing from one alternative to the next in this basic ordering, each individual monotonically rises to the peak of his preferences and then monotonically drops off in the direction of dis-preference.[5] To see that the seven rankings on p. 354 plus the ranking A_2 over A_3 over A_1 over A_4 satisfy a single-peakedness condition, consider the basic ordering from left to right: A_1, A_2, A_3, A_4. It is routine to verify that each ranking is single peaked in its preferences. For example, an individual registering the ranking A_3, A_2, A_1, A_4 goes up in his preferences as he goes from A_1 to A_2 to A_3 and then drops off. It is not difficult to see that *a profile of rankings is compatible with an underlying joint qualitative scale if and only if it satisfies a single-peakedness condition.*

Black [1948 *a, b*] has shown that, if the number of individuals is odd and if a profile meets the single-peakedness condition, then there is exactly one alternative which receives a majority over any other alternative.[5] Arrow has extended this result to show that for profiles meeting this condition simple majority rule generates a transitive ordering of the alternatives. Therefore, if Arrow's first condition is modified to the demand that a social welfare function be defined only for those profiles of individual preferences which meet a single-peakedness condition (or, equivalently, for those which can be generated from a Coombsian joint qualitative scale), then simple majority decision satisfies this and his remaining conditions 2, 3, 4, and 5, provided the number of individuals is odd.

▶To gain further insight into these results, suppose that for a profile generated by a Coombsian joint qualitative scale a majority of individuals prefer A_i to A_j, A_j to A_k and A_k to A_i. We will establish that this assumption leads to a contradiction, thereby showing that intransitivities cannot occur. The crux of the argument is the observation that, if an alternative of the joint scale lies between two others, it cannot be less preferred than both of them for any individual. Try it! There are three cases to worry about: on the joint scale either

(i) A_i lies between A_j and A_k,

or

(ii) A_k lies between A_i and A_j,

[5] In this section we are assuming no indifferences in individual values for distinct alternatives. If this assumption is weakened, some simple modifications of the conclusions have to be made.

or

(iii) A_j lies between A_i and A_k.

If, as is assumed in the hypothesis we wish to contradict, a majority prefers A_j to A_k and a (not necessarily the same) majority prefers A_k to A_i, then there is at least one person in common to the two majorities, and so he prefers A_j to A_k to A_i; but this is incompatible with case (i).   Similarly, if a majority prefers A_i to A_j and another prefers A_j to A_k, then some person must prefer A_i to A_j to A_k, which is incompatible with case (ii).   If a majority prefers A_k to A_i and A_i to A_j, then some person must prefer A_k to A_i to A_j, which is incompatible with case (iii).   This completes the demonstration.   An actual proof of Arrow's result is constructed along these lines.
◀

14.8 STRATEGIC ASPECTS OF MAJORITY RULE

As we have seen, no rule can satisfy all of Arrow's demands, but some seem to fail in better ways than others.   Many feel that simple majority rule is one of the best, its main failing being that it leads to intransitivities for some profiles.   For this reason and because of its considerable social importance, it has been discussed at length in the literature.

One major feature of the simple majority rule is that it satisfies the axiom of the independence of irrelevant alternatives.   Therefore, society's choice between a pair of alternatives depends solely upon the profile of preferences for that pair, and so any further characterization of the rule can be based simply on paired comparisons.   Under that restriction, May [1952] points out that simple majority rule has the following four properties:

i. *Decisiveness.*   For any profile of individual choices, it specifies a unique group decision for each paired comparison.

ii. *Anonymity.*   It does not depend upon the labeling of individuals.

iii. *Neutrality.*   It does not depend upon the labeling of the two alternatives.

iv. *Positive responsiveness.*   If for a given profile the rule specifies that y is not preferred to x (i.e., xPy or xIy) and if a single individual then changes his paired comparison in favor of x (i.e., changes yP_ix to either yI_ix or xP_iy, or changes yI_ix to xP_iy), while the remainder of the society maintain their former choices, then the rule requires that society prefer x to y for the new profile.

May also proves the deeper result: *simple majority rule is the only rule satisfying these four properties.*   The idea of the proof is straightforward.

▶Suppose that a specific profile of individual choices for x versus y is given.   By anonymity, the group choice will only depend upon three numbers: the number N_x of individuals who favor x, the number N_I who are indifferent, and the number

N_y who favor y. By *neutrality* the group must be indifferent if $N_x = N_y$. By repeated applications of *positive responsiveness* (starting from $N_x = N_y$) we conclude that the group favors x if $N_x > N_y$; therefore, by *neutrality*, it favors y if $N_x < N_y$. But this is simple majority rule. The *decisiveness* property is employed all along since we tacitly assumed a unique group decision.   ◄

May also points out that these four conditions are independent in the sense that rules can be devised which satisfy any three of them while not satisfying the fourth. Examples are: (i) no decision in case of ties (i.e., $N_x = N_y$); (ii) simple majority rule where one specific individual has two votes; (iii) two-thirds majority needed for x; (iv) unanimous decision required. See also May [1953].

As we know, simple majority rule can lead to intransitive group preferences for dissimilar individual orderings. Modifications of simple majority rule have been suggested by Copeland [1951] which avoid the difficulty at the expense of violating Arrow's third condition. Let $u(x)$ denote the number of alternatives which lose to x, less the number which beat x in simple majority rule. In one modification, alternatives are ranked according to this index. For example, the Copeland indices for the profile

1	2	3
x	y	z
y	z	x
z	x	y

are $u(x) = u(y) = u(z) = 0$. If a new alternative w is introduced and the profile is changed to

1	2	3
x	y	w
w	z	z
y	x	x
z	w	y

then $u(x) = 1$, $u(y) = -1$, $u(z) = -1$, $u(w) = 1$. Thus, Copeland's suggestion fails to satisfy the independence of irrelevant alternatives, since the paired comparison of x versus y depends upon whether w is present or not. It obviously meets the other axioms.

Another modification of simple majority rule which cannot generate intransitivities is this: For each alternative x, let $v(x)$ denote the number of distinct paired comparisons over *all* the individuals which involve x and which are resolved in x's favor, less the number resolved against x. Alternatives are ranked according to this index. (For example, if there are 5 alternatives and 10 individuals, there are 40 paired comparisons involving each alternative x. Should 25 be resolved in favor of x, 7 be against x, and 8 be ties, then $v(x) = 25 - 7 = 18$.)

Political scientists may not feel too disturbed by the possible intransitivities of simple majority rule, since in most legislative bodies individuals are asked to select a single alternative, not to order them by preference. Often a single alternative to an existing law is suggested, and when there are several alternatives they are forwarded in succession, each being pitted against the current status quo situation. Some may, therefore, feel that intransitivities cannot arise under present practice. This is a naive view, however. Let x be the existing law, and let y and z be possible replacements. Suppose the legislative body divides into three equal groups, which, if called upon, would register the profile:

Group 1	Group 2	Group 3
x	y	z
v	z	x
z	x	y

Suppose y is first pitted against x and then z against the winner. This ultimately leads to z, since, by majority rule, y loses to x and x to z. But, suppose z is first pitted against x and then y against the winner. This ultimately leads to y since x loses to z and z to y. Thus, as is well known to practical politicians, the final outcome can depend upon the order of presentation of the bills. If a defeated bill is reintroduced and then wins, the interpretation often made is that some people have changed their minds. This may be so, but it need not be. The interpretation usually made ignores the different status quo's in the two cases. Thus, given the usual application of the simple majority rule in legislative bodies, observers may be quite unaware of intransitivities even when they do exist.

Arrow points out that it can benefit an individual legislator to misrepresent his true feelings in legislatures which vote on successive motions by simple majority rule. He cites the following example.

Let individual 1 have ordering x, y, z; individual 2, y, x, z; and individual 3, z, y, x. Suppose that the motions come up in the order y, z, x. If all individuals voted according to their orderings, y would be chosen over z and then over x. However, individual 1 could vote for z the first time, insuring its victory; then, in the choice between z and x, x would win if individuals 2 and 3 voted according to their orderings, so that individual 1 would have a definite incentive to misrepresent. The problem treated here is similar to, though not identical with, the majority game, and the complicated analysis needed to arrive at rational solutions there suggests strongly the difficulties of this more general problem of voting. [Arrow, 1951 *a*, pp. 80–81.]

Still another example in voting strategy is instructive. Let x be the existing law, and suppose that a three-quarters majority is required to replace it. Consider a legislature composed of four equal groups, and suppose groups 1, 2, and 3 all prefer alternative y to x, thus ensuring its passage. However, in an attempt to keep y from becoming law, group 4 suggests an alternative modification z. Suppose the profile of preferences for x, y, z is as follows:

Group 1	Group 2	Group 3	Group 4
z	z	y	x
y	y	x	z
x	x	z	y

Group 4 demands a vote between z and y to determine which will be pitted against existing law x. If the legislative rules permit this order of voting, z defeats y, but it fails to get the three-quarters majority necessary to defeat x. Alternative x remains inviolate and group 4 gets its way. *Note:* Group 4 could be perfectly sincere in suggesting z instead of y!

Still another game-theoretic aspect of the simple majority rule has been cited by Majumdar [1956]. He considers a legislature where there are two people (or parties) which may sponsor bills, and a majority vote decides which of the two bills is passed. Suppose that there are four bills M, N, O, and P of interest, that the transitive preference ordering for sponsor 1 is M over N over O over P, and that for sponsor 2 the ordering is just the opposite. When any pair is presented, it is assumed that the alternative which will prevail by majority rule is known. Suppose the outcomes are those shown in Table 14.3a. In Table 14.3b we have

TABLE 14.3

		Sponsor 2					Sponsor 2		
		M	N	O	P		M	N	O	P
	M	M	N	M	P	M	$(4, 1)$	$(3, 2)$	$(4, 1)$	$(1, 4)$
Sponsor	N	N	N	O	N	Sponsor N	$(3, 2)$	$(3, 2)$	$(2, 3)$	$(3, 2)$
1	O	M	O	O	O	1 O	$(4, 1)$	$(2, 3)$	$(2, 3)$	$(2, 3)$
	P	P	N	O	P	P	$(1, 4)$	$(3, 2)$	$(2, 3)$	$(1, 4)$

a   b

replaced the outcomes by numbers indicating the *ordinal* preferences of the two sponsors. For example, if 1 sponsors N and 2 sponsors O, then O gets a simple majority and the corresponding ordinal preference is $(2, 3)$. Note that, as shown in Table 14.3a, the legislature prefers O to N, N to P, P to M, and M to O; as we know, such intransitives are perfectly possible with the simple majority rule.

Majumdar suggests that the appropriate strategy for a sponsor in this game is to bargain to be last bidder, and, if, unhappily, one is to move first, then he should misrepresent his true tastes. We may utilize our knowledge of two-person theory to comment on these observations.

First, since the sponsors have strictly opposing preferences for the alternatives, and hence the outcomes, the game is constant-sum. From what we have seen earlier, it is therefore never advantageous for either player to disclose his strategy. It is not, however, disadvantageous to disclose (to a smart adversary) that one is going to use a maximin strategy. Since this game fails to have a saddle point in pure strategies, at least one of the maximin strategies must be randomized. It is not difficult to see that 1 should randomize between M and N and 2 between O and P if both players are to move simultaneously.[6]

If the sponsor's preferences are not strictly opposing, then the game is not constant-sum and so having the first move is not necessarily disadvantageous. Of course, problems of preplay communication, arbitration, etc., arise.

At first glance, simple majority rule appears to abstract away all individual intensities of preference. Dahl [1956, p. 90] writes:

What if the minority prefers its alternative much more passionately than the majority prefers a contrary alternative? Does the majority principle still make sense?

This is the problem of intensity. And, as one can readily see, intensity is almost a modern psychological version of natural rights. For, much as Madison believed that government should be constructed so as to prevent majorities from invading the natural rights of minorities, so a modern Madison might argue that government should be designed to inhibit a relatively apathetic majority from cramming its policy down the throats of a relatively intense minority.

Yet we can argue that in some measure intensity of preference does receive expression in actual practice through "logrolling." A senator who feels strongly about bill q and indifferent about bills r and s will trade his votes on r and s for the desired votes of senators indifferent about q. Thus, his strong preference for bill q is recorded, and it may be passed even though according to the true tastes of the senators it should have been defeated. Is this good? That depends upon the bill q—at times we grumble "Shameful!" and at others chuckle "Beautiful strategy!"

The dynamic strategic aspects of legislative voting generate lovely examples of n-person games in extensive form. If bill q would have been passed without the sale of votes on r and s, then our senator has wasted assets that could have been put to better use.

Filibustering and its threatened use is, of course, another method by which an intense minority can exert pressure on the majority. A significant difference between filibustering and logrolling, however, is that an intense minority can defeat an *intense* majority with filibustering but not with logrolling.

[6] For player 1, row P is dominated by row N, so it can be deleted. Similarly, M and N are dominated for player 2. Once M is deleted for 2, O is dominated for 1.

There are alternative legislative voting schemes which make it still easier for the legislators to express their intensity of preference. Consider the following rule: A group of bills concerned with different issues are all debated and then simultaneously voted. Each individual is given 100 units to be distributed over the bills any way he wishes. The voting for each bill and the apportionments of weights over all of them is by secret ballot. The game-theoretic aspects are manifest. In fact, it is appalling to contemplate the ensuing havoc and recriminations.

Another important voting concept in democratic theory, which has received some mathematical treatment, is proportional representation. March and Levitan [1955] specify some "reasonable" conditions on a "political representation function," which they show imply that the percentage of votes won in an election by a party must be transformed into an equivalent percentage of seats won in a legislature.

Let there be m parties, and let each of n individuals vote for a single party. Implicit in their work is the democratic assumption that all individual voters shall be treated equally, so any set of individual choices can be summarized by an m-tuple $(x_1, x_2, \cdots, x_m)$, where x_i represents the proportion of votes cast for the ith party. A *political representative function* is a rule for assigning to each such $(x_1, x_2, \cdots, x_m)$ an m-tuple $(y_1, y_2, \cdots, y_m)$, where y_i indicates the proportion of seats won in the legislature by the ith party.[7] We shall idealize the model by *not* requiring that y_i times the total number of seats be an integer. Although this procedure is hardly to be recommended in practice, academically a non-integral number of seats can be interpreted to mean that a radomization scheme will be used. March and Levitan require:

1. *Equal treatment of parties.* (That is, the procedure does not depend upon the labeling of the parties.)

2. *Party legislation strength depends only upon party voting strength.* (This condition, similar to a condition of independence of irrelevant alternatives, requires that party representation be independent of the irrelevant distribution of the remaining vote over the other parties.)

In addition, one might be tempted to add a condition of *No votes, no seats* (analogous to Arrow's condition of non-imposition), or *More votes do not mean fewer seats* (analogous to Arrow's condition of positive association of individual values). However, this need only be done if we weaken either condition 1 or condition 2, for both are implied by our two requirements

---

[7] Mathematically, the word "proportion" here implies

$$y_i \geqslant 0 \quad \text{and} \quad \sum_{i=1}^{m} y_i = 1$$

when there are more than two parties.   Indeed, if there are more than two parties, conditions 1 and 2 imply that the rule must assign $(x_1, x_2, \cdots, x_m)$ to the vote $(x_1, x_2, \cdots, x_m)$.   That is, each party is represented strictly according to the percentage of votes it receives.   This is the principle of proportional representation.

▶The idea of the proof is as follows: From condition 2 (the proportion of representation y_i of the ith party depends only upon the proportion x_i that the ith party receives), it follows that y_i is some function of x_i, i.e., $y_i = f_i(x_i)$.   By condition 1, the function f_i does not depend upon i, so we can write $y = f(x)$, where y is the proportion of representation of any party that receives the proportion x of the vote.   If u and v are any two numbers between 0 and 1 such that $u + v \leqslant 1$, it follows that

(i) $f(u) + f(v) + f(1 - u - v) = 1$,

and

(ii) $f(u + v) + f(1 - u - v) = 1$.

Subtracting (i) from (ii) yields

(iii) $f(u + v) = f(u) + f(v)$.

Let n be the total number of people voting; then both u and v must be of the form i/n, where i is one of the integers $0, 1, 2, \cdots, n$.   If we choose $u = v = 0$, then (iii) reads $f(0) = f(0) + f(0)$, and so $f(0) = 0$.   Similarly, (ii) reads $f(0) + f(1) = 1$, and so $f(1) = 1$.   If we now choose $u = v = 1/n$, then (iii) reads $f(2/n) = 2f(1/n)$.   If, next, we let $u = 2/n$, then (iii) reads $f(3/n) = 3f(1/n)$. Continuing in this manner, we see that $f(i/n) = if(1/n)$, for $i = 0, 1, \cdots, n$. But, since $1 = f(1) = f(n/n) = nf(1/n)$, we conclude that $f(1/n) = 1/n$ and $f(i/n) = i/n$, as was to be shown.
The number of parties, m, had to be greater than 2 to justify step (i) above, which required three parties getting proportions u, v, and $1 - u - v$ respectively.◀

Condition 2, which demands that party representation be independent of the way the remaining vote is distributed among the other parties, is extremely strong, and it has been and will be heatedly debated for a long time to come.   Since condition 1 is quite innocuous in most contexts, debate over condition 2 boils down to a debate over P.R. itself—which inescapably involves such dynamic game-theoretic aspects as coalition formation and the stability of coalitions in the legislature itself.   We could easily be tempted to go off in this direction if *only* we had something new to add.

14.9 GAMES OF FAIR DIVISION

All the methods so far described to amalgamate individual preferences into a social preference have this one element in common: it is supposed that explicit rankings of the several alternatives are known for each of the individuals.   If these ground rules are changed, alternative procedures must be used.   For example, a game may be devised to resolve the conflict without necessitating that each individual present his schedule of preferences.   The rules will be so concocted that when players act "rationally" in their own selfish interests the outcome is "socially desirable" or "fair" to all the participants.   On the macroscopic level it is often alleged that economic markets fulfill such a role; it is clear, however, that this need not always be the case.   The economic analogue of the prisoner's dilemma (see section 5.4) appears not to result in a socially desirable outcome.   In this section we shall be concerned with games explicitly designed to lead to fair outcomes.

First, let us reconsider and modify the two-person bargain from this point of view (see Chapter 6).   Recall that a bargain is characterized by a set $\mathfrak{I}$ of feasible trades or reapportionments of the collective sum of goods and by a special distinguished trade T^* representing the status quo point (i.e., the division which gives each player the exact bundle of goods with which he started).   We shall modify this by assuming that the players as a group are given a set of items to be divided among them, and not that they bring their own goods to be bartered.   The set of items might be any of the following: $100, a pie, a painting, a complex bundle of goods, several pieces of real estate, a set of obligations which they must jointly perform, etc.   It can be argued that in all these cases there is also a status quo point, namely, the state obtaining before they were given anything to distribute. A well-known two-person method of "fair division" is for one individual to divide the commodity into two parts and for the other to choose the part he prefers; the asymmetry of the two roles can be eliminated from the game by assigning people to roles through a toss of a fair coin.   In what sense is this a "fair" procedure?   First of all, it is *egalitarian* in that it does not depend upon the labeling of the individuals; and, second, there is a presumption that the resulting outcome will be *Pareto optimal*.   For example, if the commodity to be divided is $100, the obvious strategy for the divider is an even split, and the same applies to a pie or any other finely divisible commodity.   Actually, even in these oversimplified cases, one cannot always expect people to follow this simple procedure.   Consider a wealthy man and a pauper who together come upon $100.   The rich man may prefer to give a larger share to the pauper, and therefore as divider he may make a 60–40 split.   Even with their roles reversed, the pauper, in an obvious appeal to the conscience of the rich man, might also elect the 60–40 rather than the 50–50 split.

A commodity which is indivisible, such as a painting, presents a new problem, for the divider has no real choice.   The procedure reduces to nothing but flipping a fair coin to determine the lucky player. This resolution, although egalitarian and Pareto optimal for the problem as described, may fail on the score of Pareto optimality if the scope of action for the players is enlarged to include side compensation. We have in mind this sort of modification: Each player adds x dollars to the pot, where x is an amount clearly in excess of the worth of the indivisible commodity—a painting, say—and then the usual fair division game is played with the $2x$ dollars plus the painting. The divider splits the pot into two parts, (1) the painting plus y dollars, and (2) $2x - y$ dollars. The chooser selects the part he prefers more. Hence, the non-trivial strategic element of this game is the divider's selection of y. To be specific, suppose the coin selects 1 as divider and 2 as chooser. Furthermore, suppose 2 prefers the painting plus y dollars if and only if $y > y_2$. From 1's point of view, the unknown quantity y_2 can be thought of as the true state of nature in his decision problem under uncertainty, and his optimal choice naturally depends upon specifying his *a priori* knowledge of the true state. Suppose 1 feels indifferent between the painting plus y_1 dollars and $2x - y_1$ dollars. Then his maximin strategy is to divide the pot into these two indifferent parts. If, as dividers, both players are committed to their maximin strategies, clearly it is advantageous to be the chooser. On the other hand, if each person has precise knowledge of the other's indifference point, then it is advantageous to be the divider. Since this is not entirely obvious, let us look at it in more detail. Suppose $y_1 < y_2$. As divider, 1 can get slightly less than the painting plus y_2 dollars; as chooser, 1 can only expect slightly more than the painting plus y_1 dollars. So the advantage to 1 in being the divider is roughly $y_2 - y_1$ dollars, provided, of course, each player is aware of the true tastes of the other. A more striking and well-known case of "divider advantage" arises when the divider is indifferent about the painting but knows that the chooser is sentimentally attached to it. This knowledge of weakness—at least in a society of business men—can be exploited in the obvious way, so that when 2 chooses the painting he leaves the bulk of the money for 1. On the other hand, if 2 is the divider he can exploit 1's indifference for the painting.

In each instance, the final outcome of the above procedure is *Pareto optimal* in the sense that no alternative distribution can be suggested which both players prefer. However, *the procedure itself need not be Pareto optimal*, for both players may prefer an alternative procedure of "fair division." For example, if the players toss a fair coin to determine who is to receive a painting, the winner will not be willing to accept an alternate division—therefore the final outcome is Pareto optimal. Nonetheless, both players might prefer to share the painting over time (e.g., having it alternate years) than to gamble for its full possession.

Steinhaus [1948] reports a generalization of the "divide and choose" scheme to n players due to B. Knaster and S. Banach. Their solution, which pertains to an infinitely divisible homogeneous commodity, such as a cake, is as follows:[8]

> The partners being ranged 1, 2, 3, $\cdots$, n, 1 cuts from the cake an arbitrary part. 2 has now the right, but is not obliged, to diminish the slice cut off. Whatever he does, 3 has the right (without obligation) to diminish still the already diminished (or not diminished) slice, and so on up to n. Thr rule obliges the "last diminisher" to take as his part the slice he was the last to touch. This partner thus disposed of, the remaining $n - 1$ persons start the same game with the remainder of the cake. After the number of participants has been reduced to two, they apply the classical rule [one divides while the other chooses] for halving the remainder.

Knaster also suggests (cf. Steinhaus [1948]) a method of division applicable to the case where the bundle of goods to be distributed contains fairly indivisible objects. To be concrete, suppose a father leaves his single estate of four indivisible commodities to be shared "equally" among his three children. Let the four commodities A, B, C, and D have the monetary values shown in Table 14.4. For the time being, assume that the monetary worth to each of the children of any subset of the items is merely the sum of *his* monetary evaluations of the individual items.

Table 14.4

	Individuals		
Items	1	2	3
A	\$10,000	\$4,000	\$7,000
B	2,000	1,000	4,000
C	500	1,500	2,000
D	800	2,000	1,000
Total valuation	13,300	8,500	14,000
Fair share	4,425	2,833.33	4,666.67
Commodities received	A	D	B and C
Monetary worth of commodities received	10,000	2,000	6,000
Excess	+5,575	−833.33	+1,333.33
Final division	$A - 3,550$	$D + 2,858.33$	$B, C + 691.67$

Let us carry out the analysis for individual 2. His monetary evaluations for A, B, C, and D are \$4000, \$1000, \$1500, and \$2000, respectively, for a total of \$8500. In terms of 2's *own estimation*, his fair share should be $\frac{1}{3} \times \$8500 = \$2833.33$. Since he is high bidder only on commodity D, this is the only commodity he receives. Commodity D is worth \$2000

---

[8] Steinhaus represents the individuals by A, B, $\cdots$ and commodity units by 1, 2, $\cdots$.

to 2, so with it given to him he is still short (negative excess) \$833.33. A similar analysis shows that 1 and 3 have a total excess of \$5575 + \$1333.33 = \$6908.33. When 2 is paid his deficit, there remains a total excess of \$6908.33 − \$833.33 = \$6075. Each player's share of this total excess is $\frac{1}{3}$ × \$6075 = \$2025. Hence the final division should give each player \$2025 in excess of his "fair share." It is easy to verify that in general the final division will give each player at least as much as his fair share.

This procedure can be generalized to unequal shares. For example, suppose the will had stipulated $\frac{1}{2}$ share to 1, $\frac{3}{8}$ to 2 and $\frac{1}{8}$ to 3. The fair shares are then $\frac{1}{2}$ × \$13,300, $\frac{3}{8}$ × \$8500, and $\frac{1}{8}$ × \$14,000, instead of \$4425, \$2833.33, and \$4666.67, respectively. From here the analysis proceeds in a similar manner. It is also not difficult to suggest a modified procedure for situations when the monetary worth of a set of items is not the sum of the monetary worths of the items in the set.

Once such conciliation machinery has been established, players may find it profitable to misrepresent their true tastes and to enter into coalitions. For example, suppose 1 knew 2 and 3's recorded evaluations. It would then be to his advantage to value A at \$7001, B at \$3999, C at \$1999, and D at \$1999. Player 1's fair share is then $\frac{1}{3}$(7001 + 3999 + 1999 + 1999) = \$4999, and possession of A yields only an excess of \$2002. Thus, 1's final division is A − \$1135 instead of A − \$3550. If 1 does not know 2 and 3's valuations, it can be dangerous for him to misrepresent his tastes too grossly. On the other hand, a collusion of two players and collective misrepresentation of both their tastes is less dangerous.

The "divide and choose" principle yields an alternate way for sharing the estate $\{A, B, C, D\}$ which does not necessitate a prior recording of individual evaluations. To surmount the non-divisibility feature, let each player add \$10,000 to the pot giving a total commodity bundle of $\{A, B, C, D, \$30,000\}$, and then apply the n-person variant of the "divide and choose" principle to this set of goods. Again, the relative advantage or disadvantage in being the initial divider depends upon one's *a priori* knowledge of the true tastes of one's adversaries. A random selection of the order for the players eliminates this asymmetry.

A group decision (welfare) function which dictates precisely how to pass from individual values to social preferences is too cumbersome and impractical to be employed in many contexts. Often, an automatic adjustment process is needed which modifies the social choice slightly without necessitating an intricate re-evaluation each time there is a slight change in individual values. It is extremely difficult for a thoroughly planned system, which attempts to be egalitarian, to be flexible enough to cope with the dynamic vicissitudes of individual tastes. Using a flexible game mechanism to establish a "fair division" is not a novel idea, and, as mentioned before, the economic market is one such mechanism. Social planners can and do exercise indirect controls on the social outcome of a game by changing its rules and by altering various parameters of the system. The feasibility of introducing games of fair division to resolve conflicts of interest in non-economic contexts is an intriguing area for research. Ideally, such a game should yield a unique equilibrium point (assigning a "fair share" to each player) which is Pareto optimal. It would be nice if, in addition, the players would each act in accord with their true tastes when at the equilibrium point. As long as we are dreaming we might as well throw in a demand for a dynamic structure to the game such that even moderately intelligent mortals will be inexorably forced from non-equilibrium points toward equilibrium during repeated plays of the game.

14.10 SUMMARY

The social welfare problem, as Arrow formulates it, is: Given the preference rankings (ties allowed) of m alternatives by the members of a society of n individuals, define "fair" methods for aggregating this set of individual rankings into a single ranking for the society. Such a rule for transforming an n-tuple of rankings—one ranking for each individual—into a ranking for the society is called a social welfare function. Arrow has shown that five seemingly innocuous requirements of "fairness" for social welfare functions are inconsistent (i.e., *no* welfare function exists which satisfies all of them). The five conditions are: (1) universal domain (the function has to resolve all conceivable profiles of preference patterns); (2) positive association of individual values; (3) independence of irrelevant alternatives; (4) citizen's sovereignty (or non-imposition); and (5) non-dictatorship. We discussed the meaning and motivation of each of these conditions and sketched out the nature of Arrow's impossibility proof. We feel that the weakest link in the development is the axiom of independence of irrelevant alternatives, and we supported this contention by presenting counterintuitive examples.

In discussing Arrow's paradox, an analogy between the social welfare problem and the problem of individual decision making under uncertainty was cited. Actually, a result in the latter area yields Arrow's impossibility theorem as a corollary. The relationship between these two problems was also explored when the social choice problem is generalized to include lotteries as alternatives, thus allowing utilities and strengths of preferences to be introduced. If preference strengths are incorporated into the data of the problem, then we have to contend with the inter-personal comparison problem in the sense of establishing either a commensurable unit and/or base of reference (zero). Four procedures were discussed:

i. Goodman and Markowitz employ a common unit which can be thought of as a variation of either the just-noticeable-difference notion used by sensory psychologists or the minimum sensibles of Edgeworth. Their primary result is related to the criterion for decision making under uncertainty based on the principle of insufficient reason.

ii. Nash's work on the bargaining problem was generalized and interpreted as a possible resolution to the social choice problem. Although this procedure does not introduce commensurable units, a base of reference (status quo point) is required. The Chapter 6 discussion of the Nash bargaining problem translates with only minor modifications into the social welfare context.

iii. Hildreth also introduced strengths of preferences via utility assignments, and he established both a common unit and a base of reference by positing two special social states such that for each state the individuals receive the same goods and services and for which their preferences can be said to be equal.

iv. In terms of an example, we outlined a method which takes into account both strengths of preference and the asymmetries of the rôles of the members of the group whereby a group might combine the differing individual values to arrive at a group choice. In essence, the group's manifest behavior in resolving some specific problems is used to estimate certain parameters in a hypothesized model, and in turn these estimates are used, via the model, to reconcile other cases of group conflict.

One might have expected, *a priori*, that simple majority rule would satisfy Arrow's conditions. Indeed it does except when the individual rankings are very dissimilar, in which case it gives rise to intransitivities. It is natural, therefore, to search for reasonable restrictions to be placed on the profiles of individual rankings such that majority rule always leads to a consistent social ordering. The concept of a profile which is compatible with an underlying Coombsian joint *quantitative* scale was introduced, and we showed that for such profiles the median individual's ranking on the scale is the same as that induced by simple majority rule. Black's single-peakedness condition is equivalent to the existence of an underlying Coombsian joint *qualitative* scale, and Arrow has shown that, if these equivalent assumptions are met and if the number of individuals is odd, simple majority rule can never yield a non-transitive social ordering.

The following three topics which relate to various aspects of simple majority rule were discussed: (1) May's independent set of necessary and sufficient conditions for simple majority rule.   (2) As an alternative to employing simple majority rule for a restricted domain of profiles, the domain may be left unrestricted and the rule so modified that it always leads to a transitive social ordering.   The variations mentioned violated Arrow's axiom of the independence of irrelevant alternatives.   (3) Game-theoretic strategical aspects arise when simple majority rule is employed for an unrestricted domain of profiles.   This was illustrated by examples of the difference made when bills are presented to a legislature in different orders.

In the final section we reversed our tack.   Instead of suggesting different planned programs for passing from individual values to social preferences or investigating the game aspects of such plans, artificial games were concocted so as to have the property that when the players act in their own selfish interests the outcome is "fair" in the eyes of the planner.   The use of games of fair division to resolve social conflicts has the distinct advantage that prior, detailed individual preference information is not needed. This plus added flexibility allows for more decentralized planning.

A PROBABILISTIC THEORY
OF UTILITY

A1.1 INTRODUCTION

Utility theory as formulated by von Neumann and Morgenstern (see Chapter 2) assumes, among other things:

1. That, given two alternatives, a person either prefers one to the other or is indifferent between them.

2. That there are certain well-defined chance events having probabilities attached to them which are manipulated according to the rules of the probability calculus.

In criticizing that theory, we emphasized that some experimental data suggest that the latter assumption is in error; and, although we did not particularly question the first assumption, we did stress the difficulty of obtaining transitive preference reports. These two may not be unrelated, for, if we replace assumption 1 by the assumption that a person has a certain probability of expressing a preference for one alternative over another, then a single choice from each pair of alternatives cannot generally result in transitive patterns. Thus, it may really be assumption 1, not transitivity as such, which is the source of some difficulties in utility theory.

We know of no direct empirical method to decide whether assumption 1 or the assumption of probabilistic preferences is correct. Certainly, if a person expresses his preference between two alternatives only once, we cannot distinguish between them. But, if we ask him to express his preference several times for a given pair of alternatives, other effects may enter, and these, so long as they cannot be ruled out, seem to render it impossible to decide between the assumptions. First, the very act of making a choice may change the situation so that the person's second choice is not made under the same conditions as his first one. If so, the expressed preferences could be different without invalidating assumption 1. But, equally, even if the choices are prefectly consistent, we cannot conclude that assumption 1 is necessarily supported. If the choice expressed is remembered and if consistency is an overriding virtue for the person, then the chance of his making the same choice will be sharply increased from what it was originally. That is, one effect of memory may be to alter the probabilistic structure.

So far as we can see, one is forced to select between these two assumptions in terms of the overall adequacy and predictive power of the theoretical structures which are possible in each case, not in terms of direct experimental evidence. Our goal here is not to reach and defend a choice between them but to show one possible structure generated by the assumption that preferences are probabilistic. Nonetheless, the *a priori* possibilities just mentioned raise basic empirical difficulties for both models. If the very act of making a choice can alter the total situation, it is difficult to see how by using the present sequential methods one can ever be certain of obtaining appropriate data for the von Neumann-Morgenstern model. But, equally well, if a probabilistic model is assumed, it is unclear how to estimate the postulated probabilities. Our ideal would be an individual who immediately forgets his choice upon making it, for we need a series of independent trials governed by the same probabilities. Since our ideal is unreal, we will have to resort to dodges. Possibly when a subject is confronted with a large collection of pairs of alternatives—especially fairly complicated gambles—in a relatively short time, we are justified in assuming that he is unable to recall his choice for more than a few trials and that the postulated probabilities remain constant. If it can be decided what to take as "a large collection," "a relatively short time," and "more than a few trials" in order for both of these conditions to be satisfied, then empirical estimates of the postulated probabilities can be made. No doubt considerable experimentation will be required before a suitable procedure is devised. We might mention here that one of our theoretical results suggests that money, or any other commodity having a well-accepted simple ordering of worth, may give quite different results from alternatives not culturally perfectly ordered; it appears that—no matter how convenient it may be—money should not be employed in experiments to the exclusion of other commodities.

Let us turn next to the second assumption mentioned above of the von Neumann-Morgenstern theory. As we pointed out in section 2.8, there is a fair amount of evidence to suggest that most people are behaviorally innocent of the calculus of probabilities. Moreover, casual observations suggests that they are not consistently certain as to which of two chance events is more probable. Finally, many decision situations depend upon what are commonly called chance events, but for which one is very hard pressed to assign objective probabilities. Each spring a farmer must estimate the chance of another frost; from time to time most of us must decide about the risk of another plane trip; an investor must consider the likelihood of the market falling or not; and so on. It is difficult to see how to attach objective probabilities to these events in the certain way one does to a carefully manufactured and tested die. There are complicated cyclical fluctuations in the weather which are not adequately summarized in the statistics available to a farmer; it is questionable whether one plane trip compares to another in the way one flip of a coin does to another; etc. Yet subjectively we each assign some sort of fuzzy "probabilities" to such events, at least to the extent of feeling that one is more or less probable than another. The fuzziness is suggested by our inconsistencies when we are forced to make the comparison several times, especially when we do not realize we are making the same comparison or when we have a lapse of memory as to our previous choice. So the second change we propose in utility theory is to admit that we shall be dealing with fuzzy subjective probabilities, not sharp objective ones.

In sum, utility theory will be modified by assuming that people can neither discriminate perfectly between alternatives with respect to preferences nor between events with respect to likelihood. This is not a question of psychophysical, i.e., stimulus, discrimination: we shall suppose that there is not the slightest difficulty in telling one alternative from another, or one event from another, as physical stimuli. The assumed trouble is in separating alternatives consistently as to preference and events as to likelihood.

The technique of investigation we shall employ is this appendix differs somewhat from that of Chapter 2 in that we will not give a set of axioms that ensures the existence of a utility function. Rather, we shall assume that both a utility function and a subjective probability function exist and satisfy certain conditions, including the expected utility hypothesis, and we shall also assume that the discriminations people make satisfy certain plausible conditions, and then enquire into the implications of these assumptions. One of our results is more like some of those in Chapters 13 and 14 than in Chapter 2 in that it is an impossibility theorem asserting that a set of conditions, each individually more or less plausible, are mutually inconsistent.

Other studies in which preferences are assumed to be probabilistic are: Davidson and Marschak [1957], Georgescu-Roegen [1936], Marschak [1955], Papandreou [1953], Papandreou *et al.* [1954], and Quandt [1956].

A1.2 PREFERENCE DISCRIMINATION AND INDUCED PREFERENCE

As in Chapter 2, suppose that a set A of pure alternatives is under consideration by the individual and that from these a set of gambles is developed using chance events taken from a set (actually, a Boolean algebra) E of events. If a and b are any two alternatives, or gambles, and α is an event in E, then the symbol $a\alpha b$ will be used to denote the gamble in which a is the outcome if the event α occurs and b if it does not. (In Chapter 2, a somewhat different notation was used. Assuming event α has objective probability p, we denoted the gamble by $[pa, (1 - p)b]$, so the analogous notation for events would be $[\alpha a, \bar{\alpha} b]$, where $\bar{\alpha}$ denotes "not α." It seems, however, more convenient here to use the slightly more compact notation $a\alpha b$.) The set of all such gambles, including the pure alternatives of A, that can be so generated will be denoted by G.

Axiom 1.   *For every a in G, $a\alpha a = a$.*

In words, the gamble in which a is the outcome whether or not α occurs is not distinguished as different from a itself. It is hard to quarrel with this, although when combined with axiom 9 it implies that the subjective probabilities of an event and of its complement sum to 1, which Edwards [1954 c] has questioned.

If a and b are two gambles from G, we suppose that there exists an objective probability $P(a, b)$ that the given individual will prefer a to b. As we indicated earlier, it is not easy to see how to estimate such probabilities in practice, but we need not concern ourselves about that when describing the model.

Although it is true that imperfect preference discrimination has been introduced in part to avoid the strong transitivity requirements of the von Neumann and Morgenstern theory, it would be folly to ignore the empirical evidence suggesting that preferences are approximately transitive. It is easy to go astray at this point by assuming certain inequalities among the three quantities $P(a, b)$, $P(b, c)$, and $P(a, c)$; apparently this is not strong enough. Our tack is a bit different. Observe that, in an induced sense, a is "preferred or indifferent to" b if for every c in G both

$$P(a, c) \geqslant P(b, c) \qquad \text{and} \qquad P(c, b) \geqslant P(c, a);$$

whenever these two sets of inequalities hold we shall write $a \succsim b$. It is easy to see that $\succsim$ must always be transitive, but that in general there will be alternatives which are not comparable according to $\succsim$. A basic restriction we shall make about preference discrimination is that such comparisons are always possible, i.e.,

Axiom 2. *For every a and b in G, either $a \succsim b$ or $b \succsim a$.*

This a strong assumption, but we do not believe it to be nearly so strong as the corresponding ones in Chapter 2. There, comparability was operationally forced by the demand that the individual make a choice but transitivity was in doubt. Here, transitivity is certain and comparability is in doubt. Although it is plausible that axiom 2 is met in some empirical contexts, the following example strongly suggests that this is not always the case. Suppose that a and b are two alternatives of roughly comparable value to some person, e.g., trips from New York City to Paris and to Rome. Let c be alternative a plus \$20 and d be alternative b plus \$20. Clearly, in general

$$P(a, c) = 0 \quad \text{and} \quad P(b, d) = 0.$$

It also seems perfectly plausible that for some people

$$P(b, c) > 0 \quad \text{and} \quad P(a, d) > 0,$$

in which event a and b are not comparable, and so axiom 2 is violated. In one respect this example is special: c differs from a and d from b by the addition of an extra commodity which is always desirable; therefore, we may expect perfect discrimination within each of these two pairs. As we shall see, there are theoretical reasons for believing that the occurrence of perfect preference discrimination may require a somewhat different model from when it never occurs.

Let us say that a and b are indifferent in the induced sense, and write $a \sim b$, whenever both $a \succsim b$ and $b \succsim a$. We next argue that certain two-stage gambles should be indifferent.

Consider the gamble $(a\alpha b)\beta c$, where a, b, and c are pure alternatives. If one analyzes what this means, one sees that outcome a results if both α and β occur, i.e., if the event $\alpha \cap \beta$ occurs; b results if both $\bar{\alpha}$ and β occur, i.e., if $\bar{\alpha} \cap \beta$ occurs; and c results if $\bar{\beta}$ occurs. A similar analysis of the gamble $a(\alpha \cap \beta)(b\beta c)$ shows that a, b, and c occur under exactly the same conditions. Thus, there is no difference between the two gambles, and so it is reasonable to argue that a person should be indifferent between them. We shall demand that this hold not strictly but only in the weaker sense of induced preference.

Axiom 3. *If a, b, and c are in A and α and β are in E, then*

$$(a\alpha b)\beta c \sim a(\alpha\cap\beta)(b\beta c).$$

Actually, the results that we shall state depend only upon the weaker assumption

$$(a\alpha b)\beta b \sim a(\alpha\cap\beta)b,$$

which follows from axiom 3 by setting $c = b$ and then using axiom 1.

A1.3 LIKELIHOOD DISCRIMINATION AND QUALITATIVE PROBABILITY

Suppose that our subject must decide between the two gambles $a\alpha b$ and $a\beta b$. He can simplify his choice by asking himself which alternative, a or b, he prefers, and which event, α or β, he considers more likely to occur. Of the four combinations, two should lead to preference for $a\alpha b$ over $a\beta b$:

1. a is preferred to b, and α is deemed more likely to occur than β.
2. b is preferred to a, and β is deemed more likely to occur than α.

By assumption, the probability that he will prefer a to b is $P(a, b)$. If we suppose that his discrimination as to the likelihood of events is statistically independent of his preference discriminations, and that it is governed by a probability $Q(\alpha, \beta)$, then the probability that he will both prefer a to b and deem α more likely to occur than β is $P(a, b)Q(\alpha, \beta)$. Similarly, the probability that he will both prefer b to a and deem β more likely to occur than α is $P(b, a)Q(\beta, \alpha)$. Since these two cases are exclusive of each other, the sum of the two numbers should give the probability that he will prefer $a\alpha b$ to $a\beta b$.

The important assumption made in this argument is that the two discrimination processes are statistically independent. This seems reasonable when and only when the subject believes the two gambles a and b to be "independent" of the events α and β, for, if alternative a depends on α and he believes α is likely to occur, then he is really forced to compare the outcome of a which arises when α occurs with $a\beta b$, in which case his preference between $a\alpha b$ and $a\beta b$ may be different from what it would be if a were independent of α. There is at least one case when it is plausible that the subject should deem a and b to be independent of α and β, namely, when a and b are pure alternatives having nothing to do with chance events. We shall assume our conclusion holds in that case.

Axiom 4. *There is a probability $Q(\alpha, \beta)$ for every α and β in E such that, if a and b are in A,*

$$P(a\alpha b, a\beta b) = P(a, b)Q(\alpha, \beta) + P(b, a)Q(\beta, \alpha).$$

There is, as yet, no direct evidence as to whether these two discriminations actually are statistically independent. Conceptually, we clearly separate preferences among alternatives from likelihood among events, and it seems reasonable that people attempt to deal with these as distinct, independent dimensions. On the other hand, casual observation indicates that people do play long shots, and such behavior appears to violate the axiom. At the least, the axiom seems sufficiently compelling as a dictum of sensible behavior to warrant its investigation, and it can be looked on as a generalization of related, but non-probabilistic, assumptions found in other work, e.g., in Ramsey [1931] and in Savage [1954] (see the second postulate in section 13.5).

Our next axiom is comparatively innocent. Let us state it first and then discuss its import.

Axiom 5. *For every a and b in G,*

$$P(a, b) \geqslant 0 \qquad and \qquad P(a, b) + P(b, a) = 1.$$

For every α and β in E,

$$Q(\alpha, \beta) \geqslant 0 \qquad and \qquad Q(\alpha, \beta) + Q(\beta, \alpha) = 1.$$

There exist at least two alternatives a* and b* in A such that $P(a^*, b^*) > \frac{1}{2}$.

First, we have supposed that the *P*'s and *Q*'s are actually probabilities in the sense that they lie between 0 and 1 inclusive and we have supposed that the subject is forced to make choices between alternatives and between events. That is, he cannot report that he is indifferent between *a* and *b*. Experimentally, this is known as the "forced-choice" technique, and it is in standard use. It may be worth mentioning that, if one allows indifference reports in the sense of only demanding $P(a, b) + P(b, a) \leqslant 1$, then the mathematics leads to two quite distinct cases—the one we shall describe here and another one somewhat like it but apparently less realistic. The final condition simply demands that the situation be non-trivial in the sense that not all pure alternatives are equally confused with respect to preference.

From axioms 4 and 5, it is trivial to show that

$$Q(\alpha, \beta) = \frac{P(a\alpha b, a\beta b) + P(a, b) - 1}{2P(a, b) - 1},$$

for every *a* and *b* in *A* such that $P(a, b) \neq \frac{1}{2}$ [by axiom 5, at least one such pair (*a**, *b**) exists]. This expression is useful because it permits one both to determine whether a given set of preference data do satisfy the independence assumption and, if they do, to estimate $Q(\alpha, \beta)$.

In complete analogy to "induced preference," we may define a relation on the set of events E. We write $\alpha \mathrel{\underset{\sim}{\succ}} \beta$ if

$$Q(\alpha, \delta) \geqslant Q(\beta, \delta) \quad \text{and} \quad Q(\delta, \alpha) \leqslant Q(\delta, \beta)$$

for every δ in E. We shall refer to this as the "qualitative probability" (induced by Q) on E. One might expect us now to impose a comparability axiom like axiom 2 on qualitative probability, but this is unnecessary as it is a consequence of our other axioms. Rather, an entirely different assumption, peculiar to the notion of probability, is required. We shall suppose that the subject is certain that the universal event e of the Boolean algebra E will occur. For the moment, we will demand that no event have a qualitative probability in excess of e or less than its complement.

Axiom 6.   *If e is the universal event in E, then $e \mathrel{\underset{\sim}{\succ}} \alpha \mathrel{\underset{\sim}{\succ}} \bar{e}$ for every α in E.*

A1.4 THE UTILITY AND SUBJECTIVE PROBABILITY FUNCTIONS

So far, our technique of study has been similar to that exhibited in Chapter 2, but now we depart from that tradition by assuming that utility and subjective probability[1] functions exist having, among others, properties like those established in Chapter 2. Of course, neither of these two functions, however we choose them, can be a complete representation of the assumed data in the same sense that the utility functions of Chapter 2 were. We no longer have a simple transitive relation to be represented numerically but rather a set of probabilities. The role of what we shall continue to call the utility and subjective probability functions will be a partial and—as we shall see—comparatively simple representation of the probabilities. It is analogous to using a statistic such as the mean or standard deviation to give a partial description of a probability distribution.

We shall suppose that there exists at least one real-valued function u on G called the utility function and at least one real-valued function ϕ on E called the subjective probability function and that the following axioms are met.

Axiom 7.   *u preserves the induced preference relation on G, and ϕ preserves the qualitative probability on E, i.e.,*

$$u(a) \geqslant u(b) \text{ if and only if } a \mathrel{\underset{\sim}{\succ}} b, \quad \text{for } a \text{ and } b \text{ in } G,$$

and

$$\phi(\alpha) \geqslant \phi(\beta) \text{ if and only if } \alpha \mathrel{\underset{\sim}{\succ}} \beta, \quad \text{for } \alpha \text{ and } \beta \text{ in } E.$$

---

[1] The meaning of subjective probability here will be self-contained and is not exactly the same as discussed in Chapter 13. There are, however, certain important similarities.

As this sort of condition is already very familiar we need not comment on it.

Axiom 8. $\phi(e) = 1$ *and* $\phi(\bar{e}) = 0$.

This prescribes more clearly the role of the universal event e. It is an event which is subjectively certain to occur, and its complement is subjectively certain not to occur.

Given a subjective probability function ϕ, we may follow the usual terminology for objective probabilities and say that two events α and β are (subjectively) *independent* if and only if $\phi(\alpha \cap \beta) = \phi(\alpha)\phi(\beta)$. It is clear that we cannot ascertain which events are independent until we know the subjective probability function ϕ, and so it would appear as though we were rapidly getting ourselves into a circle. However, it turns out that all of our final conclusions can be stated without reference to independent events provided only that axiom 4 can be extended in a certain way and that there are enough independent events—so many that no exhaustive check would be possible anyhow. These conditions will be formulated as axioms 9 and 10.

Earlier, when we introduced axiom 4, describing the statistical independence of the two discrimination processes, we held that it should be met whenever the two gambles a and b are "independent" of the events α and β, without, however, specifying what we might mean by this except that it should hold for all pure alternatives. We now extend axiom 4 as follows:

Axiom 9.   *If a and b are in A and α and β are events which are subjectively independent of event γ, then*

$$P[(a\gamma b)\alpha b, (a\gamma b)\beta b] = P(a\gamma b, b)Q(\alpha, \beta) + P(b, a\gamma b)Q(\beta, \alpha).$$

Axiom 10.   *The subjective probability function ϕ shall have the property that, for all numbers x, y, and z, where $0 \leqslant x, y, z \leqslant 1$, there are events α, β, and γ in E such that*

(i) $\phi(\alpha) = x$, $\phi(\beta) = y$, *and* $\phi(\gamma) = z$.
(ii) α *and* β *are both subjectively independent of* γ.

This axiom postulates a very dense set of independent events, so dense that every conceivable subjective probability is exhibited at least twice. Put another way, we are making a continuum assumption about the individual being described via the axioms. Although we have never made it so explicit before, such an assumption was implicit in the work of Chapter 2, for there we tacitly supposed (as is reasonable) that we could deal with any objective probability.

Axiom 11. *These two subjective scales satisfy the expected-utility hypothesis in the sense that, for a and b in A and α in E,*

$$u(a\alpha b) = \phi(\alpha)u(a) + \phi(\bar{\alpha})u(b).$$

At this point there should be little reason to discuss this idea further, except to note that we have not previously restricted a and b to be pure alternatives. Although no restrictions are usually stated when the expected-utility hypothesis is made, it is always tacitly assumed that it only holds for gambles whose component events are independent of the event α of the hypothesis. In utility theory, of course, independence is meant in the usual objective sense. For our purposes, it is sufficient to assume the hypothesis only for pure alternatives which are trivially independent of events.

A1.5 CONCLUSIONS ABOUT THE SUBJECTIVE SCALES

On the basis of these eleven axioms, the following conclusions can be established as to the form of the discrimination functions and the subjective scales. First of all, Q must depend only upon the difference of the subjective probabilities of its two events. Put more formally, there exists real-valued function Q^* of one real variable such that

$$Q(\alpha, \beta) = Q^*[\phi(\alpha) - \phi(\beta)].$$

This result is interesting because of its connection with a very old problem in psychology. A century ago Fechner introduced into psychology a concept of subjective sensation, which has since played a crucial and controversial role in the development of psychophysics. Even today, his idea in somewhat generalized form continues to be debated and to be the source of experimental studies. The modern statement of his formal definition of a subjective scale of sensation is exactly the property stated above for Q. The source of controversy in psychophysics need not concern us here.

Actually, we can give a much more explicit result than that ϕ is a sensation scale: we can describe the mathematical form of Q. There are three cases. In the first, there is a positive constant ϵ and Q is of the form

$$Q(\alpha, \beta) = \begin{cases} \frac{1}{2} + \frac{1}{2}[\phi(\alpha) - \phi(\beta)]^\epsilon, & \text{if } \alpha > \beta, \\ \frac{1}{2}, & \text{if } \alpha \sim \beta, \\ \frac{1}{2} - \frac{1}{2}[\phi(\beta) - \phi(\alpha)]^\epsilon, & \text{if } \beta > \alpha. \end{cases}$$

(See Fig. 1.)   The second is the discontinuous function

$$Q(\alpha, \beta) = \begin{cases} 1, & \text{if } \alpha > \beta, \\ \frac{1}{2}, & \text{if } \alpha \sim \beta, \\ 0, & \text{if } \beta > \alpha, \end{cases}$$

which results from the first case by taking the limit as ϵ approaches 0. This represents perfect likelihood discrimination. The third is the function obtained by taking the limit as ϵ approaches infinity, and it represents almost total lack of discrimination.

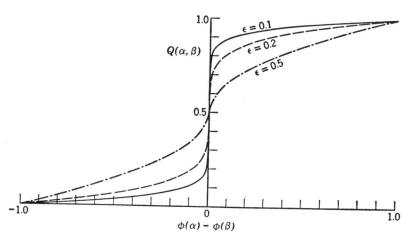

Fig. 1. The function

$$Q(\alpha, \beta) = \begin{cases} \frac{1}{2} + \frac{1}{2}[\phi(\alpha) - \phi(\beta)]^{\epsilon}, & \alpha \succ \beta \\ \frac{1}{2}, & \alpha \sim \beta \\ \frac{1}{2} - \frac{1}{2}[\phi(\beta) - \phi(\alpha)]^{\epsilon}, & \beta \succ \alpha \end{cases}$$

for $\epsilon = 0.1, 0.2,$ and 0.5.

It is easy to see that in the first case, but not in the other two, one can express ϕ in terms of Q, namely, as

$$\phi(\alpha) = [Q(\alpha, \bar{e}) - Q(\bar{e}, \alpha)]^{1/\epsilon}$$

or, more usefully, as

$$\phi(\alpha) = \begin{cases} \frac{1}{2} + \frac{1}{2}[2Q(\alpha, \bar{\alpha}) - 1]^{1/\epsilon}, & \text{if } Q(\alpha, \bar{\alpha}) > \frac{1}{2}, \\ \frac{1}{2}, & \text{if } Q(\alpha, \bar{\alpha}) = \frac{1}{2}, \\ \frac{1}{2} - \frac{1}{2}[1 - 2Q(\alpha, \bar{\alpha})]^{1/\epsilon}, & \text{if } Q(\alpha, \bar{\alpha}) < \frac{1}{2}. \end{cases}$$

Similar results hold for u and P over the set A of pure alternatives. First, P can be shown to be a function only of $u(a) - u(b)$, for a and b in A. Second, assuming a Q of the first type above and letting ϵ be the constant determined there, then

$$P(a, b) = \begin{cases} \frac{1}{2} + \frac{1}{2}[P(a^*, b^*) - P(b^*, a^*)][u(a) - u(b)]^{\epsilon}, & \text{if } a \succ b, \\ \frac{1}{2}, & \text{if } a \sim b, \\ \frac{1}{2} - \frac{1}{2}[P(a^*, b^*) - P(b^*, a^*)][u(b) - u(a)]^{\epsilon}, & \text{if } b \succ a, \end{cases}$$

and

$$
u(a) = \begin{cases} \left[\dfrac{P(a, b^*) - P(b^*, a)}{P(a^*, b^*) - P(b^*, a^*)} \right]^{1/\epsilon}, & \text{if } a \succ b^*, \\[3ex] 1 - \left[\dfrac{P(a^*, a) - P(a, a^*)}{P(a^*, b^*) - P(b^*, a^*)} \right]^{1/\epsilon}, & \text{if } b^* \succ a, \end{cases}
$$

where a^* and b^* are mentioned in axiom 5. Any positive linear transformation of u is equally acceptable.

Thus, we have the following situation. If the axioms are accepted and if it is assumed that discrimination of events is neither perfect nor totally absent, then the mathematical form of the model is completely specified except for a single parameter ϵ, which appears to reflect the individual's sensitivity of discrimination; and the two subjective scales can be inferred from the empirical estimates of the probabilities P. The subjective probability scale is unique, and the utility scale is unique except for its zero and unit. There is only one trouble with all of this: it is extremely doubtful that people satisfy all the axioms.

An example and a theorem will formulate our doubts. Although the mathematical argument used to establish our results rests heavily on steps involving independent events, the final results can be shown to hold for events whether or not they are independent, so we need not worry about independence in a counterintuitive example. Consider the two chance events: rain on Wall Street at time t, and rain on both Wall Street and 34th Street at time t. Since the locations are not widely separated, both being in New York City, it is highly likely that if it rains on Wall Street it will also rain on 34th Street, so the subjective probability of rain on Wall Street alone will only be slightly larger than rain at both places. Yet, if one is asked which is more likely, it seems silly ever to say the latter. If so, we have $\phi(\alpha)$ and $\phi(\beta)$ very close and $Q(\alpha, \beta) = 1$. If people actually behave in this way when making choices, then at least one of our axioms must be false.

A1.6 AN IMPOSSIBILITY THEOREM

Casual observation suggests that there are many situations, e.g., those involving gambles of money, in which these conditions can be satisfied: First, there are at least three prospects a, b, and c which are perfectly discriminated with respect to preference, i.e., $P(a, b) = P(b, c) = P(a, c) = 1$. This will hold, we are sure, when all other things are equal and $a = \$10$, $b = \$5$, and $c = \$1$. Second, there are at least two events, α and β, which are neither perfectly discriminated nor equally confused, i.e., such that $Q(\alpha, \beta) \neq 0, \frac{1}{2}$, or 1. The impossibility theorem asserts that these two assumptions are inconsistent with the eleven axioms we have previously stated.

This result seems disturbing, for most of the assumptions upon which it is based have, by now, acquired a considerable respectability. Yet, clearly, they cannot all be satisfied. The task of reappraising them is quite delicate, for there are numerous reasons for supposing that they are not terribly far from the truth. Some of these reasons have been given in Chapter 2. Another is that the derived form of the discrimination function for events is sufficiently similar to much discrimination data to suggest that we are not completely afield.

It would appear that six of our assumptions are subject to the greatest doubt. Of these, three (axiom 2, requiring that every pair of gambles be comparable by the induced preference relation; axiom 3, requiring that two gambles which decompose in the same way be indifferent in the induced sense; and axiom 4, requiring that the two discrimination processes be statistically independent for pure alternatives) are subject to direct experimental study. The other three (axiom 9, requiring that axiom 4 hold for certain gambles involving subjectively independent events; axiom 10, requiring that certain triples of independent events be extremely dense; and axiom 11, requiring that the expected-utility hypothesis be true for pure alternatives) are impossible to study directly. Because of this, one can expect that most attempts to get out of the bind will be concentrated on the second three.

Since all the rest of decision theory is so dependent upon the expected-utility hypothesis, special attention will undoubtedly be given to axioms 9 and 10. There is the intriguing possibility that these subjective scales are discrete rather than continuous, as has generally been assumed, which would make them more in accord with the way people seem to classify, say, events: impossible, not very likely, etc. In that case, axiom 10 might be abandoned. On the other hand, axiom 9 when coupled with our definition of independence may be the source of difficulty. As the axiom seems reasonable for one's intuitive idea of subjectively independent events, it may be the definition that should be altered.

As it stands, two conceptual features of this theory are of interest. First, by making the assumption that the two discrimination processes are statistically independent, it has been possible to deal simultaneously with both subjective value (utility) and subjective probability. Second, by using axioms which are closely related to those of traditional utility theory and the independence assumption (axiom 4), it has been possible to demonstrate that both utility and subjective probability form sensation scales in the Fechnerian sense. In psychophysics it has been argued,

though never fully accepted, that subjective experience must be repre
sented by such scales; however, the defining condition is neither simpl(
nor has it been derived from other assumptions.   The traditional practic(
has been to postulate this condition as an *a priori* definition of subjectiv(
sensation, and, of course, many have objected that it is much too sophisti
cated to be accepted as a basic axiom.   Whether a model that parallel
this one and that arrives at sensation scales as a consequence, not as (
postulate, can be developed for psychophysical problems is not known.

For a fuller statement of this theory and for proofs of the assertions, se(
Luce [1956 *b*].

THE MINIMAX THEOREM

A2.1 STATEMENT OF THE PROBLEM

The general two-person zero-sum game with finite pure strategy sets can be characterized as follows:

i. There are two players, 1 and 2.

ii. 1 has a set $A = \{\alpha_1, \alpha_2, \cdots, \alpha_m\}$ of m pure strategies.

iii. 2 has a set $B = \{\beta_1, \beta_2, \cdots, \beta_n\}$ of n pure strategies.

iv. Associated to each pair of strategies (α_i, β_j) is a payoff of $M(\alpha_i, \beta_j)$ units from player 2 to 1. $M(\alpha_i, \beta_j)$ is abbreviated by a_{ij}. Hence the values to 1 and 2 of the strategy pair (α_i, β_j) are a_{ij} and $-a_{ij}$ units respectively. Because these values sum to zero for every (α_i, β_j) pair, the game is called *zero-sum*.

v. Player 1 may adopt a *randomized* (or *mixed*) strategy by employing α_1 with probability x_1, α_2 with probability x_2, $\cdots$, α_m with probability x_m, where

$$\sum_{i=1}^{m} x_i = 1 \quad \text{and} \quad x_i \geqslant 0.$$

Such a strategy is symbolically represented by $\mathbf{x} = (x_1\alpha_1, x_2\alpha_2, \cdots, x_m\alpha_m)$. The strategy $(0\alpha_1, 0\alpha_2, \cdots, 1\alpha_i, \cdots, 0\alpha_m)$, which places all the weight on α_i, is considered to be the same as the pure strategy α_i. The set of all randomized strategies for player 1 is designated X_m (where m indicates the number of pure strategies available to 1).

vi. The generic randomized strategy for 2 is denoted by $\mathbf{y} = (y_1\beta_1, y_2\beta_2, \cdots, y_n\beta_n)$, where

$$\sum_{j=1}^{n} y_j = 1 \quad \text{and} \quad y_j \geqslant 0.$$

The pure strategy β_j is considered to be the same as the randomized strategy $(0\beta_1, 0\beta_2, \cdots, 1\beta_j, \cdots, 0\beta_n)$. The set of all randomized strategies for 2 is designated by Y_n.

vii. For each randomized strategy pair $(\mathbf{x}, \mathbf{y})$, the payoff $M(\mathbf{x}, \mathbf{y})$ to 1 is defined to be

$$\begin{aligned}
M(\mathbf{x}, \mathbf{y}) &= \sum_{i=1}^{m} \sum_{j=1}^{n} x_i a_{ij} y_j \\
&= \sum_{j=1}^{n} y_j \left(\sum_{i=1}^{m} x_i a_{ij} \right) \\
&= \sum_{i=1}^{m} x_i \left(\sum_{j=1}^{n} a_{ij} y_j \right);
\end{aligned}$$

the payoff to 2 is $-M(\mathbf{x}, \mathbf{y})$.

The symbol

$$M(\alpha_i, \mathbf{y}) = \sum_{j=1}^{n} a_{ij} y_j$$

means the payoff to 1 when 1 uses the pure strategy α_i and 2 uses $\mathbf{y}$. Quite analogously, when 1 uses $\mathbf{x}$ and 2 uses β_j, the payoff is:

$$M(\mathbf{x}, \beta_j) = \sum_{i=1}^{m} a_{ij} x_i.$$

Of course,

$$M(\alpha_i, \beta_j) = a_{ij}.$$

viii. Symbolically, we may denote the whole pure strategy game by the triplet (A, B, M), which puts into evidence the principal ingredients, namely, the two pure strategy spaces and the payoff function M. The extension of (A, B, M) to spaces of randomized strategies is denoted by the triplet (X_m, Y_n, M).

ix. Player 1's aim is to select a randomized strategy $\mathbf{x}$ from X_m so as to maximize his return or, equivalently (because of the strictly opposing nature of the game), to minimize 2's return. However, the actual outcome of the game depends upon the players joint actions. Thus, we are given the number $M(\mathbf{x}, \mathbf{y})$ for each pair $(\mathbf{x}, \mathbf{y})$, and 1 attempts to *maximize* $M(\mathbf{x}, \mathbf{y})$ by choosing $\mathbf{x}$ and, simultaneously, 2 attempts to *minimize* $M(\mathbf{x}, \mathbf{y})$ by choosing $\mathbf{y}$. The rules of the game require that each player choose his strategy (pure or randomized) in complete ignorance of his opponent's selection.

x. For each $\mathbf{x}$ belonging to X_m, player 1's security level is defined to be

$$v_1(\mathbf{x}) = \min_{\mathbf{y}} M(\mathbf{x}, \mathbf{y}).$$

Since

$$M(\mathbf{x}, \mathbf{y}) = \sum_{j=1}^{n} y_j \left(\sum_{i=1}^{m} x_i a_{ij} \right) = \sum_{j=1}^{n} y_j M(\mathbf{x}, \beta_j)$$

is a weighted average of the n payoffs $M(\mathbf{x}, \beta_j)$, $j = 1, 2, \cdots, n$, it is minimized when all of the weight is assigned to the least of these, i.e.,

$$v_1(\mathbf{x}) = \min [M(\mathbf{x}, \beta_1), M(\mathbf{x}, \beta_2), \cdots, M(\mathbf{x}, \beta_n)].$$

We may interpret $v_1(\mathbf{x})$ as the return to player 1 if he discloses to 2 that $\mathbf{x}$ is his choice and if 2 is allowed to choose his best response to $\mathbf{x}$.

If 1 wishes to maximize his security level, he must choose a strategy $\mathbf{x}^{(0)}$ such that

$$v_1(\mathbf{x}^{(0)}) \geqslant v_1(\mathbf{x}), \quad \text{for all } \mathbf{x} \text{ of } X_m.$$

Thus, if we let $v_1(\mathbf{x}^{(0)}) = v_1$, then

$$v_1 = v_1(\mathbf{x}^{(0)}) = \max_{\mathbf{x}} v_1(\mathbf{x}) = \max_{\mathbf{x}} \min_{\mathbf{y}} M(\mathbf{x}, \mathbf{y}).$$

Note that $v_1(\mathbf{x}^{(0)}) = v_1$ implies $M(\mathbf{x}^{(0)}, \mathbf{y}) \geqslant v_1$, for all y; hence, $\mathbf{x}^{(0)}$ guarantees to 1 a return of at least v_1. A strategy $\mathbf{x}^{(0)}$ which maximizes 1's security level is called a *maximin strategy* for player 1. Maximin strategies always exist, *but they need not be unique.* We let $\mathcal{O}_1$ ($\mathcal{O}$ standing for optimal) designate the set of all maximin strategies.[1] Thus, if $\mathbf{x}^*$ belongs to $\mathcal{O}_1$, then $\mathbf{x}^*$ has a security level of v_1. If $\mathbf{x}'$ does not belong to $\mathcal{O}_1$, then $\mathbf{x}'$ has a security level less than v_1.

xi. Because the game is zero-sum, we may phrase 2's aims as the minimization of 1's return rather than the maximization of his own. If 2 uses $\mathbf{y}$, 1 cannot obtain a return greater than

$$v_2(\mathbf{y}) = \max_{\mathbf{x}} M(\mathbf{x}, \mathbf{y}).$$

---

[1] The set $\mathcal{O}_1$ is a closed convex set.

In perfect analogy to 1's trying to maximize his security level, 2 tries to minimize $v_2(\mathbf{y})$. Let $\mathbf{y}^{(0)}$ be such that

$$v_2 = v_2(\mathbf{y}^{(0)}) \leqslant v_2(\mathbf{y}), \qquad \text{for all } \mathbf{y} \text{ of } Y_n.$$

Then,

$$v_2 = v_2(\mathbf{y}^{(0)}) = \min_{\mathbf{y}} \max_{\mathbf{x}} M(\mathbf{x}, \mathbf{y}),$$

and

$$M(\mathbf{x}, \mathbf{y}^{(0)}) \leqslant v_2, \qquad \text{for all } \mathbf{x}.$$

The strategy $\mathbf{y}^{(0)}$ is called a *minimax* strategy for 2. We let $\mathfrak{O}_2$ denote the set of all minimax strategies for 2. Thus, if $\mathbf{y}^*$ belongs to $\mathfrak{O}_2$, then 1 can surely be held down to at most v_2 by using $\mathbf{y}^*$. If, however, a $\mathbf{y}'$ is used which does not belong to $\mathfrak{O}_2$, then it is possible for 1 to get more than v_2.

xii. Thus, if 1 uses a maximin strategy, he guarantees himself a return of at least v_1 units. If 2 uses a minimax strategy, he guarantees that 1 cannot receive more than v_2 units. Hence, it follows that $v_1 \leqslant v_2$.

xiii. A pair $(\mathbf{x}', \mathbf{y}')$ is said to be in *equilibrium* if $\mathbf{x}'$ is good against $\mathbf{y}'$ [i.e., $M(\mathbf{x}, \mathbf{y}') \leqslant M(\mathbf{x}', \mathbf{y}')$, for all $\mathbf{x}$] and if $\mathbf{y}'$ is good against $\mathbf{x}'$ [i.e., $M(\mathbf{x}', \mathbf{y}') \leqslant M(\mathbf{x}', \mathbf{y})$, for all $\mathbf{y}$]. These conditions may be written simply as

$$M(\mathbf{x}, \mathbf{y}') \leqslant M(\mathbf{x}', \mathbf{y}') \leqslant M(\mathbf{x}', \mathbf{y})$$

for all $\mathbf{x}$ and $\mathbf{y}$, or equivalently as

$$\max_{\mathbf{x}} M(\mathbf{x}, \mathbf{y}') = M(\mathbf{x}', \mathbf{y}') = \min_{\mathbf{y}} M(\mathbf{x}', \mathbf{y}).$$

The following theorem is fundamental to a real understanding of the main result in the two-person zero-sum theory.

Theorem.   *Each of the following three conditions implies the other two:*

Condition 1.   *An equilibrium pair exists.*
Condition 2.

$$v_1 \equiv \max_{\mathbf{x}} \min_{\mathbf{y}} M(\mathbf{x}, \mathbf{y}) = \min_{\mathbf{y}} \max_{\mathbf{x}} M(\mathbf{x}, \mathbf{y}) \equiv v_2$$

(*i.e., the order of the operators* $\max_{\mathbf{x}}$ *and* $\min_{\mathbf{y}}$ *makes no difference, or, in technical jargon, they are commutative*).

Condition 3.   *There exists a real number* v, *an* $\mathbf{x}^{(0)}$ *in* X_m, *and a* $\mathbf{y}^{(0)}$ *in* Y_n *such that*

(*a*) $\displaystyle\sum_i a_{ij} x_i^{(0)} \geqslant v, \qquad$ *for* $j = 1, 2, \cdots, n,$

(*b*) $\displaystyle\sum_j a_{ij} y_j^{(0)} \leqslant v, \qquad$ *for* $i = 1, 2, \cdots, n.$

(*That is, by adopting* $\mathbf{x}^{(0)}$ *player* 1 *can guarantee himself a return of at least v, and by adopting* $\mathbf{y}^{(0)}$ *player* 2 *can guarantee that player* 1 *gets at most v.*)

Proof. *1 implies 2:* Let $(\mathbf{x}', \mathbf{y}')$ be an equilibrium pair. We then have

$$v_2 = \min_{\substack{(1) \\ \mathbf{y}}} \max_{\mathbf{x}} M(\mathbf{x}, \mathbf{y}) \underset{(2)}{\leqslant} \max_{\mathbf{x}} M(\mathbf{x}, \mathbf{y}') \underset{(3)}{=} M(\mathbf{x}', \mathbf{y}')$$

$$= \min_{\substack{(4) \\ \mathbf{y}}} M(\mathbf{x}', \mathbf{y}) \underset{(5)}{\leqslant} \max_{\mathbf{x}} \min_{\mathbf{y}} M(\mathbf{x}, \mathbf{y}) \underset{(6)}{=} v_1.$$

These equalities and inequalities are justified as follows:

(1) Definition of v_2, number xi.
(2) Definition of minimum.
(3) and (4) Definition of equilibrium pair, number xiii.
(5) Definition of maximum.
(6) Definition of v_1, number x.

But, from number xii, $v_1 \leqslant v_2$, so $v_1 = v_2$.

2 implies 3: Let $v = v_1 = v_2$; let $\mathbf{x}^{(0)}$ be maximin, and let $\mathbf{y}^{(0)}$ be minimax. We then have for all j and i

$$\sum_i a_{ij} x_i^{(0)} \underset{(1)}{=} M(\mathbf{x}^{(0)}, \beta_j) \underset{(2)}{\geqslant} \min_{\mathbf{y}} M(\mathbf{x}^{(0)}, \mathbf{y}) \underset{(3)}{=} \max_{\mathbf{x}} \min_{\mathbf{y}} M(\mathbf{x}, \mathbf{y}) \underset{(4)}{=} v$$

$$\underset{(5)}{=} \min_{\mathbf{y}} \max_{\mathbf{x}} M(\mathbf{x}, \mathbf{y}) \underset{(6)}{=} \max_{\mathbf{x}} M(\mathbf{x}, \mathbf{y}^{(0)}) \underset{(7)}{\geqslant} M(\alpha_i, \mathbf{y}^{(0)}) \underset{(8)}{=} \sum_j a_{ij} y_j^{(0)}.$$

These inequalities are justified as follows:

(1) Definition of $M(\mathbf{x}, \mathbf{y})$, number vii.
(2) Definition of minimum.
(3) Choice of $\mathbf{x}^{(0)}$.
(4) and (5) Definition of v and condition 2.
(6) Choice of $\mathbf{y}^{(0)}$
(7) Definition of maximum.
(8) Definition of $M(\mathbf{x}, \mathbf{y})$, number vii.

3 implies 1: From a and b of condition 3 it follows that

$$M(\mathbf{x}^{(0)}, \mathbf{y}) \geqslant v \geqslant M(\mathbf{x}, \mathbf{y}^{(0)}),$$

for all $\mathbf{x}$ and $\mathbf{y}$. But putting $\mathbf{x} = \mathbf{x}^{(0)}$ and $\mathbf{y} = \mathbf{y}^{(0)}$ we see $v = M(\mathbf{x}^{(0)}, \mathbf{y}^{(0)})$; hence $(\mathbf{x}^{(0)}, \mathbf{y}^{(0)})$ is an equilibrium pair by definition xiii.

Remarks. (*a*) From the proof that 1 implies 2, we see that, if $(\mathbf{x}', \mathbf{y}')$ is an equilibrium pair, then $\mathbf{x}'$ and $\mathbf{y}'$ are maximin and minimax respectively.

(b) From the proof that 2 implies 3, the common value of v_1 and v_2 is the v of condition 3.

(c) We still *do not know* whether for an arbitrary finite strategy game, (A, B, M), an equilibrium pair exists, or if $v_1 = v_2$, or if there exists a triplet $(v, \mathbf{x}^{(0)}, \mathbf{y}^{(0)})$ satisfying a and b of condition 3. The principal theorem, generally known in the literature as the minimax theorem, establishes this existence; it was first proved by von Neumann in his 1928 paper.

A2.2 HISTORICAL REMARKS

The several proofs of the minimax theorem which exist fall into two general categories: those which rest on fixed-point theorems or iterative processes and those which depend upon separation properties of convex sets. In giving some geometrical insight into the principal theorem (Appendices 3 and 4), in describing the linear-programing problem and its relation to two-person zero-sum games (Appendix 5), and in surveying the methods for solving such games (Appendix 6) we almost, but not quite, prove the theorem in several different ways.   As none of these incomplete proofs are of the fixed-point variety, a complete and elegant proof due to Nash [1950 a], based on Brouwer's fixed-point theorem, will be included in the next section of this appendix.   But first some historial remarks, which are little more than a partial summary of Kuhn's [1952, pp. 71–84] excellent survey of this literature.

The first proof of the minimax theorem was given by von Neumann [1928]; it, too, made use of Brouwer's theorem, but is quite involved. Motivated by von Neumann's 1928 proof, Kakutani [1941] presented a generalization of Brouwer's theorem which is tailor-made to prove the minimax theorem—so much so that it becomes almost a trivial corollary of his fixed-point theorem.   We have chosen to use Nash's proof rather than Kakutani's, because it depends only upon the intuitively more plausible Brouwer theorem.   In addition, Nash's proof is related to an iterative technique discussed in Appendix 6.

The first elementary, though still partially topological, proof was given by Ville [1938].   Since the statement of the minimax theorem is completely algebraic, it should be possible to give an entirely algebraic proof. The first one, and still the shortest self-contained proof, is due to Loomis [1946], who uses an induction on the total number of pure strategies available to the two players.   Weyl [1950] succeeded in developing a noninductive, completely algebraic proof, but it is complex.   Dantzig [1956] has obtained a simple, non-inductive, constructive, and completely algebraic proof which uses his simplex method for linear programing.   For other algebraic-type proofs see Shapley, Snow [1950], Gale, Kuhn, Tucker [1950a].

A2.3 NASH'S PROOF OF THE MINIMAX THEOREM

In broad outline, Nash proves the theorem in this way: He defines a transformation T which maps mixed strategy pairs $(\mathbf{x}, \mathbf{y})$ into mixed strategy pairs $T(\mathbf{x}, \mathbf{y}) = (\mathbf{x}', \mathbf{y}')$, where T has the two properties:

i. $\mathbf{x}^{(0)}$ and $\mathbf{y}^{(0)}$ are optimal strategies if and only if $T(\mathbf{x}^{(0)}, \mathbf{y}^{(0)}) = (\mathbf{x}^{(0)}, \mathbf{y}^{(0)})$, i.e., if and only if $(\mathbf{x}^{(0)}, \mathbf{y}^{(0)})$ is a fixed point under the transformation.

ii. T has at least one fixed point.

The transformation is defined in this fashion. Let

$$c_i(\mathbf{x}, \mathbf{y}) = \begin{cases} M(\alpha_i, \mathbf{y}) - M(\mathbf{x}, \mathbf{y}), & \text{if this quantity is positive,} \\ 0, & \text{otherwise;} \end{cases}$$

$$d_j(\mathbf{x}, \mathbf{y}) = \begin{cases} M(\mathbf{x}, \mathbf{y}) - M(\mathbf{x}, \beta_j), & \text{if this quantity is positive,} \\ 0, & \text{otherwise.} \end{cases}$$

Using the notation $T(\mathbf{x}, \mathbf{y}) = (\mathbf{x}', \mathbf{y}')$, we define

$$x_i' = \frac{x_i + c_i(\mathbf{x}, \mathbf{y})}{1 + \displaystyle\sum_{k=1}^{m} c_k(\mathbf{x}, \mathbf{y})}$$

and

$$y_j' = \frac{y_j + d_j(\mathbf{x}, \mathbf{y})}{1 + \displaystyle\sum_{k=1}^{n} d_k(\mathbf{x}, \mathbf{y})}.$$

It is straightforward to verify that

$$x_i' \geqslant 0, \quad \sum_{i=1}^{m} x_i' = 1, \quad y_j' \geqslant 0, \quad \text{and} \quad \sum_{j=1}^{n} y_j' = 1.$$

We first show that $(\mathbf{x}, \mathbf{y})$ is a pair of optimal strategies if and only if it is a fixed point of this T. Observe that $c_i(\mathbf{x}, \mathbf{y})$ measures the amount that α_i is better than $\mathbf{x}$, if at all, as a response against $\mathbf{y}$, and that $d_j(\mathbf{x}, \mathbf{y})$ measures the amount that β_j is better than $\mathbf{y}$ as a response against $\mathbf{x}$. Now suppose that $\mathbf{x}$ and $\mathbf{y}$ are optimal. Since $\mathbf{x}$ is good against $\mathbf{y}$, it follows that $c_i(\mathbf{x}, \mathbf{y}) = 0$ for all i, so $x_i' = x_i$, for all i. Similarly, $y_j' = y_j$. Thus, if $(\mathbf{x}, \mathbf{y})$ is a pair of optimal strategies, $T(\mathbf{x}, \mathbf{y}) = (\mathbf{x}, \mathbf{y})$.

To show the converse, suppose $(\mathbf{x}, \mathbf{y})$ is a fixed point. We first show that there must be at least one i such that both $x_i > 0$ and $c_i(\mathbf{x}, \mathbf{y}) = 0$.

Since, by definition,

$$M(\mathbf{x}, \mathbf{y}) = \sum_{i=1}^{m} x_i M(\alpha_i, \mathbf{y}),$$

we conclude that $M(\mathbf{x}, \mathbf{y}) < M(\alpha_i, \mathbf{y})$ cannot hold for all i such that $x_i > 0$. Thus, for at least one i, $c_i(\mathbf{x}, \mathbf{y}) = M(\alpha_i, \mathbf{y}) - M(\mathbf{x}, \mathbf{y}) = 0$. But for this i, the fact that $(\mathbf{x}, \mathbf{y})$ is a fixed point implies

$$x_i = \frac{x_i}{1 + \sum_{k=1}^{m} c_k(\mathbf{x}, \mathbf{y})},$$

so $\sum_{k=1}^{m} c_k(\mathbf{x}, \mathbf{y}) = 0$. But the terms $c_k(\mathbf{x}, \mathbf{y})$ are all non-negative, so they must all equal 0. Thus, $\mathbf{x}$ is at least as good a response against $\mathbf{y}$ as any α_k, so $\mathbf{x}$ is good against $\mathbf{y}$. Similarly, $\mathbf{y}$ is shown to be good against $\mathbf{x}$, and so $(\mathbf{x}, \mathbf{y})$ is an optimal pair. This concludes the proof of the first assertion.

The existence of a fixed point for T follows from the Brouwer fixed-point theorem. We shall state a particular version of that theorem and indicate how it can be used to prove that T has a fixed point. The version is this:

If a function maps each point of a sphere S (interior plus boundary) located in a Euclidean space of finite dimension into another (not necessarily distinct) point of S and if the function is continuous, then there exists at least one point which is mapped into itself.

In our case, the set of mixed strategy pairs is certainly not a sphere—but when we don the topologist's glasses it can be made to look like one. More specifically, we can find a one-to-one correspondence between our set of strategy pairs and the points of a sphere which is continuous both ways in the sense that points "close" together in one set come from or go into points "close" together in the other. The mapping T, which is clearly continuous, when iterated with this one-to-one correspondence induces a mapping of the sphere into itself that is easily shown to be continuous. By the Brouwer theorem, the induced mapping has a fixed point; hence, so does T.

▶We cannot prove the Brouwer theorem here (for a proof see Hurewicz and Wallman [1948]), but it can be made extremely plausible in 2-space, i.e., the plane. Let S be the sphere, i.e., circle in the plane, and F the mapping, where F takes $\mathbf{z}$ into $F(\mathbf{z})$ (see Fig. 1). If F had no fixed point, i.e., if the image of every point were to be distinct from the point itself, then we could perform the following trick. For each **z**, let $G(\mathbf{z})$ denote the point where the ray beginning at $F(\mathbf{z})$ and passing through **z** intersects the boundary of S. For **z** on the boundary, $G(\mathbf{z}) = \mathbf{z}$. Since $F(\mathbf{z}) \neq \mathbf{z}$ and since F is continuous, it follows that G is also continuous. But a function which maps the whole sphere onto its boundary, keeping boundary

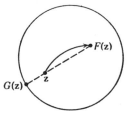

Fig. 2

points fixed, necessitates "ripping" the interior of the sphere, i.e., there must be points "close" together in the interior of S which under the mapping are shoved "far apart." Thus, the function is not continuous, contrary to what we have shown for G. The assumption that got us into this contradiction was that F had no fixed point, so we must conclude $F(\mathbf{z}) = \mathbf{z}$ for some **z**. ◀

FIRST

GEOMETRICAL INTERPRETATION

OF A TWO-PERSON

ZERO-SUM GAME

This appendix presents a complete geometric interpretation of the minimax theorem when player 1's pure strategy space consists of two elements, namely $A = \{\alpha_1, \alpha_2\}$.

Consider the game

$$\text{Player 1} \begin{array}{c} \\ \alpha_1 \\ \alpha_2 \end{array} \overset{\displaystyle \begin{array}{cc} \beta_1 & \beta_2 \end{array}}{\begin{bmatrix} a_{11} & a_{12} \\ a_{21} & a_{22} \end{bmatrix}}.$$

Any randomized strategy $(x_1\alpha_1, x_2\alpha_2)$, where $x_1 + x_2 = 1$, $x_1 \geqslant 0$, and $x_2 \geqslant 0$, can be identified with a point (x_1, x_2) on the line segment of length 1, as in Fig. 1. If player 1 chooses $(x_1\alpha_1, x_2\alpha_2)$ and player 2 chooses β_1, the return to player 1 is

$$M[(x_1\alpha_1, x_2\alpha_2), \beta_1] = a_{11}x_1 + a_{21}x_2.$$

394

Geometrically, we can exhibit the relation of $M[(x_1\alpha_1, x_2\alpha_2), \beta_1]$ to the point (x_1, x_2) as in Fig. 2.

If, however, player 2 chooses β_2, then

$$M[(x_1\alpha_1, x_2\alpha_2), \beta_2] = a_{12}x_1 + a_{22}x_2,$$

and a different, but similar diagram results. Superimposing these two lines we get a drawing of the type shown in Fig. 3.   The particular case drawn

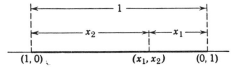

FIG. 1.   The point labeled (x_1, x_2) is x_1 units from $(0, 1)$ and x_2 units from $(1, 0)$, and $x_1 + x_2 = 1$.

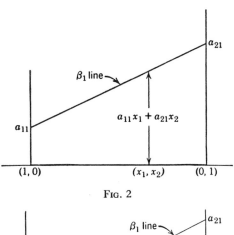

FIG. 2

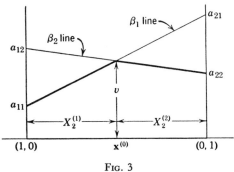

FIG. 3

supposes that $a_{21} > a_{22}$, $a_{12} > a_{11}$, and $a_{12} > a_{22}$.   If player 1 chooses the strategy (x_1, x_2), which lies in the interval marked $X_2^{(1)}$, then player 2's best response is β_1.   The vertical distance from the point (x_1, x_2) to the β_1 line represents 1's security level corresponding to $(x_1\alpha_1, x_2\alpha_2)$.   Similarly, if (x_1, x_2) is in $X_2^{(2)}$, then β_2 is the best response, and again the vertical distance from (x_1, x_2) to the heavy line represents 1's security level. Hence, the heavy line represents 1's security level. Thus, $\mathbf{x}^{(0)} = (x_1^{(0)}, x_2^{(0)})$ is 1's unique maximin strategy, and the value of the game is v.

If player 2 were to use β_1, then 1 could secure a_{21} ($> v$) by employing $(0\alpha_1, 1\alpha_2)$; and if he were to use β_2, then 1 could secure a_{12} ($> v$) by employing $(1\alpha_1, 0\alpha_2)$. Hence, to hold player 1 down to v, player 2 must use a randomized strategy $(y_1\beta_1, y_2\beta_2)$. The payoff resulting from these randomized strategies is:

$$M[(x_1\alpha_1, x_2\alpha_2), (y_1\beta_1, y_2\beta_2)]$$

$$= a_{11}x_1y_1 + a_{12}x_1y_2 + a_{21}x_2y_1 + a_{22}x_2y_2$$

$$= y_1(a_{11}x_1 + a_{21}x_2) + y_2(a_{12}x_1 + a_{22}x_2)$$

$$= y_1M[(x_1\alpha_1, x_2\alpha_2), \beta_1] + y_2M[(x_1\alpha_1, x_2\alpha_2), \beta_2].$$

So we see that $(y_1\beta_1, y_2\beta_2)$ yields a line which can be pictured on our diagram as a weighted average of the lines corresponding to β_1 and β_2. Since $y_1 + y_2 = 1$, the line associated with $\mathbf{y} = (y_1\beta_1, y_2\beta_2)$ must always lie between the β_1 line and the β_2 line, and so it must go through the point $[(x_1^{(0)}, x_2^{(0)}), v]$. Indeed, as y_1 goes from 1 to 0, a family of lines is generated which, so to speak, pivot clockwise about the point $[(x_1^{(0)}, x_2^{(0)}), v]$ from the line β_1 to β_2. For each particular line $\mathbf{y}$ chosen by 2, player 1 has a strategy choice which will maximize his return. In all cases, save when the line is horizontal, the choice is either $(1\alpha_1, 0\alpha_2)$ or $(0\alpha_1, 1\alpha_2)$, and his return exceeds v. To be certain that he will hold player 1 down to v, 2 must therefore choose the horizontal line in the family. For the horizontal line we have:

$$M[(x_1\alpha_1, x_2\alpha_2), (y_1^{(0)}\beta_1, y_2^{(0)}\beta_2)] = v, \qquad \text{for all } (x_1, x_2).$$

By setting $x_1 = 1$ and then $x_2 = 0$, we obtain the equality

$$y_1^{(0)}a_{11} + y_2^{(0)}a_{12} = y_1^{(0)}a_{21} + y_2^{(0)}a_{22} \qquad (= v).$$

Since $y_1^{(0)} + y_2^{(0)} = 1$, we can solve for $(y_1^{(0)}, y_2^{(0)})$ and for v. If v is known from player 1's analysis, then we can simplify our computation slightly since

$$y_1^{(0)}a_{11} + (1 - y_1^{(0)})a_{12} = v.$$

Lest the reader assume that all 2 by 2 games have the same structure as the one just analyzed, we present in diagrams a through i of Fig. 4 some of the different features which can occur. In c, d, and h, all of player 1's strategies are optimal (maximin). In b, e, g, and i, player 1 has a unique pure strategy [since the maximum of the (heavy) minimum function occurs in each case at the boundary—i.e., at $(1\alpha_1, 0\alpha_2)$, or $(0\alpha_1, 1\alpha_2)$]. In f, player 1 has an interval of optimal strategies which is less than the whole interval of strategies.

In a, player 2 has a unique optimum (minimax) strategy. This case was considered in detail above. In b and c, β_1 is optimal. Note that in c all lines associated with the family $(y_1\beta_1, y_2\beta_2)$, where $0 \leqslant y_1 \leqslant 1$, are

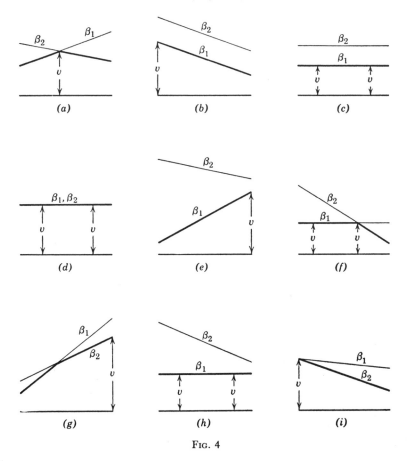

FIG. 4

horizontal, but only the lowest line is optimal. In d everything is optimal for player 2. In e and f, β_1 is optimal; in g, β_2; in h, β_1; and in i, even though all of player 2's strategies are minimax, β_2 is 2's best strategy in the sense that it not only is minimax but it is the best response against any strategy of 1.

Whether or not one constructs the diagram associated with a game (it cannot be done in two dimensions for games where both players have three or more pure strategies), one is interested in the intersections of the lines (or planes or hyperplanes when there are more strategies) with each other and with the vertical boundaries.   These points of intersection can always be found algebraically.   The above examples show that the lines or planes may not intersect; may intersect at points where an x_i is negative (cf. e and h); may intersect at a point representing a strategy, but one which is not optimal (cf. g); may intersect at the unique optimal point (cf. a); or may intersect at a non-unique optimal point (cf. f).

An extension of this analysis to games where player 2 has more than two strategies is extremely simple.   Consider, for example, the case where $A = \{\alpha_1, \alpha_2\}$ and $B = \{\beta_1, \beta_2, \cdots, \beta_5\}$.   To each $\beta_j, j = 1, 2, 3, 4, 5,$

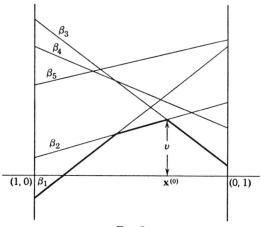

Fig. 5

there is a line in the diagram as in Fig. 5.   The security value for player 1, as a function of $\mathbf{x} = (x_1\alpha_1, x_2\alpha_2)$, is the heavy line in Fig. 5, i.e., the segments corresponding to the smallest β_i line.   Player 1 maximizes his security level by choosing $(x_1^{(0)}, x_2^{(0)})$.   If 2 wishes to hold 1 down to at most v, he must use a randomized strategy involving only β_2 and β_3.   For, if 2 places any positive weight on β_1, β_4, or β_5 then by using $(x_1^{(0)}, x_2^{(0)})$ player 1 will obtain more than v.   Of course, if player 1 fails to play optimally [i.e., fails to choose $(x_1^{(0)}\alpha_1, x_2^{(0)}\alpha_2)$], then it may benefit 2 to use β_1, but *never* β_5, since β_2 is always better than β_5, or β_4, since there are mixtures of β_2 and β_3 which are always preferable to β_4.

Without going into details, we may point out that, if $A = \{\alpha_1, \alpha_2, \alpha_3\}$, player 1's randomized strategies can be identified with the points of an equilateral triangle, and conversely.   If player 2 chooses β_j, then player 1's possibilities can be pictured as in Fig. 6.   Player 1's payoff if he uses $\mathbf{x}$ and if player 2 chooses β_j is the vertical distance from the point $\mathbf{x}$ of the horizontal equilateral triangle to the β_j plane. By superimposing the $\beta_1, \beta_2, \cdots, \beta_n$ planes (of course, the diagram is terrifically messy by now!), we can examine the *minimum function* or *security level function*, which is now a surface whose values depend upon the generic point of the equilateral triangle (i.e., of the generic strategy **x** of player 1). Player 1

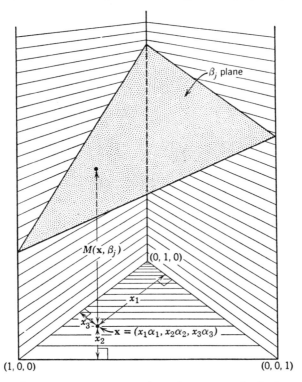

FIG. 6. There is a one-to-one correspondence between randomized strategies $\mathbf{x} = (x_1\alpha_1, x_2\alpha_2, x_3\alpha_3)$ and the points of the equilateral base triangle having an altitude of unit length. Note that $x_1 + x_2 + x_3 = 1$ for every point of this triangle. For purposes of clarity, the front face ($x_2 = 0$) of the game cylinder has been removed.

chooses $\mathbf{x}^{(0)}$ to maximize this security level. Player 2 uses a random strategy corresponding to the plane(s) which is a linear combination of the β_i planes and which is never more than v units (the value of the game) from the horizontal.

On the basis of such geometry one can develop a formal inductive proof of the minimax theorem (cf. Appendix 1 of Kuhn [1952]). We shall return to this geometrical interpretation again in Appendix 6.

SECOND

GEOMETRICAL INTERPRETATION

OF A TWO-PERSON

ZERO-SUM GAME

The following geometrical interpretation can be presented pictorially only if m, the number of pure strategies for player 1, is 2 or 3. We shall illustrate it for $m = 2$, let the reader visualize it for $m = 3$, and assert without proof that the geometry of these special cases carries over to any finite m with only minor terminological modifications. Although the concepts cannot be represented pictorially for $m > 3$, it is nonetheless extremely advantageous to employ the same geometrical terminology as developed from $m = 2$ and 3.

Let (A, B, M) be a game where $A = \{\alpha_1, \alpha_2\}$, $B = \{\beta_1, \beta_2, \cdots, \beta_n\}$, and let (X_2, Y_n, M) be its randomized strategy extension. For any randomized strategy $\mathbf{y} = (y_1\beta_1, y_2\beta_2, \cdots, y_n\beta_n)$, we can plot the pair of values $M(\alpha_1, \mathbf{y})$, $M(\alpha_2, \mathbf{y})$, where

$$M(\alpha_1, \mathbf{y}) = \sum_{j=1}^{n} a_{1j}y_j,$$

$$M(\alpha_2, \mathbf{y}) = \sum_{j=1}^{n} a_{2j}y_j,$$

as a point in the plane. Thus, the point associated with $\mathbf{y}$ has the interpretation that its ith coordinate is the return to player 1 if α_i is used. (This terminology extends to any m: if $m = 3$, the point associated with $\mathbf{y}$ is a point in 3-space; if $m > 3$, it is a point in m-space). If player 2 uses $\mathbf{y}$, then 1's best response is to choose the strategy corresponding to the largest coordinate of the point associated with $\mathbf{y}$.

Let $[m_1(\mathbf{y}), m_2(\mathbf{y})]$ be an abbreviation for $[M(\alpha_1, \mathbf{y}), M(\alpha_2, \mathbf{y})]$ and let $\mathfrak{M}$ be the set of all such points of the plane generated as $\mathbf{y}$ takes on values in Y_n. In symbolic notation,[1]

$$\mathfrak{M} = \{[m_1(\mathbf{y}), m_2(\mathbf{y})] \mid \mathbf{y} \text{ belongs to } Y_n\}.$$

Note that to each $\mathbf{y}$ belonging to Y_n there is associated a point in $\mathfrak{M}$, and that to each point in $\mathfrak{M}$ there is associated one or more elements of Y_n. If $[m_1(\mathbf{y}'), m_2(\mathbf{y}')] = [m_1(\mathbf{y}''), m_2(\mathbf{y}'')]$, then $\mathbf{y}'$ and $\mathbf{y}''$ should be considered strategically equivalent since they present identical opportunities to player 1.

We can therefore view the strategic role of player 2 as *choosing an element from the set* $\mathfrak{M}$. If player 2 chooses the point (m_1, m_2) of $\mathfrak{M}$ and player 1 chooses $(x_1\alpha_1, x_2\alpha_2)$, player 1 receives $x_1m_1 + x_2m_2$. This is a weighted average of the coordinates of the point of $\mathfrak{M}$ selected by player 2—the weights being selected, of course, by player 1.

The geometrical nature of $\mathfrak{M}$ is particularly simple, namely, a bounded, closed, convex polygon (i.e., it can be enclosed in a circle of finite radius, the boundary of $\mathfrak{M}$ belongs to $\mathfrak{M}$, if two points belong to $\mathfrak{M}$ so does the line segment joining them, and the boundary is composed of linear segments). In higher dimensional space ($m > 2$), the polygon becomes a polyhedron and the boundary is composed of (hyper)planes.

---

[1] In other words, $\mathfrak{M}$ is the set of points (m_1, m_2) where

$$m_1 = M(\alpha_1, \mathbf{y}) = \sum_{j=1}^{n} a_{1j}y_j$$

$$m_2 = M(\alpha_2, \mathbf{y}) = \sum_{j=1}^{n} a_{2j}y_j$$

and $\mathbf{y}$ is in Y_n, i.e., $\sum_{j=1}^{n} y_j = 1$ and $y_j \geqslant 0$.

To make matters a little more concrete, let us consider the following game:

Player 2

$$\begin{array}{c} & \beta_1 \; \beta_2 \; \beta_3 \; \beta_4 \; \beta_5 \\ \alpha_1 & \begin{bmatrix} 1 & 2 & 2 & 3 & 4 \\ \\ 3 & 1 & 4 & 3 & 3 \end{bmatrix} \\ \alpha_2 \end{array}$$

Player 1

The set $\mathfrak{M}$ is constructed by plotting the points in the plane (2-space) associated with the columns of the game matrix and then forming the smallest convex set containing all these points, as in Fig. 1. For example, the point **m** in Fig. 1 represents one of player 2's randomized strategies which places positive weights only on β_1 and β_2.

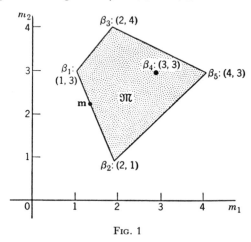

Fig. 1

In the following discussion, let us consider the four special games with the associated regions shown in Fig. 2. There is no loss of generality in assuming that $\mathfrak{M}$ is in the positive quadrant (orthant, if we are in higher dimensional space) since adding the same positive quantity to all payoffs of a game does not alter the strategic considerations involved. In all the diagrams, the 45° line is dotted and labeled l. Note that, below l, coordinate m_1 is larger than m_2, and that, above l, m_2 is larger than m_1. Thus, for example, α_2 is a good response to player 2's choice of (m_1, m_2) if and only if (m_1, m_2) is a point of $\mathfrak{M}$ on or above l.

It is 2's aim to select a point (m_1, m_2) of $\mathfrak{M}$ such that its maximum coordinate (the most player 1 can get from that point) is not greater than the maximum coordinate of any other point of $\mathfrak{M}$; in other words, he wants to choose a point (m_1, m_2) corresponding to a minimax strategy.

Since 2 wants to hold 1 down to as little as possible, it is reasonable to consider various values, such as v^*, and to ask whether 2 can hold 1 down to v^* or not. It is easily seen that this is possible if and only if the set $\mathfrak{M}$ contains a point both of whose coordinates do not exceed v^*. But this

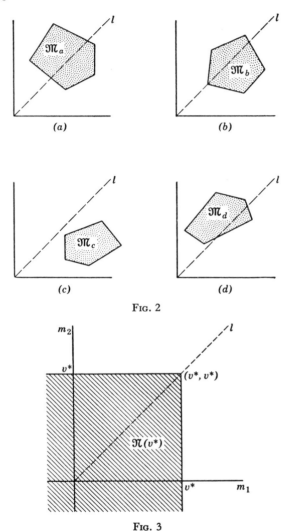

Fig. 2

Fig. 3

occurs if and only if the region labeled $\mathfrak{N}(v^*)$ in Fig. 3 contains at least one point of $\mathfrak{M}$. Formally,

$$\mathfrak{N}(v^*) = \{(m_1, m_2) \mid m_1 \leqslant v^*, m_2 \leqslant v^*\},$$

so $\mathfrak{N}(v^*)$ is the negative orthant with its origin displaced to (v^*, v^*).

There exists a value v (the value of the game) such that player 2 can hold player 1 down to v but not below v. Hence $\mathfrak{N}(v)$ must touch $\mathfrak{M}$ at some boundary point(s). See Fig. 4 for these points of contact in our four examples.

If the quantity v^* is small enough, then (in a particular game) player 2 will not be able to hold player 1 down to v^*. That is, $\mathfrak{N}(v^*)$ will be disjoint from $\mathfrak{M}$. As v^* increases, the displaced negative quandrant $\mathfrak{N}(v^*)$ gets translated in a northeasterly direction, so to speak, and there is a quantity v such that $\mathfrak{N}(v^*)$ just touches $\mathfrak{M}$ when v^* equals v.

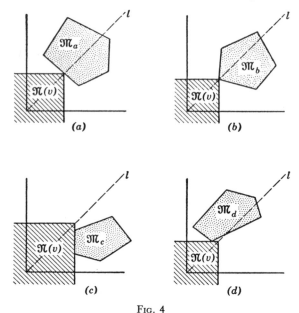

FIG. 4

Any point that player 2 chooses which is common to $\mathfrak{M}$ and $\mathfrak{N}(v)$ holds player 1 down to at most v. Hence any such point corresponds to a minimax strategy for player 2. Observe that, in the games corresponding to $\mathfrak{M}_a$, $\mathfrak{M}_b$, and $\mathfrak{M}_d$, the minimax strategy for player 2 is unique,[2] whereas for $\mathfrak{M}_c$ there is a whole segment of minimax strategies for player 2.

For any two convex sets, such as $\mathfrak{M}$ and $\mathfrak{N}(v)$, there is a line (hyperplane in higher dimension) separating them.[3] That is, a line exists touching

---

[2] Recall that strategies with equivalent payoffs are considered identical.

[3] For $v^* < v$, the sets $\mathfrak{N}(v^*)$ and $\mathfrak{M}$ are disjoint, and so the perpendicular bisector of the shortest line segment joining $\mathfrak{N}(v^*)$ and $\mathfrak{M}$ separates those bodies. It is possible to choose a sequence of values v_n^* approaching v in such a manner that the perpendicular bisectors approach a limiting line which can be shown to separate the bodies $\mathfrak{N}(v)$ and $\mathfrak{M}$. The discussion of this appendix carries over to the case when the

both $\mathfrak{M}$ and $\mathfrak{N}(v)$ such that $\mathfrak{M}$ lies on one side of the line and $\mathfrak{N}(v)$ on the other side. (As can be seen in b, this line need not be unique.)   We next show how it relates to player 1's solution of the game.

Let L be a line separating $\mathfrak{M}$ and $\mathfrak{N}(v)$.   Such a line can be thought of as the set of points (m_1, m_2) which satisfies an equation of the form

$$x_1^{(0)}m_1 + x_2^{(0)}m_2 = k,$$

where $(x_1^{(0)}, x_2^{(0)})$ determines the slope of the line and k fixes which particular line is chosen from the family of lines whose slopes are dictated by $(x_1^{(0)}, x_2^{(0)})$.

It is easily seen and is easily proved that:

i. Neither $x_1^{(0)}$ nor $x_2^{(0)}$ can be negative [for otherwise we could find an (m_1, m_2) on the line which is interior to $\mathfrak{N}(v)$, i.e., not on the boundary of $\mathfrak{N}(v)$].

ii. The point (v, v) is on the line L (check this in c and d above).

From (i), we lose no generality in assuming $x_1^{(0)} + x_2^{(0)} = 1$ (for if $x_1^{(0)} + x_2^{(0)} \neq 1$ we could consider the line

$$\frac{x_1^{(0)}}{x_1^{(0)} + x_2^{(0)}}\, m_1 + \frac{x_2^{(0)}}{x_1^{(0)} + x_2^{(0)}}\, m_2 = \frac{k}{x_1^{(0)} + x_2^{(0)}},$$

and relabeling we would get

$$x_1{}'m_1 + x_2{}'m_2 = k',$$

where $x_1{}' + x_2{}' = 1$).   Since (i) enables us to take $x_1^{(0)} + x_2^{(0)} = 1$ and since (v, v) lies on L, we may conclude that

$$x_1^{(0)}v + x_2^{(0)}v = k = (x_1^{(0)} + x_2^{(0)})v = v,$$

so $k = v$.   Hence

$$L = \{(m_1, m_2) \mid x_1^{(0)}m_1 + x_2^{(0)}m_2 = v\},$$

where $x_1^{(0)} + x_2^{(0)} = 1$, $x_1^{(0)} \geqslant 0$, and $x_2^{(0)} \geqslant 0$.   But then all points to the right of L, or above L, must satisfy

$$x_1^{(0)}m_1 + x_2^{(0)}m_2 \geqslant v.$$

Thus, for any (m_1, m_2) of $\mathfrak{M}$, we have

$$x_1^{(0)}m_1 + x_2^{(0)}m_2 \geqslant v.$$

---

number m of pure strategies of player 1 is greater than 2.   Instead of the geometry being embedded in 2-space, it is then embedded in m-space.   The separation of the two convex bodies $\mathfrak{M}$ and $\mathfrak{N}(v)$ by a hyperplane is by far the deepest mathematical result needed in the proof of the minimax theorem which results when the outline we have given here is made rigorous.

Hence, if player 1 chooses the strategy $(x_1^{(0)}\alpha_1, x_2^{(0)}\alpha_2)$, he can get at least v for *all* points (m_1, m_2) of $\mathfrak{M}$. Thus such a strategy is maximin for player 1.

In a, c, and d the separating hyperplane is unique, and so player 1's maximin strategy is unique. In b, there is not a unique strategy since there is *not* a unique separating line. In c, the line is vertical and hence of the form

$$1m_1 + 0m_2 = v,$$

i.e., α_1 is maximin for player 1. [Indeed, α_1 is best for all (m_1, m_2) of $\mathfrak{M}_c$ since all of $\mathfrak{M}_c$ is below l.] In d, the separating line is horizontal, namely:

$$0m_1 + 1m_2 = v.$$

In other words, α_2 is maximin for 1; however, it is not best for all (m_1, m_2) of $\mathfrak{M}_d$ since $\mathfrak{M}_d$ intersects l.

This completes the story as far as optimum[4] strategies are concerned. In addition, using these same geometric considerations, it is easy to see the possible effects when player 2 chooses a non-minimax strategy. To this end, let us suppose that player 1 chooses $(x_1\alpha_1, x_2\alpha_2)$. If player 2 chooses the element (m_1, m_2) of $\mathfrak{M}$ which lies on the line

$$x_1m_1 + x_2m_2 = k,$$

then player 1 receives an amount k. Consider the family of lines

$$x_1m_1 + x_2m_2 = k,$$

where k takes on different values, as shown in Fig. 5. Since player 1 chooses $(x_1\alpha_1, x_2\alpha_2)$, any of the points on the same line of the family yield the same return. Hence player 2's response to $(x_1\alpha_1, x_2\alpha_2)$ amounts to choosing a line in the family; but, of course, not all lines are available to player 2—only those which contain points of $\mathfrak{M}$. Thus, his best response to $(x_1\alpha_1, x_2\alpha_2) = \mathbf{x}$ is the line which both contains elements of $\mathfrak{M}$ and is as far left (or down) as possible in the family. For the case drawn, the heavy line represents 2's best choice.

Player 1's expected return when he uses $\mathbf{x}$ and when 2 chooses the line of best response is the security level of strategy $\mathbf{x}$, which is denoted $v_1(\mathbf{x})$. Thus, the equation for the heavy line is

$$x_1m_1 + x_2m_2 = v_1(\mathbf{x}).$$

[4] Throughout, "optimum" means either "maximin" or "minimax," and it should not be allowed to acquire the flavor of "this is what the player should do." Such statements are the meta game theoretic, not game theoretic.

This line must intersect the 45° line at the point $[v_1(\mathbf{x}), v_1(\mathbf{x})]$. Hence to represent the security level of a strategy $\mathbf{x}$, draw the family of lines associated with $\mathbf{x}$, and choose the one which is a left-sided support of $\mathfrak{M}$. The common value of the coordinates of its intersection with the 45° line, l, is the security level of $\mathbf{x}$. Player 1's security level is maximized by a

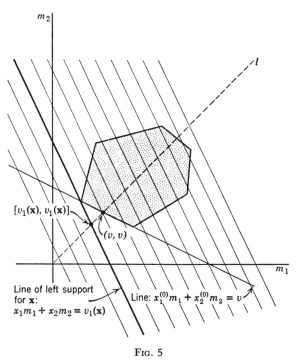

FIG. 5

strategy $\mathbf{x}^{(0)}$ such that its line of left support intersects l as far to the right as possible. Clearly, for the case shown in Fig. 5, the maximum security level is given by the line $(x_1^{(0)}\alpha_1, x_2^{(0)}\alpha_2)$. The reader will find it profitable to check through the above discussion for $\mathfrak{M}_b$, $\mathfrak{M}_c$, and $\mathfrak{M}_d$, shown in Fig. 4. Remember that, since x_1 and x_2 cannot be negative, the slopes of the lines cannot be positive.

The geometry of the minimax theorem described in this appendix was formulated by Gale, Kuhn, and Tucker in 1948. Later it was utilized by Gale [1951] and Karlin [1950].

appendix 5

LINEAR PROGRAMING AND
TWO-PERSON ZERO-SUM GAMES

This appendix is divided into three sections. In the first we will demonstrate that a two-person zero-sum game can be reduced to a special linear-programing problem. In the second we will define the general linear-programing problem and discuss its duality theory. The material in section 1 serves as motivation for the duality theory. In the third section we will employ the duality theory to show how the general linear-programing problem can also be interpreted as a two-person zero-sum game. The principal reference is Dantzig [1951 a]; however, we shall depart from Dantzig's treatment at several points in an attempt to achieve maximum clarity. We shall also use Gale, Kuhn, Tucker [1951].

A5.1 REDUCTION OF A GAME TO A LINEAR-
PROGRAMING PROBLEM

Let us assume we have a specific two-person zero-sum game (A, B, M), where $A = \{\alpha_1, \cdots, \alpha_m\}$, $B = \{\beta_1, \cdots, \beta_n\}$, and where $a_{ij} \equiv M(\alpha_i, \beta_j)$ is positive for all i and j. The last requirement does not entail any loss of generality since adding the same positive quantity to all the payoff entries does not alter the strategic structure of the game.

408

Player 1 can guarantee himself at least v^*, $(v^* > 0)$, if there exists an $\mathbf{x} = (x_1, \cdots, x_m)$, where $x_i \geqslant 0$ and $\sum_{i=1}^{m} x_i = 1$, such that

$$M(\mathbf{x}, \beta_j) \geqslant v^*, \qquad \text{for } j = 1, 2, \cdots, n, \tag{1}$$

which is equivalent to

$$\sum_{i=1}^{m} a_{ij} x_i \geqslant v^*, \qquad \text{for } j = 1, 2, \cdots, n. \tag{2}$$

By dividing eq. 2 by v^* and writing $x_i/v^* = u_i$, it is seen that player 1 can get at least v^* if there is a $\mathbf{u} = (u_1, u_2, \cdots, u_m)$, where $u_i \geqslant 0$, for $i = 1$, $2, \cdots, m$, and $\sum_i u_i = 1/v^*$, such that

$$\sum_{i=1}^{m} a_{ij} u_i \geqslant 1, \qquad \text{for } j = 1, 2, \cdots, n. \tag{3}$$

Equation 3 is equivalent to eq. 2 since multiplying by v^* and writing $u_i \cdot v^* = x_i$ yields eq. 2. Consequently, we can view the problem confronting player 1 as follows:

Player 1's problem. Let U be the set of all m-tuples $\mathbf{u} = (u_1, u_2, \cdots, u_m)$ such that

$$u_i \geqslant 0, \qquad \text{for } i = 1, 2, \cdots, m,$$

and

$$\sum_{i=1}^{m} a_{ij} u_i \geqslant 1, \qquad \text{for } j = 1, 2, \cdots, n.$$

To find those $\mathbf{u}$ belonging to U such that $\sum_{i=1}^{m} u_i$ is a minimum.

Remarks. (1*a*) If $\mathbf{u} = (u_1, \cdots, u_m)$ belongs to U, we have seen that player 1 can guarantee himself at least $\dfrac{1}{\sum_i u_i}$. In order to secure the maximum guarantee, player 1 should attempt to find a $\mathbf{u}$ in U which *maximizes* $\dfrac{1}{\sum_i u_i}$ or, equivalently, minimizes $\sum_i u_i$.

(1*b*) The problem of minimizing a linear form such as $\sum_i u_i$ (or, more generally, $\sum_i c_i u_i$) subject to restrictions involving linear inequalities such as

$\sum\limits_i a_{ij}u_i \geqslant 1$, for $j = 1, 2, \cdots, n$ (or, more generally, $\sum\limits_i a_{ij}u_i \geqslant b_j$, for $j = 1, 2, \cdots, n$), where $u_i \geqslant 0$, $i = 1, 2, \cdots, m$, is called a *linear-programing problem* (of the minimizing variety).

We next investigate the game problem from player 2's point of view and reduce that to a maximization problem involving linear inequalities.

Player 2 can guarantee that player 1 gets at most v^* ($v^* > 0$), by using $\mathbf{y} = (y_1, y_2, \cdots, y_n)$, where $\sum\limits_j y_j = 1$ and $y_j \geqslant 0$, for $j = 1, 2, \cdots, n$, provided that

$$M(\alpha_i, \mathbf{y}) \leqslant v^*, \qquad \text{for } i = 1, 2, \cdots, m, \tag{4}$$

or, equivalently, provided that

$$\sum_j a_{ij}y_j \leqslant v^*, \qquad \text{for } i = 1, 2, \cdots, m. \tag{5}$$

Equivalently, it is easily seen that player 1 can get at most v^* if there exists a $\mathbf{w} = (w_1, w_2, \cdots, w_n)$, where $w_j \geqslant 0$, for $j = 1, 2, \cdots, n$, and $\sum\limits_j w_j = 1/v^*$, such that

$$\sum_j a_{ij}w_j \leqslant 1, \qquad \text{for } i = 1, 2, \cdots, m. \tag{6}$$

Consequently, we can view the problem confronting player 2 as follows:

Player 2's problem.   Let W be the set of all n-tuples $\mathbf{w} = (w_1, w_2, \cdots, w_n)$ such that

$$w_j \geqslant 0, \qquad \text{for } j = 1, 2, \cdots, n,$$

and

$$\sum_{j=1}^{n} a_{ij}w_j \leqslant 1, \qquad \text{for } i = 1, 2, \cdots, m.$$

To find those $\mathbf{w}$ belonging to W such that $\sum\limits_{j=1}^{n} w_j$ is a maximum.

Remarks.   (2a) If $\mathbf{w} = (w_1, w_2, \cdots, w_n)$ belongs to W, we have seen that player 2 can hold player 1 down to at most $\dfrac{1}{\sum\limits_j w_j}$. In order to hold player 1 down as much as possible, player 2 should attempt to find a $\mathbf{w}$ in W which minimizes $\dfrac{1}{\sum\limits_j w_j}$, or equivalently, *maximizes* $\sum\limits_j w_j$.

(2b) The problem of maximizing a linear form such as $\sum_j w_j$ (or, more

generally, $\sum_j b_j w_j$) subject to restrictions involving linear inequalities such

as $\sum_j a_{ij} w_j \leqslant 1$, for $i = 1, 2, \cdots, m$ (or, more generally, $\sum_j a_{ij} w_j \leqslant c_i$,

for $i = 1, 2, \cdots, m$), where $w_j \geqslant 0$, for $j = 1, 2, \cdots, n$, is called a
linear-programing problem (of the maximizing variety).

(3) Player 1's problem and Player 2's problem are said to be dual
linear-programing problems.[1]

Any **u** in U guarantees player 1 *at least* $\dfrac{1}{\sum_i u_i}$.   Any **w** in W guarantees 1

at most $\dfrac{1}{\sum_j w_j}$.   Since player 1 can get *at least* $\dfrac{1}{\sum_i u_i}$ and *at most* $\dfrac{1}{\sum_j w_j}$, we must

have

$$\frac{1}{\sum_i u_i} \leqslant \frac{1}{\sum_j w_j},$$

i.e.,

$$\sum_j w_j \leqslant \sum_i u_i.$$

But we know that all zero-sum games have a value which can be interpreted as follows: Player 1 can get at least v (i.e., there is a $\mathbf{u}^{(0)}$ in U such
that

$$\frac{1}{\sum_i u_i^{(0)}} = v),$$

and player 2 can hold player 1 down to at most v (i.e., there is a $\mathbf{w}^{(0)}$ in W
such that

$$\frac{1}{\sum_j w_j^{(0)}} = v).$$

Summarizing, we have the symmetric problem.

[1] In remarks 1b and 2b following the two problems, we indicated how the linear-programing problems are generalized by introducing numbers $(b_1, \cdots, b_n)$ and $(c_1, \cdots, c_m)$ as data of the problems. This was done in such a manner that the problems given in 1b and 2b are also said to be *dual*.

Symmetric problem.[2] To find $\mathbf{u}^{(0)}$ in U and $\mathbf{w}^{(0)}$ in W such that

$$\sum_{j=1}^{n} w_j^{(0)} = \sum_{i=1}^{m} u_i^{(0)}.$$

Remarks. (4) If $\mathbf{u}^{(0)}$, $\mathbf{w}^{(0)}$ solve the symmetric problem, then

$$v = \frac{1}{\sum_j w_j^{(0)}} = \frac{1}{\sum_i u_i^{(0)}};$$

and $\mathbf{x}^{(0)} = (x_1^{(0)}, \cdots, x_m^{(0)})$, where $x_i^{(0)} = v \cdot u_i^{(0)}$, for $i = 1, 2, \cdots, m$, is maximin for player 1; and $\mathbf{y} = (y_1^{(0)}, \cdots, y_n^{(0)})$, where $y_j^{(0)} = v \cdot w_j^{(0)}$, for $j = 1, 2, \cdots, m$, is minimax for player 2.   Conversely, if $(\mathbf{x}^{(0)}, \mathbf{y}^{(0)}, v)$ constitutes a solution of the game, defining $u_i^{(0)} = x_i^{(0)}/v$, $w_j^{(0)} = y_j^{(0)}/v$ yields a solution of the symmetric problem.   Furthermore, $\mathbf{u}^{(0)}$ is a solution of 1's problem, and $\mathbf{w}^{(0)}$ is a solution of 2's problem.

A5.2 DUALITY THEORY OF THE GENERAL LINEAR-PROGRAMING PROBLEM

The data of the general linear-programing problem are an n-tuple $\mathbf{b} = (b_1, b_2, \cdots, b_n)$, an m-tuple $\mathbf{c} = (c_1, c_2, \cdots, c_m)$, and an m by n array of numbers a_{ij}, where $i = 1, 2, \cdots, m$ and $j = 1, 2, \cdots, n$. These may be any numbers—in particular, they are not assumed to be non-negative.

The minimization problem. Let U be the set of all m-tuples $\mathbf{u} = (u_1, u_2, \cdots, u_m)$ such that

(i) $u_i \geqslant 0$, for $i = 1, 2, \cdots, m$.
(ii) $u_1 a_{1j} + u_2 a_{2j} + \cdots + u_m a_{mj} \geqslant b_j$, for $j = 1, 2, \cdots, n$.

Find those $\mathbf{u}$ belonging to U such that the index (which is a linear form)

$$c_1 u_1 + c_2 u_2 + \cdots + c_m u_m$$

is a minimum.

The maximization problem. Let W be the set of all n-tuples $\mathbf{w} = (w_1, w_2, \cdots, w_n)$ such that

(i) $w_j \geqslant 0$, for $j = 1, 2, \cdots, n$.
(ii) $w_1 a_{i1} + w_2 a_{i2} + \cdots + w_n a_{in} \leqslant c_i$, for $i = 1, 2, \cdots, m$.

---

[2] More generally, if the dual problems are taken to be the general versions of remarks 1*b* and 2*b*, then the symmetric problem is to find $\mathbf{u}^{(0)}$ in U and $\mathbf{w}^{(0)}$ in W such that

$$\sum_i c_i u_i^{(0)} = \sum_j b_j w_j^{(0)}.$$

Find those $\mathbf{w}$ belonging to W such that the index (which is a linear form)

$$b_1 w_1 + b_2 w_2 + \cdots + b_n w_n$$

is a maximum.

The symmetric problem. To find those pairs $(\mathbf{u}, \mathbf{w})$, where $\mathbf{u}$ belongs to U and $\mathbf{w}$ belongs to W such that,

$$c_1 u_1 + c_2 u_2 + \cdots + c_m u_m = b_1 w_1 + b_2 w_2 + \cdots + b_n w_n.$$

Remarks. (1) The maximization, minimization, and symmetric problems stated here are obvious generalizations of the problems encountered in the previous section.

(2) The "diet problem" discussed in section 3 of Chapter 2 is in the form of the minimization problem. In that example, the following interpretations were made: a_{ij} is the amount of nutrient j per unit amount of food i; b_j is the minimum amount of nutrient j required; and c_i is the cost of a unit amount of food i. A "diet" is an m-tuple $\mathbf{u} = (u_1, \cdots, u_m)$, where u_i is the number of units of food i in the "diet."

(3) For given data $\mathbf{b}$, $\mathbf{c}$, $[a_{ij}]$ it can happen either that there are no m-tuples $\mathbf{u}$ in U or that, although there are m-tuples in U, there is no lower bound to the index $\sum_i c_i u_i$. In either case the minimization problem has no solution. Similarly, W might be the empty set or there may be no upper bound to the index $\sum_j b_j w_j$ when it is non-empty. Again, in either case there is no solution to the maximization problem.

Principal theorem of linear programing.

1. If there exists a $\mathbf{u}$ in U and a $\mathbf{w}$ in W, then

$$c_1 u_1 + \cdots + c_m u_m \geqslant b_1 w_1 + \cdots + b_n w_n.$$

2. If $(\mathbf{u}^{(0)}, \mathbf{w}^{(0)})$ is a solution to the symmetric problem, then $\mathbf{u}^{(0)}$ is a solution to the minimization problem and $\mathbf{w}^{(0)}$ is a solution to the maximization problem.

3. If $\mathbf{u}^{(0)}$ is a solution to the minimization problem and $\mathbf{w}^{(0)}$ is a solution to the maximization problem, then

$$c_1 u_1^{(0)} + \cdots + c_m u_m^{(0)} = b_1 w_1^{(0)} + \cdots + b_n w_n^{(0)},$$

i.e., $(\mathbf{u}^{(0)}, \mathbf{w}^{(0)})$ is a solution to the symmetric problem.

4. If a solution exists to one problem, then solutions exist to the other two problems.

5. If both U and W are non-empty, then all three problems have solutions.

Remark. We will outline two different proofs of this theorem. The first, given in small print in the remainder of this section, does not depend upon the minimax theorem for games, but rather upon a theorem about polyhedral cones due to Farkas. Thus, this section is self-contained in that it includes both the statement and a proof of the duality theory of linear programing. Incidently, this proof can be used to provide still another proof of the minimax theorem. In the next section, a second proof is given which rests on a slight generalization of the minimax theorem; it demonstrates clearly the intimate relationship between two-person zero-sum games and linear programing.

▶ *Proof.* 1. This follows from a chain of three inequalities:

$$\sum_j b_j w_j \leqslant \sum_j \left(\sum_i u_i a_{ij} \right) w_j = \sum_i \left(\sum_j a_{ij} w_j \right) u_i \leqslant \sum_i c_i u_i.$$

The first of these arises if we multiply the jth inequality of (ii) in the minimization problem by w_j and then sum over all j. The middle equality follows from a change in the order of summation. The last inequality arises if we multiply the ith inequality of (ii) in the maximization problem by u_i and then sum over all i.

2. If ($\mathbf{u}^{(0)}$, $\mathbf{w}^{(0)}$) is a solution of the symmetric problem, then $\mathbf{u}^{(0)}$ belongs to U and $\mathbf{w}^{(0)}$ to W. Furthermore, by the inequality of part 1, the index for $\mathbf{u}^{(0)}$ must be a minimum and the index for $\mathbf{w}^{(0)}$ must be a maximum. Hence $\mathbf{u}^{(0)}$ and $\mathbf{w}^{(0)}$ are solutions of their respective problems.

3 and 4. These two assertions are mathematically *much* deeper than the preceding ones, and the proofs are correspondingly more difficult. We shall be content merely to *outline* the nature of the proofs, which hinge on the following non-trivial lemma (first stated and proved in 1902 by J. Farkas).

Lemma. Let the following array of numbers be given:

$$\begin{bmatrix} d_{11} & d_{12} & \cdots & d_{1r} \\ d_{21} & d_{22} & \cdots & d_{2r} \\ \cdot & \cdot & & \cdot \\ \cdot & \cdot & & \cdot \\ \cdot & \cdot & & \cdot \\ d_{p1} & d_{p2} & \cdots & d_{pr} \\ d_{p+1,\,1} & d_{p+1,\,2} & \cdots & d_{p+1,\,r} \end{bmatrix},$$

where no row consists entirely of zero elements. If for any r-tuple ($\rho_1, \rho_2, \cdots \rho_r$) such that

$$\rho_1 d_{i1} + \rho_2 d_{i2} + \cdots + \rho_r d_{ir} \geqslant 0, \qquad \text{for } i = 1, 2, \cdots, p,$$

it follows that

$$\rho_1 d_{p+1,\,1} + \rho_2 d_{p+1,\,2} + \cdots + \rho_r d_{p+1,\,r} \geqslant 0,$$

then there exists a p-tuple $\boldsymbol{\lambda} = (\lambda_1, \lambda_2, \cdots, \lambda_p)$, such that

$$\lambda_i \geqslant 0, \qquad \text{for } i = 1, 2, \cdots, p,$$

and

$$\lambda_1 d_{1j} + \lambda_2 d_{2j} + \cdots + \lambda_p d_{pj} = d_{p+1,\,j}, \text{ for } j = 1, 2, \cdots, r.$$

Remarks. The lemma asserts that, if whenever a ρ vector makes a non-obtuse angle with each of the first p row vectors it also makes a non-obtuse angle with the $(p + 1)$st row vector, then the $(p + 1)$st row vector is a non-negative linear combination of the first p row vectors. When we say that $\varrho = (\rho_1, \cdots, \rho_r)$ forms a non-obtuse angle with $\mathbf{d}_i = (d_{i1}, \cdots, d_{ir})$, we merely mean that

$$\rho_1 d_{i1} + \cdots + \rho_r d_{ir} \geqslant 0.$$

Some geometrical insight into this lemma can be gained if we examine the case $r = 3$ and $p = 3$. Then the three rows can be identified with three points in 3-space. (See Fig. 1.) The row vector $\mathbf{d}_4 = (d_{41}, d_{42}, d_{43})$ is a non-negative linear combination of row vectors $\mathbf{d}_1 = (d_{11}, d_{12}, d_{13})$, $\mathbf{d}_2 = (d_{21}, d_{22}, d_{23})$, and $\mathbf{d}_3 = (d_{31}, d_{32}, d_{33})$ if and only if $\mathbf{d}_4$ is a point in the polyhedral cone in Fig. 1. To illustrate the

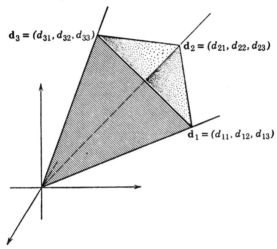

FIG. 1

plausibility of the lemma, it suffices to show that, if $\mathbf{d}_4$ does *not* belong to the cone generated by the other row vectors, then there exists a vector $\varrho = (\rho_1, \rho_2, \rho_3)$, which forms an obtuse angle with $\mathbf{d}_4$ (i.e., $\mathbf{d}_4 \cdot \varrho = d_{41}\rho_1 + d_{42}\rho_2 + d_{43}\rho_3 < 0$) and a non-obtuse angle with $\mathbf{d}_1$, $\mathbf{d}_2$, $\mathbf{d}_3$ (i.e., $\mathbf{d}_i \cdot \varrho = d_{i1}\rho_1 + d_{i2}\rho_2 + d_{i3}\rho_3 \geqslant 0$ for $i = 1, 2, 3$). If $\mathbf{d}_4$ does not belong to the cone, then intuitively it seems clear that there is a hyperplane going through the origin which separates $\mathbf{d}_4$ from the cone, where by "separates" we mean that $\mathbf{d}_4$ is on one side of the hyperplane and the polyhedral cone is on the other side. The deepest mathematical aspect of the lemma is the proof of this separation property. Assuming the existence of the separating hyperplane, we now show that its algebraic interpretation yields a proof of the lemma (in this special case). A hyperplane passing through the origin is the locus of points $\mathbf{d} = (d_1, d_2, d_3)$ such that

$$d_1\rho_1 + d_2\rho_2 + d_3\rho_3 = 0,$$

for some suitable 3-tuple $\varrho = (\rho_1, \rho_2, \rho_3)$. Put in other terms, the hyperplane is the locus of all points orthogonal to the vector ϱ. Since $\mathbf{d}_1$, $\mathbf{d}_2$, and $\mathbf{d}_3$ lie on one side of the hyperplane and $\mathbf{d}_4$ lies on the other, the quantities

$$\mathbf{d}_i \cdot \varrho = d_{i1}\rho_1 + d_{i2}\rho_2 + d_{i3}\rho_3, \qquad i = 1, 2, 3,$$

are of one sign and

$$\mathbf{d}_4 \cdot \boldsymbol{\varrho} = d_{41}\rho_1 + d_{42}\rho_2 + d_{43}\rho_3$$

is of the opposite sign. Thus, the orientation of $\boldsymbol{\varrho}$ can be so chosen that $\mathbf{d}_4 \cdot \boldsymbol{\varrho} < 0$ and $\mathbf{d}_i \cdot \boldsymbol{\varrho} > 0$ for $i = 1, 2, 3$, as was to be shown. This proof, once the separation property is proved in general, can be extended to arbitrary r and p.

Having paid tribute to the lemma, let us return to our main task of proving parts 3 and 4 of the principal theorem. To this end, we will show that *if $\mathbf{u}^{(0)}$ is a solution of the minimization problem, then there exists a $\mathbf{w}^{(0)}$ in W such that*

$$c_1 u_1^{(0)} + \cdots + c_m u_m^{(0)} \leqslant b_1 w_1^{(0)} + \cdots + b_n w_n^{(0)}.$$

Once this has been established, then we know by part 1 that the equality sign must hold, and so, by part 2, $\mathbf{w}^{(0)}$ must solve the maximization problem.

Consider the following array:

	u_1	u_2	$\cdots$	u_m	z
$z_1^{(0)}$	1	0	$\cdots$	0	0
$z_2^{(0)}$	0	1	$\cdots$	0	0
.	.	.		.	.
.	.	.		.	.
.	.	.		.	.
$z_m^{(0)}$	0	0	$\cdots$	1	0
$z_{m+1}^{(0)}$	0	0	$\cdots$	0	1
$w_1^{(0)}$	a_{11}	a_{21}	$\cdots$	a_{m1}	$-b_1$
$w_2^{(0)}$	a_{12}	a_{22}	$\cdots$	a_{m2}	$-b_2$
.	.	.		.	.
.	.	.		.	.
.	.	.		.	.
$w_n^{(0)}$	a_{1n}	a_{2n}	$\cdots$	a_{mn}	$-b_n$
	c_1	c_2	$\cdots$	c_m	$-\mu$

The quantity μ is defined to be $c_1 u_1^{(0)} + c_2 u_2^{(0)} + \cdots + c_m u_m^{(0)}$, where $\mathbf{u}^{(0)}$ is a solution of the minimization problem. The vector $(u_1, \cdots, u_m, z)$ will play the role of the $\boldsymbol{\varrho}$ vector in the lemma, the number $m + 1 + n$ the role of p, and $(c_1, c_2, \cdots, c_m, -\mu)$ is the $(p + 1)$st row. To apply the lemma, we must show that, if $(u_1, \cdots, u_m, z)$ forms a non-obtuse angle with each of the first $m + 1 + n$ row vectors, it also forms a non-obtuse angle with the last row (i.e., the $(p + 1)$st row). Once this is shown, the lemma establishes the existence of non-negative numbers $z_1^{(0)}, \cdots, z_{m+1}^{(0)}, w_1^{(0)}, \cdots, w_n^{(0)}$ such that:

$$(1) \quad z_1^{(0)} + w_1^{(0)} a_{11} + \cdots + w_n^{(0)} a_{1n} = c_1$$
$$(2) \quad z_2^{(0)} + w_1^{(0)} a_{21} + \cdots + w_n^{(0)} a_{2n} = c_2$$
$$\cdot \quad \cdot$$
$$\cdot \quad \cdot$$
$$\cdot \quad \cdot$$
$$(i) \quad z_i^{(0)} + w_1^{(0)} a_{i1} + \cdots + w_n^{(0)} a_{in} = c_i$$
$$\cdot \quad \cdot$$
$$\cdot \quad \cdot$$
$$\cdot \quad \cdot$$
$$(m) \quad z_m^{(0)} + w_1^{(0)} a_{m1} + \cdots + w_n^{(0)} a_{mn} = c_m$$
$$(m + 1) \quad z_{m+1}^{(0)} - (w_1^{(0)} b_1 + \cdots + w_n^{(0)} b_n) = -\mu$$

But, since $z_i^{(0)} \geqslant 0$, the ith equation gives

$$w_1^{(0)} a_{i1} + \cdots + w_n^{(0)} a_{in} \leqslant c_i, \quad \text{for } i = 1, 2, \cdots, m,$$

and since $z_{m+1}^{(0)} \geqslant 0$, the $(m + 1)$st equation yields

$$w_1^{(0)} b_1 + \cdots + w_n^{(0)} b_n \geqslant \mu \quad (= c_1 u_1^{(0)} + \cdots + c_m u_m^{(0)}).$$

But this means that $\mathbf{w}^{(0)} = (w_1^{(0)}, w_2^{(0)}, \cdots, w_n^{(0)})$ is a solution of the maximization problem, as was to be shown!

To finish off the job, we still must show that, if $(u_1, \cdots, u_m, z)$ forms a non-obtuse angle with the first $m + 1 + n$ rows, i.e.,

(i) $u_i \geqslant 0, \quad \text{for } i = 1, 2, \cdots, m;$
      $z \geqslant 0;$
(ii) $u_1 a_{1j} + \cdots + u_m a_{mj} - z b_j \geqslant 0, \text{ for } j = 1, 2, \cdots, n;$

then $(u_1, \cdots, u_m, z)$ forms a non-obtuse angle with the last row, i.e.,

(iii) $u_1 c_1 + u_2 c_2 + \cdots + u_m c_m - z\mu \geqslant 0.$

We will consider two cases, namely, $z > 0$ and $z = 0$.

Case 1 $(z > 0)$. If $(u_1, \cdots, u_m, z)$ satisfies inequalities (i) and (ii), then $(u_1/z, u_2/z, \cdots, u_m/z)$ belongs to U and therefore

$$\frac{u_1}{z} c_1 + \frac{u_2}{z} c_2 + \cdots + \frac{u_m}{z} c_m - \mu \geqslant 0,$$

since μ is defined as the minimum of the $\mathbf{u}$ indices. Statement (iii) follows by multiplying the last inequality by z.

Case 2 $(z = 0)$. Suppose $(u_1, \cdots, u_m, 0)$ is such that

$$u_i \geqslant 0, \quad \text{for } i = 1, 2, \cdots, m,$$

and

$$u_1 a_{1j} + \cdots + u_m a_{mj} \geqslant 0, \quad \text{for } j = 1, 2, \cdots, n;$$

then we must show

$$u_1 c_1 + \cdots + u_m c_m \geqslant 0.$$

If $\mathbf{u}^{(0)} = (u_1^{(0)}, \cdots, u_m^{(0)})$ is a solution of the minimization problem, then we assert $\mathbf{u}^* = (u_1^{(0)} + \lambda u_1, u_2^{(0)} + \lambda u_2, \cdots, u_m^{(0)} + \lambda u_m)$ belongs to U for all $\lambda \geqslant 0$, since:

(a) $u_i^{(0)} + \lambda u_i \geqslant 0, \quad \text{for } i = 1, 2, \cdots, m,$

and

(b) $(u_1^{(0)} + \lambda u_1) a_{1j} + (u_2^{(0)} + \lambda u_2) a_{2j} + \cdots + (u_m^{(0)} + \lambda u_m) a_{mj}$
$= \{u_1^{(0)} a_{1j} + u_2^{(0)} a_{2j} + \cdots + u_m^{(0)} a_{mj}\} + \{\lambda(u_1 a_{1j} + \cdots + u_m a_{mj})\}$
$\geqslant u_1^{(0)} a_{1j} + \cdots + u_m^{(0)} a_{mj}$ (since the second bracketed expression $\geqslant 0$
                                                          by hypothesis)
$\geqslant b_j, \quad \text{for } j = 1, 2, \cdots, n.$

Furthermore, the index of $\mathbf{u}^*$ must be at least μ, so

$$\mu \leqslant (u_1^{(0)} + \lambda u_1)c_1 + \cdots + (u_m^{(0)} + \lambda u_m)c_m$$
$$= \{u_1^{(0)}c_1 + \cdots + u_m^{(0)}c_m\} + \{\lambda(u_1c_1 + \cdots + u_mc_m)\}$$
$$= \mu + \lambda(u_1c_1 + \cdots + u_mc_m).$$

Hence, it follows that

$$u_1c_1 + \cdots + u_mc_m \geqslant 0,$$

as was to be shown.

Summarizing, we have shown that, if $\mathbf{u}^{(0)}$ is a solution to the minimization problem, then a solution $\mathbf{w}^{(0)}$ to the maximization problem exists and

$$c_1u_1^{(0)} + \cdots + c_mu_m^{(0)} = b_1w_1^{(0)} + \cdots + b_nw_n^{(0)}.$$

In a parallel fashion we can show that, if $\mathbf{w}^{(0)}$ is a solution to the maximization problem, then a solution $\mathbf{u}^{(0)}$ to the minimization problem exists. This establishes 3 and 4.

5. Part 1 asserts that, if $\mathbf{u}$ belongs to U and $\mathbf{w}$ to W,

$$\sum_i c_iu_i \geqslant \sum_j b_jw_j.$$

Hence, the set of numbers $\left\{\sum_i c_iu_i\right\}$, for $\mathbf{u}$ in U, is bounded from below. Let μ be the greatest lower bound of these numbers. We wish to show that there exists a $\mathbf{u}^{(0)}$ in U such that

$$\sum_i c_iu_i^{(0)} = \mu.$$

Instead, we shall show there exists a $\mathbf{w}^{(0)}$ in W such that

$$\sum_j b_jw_j^{(0)} = \mu,$$

which will prove that the maximization problem has a solution; and therefore, by 4, the minimization and symmetric problems have solutions. The existence of such a $\mathbf{w}^{(0)}$ is established by making some minor modifications in the proof of 3 and 4. The interested reader should check to see that all steps encountered in the proof of 3 and 4 remain valid except for the following points:

i. μ is not defined as $\sum_i c_iu_i^{(0)}$ since in the present proof it is not initially known that such a $\mathbf{u}^{(0)}$ exists. Instead, μ is defined to be the greatest lower bound of $\sum_i c_iu_i$ for all $\mathbf{u}$ in U.

ii. In the proof for the case $z = 0$, we can no longer take $\mathbf{u}^* = \mathbf{u}^{(0)} + \lambda\mathbf{u}$, where $\mathbf{u}^{(0)}$ is a solution to the minimization problem. Rather, we use a $\mathbf{u}'$ where $\mathbf{u}'$ belongs to U and

$$\mu \leqslant \sum_i c_iu_i' < \mu + \epsilon$$

for some preassigned positive ϵ, however small. This leads to the inequality

$$\mu \leqslant \mu + \epsilon + \lambda \sum_i u_i c_i.$$

But since ϵ is arbitrarily small, we can conclude that

$$\sum_i u_i c_i \geqslant 0,$$

which was to be shown in that part of the proof.

This concludes the first proof of the principal theorem. ◀

A5.3 REDUCTION OF A LINEAR-PROGRAMING PROBLEM TO A GAME

We have a dual aim in this section: First, as the heading advertises, we will show how a linear-programing problem can be reduced to a game. Second, we will prove the principal theorem of linear programing by means of the minimax theorem.

Consider any linear-programing problem of the minimizing or maximizing variety described earlier. We shall now exhibit a two-person zero-sum game whose solutions provide solutions to the linear-programing problems, provided solutions exist at all. The appropriate game matrix is:

$$
\begin{array}{c c}
 & \begin{array}{c c c c c c c c c c}
\beta_1 & \beta_2 & \cdots & \beta_n & \beta_{n+1} & \beta_{n+2} & \cdots & \beta_{n+m} & \beta_{n+m+1}
\end{array} \\
\begin{array}{c}
\alpha_1 \\ \alpha_2 \\ \cdot \\ \cdot \\ \cdot \\ \alpha_n \\ \alpha_{n+1} \\ \alpha_{n+2} \\ \cdot \\ \cdot \\ \cdot \\ \alpha_{n+m} \\ \alpha_{n+m+1}
\end{array} &
\left[
\begin{array}{c c c c c c c c c}
0 & 0 & \ldots & 0 & -a_{11} & -a_{21} & \ldots & -a_{m1} & b_1 \\
0 & 0 & \ldots & 0 & -a_{12} & -a_{22} & \ldots & -a_{m2} & b_2 \\
\cdot & \cdot & & \cdot & \cdot & \cdot & & \cdot & \cdot \\
\cdot & \cdot & & \cdot & \cdot & \cdot & & \cdot & \cdot \\
\cdot & \cdot & & \cdot & \cdot & \cdot & & \cdot & \cdot \\
0 & 0 & \ldots & 0 & -a_{1n} & -a_{2n} & \ldots & -a_{mn} & b_n \\
a_{11} & a_{12} & \ldots & a_{1n} & 0 & 0 & \ldots & 0 & -c_1 \\
a_{21} & a_{22} & \ldots & a_{2n} & 0 & 0 & \ldots & 0 & -c_2 \\
\cdot & \cdot & & \cdot & \cdot & \cdot & & \cdot & \cdot \\
\cdot & \cdot & & \cdot & \cdot & \cdot & & \cdot & \cdot \\
\cdot & \cdot & & \cdot & \cdot & \cdot & & \cdot & \cdot \\
a_{m1} & a_{m2} & \ldots & a_{mn} & 0 & 0 & \ldots & 0 & -c_m \\
-b_1 & -b_2 & \ldots & -b_n & c_1 & c_2 & \ldots & c_m & 0
\end{array}
\right].
\end{array}
$$

Because the game matrix is skew symmetric, one conjectures that the value of the game must be *zero*. This is easily shown: if both players use identical mixed strategies (i.e., put the same probability weight on their jth pure strategy for $j = 1, 2, \cdots, n + m + 1$) the payoff to each is zero. Thus, there is no strategy which will guarantee player 1 a positive return.

Since the value is zero, the mixed strategy

$$[z_1^{(0)}\beta_1, \cdots, z_j^{(0)}\beta_j, \cdots, z_n^{(0)}\beta_n, z_{n+1}^{(0)}\beta_{n+1}, \cdots, z_{n+i}^{(0)}\beta_{n+i}, \cdots,$$
$$z_{n+m}^{(0)}\beta_{n+m}, z_{n+m+1}^{(0)}\beta_{n+m+1}]$$

is minimax for player 2 if and only if it holds player 1 down to 0, that is, if and only if

$$z_k^{(0)} \geqslant 0, \quad \text{for } k = 1, 2, \cdots, n + m + 1, \quad \sum_{k=1}^{n+m+1} z_k^{(0)} = 1,$$

and

(i) $-[z_{n+1}^{(0)}a_{1j} + \cdots + z_{n+i}^{(0)}a_{ij} + \cdots + z_{n+m}^{(0)}a_{mj}] + z_{n+m+1}^{(0)}b_j$
$$\leqslant 0, j = 1, \cdots, n,$$

(ii) $[z_1^{(0)}a_{i1} + \cdots + z_j^{(0)}a_{ij} + \cdots + z_n^{(0)}a_{in}] - z_{n+m+1}^{(0)}c_i \leqslant 0,$
$$i = 1, \cdots, m,$$

(iii) $-[z_1^{(0)}b_1 + \cdots + z_n^{(0)}b_n] + [z_{n+1}^{(0)}c_1 + \cdots + z_{n+m}^{(0)}c_m] \leqslant 0.$

The principal theorem of linear programing, as we have stated it, contains five assertions. The first two are easy and we will assume them proved (cf. the proofs on p. 414). In establishing the other three assertions, there are two cases to consider.

Case 1. There exists a minimax strategy for player 2 with $z_{n+m+1}^{(0)} > 0$. Dividing each inequality of (i), (ii), and (iii) by $z_{n+m+1}^{(0)}$ and denoting

$$z_{n+i}^{(0)}/z_{n+m+1}^{(0)} \text{ by } u_i^{(0)}, \quad \text{for } i = 1, 2, \cdots, m,$$

$$z_j^{(0)}/z_{n+m+1}^{(0)} \text{ by } w_j^{(0)}, \quad \text{for } j = 1, 2, \cdots, n,$$

we find that

$$\mathbf{u}^{(0)} = (u_1^{(0)}, u_2^{(0)}, \cdots, u_m^{(0)}) \text{ belongs to } U,$$

$$\mathbf{w}^{(0)} = (w_1^{(0)}, w_2^{(0)}, \cdots, w_n^{(0)}) \text{ belongs to } W,$$

and

$$w_1^{(0)}b_1 + \cdots + w_n^{(0)}b_n \geqslant u_1^{(0)}c_1 + \cdots + u_m^{(0)}c_m.$$

From assertion 1 of the principal theorem of linear programing and the above inequality we conclude that $(\mathbf{u}^{(0)}, \mathbf{w}^{(0)})$ is a solution of the symmetric problem, and therefore by assertion 2 of this same theorem $\mathbf{u}^{(0)}$ and $\mathbf{w}^{(0)}$ are the solutions of the maximizing and minimizing problems respectively.

Case 2. There does not exist a minimax strategy for player 2 with $z_{n+m+1}^{(0)} > 0$. For this case we will first show three things:

(a) Either U or W is empty.

(b) If W is non-empty, then the index

$$w_1 b_1 + w_2 b_2 + \cdots + w_n b_n,$$

where $\mathbf{w}$ is in W, can be made arbitrarily large (i.e., the maximization problem has no solution).

(c) If U is non-empty, then the index

$$u_1 c_1 + u_2 c_2 + \cdots + u_m c_m,$$

where $\mathbf{u}$ is in U, can be made arbitrarily small (i.e., the minimization problem has no solution).

Once a, b, and c, are demonstrated, the three remaining parts of the principal theorem follow easily.   For, if U and W are both non-empty, or, if solutions exist to either the maximizing or minimizing problems, then case 2 does not hold; but when case 1 holds there is a minimax solution of the game which yields solutions to all three versions of the linear-programing problem.

Before we establish assertions a, b, and c we will prove three preliminary remarks which are valid for case 2.

i. The crux of the proof depends upon the following assertion about the game with the payoff matrix given on p. 419.   If *all* minimax strategies of player 2 yield a return of exactly zero against α_{n+m+1}, then player 1 has a maximin strategy which puts positive weight on α_{n+m+1}.   But, by the symmetry of the problem, this would mean that player 2 has a minimax strategy which puts positive weight on β_{n+m+1} and, under the assumption of case 2, this cannot be.   Consequently there is a minimax strategy $\mathbf{z}^{(0)}$ for 2 which gives 1 a return *less* than zero against α_{n+m+1}.   This means

$$- \sum_{j=1}^{n} z_j^{(0)} b_j + \sum_{i=1}^{m} z_{n+i}^{(0)} c_i < 0,$$

i.e.,

$$\sum_{i=1}^{m} z_{n+i}^{(0)} c_i < \sum_{j=1}^{n} z_j^{(0)} b_j.$$

(ii) If there exists a $\mathbf{w}'$ in W (i.e., W is non-empty) then for $\mathbf{z}^{(0)}$ of (i) we show

$$\sum_{j=1}^{n} z_j^{(0)} b_j > 0.$$

We establish this by first observing

$$\underset{(1)}{\sum_{j=1}^{n} z_j^{(0)} b_j} > \underset{(2)}{\sum_{i=1}^{m} z_{n+i}^{(0)} c_i} \geqslant \sum_{i=1}^{m} z_{n+i}^{(0)} \left(\sum_{j=1}^{n} a_{ij} w_j' \right) = \underset{(3)}{\sum_{j=1}^{n} w_j' \left(\sum_{i=1}^{m} z_{n+i}^{(0)} a_{ij} \right)},$$

where (1) follows from (i) above, (2) from the requirement that $\mathbf{w}'$ is in W, and (3) by interchanging summation signs.   Now, since

$$\sum_{i=1}^{m} z_{n+i}^{(0)} a_{ij} \geqslant 0, \quad \text{for all } j,$$

by the inequalities (i) on p. 420 (remember $z_{n+m+1}^{(0)} = 0$), assertion (ii) follows.

iii. If there exists a $\mathbf{u}'$ in U (i.e., U is non-empty) then for $\mathbf{z}^{(0)}$ of (i) we show

$$\sum_{j=1}^{n} z_j^{(0)} b_j \leqslant 0.$$

Observe that,

$$\sum_{j=1}^{n} z_j^{(0)} b_j \leqslant \sum_{i=1}^{n} z_j^{(0)} \left(\sum_{i=1}^{m} a_{ij} u_i' \right) = \sum_{i=1}^{m} u_i' \left(\sum_{j=1}^{n} z_j^{(0)} a_{ij} \right),$$

where the inequality follows from the requirement that $\mathbf{u}'$ is in U and the equality follows from a summation interchange.   Since

$$\sum_{j=1}^{n} z_j^{(0)} a_{ij} \leqslant 0, \quad \text{for all } i,$$

by the inequalities (ii) on p. 420 (remember $z_{m+n+1}^{(0)} = 0$) assertion (iii) follows.

Now back to assertions a, b, and c.

Assertion a, that U and W cannot both be non-empty, follows because (ii) and (iii) are then in contradiction.

Assertion b, that even if W is non-empty no maximum exists, is proved as follows [*note:* The $\mathbf{z}^{(0)}$ used in this proof is the same $\mathbf{z}^{(0)}$ used in preliminary remarks, (i), (ii), and (iii)]:

If $\mathbf{w}' = (w_1', \cdots, w_j', \cdots, w_n')$ lies in W, then so does

$$(w_1' + \lambda z_1^{(0)}, \cdots, w_j' + \lambda z_j^{(0)}, \cdots, w_n' + \lambda z_n^{(0)}) \quad \text{for } \lambda > 0,$$

since

$$\sum_{j=1}^{n} a_{ij}(w_j' + \lambda z_j^{(0)}) = \sum_{j=1}^{n} a_{ij} w_j' + \lambda \sum_{j=1}^{n} a_{ij} z_j^{(0)} \geqslant c_i + \lambda \cdot 0 = c_i$$

for all i.   But the index for this point is

$$\sum_{j=1}^{n} b_j(w_j' + \lambda z_j^{(0)}) = \sum_{j=1}^{n} b_j w_j' + \lambda \left(\sum_{j=1}^{n} b_j z_j^{(0)} \right),$$

and since by (ii) of p. 421,

$$\sum_{j=1}^{n} b_j z_j^{(0)} > 0,$$

this index can be made arbitrarily large by making λ large enough.

Assertion c, that even if U is non-empty no minimum exists, is proved similarly. It requires the dual of assertion (ii), namely: If there exists a $\mathbf{u}'$ in U then

$$\sum_{i=1}^{m} z_{n+i}^{(0)} c_i < 0.$$

This completes the demonstration.

A much simpler reduction of the linear-programing problem to a game can be given provided all the components of $\mathbf{b}$, $\mathbf{c}$, and the matrix $[a_{ij}]$ are positive. In the minimization problem, given on p. 412, make the change of variables $u_i' = c_i u_i$ $(i = 1, 2, \cdots, m)$ and $a_{ij}' = a_{ij}/(b_j c_i)$. The problem then reduces to player 1's problem on p. 409. Observe that the inequality

$$\sum_i u_i a_{ij} \geqslant b_j$$

becomes, on division by b_j,

$$\sum_i u_i' a_{ij}' \geqslant 1.$$

Similarly the maximization problem on p. 412 reduces to player 2's problem on p. 410 if we let $w_j' = b_j w_j$ and $a_{ij}' = a_{ij}/(b_j c_i)$. Thus we are led to the study of the game $\{A, B, M'\}$ where

$A = \{\alpha_1, \cdots, \alpha_m\}$, $B = \{\beta_1, \cdots, \beta_n\}$, $M'(\alpha_i, \beta_j) = a_{ij}' = a_{ij}/(b_j c_i)$.

If $\mathbf{x}^{(0)} = (x_1^{(0)}, \cdots, x_m^{(0)})$ is maximin for player 1, if $\mathbf{y}^{(0)} = (y_1^{(0)}, \cdots, y_n^{(0)})$ is minimax for 2, and if v is the value of this game, then $\mathbf{u}^{(0)} = (u_1^{(0)}, \cdots, u_m^{(0)})$, where

$$u_i^{(0)} = x_i^{(0)}/(c_i v), \qquad i = 1, 2, \cdots, m,$$

is a solution of the maximization problem; and $\mathbf{w}^{(0)} = (w_1^{(0)}, \cdots, w_n^{(0)})$, where

$$w_j^{(0)} = y_j^{(0)}/(b_j v), \qquad j = 1, 2, \cdots, n,$$

is a solution of the minimization problem.

In this case the equivalent game problem is m by n instead of $(m + n + 1)$ by $(m + n + 1)$. However, as we shall see in the next appendix, for certain computational procedures, it is advantageous to have the game in symmetric form.

appendix 6

SOLVING TWO-PERSON
ZERO-SUM GAMES

A6.1 INTRODUCTION

To render this appendix relatively self-contained, we repeat (with minor modifications) some observations made earlier in the body of the book (see section 4.12).

▶ Now that we know that all two-person zero-sum games with a finite number of pure strategies have solutions, our attention turns to methods of finding these solutions. Here, at best, the story is quite discouraging. Although several methods are known for solving games, these algorithms usually require a fantastic amount of work, at least for games which purport to be realistic replicas of actual conflicts of interest. The realism is achieved only at the expense of introducing a fabulous number of pure strategies. One might hope that cases involving such a large number of strategies could be idealized by a continuous model and that refined analytical methods could be brought to bear on the idealization. In part, this is possible as we shall see in our discussion of infinite games in the next appendix. However, in all honesty, we must admit that the number of existing techniques for the infinite cases is small, and even in examples that have their "natural" description in the infinite case the usual hope is to reduce them to approximate finite games (cf. discussion on polynomial and polynomial-like games in Appendix 7). There are two saving features in the solution of many games which arise in practice. First, although a game may involve a huge number of strategies, one can often use the practical context to help reduce the model down to its bare essentials by discarding many of the inadmissible strategies. Second, the context of the game often leads one to shrewd guesses about solutions or, in iterative procedures, about intelligent starting points.

We believe, and probably most of our colleagues will agree, that many important and interesting games will *never* be solved. This does not imply that game theory will never contribute anything to these realistic games. Often, a *modus operandi* for a complicated case is to consider an auxiliary game which is motivated and related to the original one in such a way that many of the important phenomena of the original are retained while the auxiliary remains solvable. From the solution of the auxiliary game one speculates informally how the results are modified in the original game. Thus, for example, there are simplified variants of both poker and bridge in the literature. Such studies are in much the same spirit as economic analyses of idealized *Robinson Crusoe* or *Swiss Family Robinson* economies which, by means of a lot of hand waving, are used "to explain" economic phenomena and to reach policy decisions concerning the economy at large. This is dangerous, yes! Yet it is quite stimulating to our creative intuitions and often helpful in purely literary, pseudological (not said deprecatingly, but rather pragmatically) theorizing. ◀

A6.2 TRIAL AND ERROR

The most common method of solving games which arise in practice is to guess at the solution and then to check that the proposed strategies are in equilibrium. Since a pair of strategies provides a solution to the problem, if and only if they are in equilibrium, the method is foolproof—provided, of course, that one either can guess phenomenally well or is undaunted by failure. When one comes across a statement to the effect that " . . . if we try the following pair of strategies, we see they solve the game," generally it is not known whether the solution was arrived at by brilliant mathematical insight or by sheer hack work. Usually, it is a combination of the two.

There are a few guide posts, however, for one who indulges in this guessing game. Let us suppose that $(\mathbf{x}^{(0)}, \mathbf{y}^{(0)}, v)$ represents a solution to a given game ($\mathbf{x}^{(0)}$ is player 1's maximin strategy, $\mathbf{y}^{(0)}$ is player 2's minimax strategy, and v is the value of the game). Since $M(\mathbf{x}^{(0)}, \beta_j) \geqslant v$, for $j = 1, 2, \cdots, n$, the strategy $\mathbf{y}^{(0)}$ can utilize with positive probability only those β_j for which $M(\mathbf{x}^{(0)}, \beta_j) = v$. For if $\mathbf{y}^{(0)}$ uses a β_j with positive probability for which $M(\mathbf{x}^{(0)}, \beta_j) > v$, then $M(\mathbf{x}^{(0)}, \mathbf{y}^{(0)})$ would have to be greater than v, which contradicts the optimality of $\mathbf{y}^{(0)}$.[1] Hence, if $\mathbf{y}^{(0)}$ uses each $\beta_j, j = 1, 2, \cdots, n$, with positive probability, then of necessity $\mathbf{x}^{(0)}$ must have the property that $M(\mathbf{x}^{(0)}, \beta_j) = v$, for $j = 1, 2, \cdots, n$. In many games, $\mathbf{y}^{(0)}$ uses each β_j with positive probability, so it

---

[1] This follows from the fact that $M(\mathbf{x}^{(0)}, \mathbf{y}^{(0)})$ is an *average* of the numbers $M(\mathbf{x}^{(0)}, \beta_j)$, all of which are not less than v. So if any of them is greater than v and if it is weighted positively, then the weighted average must also be greater than v.

is usually wise for the guesser to seek an $\mathbf{x}^*$, say, such that $M(\mathbf{x}^*, \beta_j)$ does not depend upon j. (Such an $\mathbf{x}^*$ can be found by solving a system of simultaneous equations.) The procedures for finding a strategy which equalizes a player's expected payoff over all of his opponent's strategies have been carried to artistic zeniths in some modern work on statistical decision theory. Even if an $\mathbf{x}^*$ is found such that $M(\mathbf{x}^*, \beta_j)$ is independent of j, one must still verify that $\mathbf{x}^*$ is maximin. Finding a strategy $\mathbf{y}^*$ such that $\mathbf{x}^*$ is player 1's best response against $\mathbf{y}^*$ is a foolproof check, since, if $\mathbf{x}^*$ equalizes the possible returns to player 2, then certainly $\mathbf{y}^*$ is good against $\mathbf{x}^*$ (as are all strategies for player 2). Thus the pair is in equilibrium since $\mathbf{x}^*$ and $\mathbf{y}^*$ are good against each other. How to find $\mathbf{y}^*$ is left as an exercise in mathematical ingenuity only for the expert and the lucky.

A6.3 CHECKING ALL CRITICAL POINTS

In Appendix 3, we introduced a geometrical model which is useful in solving games. There we considered diagrams such as those shown in Figs. 1 and 2. We are mostly confronted with games where player 1 has

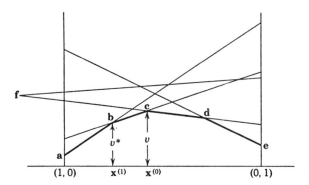

FIG. 1.   Diagram for a game where

$$A = \{\alpha_1, \alpha_2\} \text{ and } B = \{\beta_1, \beta_2, \beta_3, \beta_4, \beta_5\}.$$

more than three pure strategies, so we cannot usually depend upon having a two- or three-dimensional pictorial guide. Nevertheless, it is convenient to explain what we are about in terms of these simple cases.

One can see in these cases, and "imagine" in more complicated ones, what the minimum function (player 1's security level) looks like. It is a concave polygonal figure composed of pieces of "lines" ("planes" for $m = 3$, "hyperplanes" for $m > 3$), each generated either by a β_j or by a bounding (vertical) line (plane or hyperplane). A typical point on this minimum function is characterized by a pair of coordinates $(\mathbf{x}, z)$, say, where $\mathbf{x}$ is a m-tuple $(x_1, x_2, \cdot\cdot\cdot, x_m)$, such that $x_i \geqslant 0$ and $\sum_i x_i = 1$, and where z is the height of the point from the base plane. The set of points $(\mathbf{x}, z)$ of this minimum surface for which z is a maximum (i.e., for which $z = v$) can be "visualized" as a convex polygonal (polyhedron) set which in turn is characterized by its extreme points. For example, the

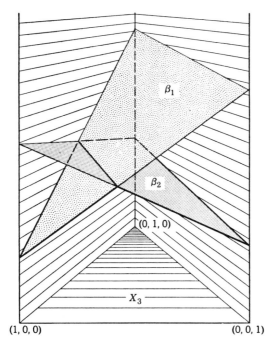

FIG. 2. Diagram for a game where

$$A = \{\alpha_1, \alpha_2, \alpha_3\} \text{ and } B = \{\beta_1, \beta_2\}.$$

For purposes of clarity, the front face $(x_2 = 0)$ of the game cylinder has been removed.

set shown in Fig. 3 is a convex polyhedron with extreme points, **a, b, c, d, e, f.** We also note that, when $m = 2$, an extreme point arises as the intersection of two lines; when $m = 3$, it arises as the intersection of 3 planes; and in general it arises as the intersection of m hyperplanes. Hence, a plan of attack is to find all the points, known as *critical points*, which arise from an intersection of *any* m hyperplanes. (*Note:* We also have to consider the bounding hyperplanes. Thus, if player 1 has m strategies and player 2 has n strategies, there is one hyperplane for each pure strategy of player 2 and m bounding hyperplanes, or a total of $m + n$ hyperplanes.)

Finding the critical point associated with any given set of m hyperplanes is equivalent to solving m linear equations, and thus it is equivalent to inverting a square matrix of order m.   Once we have found all the critical points, we can eliminate many of them as intersections outside our field of interest (e.g., f in Fig. 1 above)—that is, one or more components of the $(x_1, x_2, \cdots, x_m)$ part of the point are negative.   Referring to Fig. 1, we see that

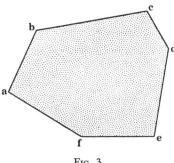

FIG. 3

the critical point c has coordinates $\mathbf{x}^{(0)}$ and v, and that $M(\mathbf{x}^{(0)}, \beta_j) \geqslant v$, for all j.   Similarly, the critical point b, with coordinates $(\mathbf{x}^{(1)}, v^*)$, is such that $M(\mathbf{x}^{(1)}, \beta_j) \geqslant v^*$, for all j.   In this manner, we can verify that only the points of intersection a, b, c, d, and e are on the minimum function, and by enumeration we see that c, i.e., $(\mathbf{x}^{(0)}, v)$, gives player 1's maximin strategy and the value of the game.

Kaplansky [1945], who made the first formal contribution to finding solutions of zero-sum games, presented an inductive procedure to determine the value of a game in a finite number of steps.   Later Shapley and

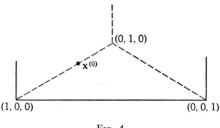

FIG. 4

Snow [1950] gave a constructive procedure to determine all solutions (i.e., all minimax and maximin strategies) of a game; for a very lucid account of this procedure see Kuhn [1952].   We shall describe it only for $m = 3$, but this case illustrates the basic idea.   Suppose that $v^{(0)}$ is the value of the game and that $\mathbf{x}^{(0)} = (x_1^{(0)}, x_2^{(0)}, 0)$ is an extreme maximin solution, (see Fig. 4).   Thus $(\mathbf{x}^{(0)}, v^{(0)})$ is a critical point on the bounding hyperplane

$x_3 = 0$. Restricting ourselves to this bounding hyperplane, the point $(x_1^{(0)}, x_2^{(0)}; v^{(0)})$ is the intersection of two β_j planes, say β_{j_1} and β_{j_2}. Thus $(x_1^{(0)}, x_2^{(0)}, v^{(0)})$ is the *unique* solution of the system of equations

$$a_{1j_1}x_1 + a_{2j_1}x_2 = v$$

$$a_{1j_2}x_1 + a_{2j_2}x_2 = v$$

$$x_1 + x_2 = 1.$$

If a solution exists to this system, it can be simply expressed in terms of the determinant

$$\begin{vmatrix} a_{1j_1} & a_{2j_1} \\ a_{1j_2} & a_{2j_2} \end{vmatrix}$$

and its cofactors.

To take another example, an extreme minimax strategy $\mathbf{x}^{(0)} = (x_1^{(0)}, x_2^{(0)}, x_3^{(0)})$, where $x_i^{(0)} > 0$, $i = 1, 2$, and 3, is detected as part of the unique solution to the equations

$$a_{1j_1}x_1 + a_{2j_1}x_2 + a_{3j_1}x_3 = v$$

$$a_{1j_2}x_1 + a_{2j_2}x_2 + a_{3j_2}x_3 = v$$

$$a_{1j_3}x_1 + a_{2j_3}x_2 + a_{3j_3}x_3 = v$$

$$x_1 + x_2 + x_3 = 1$$

for suitable indices j_1, j_2, and j_3 from the set $\{1, 2, \cdots, n\}$ of indices. Again, the unique solution (if it exists) to this system for any specific j_1, j_2, j_3 can be expressed in terms of the determinant

$$\begin{vmatrix} a_{1j_1} & a_{2j_1} & a_{3j_1} \\ a_{1j_2} & a_{2j_2} & a_{3j_2} \\ a_{1j_3} & a_{2j_3} & a_{3j_3} \end{vmatrix}$$

and its cofactors. The extreme minimax strategies $\mathbf{y}^{(0)}$ are found in a similar way.

The algorithm consists, therefore, of isolating *all* square submatrices of the payoff matrix $[a_{ij}]$ and computing for each such submatrix the *potential* extreme maximin strategy, extreme minimax strategy, and value. All of these are expressible in terms of the determinant and cofactors of the submatrices. From the set of potential candidates it is then a simple matter to check which fulfill the prerequisite equilibrium requirements for a solution.

The Shapley-Snow procedure also applies to linear programs; see, for instance, Goldman and Tucker [1956]. However, this procedure is not currently competitive as a computational algorithm, even though it is of extreme technical interest.

A6.4 THE DOUBLE DESCRIPTION METHOD

Motzkin, Raiffa, Thompson, and Thrall [1953] have suggested a computational method, called the double description method, for determining both the value and all the solutions of a two-person zero-sum game with a finite number of pure strategies.   Their procedure is also applicable to linear-programing problems.   In explaining these results, we shall use the geometry discussed in Appendix 3.

In the double description method the *minimum function* is built up by introducing the hyperplanes associated with the β_j's (called β_j planes) one at a time.   First, we consider the minimum function, called the β_1, β_2 minimum function, generated by the β_1 plane and β_2 plane.   Next, we introduce the β_3 plane and thus generate the β_1, β_2, β_3 minimum function.

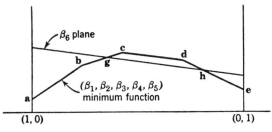

FIG. 5

We continue in this fashion until the β_1, β_2, $\cdots$, β_n minimum function is generated, from which the maximin strategy and the value of the game can be read off.

To be specific, consider a case where $A = \{\alpha_1, \alpha_2\}$.   Suppose we have carried the procedure to the point where the β_1, β_2, $\cdots$, β_5 minimum function is known, and suppose that it is the function shown in Fig. 1 above.   This minimum function is, therefore, characterized by the points **a, b, c, d, e**.   Now, we introduce the β_6 plane (actually, line in this case).   First, we compute whether the β_6 plane lies on, above, or below the critical points of the β_1, β_2, $\cdots$, β_5 minimum function.   If the β_6 plane lies above all the critical points, then it is not a part of the β_1, $\cdots$, β_6 minimum function.   However, if it lies below at least one critical point of the β_1, $\cdots$, β_5 minimum function, then the β_6 plane must be an integral part of the β_1, $\cdots$, β_6 minimum function.   Hence, new critical points must be introduced and at least one old one must be discarded to characterize the new minimum function.   To be specific, suppose that the β_6 plane is above **b** and below **c**, as in Fig. 5.   A new critical point **g** is introduced on the line segment joining **b** and **c**, so the point **c** will certainly be discarded.   To find the exact location of the new point **g**, we merely find where the line segment joining **b** and **c** pierces the β_6 plane. This simple calculation can be easily mechanized, even in higher dimensions.

There is, however, one non-trivial complication. Suppose, for example, that the β_6 plane is above **b** and below **d**. The point where the line segment connecting **b** and **d** intersects the β_6 plane is extraneous, since that line segment is not a part of the $\beta_1, \cdots, \beta_5$ minimum function. In Fig. 5 the $\beta_1, \cdots, \beta_5$ minimum function is characterized by points **a, b, c, d,** and **e**, whereas the $\beta_1, \cdots, \beta_6$ minimum function is characterized by points **a, b, g, h,** and **e**. For $A = \{\alpha_1, \alpha_2\}$, it is extremely simple to check whether the line segment joining two critical points of the minimum function is in fact a part of the minimum function. Such pairs of critical points are said to be *adjacent*. But, if player 1 has a large number of pure strategies, there is no picture to help visualize the procedure, and so it is difficult to know whether or not two critical points of a minimum function are adjacent. In m-space, two critical points are *adjacent* if and only if there are $m - 1$ planes common to their characterizations.

This difficulty can be overcome by a bookkeeping scheme. To each critical point of the minimum function (for the stage under consideration) associate not only its coordinates, but also the β_j planes which pass through the point (and thus implicitly characterize it). This *double description* of the point (coordinates plus a recording of the planes which define it) enables one to keep track of which critical points are adjacent, and so precludes introducing false critical points.

To illustrate the next point specifically, let us suppose that player 2 has ten pure strategies, that we have determined the critical points of the $\beta_1, \beta_2, \cdots, \beta_5$ minimum function, and that the critical point $(\mathbf{x}^*, v^*)$ is highest on this minimum function. We can check whether any of the planes β_6 to β_{10} lie below $(\mathbf{x}^*, v^*)$. If not, then $\mathbf{x}^*$ is maximin and v^* is the value of the game. So, in general, *it is not always necessary to compute the entire minimum function to find the maximin strategy and the value of the game.*

This raises another important issue: since the labeling of player 2's pure strategies is irrelevant, in what order shall we consider the planes in this sequential procedure so as to minimize the computational effort? One suggestion is this: at any stage find the maximum critical point of the corresponding minimum function and introduce the β_j plane which is furthest below it. This plane is easy to ascertain, and, if there is no plane below the maximum critical point, we might as well stop for we have what we are looking for.

For some purposes it may be extremely advantageous to know, in whole or in part, the final minimum function. For example, suppose $\mathbf{x}^{(0)}$ is the unique maximin strategy of a game with value v, and suppose it has the special property that $M(\mathbf{x}^{(0)}, \mathbf{y}) = v$ for all $\mathbf{y}$. Thus, $\mathbf{x}^{(0)}$ equalizes player 2's opportunities, and so, if player 1 uses $\mathbf{x}^{(0)}$, he will only get v even when player 2 does not play minimax. If player 1 suspects that player 2 is not going to play a minimax strategy, then he could try for more than v. This is only possible, however, if he avoids playing maximin. But player 1, although willing to take a risk to get more than v, may not want to expose himself to excessive dangers. For example, he might set up a safe value $v^* < v$ and decide to exploit 2's stupidity or non-conformity only to the extent of using strategies which have a security level of at least v^*. To do this, 1 can refer to his minimum function to determine the set of $\mathbf{x}$ values for which the minimum function is above v^*; his deliberations will then be confined to that set. If player 2 is not a strictly opposing player, setting up a security level $v^* < v$ might be quite realistic. See the discussion in Chapter 13 of the Hodges-Lehmann criterion and also Hodges and Lehmann [1952].

Given the minimum function of a game, which presents an analysis from player 1's point of view, techniques are known which shorten considerably the parallel analysis for player 2.

A6.5 THE SIMPLEX METHOD

The simplex method is a computational technique, devised by Dantzig [1951 b], to solve linear-programing problems. Since two-person zero-sum games can be reduced to programing problems (of a very special form), the simplex method also yields a computational procedure for solving games.

The Simplex problem. Let U' be the set of all m-tuples $\mathbf{u} = (u_1, u_2, \cdots, u_m)$ such that

(i) $u_i \geqslant 0$, for $i = 1, 2, \cdots, m$.

(ii) $u_1 a_{1j} + u_2 a_{2j} + \cdots + u_m a_{mj} = b_j$, for $j = 1, 2, \cdots, n$.

Find those $\mathbf{u}$ belonging to U' which minimize the index

$$c_1 u_1 + c_2 u_2 + \cdots + c_m u_m.$$

The simplex problem is a slight modification of the minimization problem of section A5.2. The inequalities (ii) of the minimization problem are replaced by exact equalities. At first glance, one might think that this restriction would result in a loss of generality, but this is not so because an inequality can always be changed into an equality by introducing dummy variables. For example, the inequality

$$u_1 a_{1j} + \cdots + u_m a_{mj} \geqslant b_j,$$

is changed to an equality by introducing the non-negative quantity z_j, where

$$u_1 a_{1j} + \cdots + u_m a_{mj} - z_j = b_j.$$

A simplex problem also has its dual maximization problem, but there is no longer a nice symmetry between the dual problems.

The dual of the simplex problem. Let W' be the set of all n-tuples $\mathbf{w} = (w_1, \cdots, w_n)$ such that:

$$w_1 a_{i1} + w_2 a_{i2} + \cdots + w_n a_{in} \leqslant c_i, \qquad \text{for } i = 1, 2, \cdots, m.$$

Find those $\mathbf{w}$ belonging to W' which maximize the index

$$w_1 b_1 + \cdots + w_n b_n.$$

Note that the dual problem does *not* require the w_j's to be non-negative! Also, the m constraints are given by inequalities, not by equalities.

The duality theorem for the simplex problem asserts: A solution of the simplex problem exists if and only if a solution to its dual problem exists, in which case,

$$\min_{\mathbf{u} \text{ in } U'} \sum_{i=1}^{m} c_i u_i = \max_{\mathbf{w} \text{ in } W'} \sum_{j=1}^{n} b_j w_j.$$

If $\mathbf{u}$ belongs to U', we shall say that $\mathbf{u}$ is a *feasible solution*. If $\mathbf{u} = (u_1, u_2, \cdots, u_m)$ is such that $u_i > 0$, then we shall say that $\mathbf{u}$ *uses* coordinate i. Finally, $\mathbf{u}$ will be called a *basic feasible* solution if $\mathbf{u}$ belongs to U' and if $\mathbf{u}$ uses at most n coordinates, where n refers to the n equations of (ii) in the statement of the simplex problem. It can be shown that:

1. If a feasible solution exists (which it certainly does for the linear-programing problem derived from a zero-sum game, but need not in general), then a basic feasible solution exists.

2. If a solution to the simplex problem exists, that is, if there is a minimal feasible solution (cf. the parenthetical remark in 1), then a minimal feasible solution exists which is also basic.

One final definition: two basic feasible solutions are said to be *adjacent* if there are $n - 1$ coordinates which they both use. Hence, if a basic feasible solution is modified by eliminating one used coordinate and by using a new coordinate, then the modification is a basic feasible solution adjacent to the original solution.

Two assumptions will be made about the simplex problem. The first, known as a non-degeneracy assumption,[2] implies among other things that

[2] The non-degeneracy assumption actually asserts that all n by n submatrices of the augmented coefficient matrix (n by $m + 1$) of eqs. (ii) are non-singular. If this assumption is not fulfilled, mathematical difficulties are encountered in the formal proofs which may be surmounted by perturbating the coefficients slightly so that the assumption is met.

two feasible solutions which use the same n coordinates are in fact identical. Stated alternatively, we assume that stipulating a specific set of n coordinates which are to be used uniquely determines the solution to eqs. (ii).

Second, we shall assume that a minimal feasible solution does exist. This is always so for the problem associated with a game. Actually, for any linear-programing problem, the simplex method detects in due course whether or not a minimal feasible solution exists.

The simplex technique establishes the following:

1. How to find a basic feasible solution.

2. Given a basic feasible solution, how to find an adjacent basic feasible solution with a smaller index—provided one exists.

3. If no basic feasible solution adjacent to a basic feasible solution $u^{(0)}$ has a smaller index than $u^{(0)}$, then $u^{(0)}$ is a minimal basic feasible solution. In other words, a local minimum is always a global minimum.

Knowing this, the procedure is now straightforward. Beginning with any basic feasible solution, we proceed along some path from one adjacent basic feasible solution to the next in such a manner as to decrease the index at each stage. Since, by the non-degeneracy assumption, there are only a finite number of basic feasible solutions, the process must terminate at a minimal feasible solution.

In a linear-programing problem arising from a game problem, it is always easy to find a basic feasible solution; however, in a general linear-programing problem, finding a basic feasible solution is non-trivial. It is accomplished by an iterative procedure which leads from a feasible to a basic feasible solution and which is analogous to the procedure outlined above for going from a basic to a minimum basic feasible solution.

A point of crucial concern is the speed with which this iterative procedure converges. This depends both upon the initial basic feasible solution chosen and upon the particular path taken. The path is not unique since several solutions adjacent to a given basic feasible solution may each have a lower index. There are several *ad hoc* rules for selecting among the alternatives which, in empirical tests, have proved efficient, but mathematical proofs of their optimality are still lacking.

In practice, approximately n iterates are required to go from a non-minimal to a minimal basic feasible solution. Computational techniques exist which require $2n^2 + n + m$ multiplications per step, or, roughly, $2n^3 + n^2 + mn$ multiplications in all. (With modern machine techniques, the number of multiplications required gives a good indication of the total computation time.) If a basic feasible solution is not immediately apparent, the total number of multiplications may jump to $5n^3 + 2n^2 + 2mn$.

The final part of Cowles Commission Monograph 13, on activity analysis (see Koopmans [1951]) is devoted to computational procedures in linear programing.   There, following Dantzig's presentation of the theory of the simplex technique, Dorfman [1951] illustrates its use for a simple explicit game problem.

A6.6 A GEOMETRIC INTERPRETATION OF THE SIMPLEX AND DUAL SIMPLEX PROCEDURES

Lemke [1954] offers neat geometrical insights into Dantzig's technique, as well as into his own variation of the simplex technique.   These can be

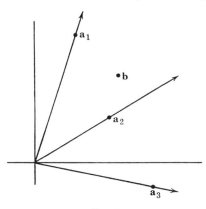

FIG. 6

presented graphically when $n = 2$, in which case the equalities (ii) of the simplex problem take the form:

$$a_{11}u_1 + a_{21}u_2 + \cdots + a_{m1}u_m = b_1$$

$$a_{12}u_1 + a_{22}u_2 + \cdots + a_{m2}u_m = b_2.$$

Denote the points (a_{i1}, a_{i2}), for $i = 1, 2, \cdots, m$, by $\mathbf{a}_i$; then a basic feasible solution exists if and only if the point $\mathbf{b} = (b_1, b_2)$ is a linear combination, with non-negative weights of the points $\mathbf{a}_1, \mathbf{a}_2, \cdots, \mathbf{a}_m$, i.e.,

$$\mathbf{b} = u_1\mathbf{a}_1 + u_2\mathbf{a}_2 + \cdots + u_m\mathbf{a}_m,$$

where $u_i \geqslant 0$, for $i = 1, 2, \cdots, m$.   Geometrically, this means that a feasible solution exists if and only if the point $\mathbf{b}$ lies in the convex polyhedral cone generated by the points $\mathbf{a}_1, \mathbf{a}_2, \cdots, \mathbf{a}_m$.   For example, feasible solutions exist for the case shown in Fig. 6, but not for that shown in Fig. 7.   Indeed, in Fig. 6 a basic feasible solution exists which uses only coordinates 1 and 2 (i.e., points $\mathbf{a}_1$ and $\mathbf{a}_2$) and another which uses only 1 and 3; however, none exist using only 2 and 3, since $\mathbf{b}$ does not belong to the cone generated by $\mathbf{a}_2$ and $\mathbf{a}_3$.

Let us next look into the geometry of the dual to the simplex problem. Each of the inequalities

$$w_1 a_{i1} + w_2 a_{i2} \leqslant c_i, \qquad \text{for } i = 1, 2, \cdots, m,$$

can be depicted by drawing the line $w_1 a_{i1} + w_2 a_{i2} = c_i$ in 2-space, as in Fig. 8. The points (w_1, w_2), which satisfy the ith inequality, lie on or below this line. The set of points satisfying the ith inequality forms a

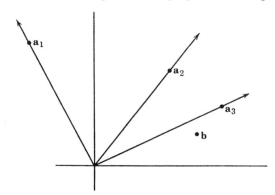

Fig. 7

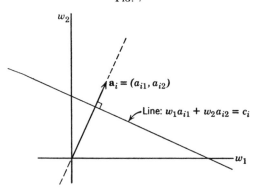

Fig. 8

half-space, and the set of points satisfying all m of the inequalities is the intersection of (i.e., points common to) all m half-spaces.

The normal to the ith line has direction numbers a_{i1} and a_{i2}. Thus, we may loosely identify the point $\mathbf{a}_i$ with the normal to this line.

Suppose we examine the simplex problem with $n = 2$ and $m = 6$ and having the geometry shown in Fig. 9. The convex region W' (speckled on the diagram) is characterized by its extreme points, $\mathbf{g}$, $\mathbf{h}$, $\mathbf{i}$, $\mathbf{j}$, and $\mathbf{k}$. According to the indicated direction of increasing indices, $\mathbf{g} = (w_1^{(0)},$

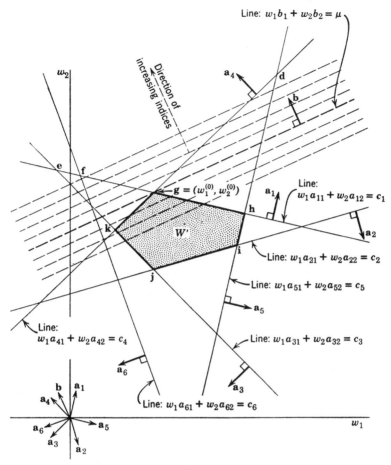

Fig. 9

$w_2^{(0)})$ is the solution of the dual of the simplex Problem, i.e., there exists a number μ such that

$$w_1^{(0)}b_1 + w_2^{(0)}b_2 = \mu,$$

and

$$w_1 b_1 + w_2 b_2 \leqslant \mu, \qquad \text{for all } (w_1, w_2) \text{ in } W'.$$

We observe in Fig. 9 that the point **g** is characterized as the intersection of the lines associated with $\mathbf{a}_1$ and $\mathbf{a}_4$ and that **b** lies in the cone generated by $\mathbf{a}_1$ and $\mathbf{a}_4$ (i.e., there is a basic feasible solution using coordinates 1 and 4). The other extreme points of W', **h**, **i**, **j**, and **k**, do not have the latter property. Specifically, **h** is characterized by $\mathbf{a}_1$ and $\mathbf{a}_5$ but **b** is *not* in the cone generated by $\mathbf{a}_1$ and $\mathbf{a}_5$; **i** is characterized by $\mathbf{a}_2$ and $\mathbf{a}_5$ but **b** is not in the cone generated by a_2 and a_5; etc. This unique property of the extreme point g of W' is algebraically related to the duality theorem.

The point d is characterized by a_4 and a_5, and b lies in the cone generated by a_4 and a_5. Thus there exists a basic feasible solution using a_4 and a_5, but it is not minimal since d is not in W'. There also exist basic feasible solutions using a_3 and a_1 (characterizing e), using a_6 and a_1 (characterizing f), and using a_4 and a_1 (characterizing g).

The simplex technique might start, for example, with the basic feasible solution which uses a_1 and a_3, i.e., with the point e. Since e does not belong to W', it is not minimal. The technique instructs us to move to an adjacent basic feasible solution with a lower index, i.e., to either f or g. If we go to g, we are done; if we go to f, then one more step takes us to g, since from f the index is only lowered by moving to g.

Lemke suggests a variation of the simplex technique, known as the "dual simplex method," which involves moving from one extreme point of W' to another in a manner that increases the index at each stage. For example, suppose we begain at the extreme point j of W'. The point j is not a solution since the cone generated by a_2 and a_3 (which characterizes j) does not contain b. We move from j in a direction perpendicular either to a_2 or a_3 until we come to the next adjacent extreme point—either i or k. (In m-space, we move in a direction perpendicular to $m - 1$ of the normals characterizing the extreme point.) Suppose we move to k; then we can go either to j or g. Since we have just come from j, we would surely go to g. But suppose we had begun at k; then we would want to make sure to go to g, not to j. This is done as follows: Change the sign of one of the normals characterizing the point, in this case a_3, so that b is in the cone generated by $-a_3$ and a_4. Then, move in the direction orthogonal to the normal whose sign was not changed, i.e., orthogonal to a_4. We see that this takes us to g. Had we been at h, then b is in the cone of $-a_5$ and a_1, so we move orthogonally to a_1 and again move to g. In m dimensions the process is similar. If b is in the cone generated by the normals characterizing our point, we have a minimal basic feasible solution. If not, changing the sign of some normals will yield a cone containing b, and, after deleting one of the normals whose sign was changed, we move in the direction orthogonal to the remaining $m - 1$ normals. The process is repeated until we arrive at a point characterized by a set of normals which, without reversal of signs, generates a cone containing b.

A6.7 DIFFERENTIAL EQUATION SOLUTIONS OF SYMMETRIC GAMES

Brown and von Neumann [1950] have given a proof of the minimax theorem which is " 'constructive' in a sense that lends itself to utilization when actually computing the solutions of specific games. The procedure could be 'mechanized' with relative ease, both for 'digital' and for 'analogy' methods." (Brown and von Neumann [1950] p. 73.)

Their procedure applies to games which are *symmetric* in the sense that the two players have the same number m of pure strategies and $M(\alpha_i, \beta_j) = -M(\alpha_j, \beta_i)$, for $i, j = 1, 2, \cdots, m$. Although the method is not directly applicable to non-symmetric games, it would be if we could symmetrize them, and then interpret the Brown-Neumann solution of the symmetrized games in terms of the original games. Hence, after discussing the procedure for symmetric games, we shall describe two procedures for symmetrizing other games.

The value of a symmetric game is zero, for, if player 1 had a strategy guaranteeing at least $v > 0$, then by adopting the corresponding strategy, player 2 could hold 1 down to at most $-v$, which is a contradiction.

The Brown and von Neumann procedure goes as follows: Begin with an initial (time 0) mixed strategy $[y_1(0), y_2(0), \cdots, y_m(0)]$ for player 2. If it does not hold player 1 down to at most 0, a "force" (to be formalized by a system of differential equations) is exerted on the strategy which tends to bring it "closer" to equilibrium. Thus we conceive of a continuous time path $\mathbf{y}(t) = [y_1(t), y_2(t), \cdots, y_m(t)]$, where $y_i(t) \geqslant 0$ and $\sum_i y_i(t) = 1$, of mixed strategies, which will be constrained so as to move toward a solution. That is, if $v(t)$ denotes the most that player 1 can obtain when player 2 uses $\mathbf{y}(t)$, then we want to constrain $\mathbf{y}(t)$ so that

1. $v(t)$ approaches zero as time t increases indefinitely.

2. If, for any t_0, $\mathbf{y}(t_0)$ guarantees that player 1 gets at most zero, then $\mathbf{y}(t_0)$ is in equilibrium in the sense that there are no forces acting on $\mathbf{y}(t_0)$ to move it.

Our problem is to set up a differential equation whose solution has such properties.

For the strategy pair $[\alpha_i, \mathbf{y}(t)]$, let $u_i(t) = M[\alpha_i, \mathbf{y}(t)]$ (= the return to player 1 corresponding to this pair of strategies). We observe that $\mathbf{y}(t)$ is a minimax strategy if and only if $u_i(t) \leqslant 0$, for $i = 1, 2, \cdots, m$. Define $\phi_i(t) = \max [0, u_i(t)]$. That is, $\phi_i(t)$ is 0 if α_i gives a non-positive return to player 1 when used against $\mathbf{y}(t)$ and it is 1's return when that return is positive. So player 2 wishes to find a strategy $\mathbf{y}(t)$ such that $\phi_i(t)$ is zero (or very small) for all i. *One* index of $\mathbf{y}(t)$'s ability to hold down player 1's payoff is $\phi(t) = \sum_{i=1}^{m} \phi_i(t)$, for, if $\phi(t)$ is very small, $\phi_i(t)$ is small for all i, and conversely. Brown and von Neumann require that $\mathbf{y}(t)$ move along a path dictated by the following set of differential equations:

$$\frac{dy_i(t)}{dt} = \phi_i(t) - \phi(t)y_i(t), \qquad i = 1, 2, \cdots, m.$$

They show that:

(i) $y_i(t) \geqslant 0$, for all i and all t.

(ii) $\displaystyle\sum_{i=1}^{m} y_i(t) = 1$, for all t.

(iii) $\phi_i(t) \leqslant c/(1 + tc)$, where $c = \sqrt{\displaystyle\sum_i \phi_i^2(0)}$.

Properties (i) and (ii) show that the path $\mathbf{y}(t)$ can always be interpreted as a mixed strategy, and (iii) says, among other things, that by using $\mathbf{y}(t)$ player 2 can hold 1 down to at most $c/(1 + tc)$. This quantity approaches zero as t increases indefinitely. The number c reflects the efficacy of the initial guess $\mathbf{y}(0)$. The path $\mathbf{y}(t)$ need not necessarily approach a limit but may oscillate among limiting points, all of which are minimax strategies for player 2. Since the game is symmetric, optimal solutions for player 1 are identical to those for player 2.

Bellman [1953] presents a variant of the Brown and von Neumann differential equation approach to symmetric games; he claims this yields an exponential rate of convergence, which of course is much faster than the rate of $1/t$ of the Brown-von Neumann process; however, a question has been raised and not yet resolved about a crucial step in Bellman's argument.

H. N. Shapiro informs us that he has also obtained some new results on rates of convergence, but these were not available to us in written form at the time of writing.

A6.8 SYMMETRIZATION OF A GAME

Perhaps the simplest symmetrization of a non-symmetric game (A, B, M) is due to von Neumann (cf. Brown and von Neumann [1950]). It has the following verbal description. The opponents, Mr. Jones and Mr. Smith, will be assigned the roles of players 1 and 2 in (A, B, M) according to the toss of a coin: if it turns up heads, Jones plays 1's role and Smith 2's role; if tails, Jones plays 2's role and Smith 1's role. This is just the symmetrization used to make chess a fair game. Certainly, the overall game generated in this manner from (A, B, M) is symmetric, and an optimal solution for Mr. Jones in the auxiliary game must provide optimal solutions for both players 1 and 2 in (A, B, M). A pure strategy for Mr. Jones in the auxiliary game can be identified with a pair (α_i, β_j) of pure strategies in (A, B, M). The interpretation is that Jones will use α_i if he is player 1 (i. e., if the coin turns up heads) and β_j if he is player 2 (i.e., if the coin turns up tails). Hence, both Jones and Smith have mn pure strategies. This symmetrization appears to increase excessively the number of pure strategies involved; however, Brown and von Neumann showed that the resulting system of mn simultaneous equations can be reduced to an auxiliary system of only $m + n$ simultaneous differential equations.

Gale, Kuhn, and Tucker [1950 a] have given a direct symmetrization involving only $m + n + 1$ pure strategies; it is intimately related to the reduction of a game to the symmetric form of its associated linear-programing problem (cf., Appendix 5). This is their procedure: For the given game (A, B, M), form player 1's associated linear-programing problem. As shown in Appendix 5 (cf. p. 409), this will be a problem of the minimizing variety where the parameters b_j's and c_i's are all 1. But associated in turn to this linear-programing problem is a two-person zero-sum game (cf. p. 419), which takes the form:

$$
\begin{bmatrix}
& & & & -a_{11} & -a_{21} & \cdots & -a_{m1} & 1 \\
& & & & -a_{12} & -a_{22} & \cdots & -a_{m2} & 1 \\
& & & & \cdot & \cdot & & \cdot & \cdot \\
& 0 & & & \cdot & \cdot & & \cdot & \cdot \\
& & & & \cdot & \cdot & & \cdot & \cdot \\
& & & & -a_{1n} & -a_{2n} & \cdots & -a_{mn} & 1 \\
a_{11} & a_{12} & \cdots & a_{1n} & & & & & -1 \\
a_{21} & a_{22} & \cdots & a_{2n} & & & & & -1 \\
\cdot & \cdot & & \cdot & & & & & \cdot \\
\cdot & \cdot & & \cdot & & & 0 & & \cdot \\
\cdot & \cdot & & \cdot & & & & & \cdot \\
a_{m1} & a_{m2} & \cdots & a_{mn} & & & & & -1 \\
-1 & -1 & \cdots & -1 & 1 & 1 & \cdots & 1 & 0
\end{bmatrix}
$$

Clearly, the induced game is symmetric, and an optimal strategy for either player yields optimal strategies for *both* players in the original game.

A verbal interpretation offered by Gale, Kuhn, and Tucker of their symmetrization is quite interesting.

We consider a game in which the players are denoted by *white* and *black*. We assume that white has an advantage (i.e., if white is the first player, $v > 0$). Then the symmetrized game is given by the following rules.

The players choose independently to play white or black, or to *hedge*. If they choose the same colors or both hedge, the play is a draw. If they choose different colors, a play of the original game ensues. As for the remaining possibilities, a hedge wins one unit from white and loses one unit to black.

It is evident that this is a symmetric game. That we learn how to play the original game by playing the symmetrized version follows from the fact that an optimal strategy for the symmetric game must include playing both white and black with positive probability. The cyclic nature of the possibilities (white, black, and hedge), reminiscent of stone, scissors, and paper, make this intuitively plausible. (Gale, Kuhn, and Tucker [1950 a], p. 83.)

A6.9 ITERATIVE SOLUTION OF GAMES BY FICTITIOUS PLAY

Brown [1951] gives

a very simple iterative method for approximating to solutions of discrete zero-sum games. This method is related to some particular systems of differential equations . . . whose steady-state solutions correspond to solutions of a game . . . . The iterative method in question can be loosely characterized by the fact that it rests on the traditional statistician's philosophy of basing future decisions on the relevant past history. Visualize two statisticians, perhaps ignorant of minimax theory, playing many plays of the same discrete zero-sum game. One might naturally expect a statistician to keep track of the opponent's past plays and, in the absence of a more sophisticated calculation, perhaps to choose at each play the optimal pure strategy against the mixture represented by all the opponent's past plays. [Page 374.]

Before describing this process in detail, we define the auxiliary notion of an *empirical mixed strategy*. Suppose that, in k iterations of a game, player 1 used the pure strategies $(\alpha^{(1)}, \alpha^{(2)}, \cdots, \alpha^{(k)})$, where $\alpha^{(i)}$ denotes the pure strategy he used on the ith iteration. If r denotes the number of times pure strategy α_i appears in the set $(\alpha^{(1)}, \alpha^{(2)}, \cdots, \alpha^{(k)})$, then let $\mathbf{x}^{(k)}$ denote the randomized strategy in which pure strategy α_i occurs with probability r/k. For example, if in five iterations player 1 uses α_3, α_2, α_4, α_2, α_3, then $\mathbf{x}^{(5)} = (\tfrac{2}{5}\alpha_2, \tfrac{2}{5}\alpha_3, \tfrac{1}{5}\alpha_4)$. The strategy $\mathbf{x}^{(k)}$ is called the *empirical mixed strategy* for player 1, which results from the sequence of pure strategies $(\alpha^{(1)}, \alpha^{(2)}, \cdots, \alpha^{(k)})$. The empirical mixed strategy $\mathbf{y}^{(k)}$ for player 2 which results from the sequence of pure strategies $(\beta^{(1)}, \beta^{(2)}, \cdots, \beta^{(k)})$ is defined similarly.

Brown's procedure, which applies to iterations of a fixed game, is described by the following steps:

Step 1. (a) Player 1 chooses a pure strategy which is called $\alpha^{(1)}$. Then $\mathbf{x}^{(1)} = \alpha^{(1)}$.

(b) Player 2 chooses a pure strategy $\beta^{(1)}$ which is best against $\mathbf{x}^{(1)}$ (i.e., against $\alpha^{(1)}$). If this instruction, or any of the instructions to follow, is ambiguous because of non-uniqueness, choose any one of the possible pure strategies compatible with it.

Step 2. (a) Choose $\alpha^{(2)}$ best against $\mathbf{y}^{(1)}$ (i.e., against $\beta^{(1)}$).

(b) Choose $\beta^{(2)}$ best against $\mathbf{x}^{(2)}$ corresponding to $\{\alpha^{(1)}, \alpha^{(2)}\}$.

Step 3. (a) Choose $\alpha^{(3)}$ best against $\mathbf{y}^{(2)}$ corresponding to $\{\beta^{(1)}, \beta^{(2)}\}$.

(b) Choose $\beta^{(3)}$ best against $\mathbf{x}^{(3)}$ corresponding to $\{\alpha^{(1)}, \alpha^{(2)}, \alpha^{(3)}\}$.

. . .

Step k. (a) Choose $\alpha^{(k)}$ best against $\mathbf{y}^{(k-1)}$ corresponding to $\{\beta^{(1)}, \cdots , \beta^{(k-1)}\}$.

(b) Choose $\beta^{(k)}$ best against $\mathbf{x}^{(k)}$ corresponding to $\{\alpha^{(1)}, \cdots , \alpha^{(k)}\}$.

. . .

Note that the process possesses only one completely arbitrary choice, namely, the starting point $\alpha^{(1)}$. Again, let us denote by $v_1(\mathbf{x})$ player 1's security level with the strategy $\mathbf{x}$, and by $v_2(\mathbf{y})$ the maximum that player 1 can get if player 2 uses $\mathbf{y}$. Naturally, for each k

$$v_1(\mathbf{x}^{(k)}) \leqslant v \leqslant v_2(\mathbf{y}^{(k)}),$$

where v is the value of the game. Clearly, if $\lim_{k \to \infty} v_1(\mathbf{x}^{(k)}) = \lim_{k \to \infty} v_2(\mathbf{y}^{(k)})$, this common value must be the value of the game, and so $\mathbf{x}^{(k)}$ and $\mathbf{y}^{(k)}$ must be "nearly" optimal for large k. Julia Robinson [1951] showed that iterative procedures of this type must converge in the limit, so the Brown procedure yields the value v in the limit.

In light of these results, finding the approximate value of a game is very easy. One merely iterates the game, as we have described, using fictitious players, and at each stage the pair of numbers $[v_1(\mathbf{x}^{(k)}), v_2(\mathbf{y}^{(k)})]$ are calculated. Since v lies in between them, the process is terminated when the desired degree of accuracy is attained.

Brown's results are not only computationally valuable but also quite illuminating from a substantive point of view. Imagine a pair of players repeating a game over and over again. It is plausible that at every stage a player attempts to exploit his knowledge of his opponent's past moves. Even though the game may be too complicated or too nebulous to be subjected to an adequate initial analysis, experience in repeated plays may tend to a statistical equilibrium whose (time) average return is approximately equal to the value of the game.

Note, however, Brown's procedure requires each player to use pure strategies which are best against the empirical mixed strategy defined for *all preceding pure strategies* of the other player. It is natural, therefore, to ask: what happens if the players have only finite memories (i.e., can only remember the preceding p moves, say)? For example, if $p = 1$, each player chooses the best strategy against the pure strategy just used by his opponent. This procedure need not converge, and it is conjectured that no finite memory procedure will work for all games. To see that a memory of order 1 will not work, consider the following game:

$$
\begin{array}{c c c c}
 & \beta_1 & \beta_2 & \beta_3 \\
\alpha_1 & \begin{bmatrix} 5 & 2 & 0 \\ \alpha_2 & 4 & 3 & 4 \\ \alpha_3 & 0 & 2 & 5 \end{bmatrix}
\end{array}
$$

Starting with α_1, we get β_3, since it is best against α_1, then α_3 (best against β_3), then β_1, then α_1, and the process repeats. However, (α_2, β_2) is in equilibrium, and α_2 is uniformly better than the empirical mixed strategy $(\frac{1}{2}\alpha_1, \frac{1}{2}\alpha_3)$ which arises from the iterative process.

A related criticism of Brown's procedure is that it gives the initial pure strategies as much weight as the later ones. Since the earlier strategies were not pitted against as much accumulated knowledge of the opponent's strategies, it would seem that they should be minimized. What is called for is a damped memory—but not so precipitous a damping as a finite memory. The discrete analogue of the Bellman differential equation method for solving games (mentioned on p. 440), which is obtained if the differential equation is replaced by a difference equation, has a damped memory effect and furnishes a new iterative algorithm which converges faster than Brown's statistical iterative procedure. The Bellman procedure is applied directly to symmetric games; therefore other games must be symmeterized first.

Von Neumann [1954] describes still another numerical method for solving two-person zero-sum games. Given the matrix $[a_{ij}]$, $i = 1, 2,$ $\cdots, m$ and $j = 1, 2, \cdots, n$, it is desired to find: (i) an m-tuple $\mathbf{x} = (x_1, x_2, \cdots, x_m)$, $x_i \geqslant 0$, for all i, and $\sum_{i=1}^{m} x_i = 1$; (ii) an n-tuple $\mathbf{y} = (y_1, y_2, \cdots, y_n)$, $y_j \geqslant 0$, for all j, and $\sum_{j=1}^{n} y_j = 1$; and a value v such that

(iii) $\displaystyle\sum_i a_{ij} x_i \geqslant v$,   for $j = 1, 2, \cdots, n$,

and

(iv) $\displaystyle\sum_j a_{ij} y_j \leqslant v$ for $i = 1, 2, \cdots, m$.

The von Neuman procedure begins with a trial triplet $(\mathbf{x}', \mathbf{y}', v')$ satisfying (i) and (ii), and an iterative procedure is prescribed for going from triplet to triplet such that a limit is monotonically approached which satisfies (iii) and (iv). To each triplet $(\mathbf{x}', \mathbf{y}', v')$ the following index is assigned to indicate how "close" it is to a solution:

$$\phi' = \sum_j{}' \left(v' - \sum_i a_{ij}x_i'\right)^2 + \sum_i{}' \left(\sum_j a_{ij}y_j' - v'\right)^2,$$

where the sum $\sum_j{}'$ is taken over all j-indices for which the inequality (iii)

fails to hold, and $\sum_i{}'$ is taken over all i-indices for which the inequality (iv)

fails to hold. It can also be written

$$\phi' = \sum_j v_j'^2 + \sum_i u_i'^2,$$

where

$$v_j' = \begin{cases} 0, & \text{if } \sum_i a_{ij}x_i' \geqslant v' \\[2mm] v' - \sum_i a_{ij}x_i', & \text{if } \sum_i a_{ij}x_i' \leqslant v', \end{cases}$$

and

$$u_i' = \begin{cases} 0, & \text{if } \sum_j a_{ij}y_j' \leqslant v' \\[2mm] \sum_j a_{ij}y_j' - v', & \text{if } \sum_j a_{ij}y_j' \geqslant v'. \end{cases}$$

Starting with the triplet $(\mathbf{x}', \mathbf{y}', v')$, von Neumann defines the triplet $(\tilde{\mathbf{x}}', \tilde{\mathbf{y}}', \tilde{v}')$, where

$$\tilde{x}_i' = \frac{u_i'}{\sum_k u_k'},$$

$$\tilde{y}_j' = \frac{v_j'}{\sum_k v_k'},$$

$$\tilde{v}' = \sum_{i,j} a_{ij}\tilde{x}_i'\tilde{y}_j'.$$

Observe that the constant relating u_i' to $\tilde{x}_i'$ is so chosen that $\tilde{\mathbf{x}}' = (\tilde{x}_1',$ $\cdots, \tilde{x}_m')$ is a probability vector. Note also that the size of u_i' depends upon how good player 1's ith strategy is against $\mathbf{y}'$ relative to an "aspiration level" of v', so to speak. The von Neumann recursive procedure then defines $(\mathbf{x}'', \mathbf{y}'', v'')$ as a suitable weighted average of $(\mathbf{x}', \mathbf{y}', v')$ and

$(\tilde{\mathbf{x}}', \mathbf{y}', \tilde{v}')$, the weights being so chosen that the index ϕ'' attached to the triplet $(\mathbf{x}'', \mathbf{y}'', v'')$ is small.

The procedure thus generates a sequence of triplets, namely,

$$(\mathbf{x}', \mathbf{y}', v') \rightarrow (\mathbf{x}'', \mathbf{y}'', v'') \rightarrow (\mathbf{x}^{(3)}, \mathbf{y}^{(3)}, v^{(3)})$$
$$\rightarrow \cdots (\mathbf{x}^{(h)}, \mathbf{y}^{(h)}, v^{(h)}) \rightarrow \cdots,$$

and if we let $\phi^{(h)}$ be the index associated with the hth iterate, it can be shown by elementary means that:

(i) $\phi^{(1)} \geqslant \phi^{(2)} \geqslant \phi^{(3)} \geqslant \cdots$

and

(ii) $\phi^{(h)} \leqslant \dfrac{m+n}{h} [\max_{i,j} a_{ij} - \min_{i,j} a_{ij}]^2.$

Hence, $\phi^{(h)} \rightarrow 0$ as h increases.

Von Neumann [1954] indicates how the above technique can be programed for a machine, and he calculates the operation time of each iterate. We quote:

> In evaluating the method it ought to be compared with G. Dantzig's "simplex method." In the latter method the a priori guarantees for the length of calculation and size of numbers are considerably less favorable than ours, but the available practical experience with the "simplex method" indicates that its actual performance—at least under the conditions under which it has been so far tested—is much better than the limits that can be guaranteed. Limited comparisons of our method with the "simplex method" again indicate that the latter converges faster, but there is reason to believe that our method can be accelerated by various tricks which amount to smoothing the iterative recursive sequence which is involved, and making this recursion dependent on several predecessors. The description of the method is offered here as a first step in this direction, i.e., in order to illustrate the new method, and to furnish a basis for the possible improvements referred to above. (Von Neumann [1954], p. 115.)

appendix 7

GAMES WITH
INFINITE PURE STRATEGY SETS

A7.1 INTRODUCTION

An extension of the minimax (equilibrium) theorem to special classes of games involving an infinite number of pure strategies was first accomplished by Ville [1938]. However, particular examples of partitioning-like games (deployment of military forces) with infinitely many pure strategies were treated by Borel in 1921 long before Ville's systematic treatment of a class of infinite games. In the last decade, a great deal of research has been concentrated on infinite games, the primary motivation being games of timing (e.g., when to fire in an air duel) and games of partitioning (e.g., what proportion of one's resources to allot to a given endeavor). In many of the examples, it is suitable to identify the pure strategies with real numbers in the unit interval and pairs of strategies with points in the unit square. A sizeable literature now exists for such games over the unit square.

Independently of Ville, Wald published a series of papers [1945 *a*, 1945 *b*, 1950 *a*] on statistical decision theory in which he developed an extensive theory for two-person games having infinitely many pure strategies. Much that Wald did in statistical decision theory using the game theory he so ably developed can now be accomplished, perhaps more elegantly, without it; but in all likelihood, today's formulations would not have been achieved so readily had Wald's pioneering game-theoretic framework been lacking.

A7.2 GAMES WITH NO VALUE

Provided each player's set of pure strategies is finite, every zero-sum (strictly competitive) two-person game has

i. A value v.
ii. An optimal (maximin) strategy for player 1.
iii. An optimal (minimax) strategy for player 2.

To see that this need not be so when strategy sets are infinite, consider the following game: Players 1 and 2 each choose a positive integer, say α and β respectively. Player 1 receives one unit from player 2 if $\alpha > \beta$, zero units if $\alpha = \beta$, and he gives one unit to player 2 if $\alpha < \beta$. The whole idea of the game is to pick a "large" number. This game, as we will show, has no value and the players do not have optimal strategies. The pure strategy set for each player is the set of positive integers $\{1, 2, 3, \cdots\}$, and a typical randomized strategy for player 1 is a sequence $\{x_1, x_2, \cdots\}$, where $\sum_{i=1}^{\infty} x_i = 1$ and $x_i \geq 0$. This strategy simply requires player 1 to choose the integer i with probability x_i, $i = 1, 2, \cdots$. We now show that, for any strategy $\mathbf{x}$, player 1's security level, $v_1(\mathbf{x})$, is -1. Since the sum $\sum_{i=N}^{\infty} x_i$ can be made arbitrarily small by choosing N sufficiently large, 2 can always render the chance of 1's winning arbitrarily small. So, if 1's choice is known, 2 can hold him down to as near to -1 as 2 desires. Thus, every $\mathbf{x} = \{x_1, x_2, \cdots\}$ has a security value $v_1(\mathbf{x}) = -1$. Similarly, for any of player 2's strategies, player 1 can ensure a return as near to 1 as he desires. Consequently,

$$v_1 = \max_{\mathbf{x}} v_1(\mathbf{x}) = -1 < v_2 = \min_{\mathbf{y}} v_2(\mathbf{y}) = +1;$$

hence the game has no value.

Furthermore, for player 1 the strategy in which the integers $1, 2, \cdots, k$ are selected, with probability zero, and the integers $k + i$, with probability x_i, for $i = 1, 2, \cdots$, i.e., $\{0, 0, \cdots, 0, x_1, x_2, \cdots\}$, is uniformly better than $\{x_1, x_2, \cdots\}$ in the sense that the former is at least as good a response as the latter for any $\mathbf{y}$ and strictly better for some. We say the latter is *inadmissible* (because it is dominated by the former). Thus, no strategy can be admissible.

For such a game, therefore, one can reasonably take the view (as we shall) that there is no "value" in ascribing a value (a number) to it. Although one might take the opposite view, arguing that the perfect symmetry of the game should result in a (generalized) "value" of zero, this has not proved to be a useful definition.

We still wish to establish a suitable abstract definition of value for some games with infinite strategy sets. A game is, as before, a triplet (A, B, M), where A and B are players 1 and 2's sets of pure strategies, respectively, and M is a real-valued function defined for all pairs (α, β), where α is in A and β in B. If (α, β) is chosen, player 1 receives $M(\alpha, \beta)$ and 2 receives $-M(\alpha, \beta)$. No assumption is made that A and B are finite. We further assume that from $A[B]$ a set $X[Y]$ of mixed or randomized strategies is generated[1] having $\mathbf{x}[\mathbf{y}]$ as a generic element. Corresponding to the pair $(\mathbf{x}, \mathbf{y})$, the expected return to 1 is denoted by $M(\mathbf{x}, \mathbf{y})$.

For the game (A, B, M) with the mixed extension (X, Y, M), the following terminology is employed:

i. The game is said to have a value v if, for every positive ϵ, however small, player 1 has a strategy in X which guarantees him a return of at least $v - \epsilon$ and player 2 has a strategy in Y which limits 1 to a return of at most $v + \epsilon$. Such a game is said to be *strictly determined*.

ii. A strategy from X is said to be *maximin* if it guarantees player 1 an amount v_1 and if no other strategy will guarantee him more.

iii. A strategy from Y is said to be *minimax* if it guarantees that player 1 receives no more than v_2 and if no other strategy in Y will hold 1 down to less.

There are, by now, classic examples of games, both with and without a value, where neither, one, or both players have optimal strategies. If players 1 and 2 have maximin and minimax strategies, respectively, then $v_1 \leqslant v_2$; the equality holds if and only if the game has a value.

---

[1] In the infinite case the set of *all* distributions need not be a meaningful concept, and so some further specification is needed. A number of levels of generality are possible, including:

i. $\mathbf{x}$ uses at most a finite number of pure strategies (i.e., for each $\mathbf{x}$ in X there is a finite subset $w_{\mathbf{x}}$ of A such that the probability, according to $\mathbf{x}$, of choosing an α in $w_{\mathbf{x}}$ is 1).

ii. $\mathbf{x}$ uses at most a denumerable subset of pure strategies.

iii. All measures $\mathbf{x}$ are densities (with respect to some specified dominating measure over a suitably chosen field or σ-algebra of sets).

iv. $\mathbf{x}$ is a completely additive measure over some σ-algebra.

v. $\mathbf{x}$ is a finitely additive measure with all sets measurable. [With this convention, the value of the game "to select the larger integer" (discussed on p. 448) is zero.]

Naturally, whether or not the game has a value or player 1 has a maximin strategy depends upon the definition of X we choose.

The rest of this appendix is devoted to classes of infinite games for which results have been published. *Throughout the discussion we shall assume that the payoff function M is bounded.*

A7.3 GAMES WHERE A (OR B) IS FINITE

Let us assume that $A = \{\alpha_1, \alpha_2 \cdot \cdot \cdot, \alpha_m\}$ and B is arbitrary. To each $\mathbf{y}$ of Y, we may associate the point $[M(\alpha_1, \mathbf{y}), \cdot \cdot \cdot, M(\alpha_m, \mathbf{y})]$ in

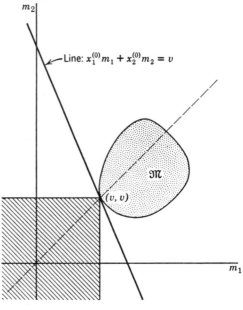

FIG. 1

m-space (cf. Appendix 4). As $\mathbf{y}$ ranges over Y, the associated points generate a region $\mathfrak{M}$ which is convex in the sense that the line segment joining any two points of $\mathfrak{M}$ is in $\mathfrak{M}$. Figure 1 is illustrative for $A = \{\alpha_1, \alpha_2\}$.

Note that the set $\mathfrak{M}$ need neither be polygonal (as when B is also finite) *nor contain all of its boundary points.* If the point (v, v) belongs to $\mathfrak{M}$ (see Fig. 1), then, by choosing this point, player 2 can hold 1 down to v. If, however, (v, v) does not belong to $\mathfrak{M}$ (located as in Fig. 1), as is possible, then player 2 cannot hold 1 down to v, but he can hold 1 down to $v + \epsilon$, for each small positive ϵ. On the other hand, player 1 can guarantee himself at least v by using $(x_1^{(0)}\alpha_1, x_2^{(0)}\alpha_2)$. Hence we see that *the game has a value and player 1 has a maximin strategy.* This result holds for all games where A is finite. If the game is such that the set $\mathfrak{M}$ contains its boundary points (at least in the lower left part of $\mathfrak{M}$), then player 2 has a minimax strategy. Note that the dashed 45° line in Fig. 1 does not necessarily have to intersect the region $\mathfrak{M}$ (cf. Fig. 2c p. 403).

A7.4 GAMES WHERE A IS "ALMOST" FINITE

Of considerable significance are those games where player 1 can restrict his randomization to a fixed finite set of pure strategies with the knowledge that he will sacrifice less than ϵ, say, regardless of what player 2 does. To be specific, suppose ϵ is given (e.g., it might be $\epsilon = 10^{-6}$). Then, we assume there is a finite set $A_\epsilon = \{\alpha_{i_1}, \alpha_{i_2}, \cdots, \alpha_{i_r}\}$, where r depends upon ϵ, such that for each $\mathbf{x}$ of X there is a "matching" randomization $\mathbf{x}^*$ over A_ϵ, with the property that

$$|M(\mathbf{x}, \mathbf{y}) - M(\mathbf{x}^*, \mathbf{y})| < \epsilon, \qquad \text{for all } \mathbf{y} \text{ in } Y.$$

We can think of A_ϵ as a finite ϵ-approximation to A.

Now, since A_ϵ is finite, the mixed extension of the game (A_ϵ, B, M) has a value and player 1 has an optimal strategy. By letting ϵ approach 0, it can be shown that the mixed extension of (A, B, M) also has a value. However, player 1 need not have a maximin strategy—its existence depends, in part, upon how general we choose to make X.

Speaking very roughly, if A is finite, then a value and a maximin strategy both exist; if it is not finite but can be approximated by a finite set in such a manner that the sacrifice is arbitrarily small, then a value again exists, but a maximin strategy need not.

A7.5 GAMES OVER THE UNIT SQUARE

The infinite games which have received most attention are those whose pure strategy sets are the real numbers in the closed interval from 0 to 1, i.e.,

$$A = \{\alpha \mid 0 \leqslant \alpha \leqslant 1\}$$
$$B = \{\beta \mid 0 \leqslant \beta \leqslant 1\}.$$

If M is a *continuous function* on the unit square $\{(\alpha, \beta) \mid 0 \leqslant \alpha \leqslant 1, 0 \leqslant \beta \leqslant 1\}$, then, for every $\epsilon > 0$, a finite ϵ-approximation to A can be found by choosing sufficient points scattered uniformly over the unit interval. Thus, all such games have a value. Players 1 and 2 have optimal strategies provided X and Y are taken to be the sets of all cumulative distribution functions over the unit interval [case (iv) in footnote 1].

More refined problems are: (i) Is it possible to characterize the optimal strategies for games with a continuous payoff over the unit square? (ii)

For which subclasses of such games are there optimal pure strategies? (iii) For which classes of games are there optimal strategies which use only a finite number of pure strategies?

So far as we know, no general characterization in answer to (i) has been given, and, unhappily, a number of extremely complicated and patho-logical examples have been exhibited.

As regards (ii), the important *strictly convex*, or *convex-concave*, games have pure strategy solutions. In particular, if, for each α, M is a convex func-tion of β, i.e., $\lambda M(\alpha, \beta') + (1 - \lambda)M(\alpha, \beta'') \geqslant M[\alpha, \lambda\beta' + (1 - \lambda)\beta'']$, for all β', β'', and $0 \leqslant \lambda \leqslant 1$, then player 2 has a minimax pure strategy and player 1 has a maximin strategy which uses at most two pure strate-gies. If M is concave in α for each β, i.e., $\lambda M(\alpha', \beta) + (1 - \lambda)M(\alpha'', \beta)$ $\leqslant M[\lambda\alpha' + (1 - \lambda)\alpha'', \beta]$, for all α', α'', and $0 \leqslant \lambda \leqslant 1$, and convex in β for each α, then both players have optimal pure strategies. See Bohnen-blust, Karlin, and Shapley [1950 b] for further results.

As to question (iii), the first important class of games which have optimal solutions involving only a finite number of pure strategies are called *polynomial games*. They are defined by the condition that there exist constants m, n, and a_{ij} ($i = 1, 2, \cdots, m; j = 1, 2, \cdots, n$) such that $M(\alpha, \beta) = \sum_{i=1}^{m} \sum_{j=1}^{n} a_{ij}\alpha^i\beta^j$. A great deal is known about solutions to these games; see Dresher, Karlin, and Shapley [1950]. It was originally hoped that polynomial games would serve as a bridge from the finite case to infinite games with continuous payoffs, since any continuous payoff can be uniformly approximated by a polynomial. Unfortunately, this pro-gram has not been too successful so far.

▶ The analysis of polynomial games is intrinsically related to the moment problem of statistics. If F and G are mixed strategies (cumulative distribution functions) for players 1 and 2, respectively, then

$$M(F, G) = \int_0^1 \int_0^1 M(\alpha, \beta) \, dF(\alpha) \, dG(\beta)$$

$$= \int_0^1 \int_0^1 \sum_{i=1}^{m} \sum_{j=1}^{n} a_{ij}\alpha^i\beta^j \, dF(\alpha) \, dG(\beta)$$

$$= \sum_{i=1}^{m} \sum_{j=1}^{n} \nu_i(F)a_{ij}\nu_j(G),$$

where $\nu_i(F) = \int_0^1 \alpha^i \, dF(\alpha)$ is the ith moment of F, and $\nu_j(G) = \int_0^1 \beta^j \, dG(\beta)$ is the jth moment of G. Consequently, player 1's strategic problem reduces to the judicious selection of an ordered set of m real numbers—these being the first m

moments of a cumulative distribution function F over the unit interval. By changing 1 to 2, m to n, and F to G, an analogous statement holds for player 2. ◀

Because of its increased versatility, an expression of the following form

$$\sum_{i=1}^{m} \sum_{j=1}^{n} a_{ij} r_i(\alpha) s_j(\beta),$$

where r_i and s_j are continuous functions, seems to be a more promising approximation to an arbitrary continuous payoff. Games with payoffs of this type are called *polynomial-like*. For given m and n they can usually approximate an arbitrary continuous game more closely than can the polynomial games, but this advantage is offset by the fact that at present much sharper results are known for polynomial games. Dresher, Karlin, and Shapley [1950] show that in every polynomial-like game both players have optimal mixed strategies which use at most min (m, n) pure strategies.

Finally, Blackwell and Girshick [1954, p. 54] have given an example of a continuous game over the unit square in which *every* optimal strategy uses *all* of the pure strategies. If X and Y are restricted to randomizations over at most a denumerable number of pure strategies, then their example has a value, but no optimal strategies.

A7.6 GAMES INVOLVING TIMING OR PARTITIONING

Games involving either timing or partitioning often can be reduced to games over the unit square, but in contrast to those we have examined previously the payoff function is not continuous. To illustrate how the discontinuities arise, consider the game in which the two players—duelists —are separated by a distance of two units. At a signal, they begin to approach one another at the same constant uniform rate. Each is at liberty to fire a single shot at any time he desires. A pure strategy for player 1 [player 2] is a number α, $0 \leqslant \alpha \leqslant 1$ [β, $0 \leqslant \beta \leqslant 1$]. This is interpreted to mean that 1 [2] fires when he has traveled α [β] units, unless 2 [1] has fired earlier and missed, in which case 1 [2] holds his fire until they are together. It is postulated that each player knows his own and his antagonist's probability of a kill as a function of the distance between them—these probability functions are assumed to increase monotonically and continuously as the distance between the players decreases. If $P_1(\gamma)$ [$P_2(\gamma)$] denotes the probability that 1 [2] hits 2 [1] if he fires when they are a distance γ apart, then *a possible* payoff M is:

(a) $\quad M(\alpha, \beta) = (1)P_1(2\alpha) + (-1)[1 - P_1(2\alpha)], \quad$ if $\alpha < \beta$,

(b) $\quad = (1)[1 - P_2(2\beta)] + (-1)P_2(2\beta), \quad$ if $\alpha > \beta$,

(c) $\quad = (1)P_1(2\alpha) + (-1)P_2(2\alpha), \quad$ if $\alpha = \beta$.

For example, in case a 1 shoots first. The probability that he hits 2 is $P_1(2\alpha)$, and so the probability of his being hit by 2 is $1 - P_1(2\alpha)$, since, if 1 misses, 2 will hold his fire until he cannot miss. We have no wish to defend the realism of this payoff function; it is merely presented to illustrate that a natural payoff in a duel is not continuous. There is a line of discontinuity along $\alpha = \beta$, since if $\alpha < \beta$ 1 shoots first, and if $\alpha > \beta$ 2 shoots first. The payoffs on the two sides of the line can differ appreciably.

Such games with a line of discontinuity in the payoff function have a value and the players have optimal strategies[2] (Karlin [1950], p. 141).

Shiffman [1953] generalized this example to what he calls a *symmetrical game of timing:* a game over the unit square where M is:

i. Strictly increasing in α, for $\alpha < \beta$ and β fixed (i.e., the payoff to 1 improves the longer he waits—so long as he fires before 2).

ii. Strictly decreasing in β, for $\beta < \alpha$ and α fixed (i.e., the payoff to 1 diminishes the longer 2 waits—so long as 2 fires before 1).

iii. $M(\alpha, \beta) = -M(\beta, \alpha)$ (i.e., the player's capabilities are symmetric).

He shows that, aside from some trivial cases, the optimal strategy is a randomization of a density type. This density function can be expressed as the solution to a certain integral equation which, in turn, often can be shown to be equivalent to a system of ordinary linear differential equations.

Karlin [1953] indicated that the salient feature of symmetrical games of timing which leads to an integral equation for the optimal strategies is the diagonal discontinuity in the payoff function arising from the order in which the players act.

The history of games of partitioning dates back at least to Borel, who, in a paper published in 1921 (a translation appears in *Econometrica;* see Borel [1953]), posed the following problem:

---

[2] Some interesting variants of the simple duel are:

i. Each player is unaware when his opponent has fired except when he is hit. (The pistols have silencers.)

ii. Only one player has a silencer.

iii. Player 1 has m bullets and 2 has n bullets.

iv. Combinations of (iii) and (i) or (ii).

In cases (i)–(iv), the players know all the data of the problem, including the hit probabilities. To be really complicated, we can suppose in case (iii) that 1 does not know the initial n and 2 does not know the initial m. Blackwell and Girshick [1954, pp. 69–73] discuss (iii) in detail under the assumption that the players have identical hitting accuracies and m and n are known.

· · · two players A and B each choose three positive numbers the sum of which is equal to 1, viz:

$$x + y + z = 1,$$

$$x' + y' + z' = 1;$$

and each player arranges the numbers he has chosen in a determined order. A wins if two of the numbers chosen by him are superior to the corresponding numbers of B.

For example, we can think of two opposing generals with equal forces each of whom must partition his own forces among three battle areas without knowing how the other will deploy his. Each aims to have a numerical superiority at two of the three sites.

Tukey [1949] and Blackett [1954] studied a more general class of games of partitioning called "Blotto games." Tukey analyzed Blotto games of a symmetric type, and Blackett examined a specific asymmetric form. Without actually defining these games, their nature and possible applications are suggested by the quotation from Blackett:

The particular problem of Colonel Blotto illustrates a general class of "Blotto" games in which:

Two players (A and B) contending on N independent battlefields (labeled 1, 2, · · · , N) must distribute their forces (F and G units, respectively) to the battlefields before knowing the opposing deployment. The payoff (a numerical measure of the gain of A or equivalently of the loss of B) on the ith battlefield is given by a function $P_i(x, y)$ depending only on the battlefield and the opposing forces x and y committed to that battlefield by A and B. The payoff of the game as a whole is the sum of the payoffs on the individual battlefields.

An interesting mathematical problem connected with these games is the determination of their solutions from the payoff functions of the individual battlefields. Instead of studying this in general, the present paper illustrates the possible applicability of Blotto games to problems of logistics and tactics by analyzing a particular problem which may be considered as an especially simple Blotto game.

Suppose a supply system is to deliver a shipment of material from a rear area to an advanced area by one of N independent routes subject to interdiction by enemy assailants. (By "independent routes" is meant routes such that a single assailant cannot interdict more than one.) If the route must be selected in ignorance of the interdiction plans of the enemy and the enemy must station his assailants without knowing the route the shipment will travel, the situation described may be regarded as a Blotto game in which A selects the route (battlefield) for the shipment (A's forces) while B distributes his assailants (B's forces) among the different possible routes (battlefields). In this game, $P_i(1, y)$ represents the gain of A (the loss of B) on the ith route when the shipment travels the ith route and y of B's G assailants interdict this route. The analogous quantity when the shipment travels some route other than the ith one is $P_i(0, y)$.

In one military interpretation of this situation the shipment is a naval convoy and the assailants are submarines. In another the shipment is a truck convoy and the assailants are attack aircraft. In a third the shipment is a bombing strike against an enemy target and the assailants are interceptors. [1954, p. 55.]

A7.7 A MODEL OF POKER DUE TO BOREL

Another class of games on the square is described in the following quotation from Kuhn's extremely fine *Lectures on the Theory of Games* [1952, p. 139]:

A fertile source of examples of infinite games is provided by models of common card games. Indeed, they invite treatment by a continuous variable on two grounds. First, the combinatorial complexity of finite models precludes consideration of any but the simplest cases. Second, the natural linear ordering of a large number of hands in games such as poker virtually invites the passage to a continuum of hands. Consider, for example, the following model of poker due to Borel:

"*La relance.* An ante of a units is required by each of the two players. At the beginning of a play they receive fixed hands, s and t, chosen at random from the unit intervals $0 \leq s \leq 1$ and $0 \leq t \leq 1$. Then 1 either bets an amount $b - a$ or drops out, losing his ante. If 1 bets, then 2 can either see the bet or drop out, losing his ante. If 2 sees a bet, the hands are compared with the higher card winning the total wager b."

We shall assume that 1 uses pure strategies of the following form: he chooses a number α in the unit interval, $0 \leq \alpha \leq 1$, and decides to bet when his hand exceeds α and to drop out otherwise. Correspondingly, a pure strategy for 2 will consist of the choice of a number β in the unit interval, $0 \leq \beta \leq 1$, and the decision to see any possible bet when his hand exceeds β and to drop out otherwise. (The careful reader will want to verify that every other pure strategy is dominated by one of these. He will also remark that the original game had too many strategies to be a game on the square.)

If $M(\alpha, \beta)$ is the game matrix, it can be shown (cf. Kuhn [1952, p. 140]) that M is continuous and convex in β, for each fixed α, and concave in α, for each fixed β. Thus, by the theory of convex-concave games, the game has a value $\left[\text{in this case } v = -a\left(\dfrac{b - a}{a + b}\right)^2 \right]$, player 1 has a pure maximin strategy $\alpha^{(0)} = \left(\dfrac{b - a}{a + b}\right)^2$, and player 2 has a *unique* minimax strategy, which is pure, $\beta^{(0)} = \dfrac{b - a}{a + b}$. Player 1 also has randomized maximin strategies—he can use any randomized strategy whose mean value is $\left(\dfrac{b - a}{a + b}\right)^2$. Thus, although 1 can "bluff" in an optimal fashion, he cannot "bluff" so as to guarantee more than that guaranteed by his pure maximin strategy. In other variants of poker, this is *not* always the case!

appendix **8**

SEQUENTIAL COMPOUNDING
OF TWO-PERSON GAMES

A8.1 INTRODUCTION

We first encountered sequentially compounded (i.e., temporally repeated) two-person games in Chapter 5, where it will be recalled compounded zero-sum games seemed to be reasonably well behaved but compounded non-zero-sum games exhibited certain anomalies (e.g., the prisoner's dilemma). Here we shall explore the structure of compounded games more fully, using more complex compounding rules than in Chapter 5, but confining ourselves primarily to component games which lead to a zero-sum overall game.

The relevant literature which dates back, even if we are generous; only to 1949, is already extensive. Apparently, the time was ripe in the early '50's, for independently a number of workers attacked variations on the theme of compounding. Unfortunately,·this has led to a good deal of redundancy, both conceptual and technical, in the literature.

The central ideas are these: In one class of games (recursive and stochastic) a normalized game is played at each stage, and the player's strategies control not only the (monetary) payoff but also the transition probabilities which govern the game to be played at the next stage. In another class (survival and attrition games) there is but one component

457

game and it is repeated.   The players have limited initial resources, and these fluctuate in time according to the outcomes of repeated plays of the given game.   The overall game is concluded when one of the players is bankrupt.   In still another class (compound decision problems) a given game is repeated, and each player attempts to control the average payoff by exploiting the statistical records of his adversary's previous choices. The final class (economic ruin games), which we discuss only briefly, is typified by the problem of corporate dividend policy: The more generous the dividend policy of the corporation, the less secure it is against future exigencies; however, in opposition to this platitude is the truth, imposed by interest rates, that a dollar today is worth more than the present value of a dollar to be delivered in the future.

As we explore these topics we will also indicate some of the interrelations: how the theory of stochastic games suggested that of recursive games which, in turn, is related to the theory of survival and attrition games; how Blackwell's approachability theory, which was motivated by attrition games, can be used to analyze compound decision problems; and how approachability theory is technically similar to a generalization of the theory of survival games.

A8.2 STOCHASTIC GAMES

The first game to be described is a specific stochastic game which involves these two payoff matrices:

Component Game Γ^1

$\beta_1{}^1 \qquad\qquad\qquad\qquad \beta_2{}^1$

$\begin{array}{c} \alpha_1{}^1 \\ \alpha_2{}^1 \end{array} \left[\begin{array}{cc} 4 \;\&\; (0.4S,\; 0.5\Gamma^1,\; 0.1\Gamma^2) & 0 \;\&\; (0.2S,\; 0.5\Gamma^1,\; 0.3\Gamma^2) \\ -2 \;\&\; (0.6S,\;\;\; 0\Gamma^1,\; 0.4\Gamma^2) & 2 \;\&\; (0.8S,\; 0.2\Gamma^1,\;\;\; 0\Gamma^2) \end{array} \right],$

Component Game Γ^2

$\beta_1{}^2 \qquad\qquad\qquad\qquad \beta_2{}^2$

$\begin{array}{c} \alpha_1{}^2 \\ \alpha_2{}^2 \end{array} \left[\begin{array}{cc} -1 \;\&\; (\;\; 1S,\;\;\; 0\Gamma^1,\;\;\; 0\Gamma^2) & 2 \;\&\; (0.6S,\; 0.2\Gamma^1,\; 0.2\Gamma^2) \\ 2 \;\&\; (0.1S,\; 0.5\Gamma^1,\; 0.4\Gamma^2) & 5 \;\&\; (0.3S,\; 0.6\Gamma^1,\; 0.1\Gamma^2) \end{array} \right].$

The meaning of the entries will become apparent in a minute.   Suppose, as an initial condition, the stochastic game starts at component game Γ^1 with the players making simultaneous choices.   Suppose 1 selects $\alpha_1{}^1$ and 2 selects $\beta_1{}^1$, then the resulting payoff 4 & $(0.4S, 0.5\Gamma^1, 0.1\Gamma^2)$ means that player 1 receives 4 units from 2 *and* a lottery is performed in which the outcomes "Stop," "Play Γ^1 next," and "Play Γ^2 next" occur with probabilities 0.4, 0.5, and 0.1, respectively.   If, for example, the lottery yields outcome Γ^2, i.e., "Play Γ^2 next," then the players do just that—they play the component game Γ^2. Suppose 1 chooses $\alpha_2{}^2$ and 2 chooses $\beta_1{}^2$; then the payoff is 2 & $(0.1S, 0.5\Gamma^1, 0.4\Gamma^2)$. As before, this means that player 2 gives 2 units to player 1 *and* a lottery is conducted in which the alternatives "Stop," "Play Γ^1 next," and "Play Γ^2 next" occur with probabilities 0.1, 0.5, and 0.4, respectively. The play continues until a lottery yields a "Stop" outcome, and the overall payoff to each player is the sum of his payoffs in the component games of the play.

The play proceeds from component game to component game according to transition probabilities controlled jointly by the players. At each trial, each player must consider not only the probable effect his choice has on his positional payoff at that trial but also its effect on his chances of playing the several component games in the future. Since each lottery has a *positive* "stop" probability, the play is "almost certain" (i.e., with probability 1) to terminate in a finite number of steps.

The generalization is straightforward. The r component games Γ^1, Γ^2, $\cdots$, Γ^r are given. In game Γ^k, player 1 has the pure strategies $\alpha_1{}^k$, $\alpha_2{}^k$, $\cdots$, $\alpha_{m_k}{}^k$ and 2 has the pure strategies $\beta_1{}^k$, $\beta_2{}^k$, $\cdots$, $\beta_{n_k}{}^k$. If 1 uses $\alpha_i{}^k$ and 2 uses $\beta_j{}^k$, the payoff is

$$a_{ij}{}^k \,\&\, (p_{ij}{}^{k0}S, p_{ij}{}^{k1}\Gamma^1, p_{ij}{}^{k2}\Gamma^2, \cdots, p_{ij}{}^{kr}\Gamma^r),$$

where

$$p_{ij}{}^{k0} > 0, p_{ij}{}^{kl} \geqq 0, \text{ for } l = 1, 2, \cdots, r,$$

and

$$p_{ij}{}^{k0} + p_{ij}{}^{k1} + \cdots + p_{ij}{}^{kr} = 1.$$

These payoffs are interpreted to mean that player 2 gives 1 $a_{ij}{}^k$ units *and* a lottery is performed in which the play terminates with the *positive* probability $p_{ij}{}^{k0}$ and the component game Γ^l is played next with probability $p_{ij}{}^{kl}$, $l = 1, 2, \cdots, r$.

If we let $\vec{\Gamma}$ stand for the collection $\{\Gamma^1, \Gamma^2, \cdots, \Gamma^r\}$, then the specific game which begins with Γ^k may be denoted by the pair $(\vec{\Gamma}; \Gamma^k)$.

Shapley [1953 *d*] first defined stochastic games, and he characterized their solutions in this sense: for each initial condition, he gave a method for finding the value of the game, 1's maximin strategy, and 2's minimax strategy. We illustrate his procedure for our specific example.

First, the given game is truncated at trial n as follows. It is played without any modification as long as it terminates prior to the nth trial. But, should it last n trials, then, instead of playing a component game at trial $n + 1$, player 2 gives 1 a fixed amount $w_1^{(0)}$, if Γ^1 was to be played, and $w_2^{(0)}$, if Γ^2 was to be played. Elliptically, the game is said to be "truncated at trial n by means of the payoffs $(w_1^{(0)}, w_2^{(0)})$." If n is large, it is intuitively clear that the truncated game is not very different from the original game, and the particular values $w_1^{(0)}$ and $w_2^{(0)}$ which are used should not critically affect the overall value of the truncated game. If this is so, it is important since the truncated game is easy to analyze. Let us see why. At trial n (if the game lasts that long) the payoffs are:

$$\text{Game } \Gamma^1 \ (w_1^{(0)} \ w_2^{(0)})$$

$$
\begin{array}{c}
\quad\quad \beta_1^{\ 1} \quad\quad\quad\quad\quad\quad \beta_2^{\ 1} \\
\begin{array}{c} \alpha_1^{\ 1} \\ \alpha_2^{\ 1} \end{array}
\left[
\begin{array}{cc}
4 + 0.5w_1^{(0)} + 0.1w_2^{(0)} & 0.5w_1^{(0)} + 0.3w_2^{(0)} \\
-2 + 0.4w_2^{(0)} & 2 + 0.2w_1^{(0)}
\end{array}
\right],
\end{array}
$$

$$\text{Game } \Gamma^2(w_1^{(0)}, \ w_2^{(0)})$$

$$
\begin{array}{c}
\quad\quad \beta_1^{\ 2} \quad\quad\quad\quad\quad\quad \beta_2^{\ 2} \\
\begin{array}{c} \alpha_1^{\ 2} \\ \alpha_2^{\ 2} \end{array}
\left[
\begin{array}{cc}
-1 & 2 + 0.2w_1^{(0)} + 0.2w_2^{(0)} \\
2 + 0.5w_1^{(0)} + 0.4w_2^{(0)} & 5 + 0.6w_1^{(0)} + 0.1w_2^{(0)}
\end{array}
\right].
\end{array}
$$

Note the labeling: the payoff matrix of component game Γ^k at trial n is denoted $\Gamma^k(w_1^{(0)}, w_2^{(0)})$, $k = 1, 2$.

Let

$$w_1^{(1)} = \text{value of zero-sum game } \Gamma^1(w_1^{(0)}, w_2^{(0)}) = \text{val } \Gamma^1(w_1^{(0)}, w_2^{(0)})$$

$$w_2^{(1)} = \text{value of zero-sum game } \Gamma^2(w_1^{(0)}, w_2^{(0)}) = \text{val } \Gamma^2(w_1^{(0)}, w_2^{(0)}).$$

We now work backwards. At trial $n - 1$, the outcome "Play Γ^k next" has the value of $w_k^{(1)}$ to player 1, $k = 1, 2$. Thus at that trial, the players should behave as if they are playing $\Gamma^1(w_1^{(1)}, w_2^{(1)})$ and $\Gamma^2(w_1^{(1)}, w_2^{(1)})$. Continuing our backward induction, for any integer s, $0 \leqslant s \leqslant n - 1$, let

$$w_k^{(s+1)} = \text{val } \Gamma^k(w_1^{(s)}, w_2^{(s)}), \quad \text{for } k = 1, 2.$$

In particular, at the first trial we have

$$w_k^{(n)} = \text{val } \Gamma^k(w_1^{(n-1)}, w_2^{(n-1)}), \quad \text{for } k = 1, 2.$$

Thus we see that $w_1^{(n)}$ and $w_2^{(n)}$ are the values of the games $(\vec{\Gamma}; \Gamma^1)$ and $(\vec{\Gamma}; \Gamma^2)$, respectively, when they are truncated at n by $(w_1^{(0)}, w_2^{(0)})$.

These considerations suggest that we define what may be called the *value transformation:* the function T which maps the pair (w_1, w_2) into the pair $[\text{val } \Gamma^1(w_1, w_2), \text{val } \Gamma^2(w_1, w_2)] \equiv T(w_1, w_2)$. In terms of T, the above induction can be written:

$$T(w_1^{(0)}, w_2^{(0)}) = (w_1^{(1)}, w_2^{(1)})$$

$$T^2(w_1^{(0)}, w_2^{(0)}) = T(w_1^{(1)}, w_2^{(1)}) = (w_1^{(2)}, w_2^{(2)})$$

$$\cdot$$
$$\cdot$$
$$\cdot$$

$$T^n(w_1^{(0)}, w_2^{(0)}) = (w_1^{(n)}, w_2^{(n)}).$$

It is intuitively plausible (and Shapley proves it) that:

1. As n increases, $T^n(w_1^{(0)}, w_2^{(0)})$ has a limit which is independent of $(w_1^{(0)}, w_2^{(0)})$. We denote it by (w_1^*, w_2^*), i.e.,

$$\lim_{n \to \infty} T^n(w_1^{(0)}, w_2^{(0)}) = (w_1^*, w_2^*).$$

2. (w_1^*, w_2^*) is the unique pair with the property that

$$T(w_1^*, w_2^*) = (w_1^*, w_2^*),$$

i.e., (w_1^*, w_2^*) is the unique solution of the two equations in two unknowns:

$$w_1 = \text{val } \Gamma^1(w_1, w_2)$$

$$w_2 = \text{val } \Gamma^2(w_1, w_2).$$

In the general case, $(w_1^*, w_2^*, \cdots, w_r^*)$ is the unique solution of the system

$$w_k = \Gamma^k(w_1, w_2, \cdots, w_r), \qquad \text{for } k = 1, 2, \cdots, r.$$

3. Player 1 has a maximin strategy for the stochastic game which consists of playing a maximin strategy for $\Gamma^k(w_1^*, w_2^*)$ whenever the component game Γ^k arises. This guarantees player 1 an expected return of at least w_k^* for the game $(\vec{\Gamma}; \Gamma^k)$. Changing 1 to 2, maximin to minimax, and w_k^* to $-w_k^*$, an analogous statement holds for player 2.

Shapley points out that, if player 2 has only one pure strategy in each component game, he is really a dummy player and this degenerate stochastic game amounts to a dynamic programing problem for player 1.

A8.3 RECURSIVE GAMES

Recursive games and stochastic games are closely related, the only difference being the form of the payoff functions in the component games. Recall that, when 1 chooses his ith strategy and 2 his jth strategy in the component game Γ^k of a stochastic game, then the payoff is

$$a_{ij}^k \,\&\, (p_{ij}^{k0}S, \, p_{ij}^{k1}\Gamma^1, \, \cdots, \, p_{ij}^{kr}\Gamma^r),$$

where $p_{ij}^{k0} > 0, p_{ij}^{kl} \geqslant 0,$ for $l = 1, 2, \cdots, r,$ and $p_{ij}^{k0} + \displaystyle\sum_{l=1}^{r} p_{ij}^{kl} = 1.$

In a recursive game two small, but important, modifications are made:

1. A payoff of actual units (such as a_{ij}^k) only occurs when the play terminates.

2. The probability of stopping, p_{ij}^{k0}, is not necessarily positive.

A typical payoff in a recursive game is of the form:

$$[p_{ij}^{k0}(a_{ij}^{k} \text{ and } S), \; p_{ij}^{k1}\Gamma^1, \; p_{ij}^{k2}\Gamma^2, \; \cdots, \; p_{ij}^{kr}\Gamma^r],$$

where $p_{ij}^{kl} \geqslant 0$, for $l = 0, 1, 2, \cdots, r$, and $\sum_{l=0}^{r} p_{ij}^{kl} = 1$. This is the interpretation: With probability p_{ij}^{k0} (not necessarily positive!), the play stops and 2 pays 1 a_{ij}^{k} units, and with probability p_{ij}^{kl} the game Γ^l is played next (and no units are exchanged at this trial!), for $l = 1, 2, \cdots, r$. The basic reference for recursive games is Everett [1954].

The conceptual differences between stochastic and recursive games are illustrated in the following simple examples discussed by Everett:

Example 1.

$$\Gamma^1: \alpha_1^1 \begin{matrix} \beta_1^1 & \beta_2^1 \\ [\Gamma^1 & 1] \end{matrix}.$$

There is only one component game, and player 1 is strategically a dummy. If β_1^1 is always played, then the component game is repeated indefinitely, and so the recursive game never terminates. This real possibility must be taken into account by the rules of recursive games. We shall say that the payoff to each player is zero for any non-terminating play of a recursive game.

Example 2.

$$\Gamma^1: \begin{matrix} & \beta_1^1 & \beta_2^1 \\ \alpha_1^1 & \begin{bmatrix} \Gamma^1 & 1 \\ 1 & 0 \end{bmatrix} \\ \alpha_2^1 & \end{matrix}.$$

By using the mixed strategy $[(1 - \epsilon)\alpha_1^1, \; \epsilon\alpha_2^1]$ repeatedly, where $0 < \epsilon < 1$, player 1 can force the play to terminate with probability one, yielding him an expected return of at least $1 - \epsilon$, regardless of 2's strategy. For, if 2 uses β_1^1 at any stage, then the play terminates with probability ϵ, yielding player 1 a return of one unit; and, if 2 uses β_2^1, then the play surely terminates yielding 1 an expected return of $1 - \epsilon$ units. Therefore, player 1 can guarantee himself an expected return arbitrarily close to 1; but he cannot reach that value, and so he is said to have ϵ-maximin strategies.

For this recursive game, the analogue of the important functional equation which arose in the analysis of stochastic games is:

$$w_1 = \text{val} \begin{bmatrix} w_1 & 1 \\ 1 & 0 \end{bmatrix},$$

which has the unique solution $w_1{}^* = 1$. However, although the unique maximin strategy for the game

$$\begin{bmatrix} w_1{}^* & 1 \\ 1 & 0 \end{bmatrix}$$

is $\alpha_1{}^1$, its repeated choice leads to a security level of 0, not 1, since player 2 can prevent the game from terminating by choosing $\beta_1{}^1$ at each trial.

Example 3.

$$
\begin{array}{c}
\Gamma^1 \\[4pt]
\begin{array}{c}
\phantom{\alpha_1{}^1} \quad \beta_1{}^1 \;\; \beta_2{}^1 \\
\begin{array}{c} \alpha_1{}^1 \\ \alpha_2{}^1 \\ \alpha_3{}^1 \end{array}
\begin{bmatrix} \Gamma^1 & \Gamma^1 \\ \Gamma^2 & 20 \\ 20 & \Gamma^2 \end{bmatrix}
\end{array}
\end{array}
\qquad
\begin{array}{c}
\Gamma^2 \\[4pt]
\beta_1{}^2 \\
\alpha_1{}^2\,[-10]
\end{array}
$$

This third example is simple to analyze, since the payoff Γ^2 in component game Γ^1 is trivially worth -10 to player 1. Player 1's maximin strategy is always to use $(0\alpha_1{}^1, \tfrac{1}{2}\alpha_2{}^1, \tfrac{1}{2}\alpha_3{}^1)$ in the game Γ^1, player 2's minimax strategy is $(\tfrac{1}{2}\beta_1{}^1, \tfrac{1}{2}\beta_2{}^1)$ for component game Γ^1, and the value is $+5$. By denoting the payoff for $(\alpha_3{}^1, \beta_2{}^1)$ and $(\alpha_2{}^1, \beta_1{}^1)$ as Γ^2, not -10, several points about recursive games are easily demonstrated:

i. In general, the solution of a recursive game cannot be obtained as the limit of solutions of truncated games. For suppose that this recursive game begins with Γ^1 and that, when the play does not terminate by the nth trial, the payoff is zero to both players. This truncated game has value $+10$ to player 1, since he can play $\alpha_1{}^1$ for the first $n-1$ trials and $(0\alpha_1{}^1, \tfrac{1}{2}\alpha_2{}^1, \tfrac{1}{2}\alpha_3{}^1)$ at trial n. Regardless of player 2's strategy at trial n, 1's payoff is the lottery $(\tfrac{1}{2}\,20, \tfrac{1}{2}\Gamma^2)$, which is worth 10 units to him since the outcome Γ^2 on trial n is worth only zero when the game is truncated at n. But, as we noted, the value of the recursive game is 5, not 10.

ii. A fixed point of the "value transformation" does not necessarily yield the value of the game. Also, there is not necessarily a unique fixed point, and, therefore, we cannot conclude (as for stochastic games) that, starting with any initial point, repetitions of the value transformation will necessarily yield a sequence converging to the value of the game. As an illustration, observe that the system

$$
w_1 = \mathrm{val}\begin{bmatrix} w_1 & w_1 \\ w_2 & 20 \\ 20 & w_2 \end{bmatrix}, \qquad w_2 = \mathrm{val}\,[-10] = -10,
$$

has $(w_1, -10)$ as a solution for every $w_1 \geqslant 5$. Furthermore, if we start with $(w_1^{(0)}, w_2^{(0)}) = (0, 0)$, then we get $T(w_1^{(0)}, w_2^{(0)}) = (10, -10)$ and $T^2(w_1^{(0)}, w_2^{(0)}) = T(10, -10) = (10, -10)$, etc. Thus $T^n(0, 0)$ converges to $(10, -10)$ instead of the value $(5, -10)$.

These examples establish that, mathematically, there must be a considerable divergence between the theory of stochastic and recursive games. Nevertheless, Everett [1954] shows that every recursive game has a value and that value is *a* (not *the!*) fixed point of the value transformation. The players do not necessarily have maximin and minimax strategies, but Everett characterizes what may be called their ϵ-optimal strategies. His results, although more complicated than those for stochastic games, can be outlined as follows:

1. Suppose a recursive game has the components $(\Gamma^1, \Gamma^2, \cdots, \Gamma^r)$. An r-tuple $(w_1, w_2, \cdots, w_r)$ is *obtainable* by player 1 if the "value transformation" T,

$$(w_1, w_2, \cdots, w_r) \xrightarrow{T} (w_1', w_2', \cdots, w_r'),$$

where

$$w_k' = \text{val } \Gamma^k(w_1, w_2, \cdots, w_r), \qquad \text{for } k = 1, 2, \cdots, r,$$

has the properties[1]

(i) $w_k' > w_k$,   whenever $w_k > 0$,
(ii) $w_k' \geqq w_k$,   whenever $w_k \leqq 0$,

for $k = 1, 2, \cdots, r$. If $(w_1, \cdots, w_r)$ is obtainable and the game begins with Γ^i, then by playing a maximin strategy of $\Gamma^k(w_1, w_2, \cdots, w_r)$ whenever the component game Γ^k occurs, $k = 1, 2, \cdots, r$, player 1 can guarantee himself an expected return of at least w_i. The analogue for player 2 is: the r-tuple $(w_1, w_2, \cdots, w_r)$ is *obtainable* by 2 if

(i) $w_k' < w_k$ whenever $w_k < 0$,
(ii) $w_k' \leqq w_k$ whenever $w_k \geqq 0$,

for $k = 1, 2, \cdots, r$. If $(w_1, \cdots, w_r)$ is obtainable and the game begins with Γ^i, then, by playing a minimax strategy of $\Gamma^k(w_1, \cdots, w_r)$ whenever the component game Γ^k occurs, $k = 1, 2, \cdots, r$, player 2 can guarantee that player 1 gets an expected return of at most w_i.

Definition. The r-tuple $(w_1^{(0)}, w_2^{(0)}, \cdots, w_r^{(0)})$ is said to be a *critical* vector if, for each $\epsilon > 0$ however small,

(*a*) There exists another r-tuple which is obtainable by 1 and is (componentwise) within ϵ of $(w_1^{(0)}, \cdots, w_r^{(0)})$,

(*b*) There exists another r-tuple which is obtainable by 2 and is (componentwise) within ϵ of $(w_1^{(0)}, \cdots, w_r^{(0)})$.

---

[1] To appreciate the delicacy of requirements (i) and (ii), the reader can check that w_1 satisfies (i) and (ii) in example 1 if and only if $w_1 \leqq 0$, and in example 2 if and only if $w_1 < 1$.

A critical vector (if it exists) is unique, and it is interpreted as the "value" of the recursive game.

2. If *the* critical vector exists, then it is *a* fixed point of the "value transformation" T, i.e., $T(w_1^{(0)}, w_2^{(0)}, \cdots, w_r^{(0)}) = (w_1^{(0)}, w_2^{(0)}, \cdots, w_r^{(0)})$.

3. *Every* recursive game possesses a critical vector.

▶ Although formulating the definition of an obtainable vector is delicate, once it is done correctly proving the assertions in 1 is a straightforward analytical matter. Assertions 2 and 3 are somewhat deeper.   First, consider a recursive game with just one component Γ^1, i.e., Γ^1 is played repeatedly until termination occurs (if it does). Let m and M be, respectively, the minimum and maximum payoff entries in Γ^1. We shall consider val $\Gamma^1(w)$ as a function of w, which can be plotted as in Fig. 1.

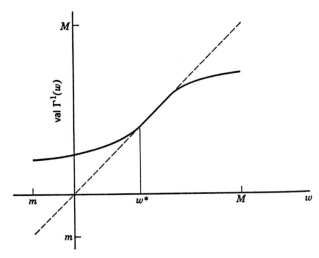

Fig. 1

The following can be shown: val $\Gamma^1(w)$ is monotonically non-decreasing in w, it is continuous, val $\Gamma^1(m) \geqslant m$, and val $\Gamma^1(M) \leqslant M$.   These four properties imply that the graph of the function for $m \leqslant w \leqslant M$ must cut the 45° line which passes through the origin, i.e., there are solutions to the equation $w = $ val $\Gamma^1(w)$.   In Fig. 1, w^* is a critical vector since, for any $\epsilon > 0$, $w^* - \epsilon$ is obtainable by 1, i.e., val $\Gamma^1(w^* - \epsilon) > w^* - \epsilon$, and w^* is itself obtainable by 2 (i.e., val $\Gamma^1(w^*) \leqslant w^*$ and $w^* > 0$).   So the recursive game has value w^*, 2 has a minimax strategy, and 1 has ϵ-maximin strategies.   (A set of inequalities, which we have not presented, rules out the possibility that the graph again crosses the 45° line as w increases. Thus, certain potential pathological examples do not really plague us.)

Next, we consider a recursive game having two components, Γ^1 and Γ^2, and now we let m and M be the minimum and maximum entries of the two payoff tables. If we were arbitrarily to substitute the number w_2 for Γ^2, then the given two component recursive game would collapse into the one component recursive game Γ^1.   As we know, this game has a critical value which, of course, depends upon w_2.   To show this dependence, we shall denote the critical value by $w_1^*(w_2)$.

So val $\Gamma^2[w_1*(w_2), w_2]$ is a function of w_2. As the reader will have anticipated, it can be shown that in the interval $m \leqslant w_2 \leqslant M$ this function is monotically non-decreasing and continuous, it neither lies below the 45° line at $w_2 = m$ nor above that line at $w_2 = M$, and it never crosses the 45° line from below to above as w_2 increases. A plot of val $\Gamma^2[w_1*(w_2), w_2]$ versus w_2 is similar in form to Fig. 1. Let w_2* be the point closest to the origin where the graph crosses the 45° line (such a point exists because of the properties we have just stated). Our claim is that $[w_1*(w_2*), w_2*]$ is a critical vector of the recursive game.

To prove this assertion, one must show that both players have (different) vectors which are obtainable and which lie arbitrarily close to $[w_1*(w_2*), w_2*]$. As an illustration of how this can be done, we will take the case where $w_1*(w_2*) > 0$ and $w_2* > 0$, and we will only worry about a vector obtainable by 1. Choose w_2' be-low and "very close"[2] to w_2*; then by continuity $w_1*(w_2')$ is very close to $w_1*(w_2*)$ and, since w_2* is the point nearest the origin where the curve cuts the 45° line,

$$\text{val } \Gamma^2[w_1*(w_2'), w_2'] > w_2'.$$

Now, $w_1*(w_2')$ is critical for the reduced game Γ^1 when w_2' is substituted for Γ^2; therefore, there exists a w_1' arbitrarily close to $w_1*(w_2')$ which is obtainable in the reduced game, namely:

$$\text{val } \Gamma^1(w_1', w_2') > w_1'.$$

Finally, we note that w_1' is "close" to $w_1*(w_2*)$, and, since w_1' is also close to $w_1*(w_2')$, continuity implies

$$\text{val } \Gamma^2(w_1', w_2') > w_2'.$$

Therefore, (w_1', w_2') is obtainable by player 1.

Everett gives a rigorous inductive proof that a critical vector exists for every recursive game which follows along the above lines.   ◀

Everett [1954] actually proves results more general than the ones we have stated. For example, the component games need not have a finite number of pure strategies provided that each game has a value when any r-tuple $(w_1, w_2, \cdots, w_r)$ is substituted, and maximin and minimax strategies do not have to exist.

As Everett points out, his results include Shapley's existence theorem for stochastic games as a special case. We elected to outline Shapley's work first because the importance of the value transformation is readily demon-strated and because at present the results for stochastic games are stronger. Let the stochastic game have a payoff of

$$a_{ij}{}^k \ \& \ (p_{ij}{}^{k0}S, \ p_{ij}{}^{k1}\Gamma^1, \ \cdots, \ p_{ij}{}^{kr}\Gamma^r)$$

assigned to the strategy pair $(\alpha_i{}^k, \beta_j{}^k)$. A related recursive game has the payoff

$$\left[p_{ij}{}^{k0} \left(\frac{a_{ij}{}^k}{p_{ij}{}^{k0}} \ \& \ S \right), \ p_{ij}{}^{k1}\Gamma^1, \ \cdots, \ p_{ij}{}^{kr}\Gamma^r \right]$$

[2] The rigorous version of this proof uses the familiar (ϵ, δ)-method.

for the same strategy pair. Since $p_{ij}^{k0} > 0$ (a basic assumption for stochastic games), the quantity a_{ij}^{k}/p_{ij}^{k0} is well defined. We note that in the stochastic game the payment of a_{ij}^{k} units is *certain* to occur; whereas, in the related recursive game, the *expected* value of the payment is a_{ij}^{k} units. Also, the transition probabilities for the outcomes "Stop" and "Play Γ^{k} next" are *identical* in the two games. Consequently, the strategic analyses of the stochastic game and its related recursive game are the same.

In the theory of stochastic games, if the assumption that p_{ij}^{k0} is strictly positive for all i, j, and k is dropped, a value does not always exist. This can be illustrated by a simple example. In the game Γ^{1}:

$$\begin{array}{c} \beta_1{}^1 \\ \alpha_1{}^1 \left[1 + \Gamma^1\right], \end{array}$$

player 2 gives player 1 one unit at each stage and the game is repeated. So the "value" (if it can be said to exist) must be infinite. If the meaning of "value" is extended to include any finite number plus "numbers" $+ \infty$ and $- \infty$, then it is easily shown that for *one*-component stochastic games an "extended value" always exists—even if $p_{ij}^{10} = 0$ for some i, j; however, this result is not generally true when there are two or more components. For example, in the game:

$$\begin{array}{cc} \text{Component } \Gamma^1 & \text{Component } \Gamma^2 \\ \beta_1{}^1 & \beta_1{}^2 \\ \alpha_1{}^1 \left[1 + \Gamma^2\right] & \alpha_1{}^2 \left[-1 + \Gamma^1\right] \end{array}$$

players 1 and 2 transfer one unit back and forth, and so a "value" does not exist in the usual sense. However, if all the a_{ij}^{k} are non-negative (or non-positive), an extended value does exist. These games Everett calls *univalent*, and one-component games he calls *simple*.

A8.4 GAMES OF SURVIVAL

Games of survival are one of the possible generalizations of the classical gambler's ruin problem: Two gamblers initially have r and $R - r$ dollars, and at each flip of a (not necessarily fair) coin the loser pays the winner a dollar. The game is terminated when one of the gamblers' capital is exhausted—when he is ruined. Centering attention on the gambler with initial capital r, let p be his (constant) probability of winning a dollar and $q = 1 - p$ his probability of losing a dollar when the coin is tossed. Let q_r denote the probability that he is ultimately ruined. It can be shown (cf. Feller [1950, p. 283]) that

$$q_r = \begin{cases} \dfrac{(q/p)^R - (q/p)^r}{(q/p)^R - 1}, & \text{if } p \neq \dfrac{1}{2}, \\[2ex] \dfrac{R - r}{R}, & \text{if } p = \dfrac{1}{2}. \end{cases}$$

Feller also discusses the distribution of the time duration of the game.

Once the gamblers are committed to playing this ruin game, no strategy problem is involved. But one can be introduced by the following modification. As before, assume two gamblers, players 1 and 2, enter into a game of survival with r and $R - r$ dollars, respectively, but, instead of a chance device determining the payment at each trial, they play a given zero-sum (in monetary units) game. If player 1 uses strategy α_i ($i = 1$, 2, $\cdots$, m) and 2 uses β_j ($j = 1, 2, \cdots, n$) at any trial, then 1 receives a_{ij} dollars from 2. The game is repeated until one of the players is ruined. As an example (Hausner [1952a, b]) suppose:

(a) Player 1 has r dollars, where $r = 1$ or 2 or 3.

(b) Player 2 has $R - r$ dollars, where $R = 4$.

(c) At each trial the players play the zero-sum game

$$\begin{array}{c} & \begin{array}{cc} \beta_1 & \beta_2 \end{array} \\ \begin{array}{c} \alpha_1 \\ \alpha_2 \end{array} & \left[\begin{array}{cc} 2 & -1 \\ -2 & 1 \end{array}\right]. \end{array}$$

This ruin game is equivalent to the following three-component recursive game:

$$\begin{array}{ccc} \text{Component} & \text{Component} & \text{Component} \\ \text{Game } \Gamma^1 & \text{Game } \Gamma^2 & \text{Game } \Gamma^3 \\[1ex] \begin{array}{c} & \begin{array}{cc} \beta_1^{\,1} & \beta_2^{\,1} \end{array} \\ \begin{array}{c} \alpha_1^{\,1} \\ \alpha_2^{\,1} \end{array} & \left[\begin{array}{cc} \Gamma^3 & 0 \\ 0 & \Gamma^2 \end{array}\right] \end{array} & \begin{array}{c} & \begin{array}{cc} \beta_1^{\,2} & \beta_2^{\,2} \end{array} \\ \begin{array}{c} \alpha_1^{\,2} \\ \alpha_2^{\,2} \end{array} & \left[\begin{array}{cc} 1 & \Gamma^1 \\ 0 & \Gamma^3 \end{array}\right] \end{array} & \begin{array}{c} & \begin{array}{cc} \beta_1^{\,3} & \beta_2^{\,3} \end{array} \\ \begin{array}{c} \alpha_1^{\,3} \\ \alpha_2^{\,3} \end{array} & \left[\begin{array}{cc} 1 & \Gamma^2 \\ \Gamma^1 & 1 \end{array}\right], \end{array} \end{array}$$

where Γ^k is interpreted as the game faced by player 1 when his capital is k dollars. Player 1's payoffs (0 if he is ruined and 1 if his adversary is ruined) are selected so that the value of the game has a natural interpretation in terms of ruin probabilities. Suppose players 1 and 2 choose strategies; then the expected value to 1 is:

$0 \times$ (Probability that 1 is ruined + Probability of a
non-terminating play)

$+1 \times$ (Probability that 2 is ruined).

Hence, the expected payoff equals the probability that 2 is ruined.

The value transformation for recursive games applied to w_1, w_2, w_3 yields

$$w_1' = \text{val} \begin{bmatrix} w_3 & 0 \\ 0 & w_2 \end{bmatrix} = \frac{w_3 w_2}{w_3 + w_2},$$

$$w_2' = \text{val} \begin{bmatrix} 1 & w_1 \\ 0 & w_3 \end{bmatrix} = \frac{w_3}{w_3 + 1 - w_1},$$

$$w_3' = \text{val} \begin{bmatrix} 1 & w_2 \\ w_1 & 1 \end{bmatrix} = \frac{1 - w_1 w_2}{2 - w_2 - w_1}.$$

It can be shown that this value transformation has only one fixed point $(w_1{}^*, w_2{}^*, w_3{}^*)$, where

$$w_1{}^* = 1 - \frac{\sqrt{2}}{2} = 0.293, \qquad w_2{}^* = 0.5, \qquad w_3{}^* = \frac{\sqrt{2}}{2} = 0.707.$$

Since $w_1{}^* > 0$, $w_2{}^* > 0$, $w_3{}^* > 0$, the general existence theory of recursive games implies that $(w_1{}^*, w_2{}^*, w_3{}^*)$ is obtainable for player 2 and that a minimax strategy for player 2 is to play his minimax strategy in each of the component games $\Gamma^k(w_1{}^*, w_2{}^*, w_3{}^*)$, $k = 1, 2, 3$. In this special game, the composite strategy in which player 1 uses his unique maximin strategies in the component games $\Gamma^k(w_1{}^*, w_2{}^*, w_3{}^*)$, $k = 1, 2, 3$, is maximin.

In summary: if the two players have a total capital of four units and repeatedly play the zero-sum game

$$\begin{array}{c} & \beta_1 & \beta_2 \\ \alpha_1 & \begin{bmatrix} 2 & -1 \\ \end{bmatrix} \\ \alpha_2 & \begin{bmatrix} -2 & 1 \end{bmatrix} \end{array},$$

then:

Current Capital of Player 1	Probability of 2's Ruin When 1's Current Capital Is r	Maximin Strategy for 1 When 1's Current Capital Is r	Minimax Strategy for 2 When 1's Current Capital Is r
$r = 1$	0.293	$(0.414\alpha_1, 0.586\alpha_2)$	$(0.414\beta_1, 0.586\beta_2)$
$r = 2$	0.5	$(0.5\alpha_1, \quad 0.5\alpha_2)$	$(0.293\beta_1, 0.707\beta_2)$
$r = 3$	0.707	$(0.586\alpha_1, 0.414\alpha_2)$	$(0.414\beta_1, 0.586\beta_2)$

Hausner's treatment of this game, which differs from that given here, predates Everett's work on recursive games and Shapley's work on stochastic games.

As another example, also given by Hausner [1952b], suppose the players each begin with one unit and they play a game of survival based on the zero-sum game

$$\begin{array}{c} & \begin{array}{cc} \beta_1 & \beta_2 \end{array} \\ \begin{array}{c} \alpha_1 \\ \alpha_2 \end{array} & \left[\begin{array}{cc} 0 & 1 \\ 1 & -1 \end{array}\right]. \end{array}$$

If (α_1, β_1) is used at each trial, the play does not terminate. Thus, if each player aims to guarantee his own survival (i.e., non-ruin), the game is non-zero-sum and using (α_1, β_1) at every trial can be thought of as a cooperative "solution." By "player 1's survival game" let us refer to the game in which 1 "wins" if and only if either 2 is eventually ruined or the play is non-terminating. In this game, using α_1 at each trial guarantees that 1 will "win." By "player 2's survival game" let us refer to the game in which 1 "wins" if and only if 2 is eventually ruined. There is no strategy which makes 1 certain of winning in player 2's survival game.

The induced recursive game for player 2's survival game is

Game Γ^1

$$\begin{array}{c} & \begin{array}{cc} \beta_1{}^1 & \beta_2{}^1 \end{array} \\ \begin{array}{c} \alpha_1{}^1 \\ \alpha_2{}^1 \end{array} & \left[\begin{array}{cc} \Gamma^1 & 1 \\ 1 & 0 \end{array}\right]. \end{array}$$

This one-component recursive game has already been studied (cf. p. 462). It will be remembered that $w_1{}^* = 1$ is a critical value; that 2 can obtain the value 1; and that for any ϵ, however small, 1 can obtain the value $1 - \epsilon$ in the sense that, if player 1 uses a mixed strategy which is maximin for the game

$$\left[\begin{array}{cc} 1 - \epsilon & 1 \\ 1 & 0 \end{array}\right]$$

at each trial, he can guarantee himself an expected return of more than $1 - \epsilon$ (but still *less than* 1) in the recursive game. The pure strategy $\alpha_1{}^1$ is maximin in $\Gamma^1(w_1{}^*)$, but it is not maximin in the recursive game since it only gives a security level 0 (because $\beta_1{}^1$ versus $\alpha_1{}^1$ leads to a non-terminating play).

Peisakoff [1952] extended Hausner's work on games of survival, which was restricted to those generated from two-person zero-sum games with two pure strategies for each player, to those with an arbitrary finite number of pure strategies in the component game. Again we shall reverse the historical order, arriving at Peisakoff's results via Everett's theory of recursive games. (Everett did not indicate this connection in his paper.)

Data of the problem.

i. The players have a total initial capital of R units.

ii. Player 1 has an initial capital of r_0 units $(r_0 = 1, 2, \cdots, R - 1)$.

iii. A two-person zero-sum game is given where the payoff to player 1 is a_{ij} if 1 uses α_i and 2 uses β_j, for $i = 1, 2, \cdots, m$, and $j = 1, 2, \cdots, n$. *The return a_{ij} is an integer for all i, j.*

iv. This zero-sum game is played repeatedly until one of the players is ruined (i.e., his capital is reduced to zero).

v. Player 1 "wins" if and only if player 2 is eventually ruined. (This is "player 2's survival game.")

The above game induces a recursive game with $R - 1$ component games Γ^k, $k = 1, 2, \cdots, R - 1$. If 1 uses $\alpha_i{}^k$ and 2 uses $\beta_j{}^k$ in Γ^k, then the payoff is:

(a) $\Gamma^{(k+a_{ij})}$ if $1 \leqslant k + a_{ij} \leqslant R - 1$,
(b) 0 if $k + a_{ij} \leqslant 0$,
(c) 1 if $k + a_{ij} \geqslant R$.

From the general existence theory of recursive games we know: 1. This game has a unique critical vector—an $(R - 1)$-tuple $(w_1{}^*, w_2{}^*, \cdots, w_{R-1}{}^*)$—which is the value of the game in this sense: If 1's initial capital is r_0 units, then he can guarantee that 2 will be ruined with a probability that is arbitrarily close to $w_{r_0}{}^*$, and 2 can guarantee that he will survive (not that 1 will be ruined!) with a probability that is at least equal to $1 - w_{r_0}{}^*$.

2. The critical vector is a fixed point of the value transformation, T, which maps $(w_1, w_2, \cdots, w_{R-1})$ into $(w_1', w_2', \cdots, w_{R-1}')$, where

$$w_k' = \text{val } \Gamma^k(w_1, w_2, \cdots, w_{R-1}), \qquad k = 1, 2, \cdots, R - 1.$$

Finding this fixed point of the value transformation entails solving $R - 1$ equations in $R - 1$ unknowns.

3. From the interpretation of the problem,

$$0 \leqslant w_1{}^* \leqslant w_2{}^* \leqslant \cdots \leqslant w_{R-1}{}^* \leqslant 1;$$

therefore, $(w_1{}^*, w_2{}^*, \cdots, w_{R-1}{}^*)$ can be obtained by player 2, and his minimax strategy is to play a minimax strategy of $\Gamma^k(w_1{}^*, w_2{}^*, \cdots, w_{R-1}{}^*)$ whenever the component game Γ^k occurs.

4. Starting with the $(R - 1)$-tuple $(0, 0, \cdots, 0)$, successive iterations of the value mapping yield a componentwise monotonic non-decreasing sequence of $(R - 1)$-tuples

$$(w_1^{(1)}, w_2^{(1)}, \cdots, w_{R-1}^{(1)}), (w_1^{(2)}, w_2^{(2)}, \cdots, w_{R-1}^{(2)}), \cdots,$$
$$(w_1^{(p)}, w_2^{(p)}, \cdots, w_{R-1}^{(p)}), \cdots,$$

which converge componentwise to the fixed point $(w_1{}^*, w_2{}^*, \cdots, w_{R-1}{}^*)$. The vector $(w_1^{(p)}, w_2^{(p)}, \cdots, w_{R-1}^{(p)})$ is the value of the recursive game truncated at trial p by the vector $(0, 0, \cdots, 0)$ in the sense that, if 1's initial capital is r_0 units, his probability of ruining 2 by trial p is $w_{r_0}^{(p)}$ provided that optimal strategies for the truncated game are used.

These results are similar to, but do not follow from, Shapley's [1953 d] theorems about stochastic games; however, they do follow from Everett's results on univalent games which in turn were motivated by Shapley's results. Of course, neither Shapley's nor Everett's results were available to Peisakoff. Peisakoff's paper contains many of the ingenious tricks used later, but arrived at independently, by Shapley and Everett.

Milnor and Shapley [1955] have further generalized the scope of the theory of games of survival by assuming that the payoffs a_{ij} are not necessarily integers. Since they need not be commensurate quantities, an infinity of different distributions of capital can occur during a single play of the game.

Again, let the total capital of the players be R units, of which player 1 has r_0 units. Neither R nor r_0 need be integers. At each stage they play the m by n zero-sum game $[a_{ij}]$, where the a_{ij}'s are not necessarily integers. It is assumed that $a_{ij} \neq 0$ for all i and j. If any row has all positive entries, its repeated use would automatically ruin player 2. Similarly, if any column has all negative entries, player 2 would be certain of ruining player 1. These cases are both trivial and special, so they are excluded from the theory, i.e., we assume every row and every column has both positive and negative entries. The ultimate payoff to player 1 is

(i)   0 if player 1 is eventually ruined,
(ii)   1 if player 2 is eventually ruined,   (A)
(iii) An amount P_∞ if the play is non-terminating.

Milnor and Shapley show that such games have a value which is independent of the number P_∞ and that both players have strategies which are uniformly optimal for all P_∞. The argument leading to these conclusions is fairly subtle, but we shall sketch it for those who are interested.

►When the a_{ij}'s are integers and, therefore, when player 1's potential set of resource allocations is finite during any play, the value of a game of survival is a fixed point of the value transformation. A related result holds for the generalization, but, rather than use the notation $(w_1{}^*, \cdots, w_{R-1}{}^*)$, which would be awkward in the present context, let us employ the symbolism $[\varphi(1), \varphi(2), \cdots, \varphi(R-1)]$. Then the fixed point of the value transformation must satisfy the functional equation

$$\varphi(r) = \text{val} \begin{bmatrix} \varphi(r + a_{11}) & \cdots & \varphi(r + a_{1n}) \\ \cdot & & \cdot \\ \cdot & & \cdot \\ \cdot & & \cdot \\ \varphi(r + a_{m1}) & \cdots & \varphi(r + a_{mn}) \end{bmatrix}, \qquad (B)$$

where

$$\varphi(r) = \begin{cases} 0, & \text{if } r \leqslant 0, \\ 1, & \text{if } r \geqslant R. \end{cases} \tag{C}$$

When r is restricted to integral values, the functional equation B, with boundary conditions C, simplifies to $R - 1$ equations in $R - 1$ unknowns. In the general case, r is not restricted to integral values, but B and C still play a central role.

In a given play of the survival game, let r_k represent player 1's capital at the end of k trials. The sequence $\{r_0,\ r_1,\ r_2,\ \cdot\cdot\cdot\}$ gives a trial-by-trial record of 1's financial holdings in the given play. If $0 < r_k < R$ for all k, then play does not terminate and 1's payoff is P_∞; if $0 < r_k < R$ for $k = 0, 1, \cdot\cdot\cdot, N - 1$ and $r_N \leqslant 0$ or $r_N \geqslant R$, then the play terminates at trial N, and we assume $r_{N+p} = r_N$ for $p = 0, 1, \cdot\cdot\cdot$. If the players choose pure strategies for the survival game, then the sequence $\{r_k\}$ is uniquely determined; if they choose mixed strategies, then they jointly generate a probability measure over the set of $\{r_k\}$ sequences. In probabilists' parlance, the set of sequences plus a probability measure over them is called a stochastic process, and a particular sequence is said to be a realization of this stochastic process.

Suppose a given play results in a sequence $\{r_k\}$. If player 1 is ultimately ruined ($r_N \leqslant 0$) his payoff is zero regardless of the value r_N, i.e., the payoff does not vary with the difference $r_N - 0$. Similarly, if player 2 is ultimately ruined ($r_N \geqslant R$), 1's payoff is 1 regardless of the value of r_N, i.e., the payoff is independent of the difference $r_N - R$. Neglecting these differences, which is conceptually trivial to do, leads to some mathematical complications. Let us see why. It is plausible that, for two different initial amounts of capital, player 1 can have the same expected payoffs. Mathematically, this means that the function $\varphi(r)$, which eventually will be identified as the value of the game to player 1 when his initial capital is r, will not be *strictly increasing* in the interval $0 \leqslant r \leqslant R$. Without strict monotonicity, and therefore without a 1 to 1 relation between $\varphi(r)$ and r, we cannot make exact inferences about the sequence $\{r_1, r_2, \cdot\cdot\cdot\}$ from an analysis of the mathematically more tractable sequence $\{\varphi(r_1), \varphi(r_2), \cdot\cdot\cdot\}$.

We can eliminate the difficulty by modifying the payoffs of the survival game to take into account the excess by which the capital limits are exceeded. For example, the payoff given in A can be changed to:

(i) ϵr_N,   if $r_N \leqslant 0$,

(ii) $1 + \epsilon(r_N - R - M)$,   if $r_N \geqslant R$,   (D)

(iii) P_∞,   if $0 < r_k < R$, for all k,

where $M = \max\limits_{i,j} |a_{ij}|$ and ϵ is a small positive quantity. (Later, limits will be evaluated as ϵ approaches zero.) Similarly, the boundary condition C can be changed to

$$\varphi(r) = \begin{cases} \epsilon r, & \text{if } r \leqslant 0, \\ 1 + \epsilon(r_N - R - M), & \text{if } r \geqslant R. \end{cases} \tag{E}$$

Note that as ϵ approaches zero, the payoffs in D approach those in A and the boundary conditions in E approach those in C.

Now let us assume that we have a function φ which satisfies B and E and which is strictly increasing in $0 \leqslant r \leqslant R$. For the game with payoff D, let player 1 adopt the strategy that, if he has a capital accumulation of r_k, $0 < r_k < R$, at trial k, then he plays a maximin strategy for the zero-sum game

$$[\varphi(r_k + a_{ij})], \qquad i = 1, 2, \cdots, m \quad \text{and} \quad j = 1, 2, \cdots, n.$$

Player 1's capital accumulation at trial $k + 1$ therefore depends upon chance if either player's strategy at trial k is mixed. To stress this fact, we denote his capital at trial $k + 1$ by $\dot{r}_{k+1}$, where the dot stands for "random quantity." For example, the probability that $\dot{r}_{k+1} = r_k + a_{ij}$ is the product of the probability that 1 chooses α_i and the probability that 2 chooses β_j at trial k. The value which the function φ assumes at trial $k + 1$ depends, therefore, upon chance.

Suppose now we have a *known* past history $(r_0, r_1, \cdots, r_k)$ at trial $k + 1$. If, as we assumed, 1 plays maximin in the game $[\varphi(r_k + a_{ij})]$ and 2 plays any fixed strategy, then the random variable $\varphi(\dot{r}_{k+1})$ must, by the meaning of maximin, have an expected value at least as large as val $[\varphi(r_k + a_{ij})]$. But by the assumption that φ satisfies eq. B, this must equal $\varphi(r_k)$. Thus, if 1 follows this strategy at every k and 2 plays any fixed strategy, a measure is induced over the sequence $\{r_k\}$, and therefore over $\{\varphi(r_k)\}$, with the property that

$$E[\varphi(\dot{r}_{k+1}) \mid \varphi(r_k), \cdots, \varphi(r_0)] \geqslant \varphi(r_k) \qquad (F)$$

for each k and each partial sequence $r_0, r_1, \cdots, r_k$. That is, for any past history $r_0, r_1, \cdots, r_k$ [or equivalently, $\varphi(r_0), \varphi(r_1), \cdots, \varphi(r_k)$] the random quantity $\varphi(\dot{r}_{k+1})$ is well defined and its expected value, conditional upon the past, is at least $\varphi(r_k)$. (To make this assertion mathematically precise, one must insert some "with probability 1" qualifiers.) A stochastic process which satisfies F is said to be a *semimartingale* (Doob [1953]). Milnor and Shapley apply Doob's existence theorem for semimartingales to show:

1. The set of infinite sequences $\{\varphi(r_k)\}$ for which the limit of $\varphi(r_k)$ as $k \to \infty$ does *not* exist has probability zero. Thus, intuitively, we can think of any play of the game as generating a sequence for which $\lim\limits_{k \to \infty} \varphi(r_k)$ exists.

2. The limiting value must depend upon chance since the sequence itself depends upon chance, so, for emphasis, we write $\lim\limits_{k \to \infty} \varphi(\dot{r}_k)$. The probability distribution of these limits is such that its expectation, conditional upon knowing r_0, is at least $\varphi(r_0)$, namely:

$$E[\lim_{k \to \infty} \varphi(\dot{r}_k) \mid \varphi(r_0)] \geqslant \varphi(r_0).$$

From these facts we can conclude, as follows, that *the probability of the play not terminating is zero*. According to 1 the probability that the sequence $\{\varphi(r_k)\}$ does not converge is zero, and, since φ is assumed to be *strictly monotonic*, this implies that the probability that the sequence $\{r_k\}$ does not converge is also zero. But, since $r_{k+1} = r_k + a_{ij}$ for some i and j and $a_{ij} \neq 0$, $r_{k+1} \neq r_k$, if $0 < r_k < R$. This with the convergence of $\{r_k\}$ implies the eventual ruin of one player. Let N be the first trial where one of the players is ruined. Now noting that, if 1 is eventually ruined $r_N \leqslant 0$ and $\varphi(r_N) \leqslant \varphi(0) = 0$, and if 2 is eventually ruined $R \leqslant r_N \leqslant R + M$ and $\varphi(r_N) \leqslant \varphi(R + M) = 1$, we get

$$E[\lim_{k \to \infty} \varphi(\dot{r}_k) \mid \varphi(r_0)] \leqslant 0 \times \text{Prob}\,[1 \text{ is ruined} \mid r_0] + 1 \times \text{Prob}\,[2 \text{ is ruined} \mid r_0],$$

and therefore by 2

$$\text{Prob}\,[2 \text{ is ruined} \mid r_0] \geqslant \varphi(r_0).$$

Summary. If φ is a solution to B and E and if player 1 plays maximin in $[\varphi(r + a_{ij})]$ whenever his current capital is r, then, regardless of 2's strategy, the play terminates with probability 1 and the probability that 2 will be ruined is not less than $\varphi(r_0)$. A similar analysis, with the roles of players 1 and 2 interchanged, shows that, if player 2 plays minimax in $[\varphi(r + a_{ij})]$ whenever 1's current capital is r, then, regardless of 1's strategy, the play terminates with probability 1, and the probability that 2 will be ruined is not greater than $\varphi(r_0) + \epsilon M$. Note that the function φ depends upon the value of the ϵ in boundary condition E, so we really should write $\varphi_\epsilon(r_0)$ instead of merely $\varphi(r_0)$. The value to player 1 of the original game is therefore

$$v(r_0) = \lim_{\epsilon \to 0} \varphi_\epsilon(r_0),$$

and this is independent of P_∞.

To show that there is a strictly monotonic solution to B and E, as we assumed earlier on p. 473, is a major feat in itself. As a step in this demonstration, Milnor and Shapley use the fact that player 1's survival game (i.e., 1 loses only if 1 is ruined, i.e., $P_\infty = 1$) has a value. This was demonstrated by Scarf and Shapley [1954] in a very abstract paper, which in turn depends upon work of Glicksberg [1950].

Milnor and Shapley [1955] go on to show that each of the players actually has an optimal strategy which forces an end to the play with probability 1, and that, therefore, such a strategy is uniformly optimal for all P_∞. ◀

Milnor and Shapley's results, although proved constructively, would be terribly difficult for a player to use, so a simple approximation to the solution is desirable. They give one which is quite good provided that max $|a_{ij}|$ is small compared with the player's initial fortunes, r_0 and $R - r_0$. Let

$$\varphi^*(r) = \begin{cases} (e^{\lambda_0 r} - 1)/\lambda_0, & \text{if } \lambda_0 \neq 0, \\ r, & \text{if } \lambda_0 = 0, \end{cases}$$

where λ_0 is the unique[3] solution to

$$\text{val} \begin{bmatrix} \dfrac{e^{\lambda a_{11}} - 1}{\lambda} & \cdots & \dfrac{e^{\lambda a_{1n}} - 1}{\lambda} \\ \cdot & & \cdot \\ \cdot & & \cdot \\ \cdot & & \cdot \\ \dfrac{e^{\lambda a_{m1}} - 1}{\lambda} & \cdots & \dfrac{e^{\lambda a_{mn}} - 1}{\lambda} \end{bmatrix} = 0.$$

Now, if at each trial of the given survival game players 1 and 2 use maximin and minimax strategies of the game $[\varphi^*(a_{ij})]$, then the game terminates with probability 1. If $-m = \min_{i,j} a_{ij}$ and $M = \max_{i,j} a_{ij}$, then player 2's probability of being ruined is at least $\varphi^*(r_0)/\varphi^*(R + M)$ and at

---

[3] The solution can be proved both to exist and to be unique.

most $\varphi^*(r_0 + m)/\varphi^*(R + m)$. Therefore, if $v(r_0)$ denotes the value of the survival game to player 1, we have the bounds

$$\frac{\varphi^*(r_0)}{\varphi^*(R + M)} \leqslant v(r_0) \leqslant \frac{\varphi^*(r_0 + m)}{\varphi^*(R + m)},$$

which, of course, increase in precision as m and M are made smaller relative to r_0 and R.

▶A sketch of the proof follows. First, $\varphi^*(r)$ is clearly strictly monotonically increasing. Second, it satisfies eq. B since

$$\text{val } [\varphi^*(r + a_{ij})] = \text{val} \left[\frac{e^{\lambda_0(r + a_{ij})} - 1}{\lambda_0} \right]$$

$$= \text{val} \left[e^{\lambda_0 r} \left(\frac{e^{\lambda_0 a_{ij}} - 1}{\lambda_0} \right) + \frac{e^{\lambda_0 r} - 1}{\lambda_0} \right]$$

$$= \frac{e^{\lambda_0 r} - 1}{\lambda_0} + e^{\lambda_0 r} \text{ val} \left[\frac{e^{\lambda_0 a_{ij}} - 1}{\lambda_0} \right]$$

$$= \varphi^* (r),$$

using the definition of φ^* and the fact that λ_0 is the solution to val $[(e^{\lambda a_{ij}} - 1)/\lambda]$ $= 0$. Third, the maximin and minimax strategies for the game $[\varphi^*(r + a_{ij})]$ are completely independent of r for, by what we have just seen, $[\varphi^*(r + a_{ij})]$ and $[\varphi^*(a_{ij})]$ are strategically equivalent, i.e., differ only by a linear transformation of the entries, and the latter does not depend upon r. Now, if at each trial 1 plays his maximin strategy for the game $[\varphi^*(a_{ij})]$, then the sequence $\{\varphi^*(r_k)\}$ is a realization of a semimartingale stochastic process which converges with probability one. But in that case, as we saw before,

$$\varphi^*(r_0) \leqslant E \left[\lim_{k \to \infty} \varphi^*(r_k) \mid \varphi^*(r_0) \right] \leqslant \varphi^*(0) \text{ Prob } [1 \text{ is ruined} \mid r_0]$$
$$+ \varphi^*(R + M) \text{ Prob } [2 \text{ is ruined} \mid r_0].$$

Since $\varphi^*(0) = 0$, we conclude

$$\text{Prob. } [2 \text{ is ruined} \mid r_0] \geqslant \frac{\varphi^*(r_0)}{\varphi^*(R + M)},$$

The analysis for player 2 is similar.   ◀

A8.5 MULTICOMPONENT ATTRITION GAMES

A whimsical instance of a multicomponent attrition game, which may be titled "Women and cats versus men and mice," is due, as is the general class of games, to Blackwell [1954 a]. In each component game, team 1, which initially consists of $a_1^{(0)}$ women and $a_2^{(0)}$ cats, puts either a woman or a cat in the ring without any knowledge of what team 2 will do; in similar ignorance, team 2, whose initial composition is $b_1^{(0)}$ men and $b_2^{(0)}$ mice, sends forth a man or a mouse. The outcome is determined by the rule: a woman eliminates a man, who eliminates a cat, who eliminates a mouse, who in turn, eliminates a woman, i.e.,

$$
\begin{array}{c}
\text{Team 2} \\
\begin{array}{cc}
\text{Man} & \text{Mouse}
\end{array} \\
\text{Team 1} \quad
\begin{array}{c}
\text{Woman} \\[2em]
\text{Cat}
\end{array}
\begin{bmatrix}
\text{Team 2 loses} & \text{Team 1 loses} \\
\text{a man} & \text{a woman} \\[1em]
\text{Team 1 loses} & \text{Team 2 loses} \\
\text{a cat} & \text{a mouse}
\end{bmatrix}.
\end{array}
$$

The overall game is one of attrition in the sense that the component games are repeated until one side is decimated, which it really is, of course, whenever one of its two components is reduced to zero. Clearly, each team's mixed strategy at each engagement should depend upon the current resources of *both* teams.

Blackwell's general class of games is a fairly straightforward generalization of the example. Player (or team) 1 has R different types of commodities with an initial supply of $a_r^{(0)}$ units of type r, $r = 1, 2, \cdots, R$. Player (or team) 2 has S different types of commodities with an initial supply of $b_s^{(0)}$ units of type s, $s = 1, 2, \cdots, S$. It will be convenient to denote the initial R-tuple $(a_1^{(0)}, a_2^{(0)}, \cdots, a_R^{(0)})$ by $\mathbf{a}^{(0)}$, and, similarly, 2's initial S-tuple by $\mathbf{b}^{(0)}$. We assume that the players have m and n pure strategies, respectively, and that the effect of each play of the game is to reduce their current supply of the commodities. This reduction is given by the attrition matrices $[\alpha_r(i, j)]$, where the typical entry is the amount that player 1's rth commodity is diminished when strategies i and j are used, and $[\beta_s(i, j)]$, where the typical entry is the amount that player 2's sth commodity is diminished when strategies i and j are used. Thus, if the strategy pairs $(i_1, j_1), (i_2, j_2), \cdots, (i_k, j_k)$ are used during the first k trials, the remaining amounts of resources are:

$$
a_r^{(k)} = a_r^{(0)} - \sum_{q=1}^{k} \alpha_r(i_q, j_q), \text{ if this number is positive,}
$$

$$
= 0, \text{ otherwise,}
$$

and

$$
b_s^{(k)} = b_s^{(0)} - \sum_{q=1}^{k} \beta_s(i_q, j_q), \text{ if this number is positive,}
$$

$$
= 0, \text{ otherwise.}
$$

It is assumed that each player tries to exhaust one—any one—of his adversary's commodities without, however, allowing any of his stocks to vanish.

This is a recursive game, as we can see by defining player 1's payoff on trial k to be either the game with the resources remaining after the game on trial k, provided neither player has lost all of any one of his commodities, or 1 if any of player 2's stocks go to zero, or 0 if none of 2's stocks go to zero and at least one of 1's do.   (Observe that the convention has been made that player 1 "wins" whenever both players simultaneously exhaust a commodity.)   The play terminates when either a 0 or a 1 payoff occurs. To guard against infinite play, Blackwell requires that at least one commodity be diminished in each engagement and that no resupply ever occurs.   Stated formally, for every (i, j) pair,

$$\sum_{r=1}^{R} \alpha_r(i, j) + \sum_{s=1}^{S} \beta_s(i, j) > 0$$

and

$$\alpha_r(i, j) \geqslant 0, \qquad \text{for every } r,$$

and

$$\beta_s(i, j) \geqslant 0, \qquad \text{for every } s.$$

In sum, a multicomponent attrition game is described by two complexes of information: the initial resources $(\mathbf{a}^{(0)}, \mathbf{b}^{(0)})$, and the attrition matrix $[(\alpha(i, j), \beta(i, j)]$, which we shall abbreviate simply as (α, β).   Each entry of the matrix (α, β) is an $(R + S)$-tuple, the first R components of the (i, j) entry $[\alpha_1(i, j), \cdots, \alpha_R(i, j)]$, being designated by $\alpha(i, j)$, and the last S components, $[\beta_1(i, j), \cdots, \beta_S(i, j)]$, by $\beta(i, j)$.   Since the payoffs have been chosen to be 0 and 1, the value of the game is merely the probability, which we denote by $P[\alpha, \beta; \mathbf{a}^{(0)}, \mathbf{b}^{(0)}]$, that player 1 wins when both players use their optimal strategies.   Since multicomponent attrition games are special cases of recursive games, we know that, when P is treated as a function of $(\mathbf{a}^{(0)}, \mathbf{b}^{(0)})$, with (α, β) held constant, it must satisfy the basic functional equation of stochastic and recursive games. Of course, even in special instances, that equation is monstrous, and Blackwell does not attempt to solve it as such.   Rather, he investigates the asymptotic behavior of P as the resources $(\mathbf{a}^{(0)}, \mathbf{b}^{(0)})$ are increased indefinitely subject to the condition that their relative sizes are fixed. For example, one can look for the set of $(\mathbf{a}^{(0)}, \mathbf{b}^{(0)})$ pairs such that

$$\lim_{t \to \infty} P[\alpha, \beta; t\mathbf{a}^{(0)}, t\mathbf{b}^{(0)}] = 1,$$

where, of course, $t\mathbf{a}^{(0)} = (ta_1^{(0)}, ta_2^{(0)}, \cdots, ta_R^{(0)})$ and similarly for $t\mathbf{b}^{(0)}$. For the women and cats versus men and mice example, Blackwell shows that

$$\lim_{t \to \infty} P \left\{ \begin{bmatrix} (0, 0; 1, 0) & (1, 0; 0, 0) \\ (0, 1; 0, 0) & (0, 0; 0, 1) \end{bmatrix}, (ta_1^{(0)}, ta_2^{(0)}; tb_1^{(0)}, tb_2^{(0)}) \right\} = 1,$$

provided that

$$a_1^{(0)} a_2^{(0)} > b_1^{(0)} b_2^{(0)}.$$

Note, for example, that $(0, 0; 1, 0)$, which is the $(1, 1)$ entry of matrix (α, β), has the interpretation: team 1 loses zero women and zero cats and team 2 loses one man and zero mice.

A8.6 APPROACHABILITY-EXCLUDABILITY THEORY AND COMPOUND DECISION PROBLEMS

The asymptotic theory of multicomponent attrition games is based on Blackwell's [1956 a] analogue of the minimax theorem for games with

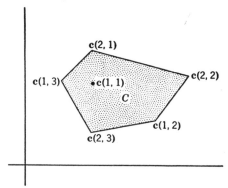

Fig. 2

vector payoffs. In such games, the players have m and n pure strategies as usual, but the payoff corresponding to the (i, j) strategy pair is a Q-tuple (or vector in Q-space) of the form $\mathbf{c}(i, j) = [c_1(i, j), c_2(i, j), \cdots, c_Q(i, j)]$. The multicomponent attrition games are of this form with $Q = R + S$ and $\mathbf{c}(i, j)$ equal to the attrition payoff, but, as we shall see below, quite different interpretations of vector games also exist.

Let us denote by C the convex hull of the set of points (in Q-space) $\mathbf{c}(i, j)$, where i and j vary over their domains. For example, if $Q = 2$, $m = 2$, and $n = 3$, then a typical region C is shown in Fig. 2. Blackwell raises this question: *If such a game is repeated in time, can player 1 force the average payoff to approach a preassigned closed subset T of C? Equally well, when can player 2 exclude the average payoff from T?*

The following notation will be useful. Let $\mathbf{x} = (x_1, x_2, \cdots, x_m)$ be one of player 1's mixed strategies on a component game; then if player 2 uses pure strategy j, the expected payoff will be

$$c(\mathbf{x}, j) = \sum_i x_i c(i, j).$$

Thus, his expected payoff when he uses $\mathbf{x}$ will lie in the smallest convex set containing the n points $c(\mathbf{x}, j)$, $j = 1, 2, \cdots, n$; we denote this set by $C(\mathbf{x}, \cdot)$. Exactly parallel notation [$\mathbf{y}$, $c(i, \mathbf{y})$, and $C(\cdot, \mathbf{y})$] is introduced for player 2. Finally, the average payoff for k trials is denoted

$$\bar{c}^{(k)} = [c(i_1, j_1) + c(i_2, j_2) + \cdots + c(i_k, j_k)]/k,$$

where (i_h, j_h) denotes the strategy pair chosen on trial h.

We observe that a sufficient condition for T to be excludable by player 2 is the existence of a strategy $\mathbf{y}^{(0)}$ such that $C(\cdot, \mathbf{y}^{(0)})$ is disjoint from T, for if $\mathbf{y}^{(0)}$ is used at each trial the average payoff will approach $C(\cdot, \mathbf{y}^{(0)})$, and so not T. Blackwell shows, in essence, that this is a necessary condition too. To be more precise: *any convex set T is either approachable by* 1 *or excludable by* 2, *and the latter is equivalent to the existence or a $\mathbf{y}^{(0)}$ such that T and $C(\cdot, \mathbf{y}^{(0)})$ are disjoint.* Furthermore, he displays a strategy for player 1 which will force the average payoff to approach T whenever such a strategy exists.

The idea is simple. If at trial k, the average payoff $\bar{c}^{(k)}$ is already in T, select any $\mathbf{x}$ on trial $k + 1$. If, however, $\bar{c}^{(k)}$ and T are disjoint, choose $\mathbf{x}$ so that $C(\mathbf{x}, \cdot)$ and T lie on the same side of the supporting hyperplane of T which both passes through the point c' of T that is closest to $\bar{c}^{(k)}$ and which is perpendicular to the line joining these two points. (See Fig. 3.) Such an $\mathbf{x}$ can be shown to exist if and only if the convex set T is not excludable by 2. (Roughly the idea is this: Suppose 1 tries to get an expected payoff which lies as far below the separating hyperplane as possible. Player 2 cannot guarantee that 1 will not get a point on or below this hyperplane since $C(\cdot, \mathbf{y})$ intersects T for all $\mathbf{y}$. Now we invoke the usual form of the minimax theorem to conclude that 1 can therefore guarantee a point on or below the hyperplane.) Since the expected payoff c^* on trial $k + 1$ will, of course, be in $C(\mathbf{x}, \cdot)$, let us, for heuristic reasons, simplify the argument by supposing that the actual payoff on trial $k + 1$ is the point c^* in $C(\mathbf{x}, \cdot)$; then the average payoff $\bar{c}^{(k+1)}$ will lie on the line joining $\bar{c}^{(k)}$ and c^*. If k is large, $\bar{c}^{(k+1)}$ will be much nearer to $\bar{c}^{(k)}$ than to c^*, and so it will be nearer to c' than $\bar{c}^{(k)}$ is. This suggests that in time the average payoff will approach T. As yet, however, the argument is not tight for we have been dealing with expected values at a given trial, whereas the approachability theorem asserts something about the time sequence $\{\bar{c}^{(k)}\}$ being true with probability 1. This gap is bridged by a probability existence theorem that we will not discuss except to remark that it is similar in spirit to the martingale theorem which arose in the section on recursive games.

Two points about approachability-excludability theory need clarification: why is it related to the study of multicomponent attrition games, and in what sense is it an analogue of the minimax theory? The first seems to be a problem since we know that multicomponent games are recursive games, whereas the present theory is not cast in that form. But recall that Blackwell confined himself to questions about ruin probabilities when the initial resources are held in fixed proportion and increased without bound. It is thus plausible that each player's ability to control the limiting behavior of the time average of the attrition payoffs will govern the outcome, and in fact it does.

Next, let us turn to the sense in which the theory generalizes the minimax theorem. Suppose that the payoffs $\mathbf{c}(i, j)$ are actually real numbers,

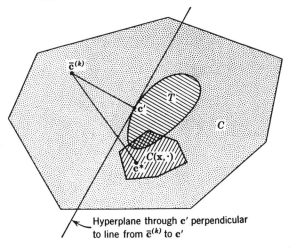

Hyperplane through $\mathbf{c}'$ perpendicular to line from $\bar{\mathbf{c}}^{(k)}$ to $\mathbf{c}'$

Fɪɢ. 3

i.e., $Q = 1$, and that they are interpreted as 1's payoffs. If we let a denote the minimum and b the maximum of these mn numbers, the set C is simply the interval of the real line from a to b inclusive. If v denotes the value of the game, player 1 can approach the interval $[v, b]$ and player 2 can approach the interval $[a, v]$. Or in more familiar words, using the law of large numbers, the expected value v of a two-person zero-sum game can be given a frequency interpretation as the limiting value of a temporal average.

Earlier we promised a second and important interpretation of the approachability-excludability theory, and it is now time to fulfill it. Let us suppose that a two-person game is to be repeated and that player 1 is solely interested in his long-term average payoff. He can certainly secure a limiting average at least equal to the maximum value of the component game by playing maximin at each stage. But, as we have pointed out previously, it has long been recognized that such a strategy is not very realistic in any of the following cases:

i. In a zero-sum game when player 2 is not a conscious minimaxer.
ii. In a non-zero-sum game.
iii. When player 2 is "nature" in the usual decision problem under uncertainty—the statistical inference problem.

Robbins [1951] has emphasized that when a (statistical) decision problem is repeated in time, e.g., when a stream of individuals must be classified by their individual test responses, the statistician can often do as well asymptotically with no prior information as when he knows the exact limiting proportion of times player 2 uses each strategy. To be more specific, suppose 1's payoffs are a_{ij} and that *a priori* he knows that the proportion of the time player 2 will use strategy j, $j = 1, 2, \cdots, n$, is y_j^*. He can, therefore, achieve the limiting average return

$$\rho(\mathbf{y}^*) = \max_i \left(\sum_j a_{ij} y_j^* \right)$$

by playing that strategy i which maximizes the right-hand expression on each trial. Hannan [1957] shows that asymptotically player 1 can do as well as $\rho(\mathbf{y}^*)$ *without knowing* $\mathbf{y}^*$ *beforehand* provided that he bases his choice at each trial on his knowledge of 2's previous choices and on chance. (Actually, he need only consider 2's empirical mixed strategy over the preceding moves.)

Blackwell [1956 b] shows that this can be concluded from approachability-excludability theory. He chooses $Q = n + 1$ and defines

$$c(i, j) = (0, 0, \cdots, 0, 1, 0, \cdots, 0, a_{ij}),$$

where the 1 appears in the jth position and a_{ij} is the (i, j) payoff of the given game to player 1. This definition may seem strange, but it is less so when one observes that the first n components of $\bar{\mathbf{c}}^{(k)}$ equal player 2's empirical mixed strategy over the first k trials and the last component is 1's average payoff during those trials. Now, let T be the set of all $(n + 1)$-tuples whose first n components represent a probability vector, call it $\mathbf{y}$, and whose last component, c_{n+1}, is at least equal to $\rho(\mathbf{y})$, i.e.,

$$T = \{\text{the set of all } (c_1, c_2, \cdots, c_n, c_{n+1}) \text{ such that } c_j \geqslant 0,$$

$$\text{for } j = 1, 2, \cdots, n, \quad \sum_{j=1}^n c_j = 1,$$

$$\text{and } c_{n+1} \geqslant \sum_{j=1}^n a_{ij} c_j, \quad \text{for } i = 1, 2, \cdots, m\}.$$

The result is proved if we can show that T is approachable by 1, for, if it is approachable, then with any limiting distribution $\mathbf{y}^*$ player 1 receives a limiting average value of at least $\rho(\mathbf{y}^*)$. Note that we do not necessarily assume that the empirical mixed strategy over the first k trials, $\mathbf{y}^{(k)}$, approaches a limit as $k \to \infty$. When the limit does not exist, the result is interpreted roughly as meaning that the average payoff for large k will be close to $\rho(\mathbf{y}^{(k)})$.

The approachability of T follows from the observation that, for each $\mathbf{y}$, the set $C(\cdot, \mathbf{y})$ just touches T. This we can see as follows: If $\mathbf{y} = (y_1, y_2, \cdots, y_n)$, then $C(\cdot, \mathbf{y})$ is the set of $(n+1)$-tuples $(y_1, y_2, \cdots, y_n, c_{n+1})$, where $\min_i \sum_j a_{ij} y_j \leqslant c_{n+1} \leqslant \max_i \sum_j a_{ij} y_j$, so it intersects T at the point $[y_1, y_2, \cdots, y_n, \rho(\mathbf{y})]$.

The choice of a strategy which leads the average payoff to approach T is far more subtle than it may seem. For example, player 1's "obvious" strategy of playing optimal on trial $k+1$ against 2's empirical mixed strategy calculated over the first k trials need not force the average payoff to converge to T. Remember that player 2 may not employ the limiting mixed strategy $\mathbf{y}^*$ at every (or indeed, any) of the trials.

Besides this asymptotic result, Hannan [1957] also has a great deal to say about the rates of convergence for certain reasonable classes of player 1's strategies. Other papers which extend the pioneering work of Robbins [1951] on compound statistical decision problems are Hannan and Robbins [1955], Laderman [1955], and Johns [1956].

A8.7 DIVIDEND POLICY AND ECONOMIC RUIN GAMES

Most of the games we have encountered in this appendix meet the following very general description: a known stochastic process is under way, but at periodic intervals two players, perhaps opposing, can exert some influence on the process. Shubik [1957] has pointed out that corporate dividend policy can be looked upon in this way, and he has begun to examine games suggested by this interpretation.

The simplest case is the degenerate single corporation game in which its assets fluctuate from period to period according to a simple chance mechanism. For example, if the capital accumulation is Z units (units in terms of thousands or tens of thousands of dollars) in one period, we might assume that in the next period it becomes $Z + 1$ with probability p, or $Z - 1$ with probability $q = 1 - p$. The corporation is ruined if at any period its capital drops below zero. Clearly, its chance of being ruined within a specified time period is less the greater the capital at the beginning of that period, but, on the other hand, money in the corporate till does not satisfy the stockholders until dividends are declared, which, of course, reduces the capital. Furthermore, a dollar delivered to the stockholders k periods from the present is assumed to be only worth ρ^k at present, where, in general, ρ depends upon the interest rate and so is of the order of 0.95. The conflicting motives are clear: should the corporation declare a dividend of s dollars at present, and thereby increase its chance of ruin, or should it wait until its financial position is more secure, knowing that money paid out in the future is of less value than if it were paid out now? If one assumes that the corporation wishes to maximize the present value of all future dividends during the period that the corporation is solvent,[4] then clearly an optimal dividend policy depends upon p and ρ, and the problem becomes one of dynamic programing of the inventory type.

As stated, the problem is naively simple, but we can complicate it in a variety of ways. First, it can be made a one-person game against nature by supposing that p is unknown. Of course, as the random walk unfolds, data will be accumulated and inferences about p can be made. But what about dividend policy in the early stages?

Second, another corporation may be assumed, also with prescribed initial assets, and the two play a competing survival game. That is, if at any trial corporation 1 plays i and corporation 2 plays j, then their assets are changed by a_{ij} and b_{ij} units, respectively. This survival game, quite likely with a non-zero-sum component game, proceeds as usual, but as in the simpler case both corporations must worry about dividend policies. The problem is not completely formulated until it is known what happens to one corporation once the other becomes insolvent; one possible assumption is that the remaining one gains δ units per period in perpetuity. This does not result in an infinite reward because of the ever-present discount rate ρ. The collusion possibilities are enormous.

Third, the entry of new funds can be introduced in the sense of allowing corporate assets to be bolstered by new shareholders. Such entry can be assumed to depend upon present assets and past dividend policy. So the model increases in complexity, if not in tractability.

In the past few years, Shubik and others have begun to take nibbles at this extremely inviting set of problems.

---

[4] To be sure, boards of directors rarely are solely interested in maximizing the present value of future dividends—if nothing else, the present value of their future salaries should have some influence on their policies. Conceptually, such features are easily included; but since we are only trying to point to a class of problems, and not to give a valid economic analysis, we shall suppress such realistic embellishments.

Bibliography

The following bibliography, although extensive, is far from being exhaustive. Its coverage of game theory proper is relatively complete, but of allied subjects it is less so. This should not create any difficulty since there are excellent bibliographies of these areas in other recent volumes. For the foundations of probability see Savage [1954]; for linear programing, convex bodies, and linear inequalities, Kuhn and Tucker [1956 *b*]; for statistical decision theory, Blackwell and Girshick [1954]; and, for utility theory, Edwards [1954 *c*] and Savage [1954].

Abramson, L. R., *Linear Conditional Utility Functions*, Department of Mathematical Statistics, Columbia University, 1956 (unpublished).

Ackoff, R. L., C. W. Churchman, and E. L. Arnoff. See Churchman.

Adams, E. W., "A survey of Bernoullian utilities and applications," Behavioral Models Project, *Technical Report 9*, Columbia University, 1954.

————, and R. D. Luce. See Luce.

Allais, Maurice, "Le comportement de l'homme rationnel devant le risque: Critique des postulats et axioms de l'école Americaine," *Econometrica*, **21**, 503–546, 1953.

Anderson, O., "Theorie der Glücksspiele und ökonomisches Verhalten," *Schweizerische Zeitschrift für Volkswirtschaft und Statistik*, **85**, 46–53, 1949.

Armstrong, W. E., "The determinateness of the utility function," *Economic Journal*, **49**, 453–467, 1939.

————, "Utility and the theory of welfare," *Oxford Economic Papers*, New Series, **3**, 259–271, 1951.

Arnoff, E. L., C. W. Churchman, and R. L. Ackoff. See Churchman.

Arrow, K. J., *Social Choice and Individual Values*, Cowles Commission Monograph 12, John Wiley & Sons, New York, 1951 (*a*).

———, "Alternative approaches to the theory of choice in risk-taking situations," *Econometrica*, **19**, 404–437, 1951 (*b*).

———, "Mathematical models in the social sciences," *The Policy Sciences*, Daniel Lerner and H. D. Lasswell, editors, pp. 129–154, Stanford University Press, Stanford, 1951 (*c*).

———, "The meaning of social welfare: a comment on some recent proposals," *Technical Report* 2, Department of Economics and Statistics, Stanford University, 1951 (*d*).

———, "Hurwicz's optimality criterion for decision-making under ignorance," *Technical Report* 6, Department of Economics and Statistics, Stanford University, 1953.

———, David Blackwell, and M. A. Girshick, "Bayes and minimax solutions of sequential decision problems," *Econometrica*, **17**, 213–243, 1949.

Baumol, W. J., "The Neumann-Morgenstern utility index—an ordinalist view," *Journal of Political Economy*, **59**, 61–66, 1951.

Bellman, Richard, "On the theory of dynamic programming," *Proceedings of the National Academy of Sciences, U. S. A.*, **38**, 716–719, 1952 (*a*).

———, "On games involving bluffing," *Rendiconti del Circolo Mathematico di Palermo*, Series 2, **1**, 139–156, 1952 (*b*).

———, "On a new iterative algorithm for finding the solutions of games and linear programming problems," *Research Memorandum* P-473, The RAND Corporation, Santa Monica, 1953.

———, "Decision making in the face of uncertainty, I, II," *Naval Research Logistics Quarterly*, **1**, 230–232, 327–332, 1954.

———, and David Blackwell, "Some two-person games involving bluffing," *Proceedings of the National Academy of Sciences, U. S. A.*, **35**, 600–605, 1949.

Bennion, E. G., "Capital budgeting and game theory," *Harvard Business Review*, **34**, 115–123, 1956.

Berge, Claude, "Sur une théorie ensembliste des jeux alternatifs," *Journal de mathématiques pures et appliquées*, **32**, 129–184, 1953 (*a*).

———, "Le problème du gain dans la théorie généralisée des jeux sans informations," *Bulletin de la société mathématique de France*, **81**, 1–8, 1953 (*b*).

Bernard, Jessie, "The theory of games of strategy as a modern sociology of conflict," *The American Journal of Sociology*, **59**, 411–424, 1954.

Bernoulli, Daniel, "Exposition of a new theory on the measurement of risk" [English translation of "Specimen theoriae novae de mensura sortis," *Commentarii academiae scientiarum imperialis Petropolitanae*, 1730 and 1731, **5**, 175–192, 1738, by Louise Sommer], *Econometrica*, **22**, 23–26, 1954.

Birch, B. J., "On games with almost complete information," *Proceedings of the Cambridge Philosophical Society*, **51**, 275–287, 1955.

Bitter, F., "The mathematical formulation of strategic problems," *Proceedings of the Berkeley Symposium*, J. Neyman, editor, pp. 223–228, University of California Press, Berkeley, 1949.

Black, Duncan, "On the rationale of group decision making," *Journal of Political Economy*, **56**, 23–24, 1948 (*a*).

———, "The decisions of a committee using a special majority," *Econometrica*, **16**, 245–261, 1948 (*b*).

———, "The elasticity of committee decisions with an altering size of majority," *Econometrica*, **16**, 262–270, 1948 (*c*).

Blackett, D. W., "Some Blotto games," *Naval Research Logistics Quarterly*, **1**, 55–60, 1954.

Blackwell, David, "On randomization in statistical games with k terminal actions," in Kuhn and Tucker, pp. 183–188 [1953].

———, "On multi-component attrition games," *Naval Research Logistics Quarterly*, **1**, 210–216, 1954 (*a*).

———, "Game theory," *Operations Research for Management*, J. F. McCloskey and F. N. Trefethen, editors, pp. 238–253, The Johns Hopkins Press, Baltimore, 1954 (*b*).

———, "An analog of the minimax theorem for vector payoffs," *Pacific Journal of Mathematics*, **6**, 1–8, 1956 (*a*).

———, "Controlled random walks," invited address, Institute of Mathematical Statistics, Seattle, August, 1956 (*b*).

———, K. J. Arrow, and M. A. Girshick. See Arrow.

———, and Richard Bellman. See Bellman.

———, and M. A. Girshick, *Theory of Games and Statistical Decisions*, John Wiley & Sons, New York, 1954.

Blau, J. H., "The existence of social welfare functions," *Econometrica*, in press.

Bohnenblust, H., M. Dresher, M. A. Girshick, T. E. Harris, O. Helmer, J. C. C. McKinsey, L. S. Shapley, and R. N. Snow, *Mathematical Theory of Zero-Sum Two-Person Games with a Finite or a Continuum of Strategies*, RAND Corp., Santa Monica, Calif., 1948.

———, and Samuel Karlin, "On a theorem of Ville," in Kuhn and Tucker, pp. 155–160 [1950].

———, Samuel Karlin, and L. S. Shapley, "Solutions of discrete two-person games," in Kuhn and Tucker, pp. 51–72 [1950] (*a*).

———, Samuel Karlin, and L. S. Shapley, "Games with continuous, convex pay-off," in Kuhn and Tucker, pp. 181–192 [1950] (*b*).

Bonnessen, T., and W. Fenchel, *Theorie der konvexen Körper*, Ergebnisse der Mathematik und ihrer Grenzgebiete, Vol. III, Part I, J. Springer, Berlin, 1934; reprinted, Chelsea Publishing Co., New York, 1948.

Borel, Emile, "Applications aux jeux de hasard," *Traité du calcul des probabilités et de ses applications*, Gauthier-Villars, Paris 1938.

———, "The theory of play and integral equations with skew symmetrical kernels"; "On games that involve chance and the skill of the players"; and "On systems of linear forms of skew symmetric determinants and the general theory of play," translated by L. J. Savage, *Econometrica*, **21**, 97–117, 1953.

Bott, Raoul, "Symmetric solutions to majority games," in Kuhn and Tucker, pp. 319–323 [1953].

Braithwaite, R. B., *Theory of Games as a Tool for the Moral Philosopher*, Cambridge University Press, Cambridge, 1955.

Bross, I. D. J., *Design for Decision*, The Macmillan Co., New York, 1953.

Brown, G. W., "Iterative solutions of games by fictitious play," in Koopmans, pp. 374–376 [1951].

———, and T. C. Koopmans, "Computational suggestions for maximizing a linear function subject to linear inequalities," in Koopmans, pp. 377–380 [1951].

———, and John von Neumann, "Solutions of games by differential equations," in Kuhn and Tucker, pp. 73–79 [1950].

Brownlee, O. H., A. G. Papandreou, O. H. Sauerlender, Leonid Hurwicz, and William Franklin. See Papandreou.

Caywood, T. E., and C. J. Thomas, "Applications of game theory in fighter versus bomber combat," *Journal of the Operations Research Society of America*, **3**, 402–411, 1955.

Champernowne, D. G., "A note on J. v. Neumann's article," *Review of Economic Studies*, **13**, 10–18, 1945–1946.

488   Bibliography

Charnes, A., W. W. Cooper, and A. Henderson, *An Introduction to Linear Programming*, John Wiley & Sons, New York, 1953.

Chernoff, Herman, *Remarks on a Rational Selection of a Decision Function*, Cowles Commission Discussion Paper, Statistics, No. 326, 1949 (unpublished).

———, "Rational selection of decision functions," *Econometrica*, **22**, 422–443, 1954.

Christie, L. S., "Information handling in organized groups, introduction," *Operations Research in Management*, **II**, J. F. McCloskey and J. M. Coppinger, editors, pp. 417–421, The Johns Hopkins Press, Baltimore, 1956.

Churchman, C. W., *Theory of Experimental Inference*, The Macmillan Co., New York, 1948.

Churchman, C. W., R. L. Ackoff, and E. L. Arnoff, *Introduction to Operations Research*, John Wiley & Sons, New York, 1957.

Coombs, C. H., "Psychological scaling without a unit of measurement," *Psychological Review*, **57**, 145–158, 1950.

———, "A theory of psychological scaling," *Engineering Research Bulletin* 34, University of Michigan Press, Ann Arbor, 1952.

———, "Social choice and strength of preference," in Thrall, Coombs, and Davis, pp. 69–86 [1954].

———, R. M. Thrall, and R. L. Davis. See Thrall.

Cooper, W. W., A. Charnes, and A. Henderson. See Charnes.

Copeland, A. H., "John von Neumann and Oskar Margenstern's theory of games and economic behavior," *Bulletin of the American Mathematical Society*, **51**, 498–504, 1945.

———, *A "Reasonable" Social Welfare Function*, University of Michigan Seminar on Applications of Mathematics to the Social Sciences, 1951 (mimeographed notes).

Dahl, Robert, *A Preface to Democratic Theory*, University of Chicago Press, Chicago, 1956.

Dalkey, Norman, "Equivalence of information patterns and essentially determinate games," in Kuhn and Tucker, pp. 217–244 [1953].

Danskin, J. M., "Fictitious play for continuous games," *Naval Research Logistics Quarterly*, **1**, 313–320, 1954.

Dantzig, G. B., "A proof of the equivalence of the programming problem and the game problem," in Koopmans, pp. 330–338 [1951] (*a*).

———, "Maximization of a linear function of variables subject to linear inequalities," in Koopmans, pp. 339–347 [1951] (*b*).

———, "Application of the simplex method to a transportation problem," in Koopmans, pp. 359–373 [1951] (*c*).

———, "Constructive proof of the min-max theorem," *Pacific Journal of Mathematics*, **6**, 25–33, 1956.

Davidson, Donald, and Jacob Marschak, *Experimental Tests of Stochastic Decision Theory*, Cowles Foundation Discussion Paper 22, Yale University, 1957 (unpublished).

———, Sidney Siegel, and Patrick Suppes, "Some experiments and related theory on the measurement of utility and subjective probability," Applied Mathematics and Statistics Laboratory, *Technical Report* 1, Stanford University, Stanford, 1955.

Davis, R. L., R. M. Thrall, and C. H. Coombs. See Thrall.

De Finetti, Bruno, "La prévision: ses lois logiques, ses sources subjectives," *Annales de l'Institut Henri Poincaré*, **7**, 1–68, 1937.

DeLeeuw, Karel, J. G. Kemeny, J. L. Snell, and G. L. Thompson. See Kemeny.

De Possel, R., "Sur la théorie mathématique des jeux de hasard et de reflexion," *Actualités scientifiques et industrielles*, No. 436, Hermann & Cie., Paris, 1936.

Deutsch, K. W., *Applications of Game Theory to International Politics: Some Opportunities and Limits*, The Center of International Studies, Princeton University, Princeton, undated (mimeographed).

Dines, L. L., "On a theorem of von Neumann," *Proceedings of the National Academy, U.S.A.*, **33**, 329–331, 1947.

Doob, J. L., *Stochastic Processes*, John Wiley & Sons, New York, 1953.

Dorfman, Robert, "Application of the simplex method to a game theory problem," in Koopmans, pp. 348–358 [1951].

Dresher, Melvin, "Methods of solution in game theory," *Econometrica*, **18**, 179–181, 1950.

———, "Games of stategy," *Mathematics Magazine*, **25**, 93–99, 1951.

———, "Solution of polynomial-like games," *Proceedings of the International Congress of Mathematicians*, **I**, 1950, American Mathematical Society, Providence, pp. 334–335, 1952.

———, H. Bohnenblust, M. A. Girshick, T. E. Harris, O. Helmer, J. C. C. McKinsey, L. S. Shapley, and R. N. Snow. See Bohnenblust.

———, and Samuel Karlin, "Solutions of convex games as fixed points," in Kuhn and Tucker, pp. 75–86 [1953].

———, Samuel Karlin, and L. S. Shapley, "Polynomial games," in Kuhn and Tucker, pp. 161–180 [1950].

———, A. W. Tucker, and Philip Wolfe, editors, *Contributions to the Theory of Games*, **III**, Annals of Mathematics Studies, 39, Princeton University Press, Princeton, 1957.

Dunne, J. J., *The Theory of Games in Extensive Form*, Ph.D. thesis, Department of Mathematics, Univeristy of Notre Dame, 1953.

———, and Richard Otter. See Otter.

Dvoretsky, Aryeh, Abraham Wald, and Jacob Wolfowitz, "Elimination of randomization in certain statistical decision problems and zero-sum two-person games," *Annals of Mathematical Statistics*, **22**, 1–21, 1951. (*a*).

———, "Recent suggestions for the reconciliation of theories of probability," in *Proceedings of the Second Berkeley Symposium on Mathematical Statistics and Probability*, Jerzy Neyman, editor, pp. 217–226, University of California Press, Berkeley, 1951. (*b*).

Edgeworth, F. Y., *Mathematical Psychics*, C. Kegan Paul and Co., London, 1881.

Edwards, Ward, "Experiments on economic decision-making in gambling situations," *Econometrica*, **21**, 349–350, 1953 (abstract) (*a*).

———, "Probability-preferences in gambling," *American Journal of Psychology*, **66**, 349–364, 1953 (*b*).

———, "Probability preferences among bets with differing expected values," *American Journal of Psychology*, **67**, 56–67, 1954 (*a*).

———, "The reliability of probability preferences," *American Journal of Psychology*, **67**, 68–95, 1954 (*b*).

———, "The theory of decision making," *Psychological Bulletin*, **5**, 380–417, 1954 (*c*).

Everett, H., "Recursive games," Princeton University, Princeton, 1954, (mimeographed).

Farkas, J., "Uber die Theorie der einfachen Ungleichungen," *Journal für reine und die angewandte Mathematik*, **124**, 1–27, 1902.

Farquharson, Robin, "Sur une généralisation de la notion d'équilibrium," *Comptes rendus hebdomadaires des séances de l'académie des sciences*, Paris, **240**, 46–48, 1955.

Feller, William, *An Introduction to Probability Theory and its Applications*, Vol. I, John Wiley & Sons, New York, 1950.

Fenchel, W., and T. Bonnessen. See Bonnessen.

Festinger, Leon, P. J. Hoffman, and D. H. Lawrence. See Hoffman.

Fisher, R. A., "Randomisation, and an old enigma of card play," *Mathematical Gazette*, **18**, 294–297, 1934.

490 Bibliography

Flood, M. M., "Some experimental games," *Research Memorandum RM*-789, The RAND Corporation, Santa Monica, 1952.

——, "Game-learning theory and some decision making experiments," in Thrall, Coombs, and Davis, pp. 139–158 [1954] (*a*).

——, "Environmental non-stationarity in a sequential decision-making experiment," in Thrall, Coombs, and Davis, pp. 287–300 [1954] (*b*).

Franklin, William, A. G. Papandreou, O. H. Sauerlender, O. H. Brownlee, and Leonid Hurwicz. See Papandreou.

Fréchet, Maurice, "Emile Borel, initiator of the theory of psychological games and its application," *Econometrica*, **21**, 95–96, 1953.

——, and John von Neumann, "Commentary on the Borel note," *Econometrica*, **21**, 118–127, 1953.

Friedman, Milton and L. J. Savage, "The utility analysis of choices involving risk," *Journal of Political Economy*, **56**, 279–304, 1948. Reprinted, with a correction, in *Readings in Price Theory*, G. J. Stigler and K. E. Boulding, editors, Richard D. Irwin, Chicago, 1952. (*a*).

——, and L. J. Savage, "The expected-utility hypothesis and the measurability of utility," *Journal of Political Economy*, **60**, 463–474, 1952 (*b*).

Gale, David, "Convex polyhedral cones and linear inequalities," in Koopmans, pp. 287–297 [1951]. (*b*).

——, "A theory of *n*-person games with perfect information," *Proceedings of the National Academy of Sciences, U. S. A.*, **39**, 496–501, 1953.

——, H. W. Kuhn, and A. W. Tucker, "On symmetric games," in Kuhn and Tucker, pp. 81–88 [1950] (*a*).

——, H. W. Kuhn, and A. W. Tucker, "Reduction of game matrices," in Kuhn and Tucker, pp. 89–96 [1950] (*b*).

——, H. W. Kuhn, and A. W. Tucker, "Linear programming and the theory of games," in Koopmans, pp. 317–329 [1951].

——, and Seymour Sherman, "Solutions of finite two-person games," in Kuhn and Tucker, pp. 37–50 [1950].

——, and F. M. Stewart, "Infinite games with perfect information," in Kuhn and Tucker, pp. 245–266 [1953].

Georgescu-Roegen, N., "The pure theory of consumer's behavior," *Quarterly Journal of Economics*, **50**, 545–593, 1936.

Gerstenhaber, Murray, "Theory of convex polyhedral cones," in Koopmans, pp. 298–316 [1951].

Gillies, D. B., "Discriminatory and bargaining solutions to a class of symmetric *n*-person games," in Kuhn and Tucker, pp. 325–342 [1953] (*a*).

——, *Some Theorems on n-Person Games*, Ph.D. thesis, Department of Mathematics, Princeton University, Princeton, 1953 (*b*).

——, J. P. Mayberry, and John von Neumann, "Two variants of poker," in Kuhn and Tucker, pp. 13–50 [1953].

Girshick, M. A., K. J. Arrow, and David Blackwell. See Arrow.

——, and David Blackwell. See Blackwell.

——, H. Bohnenblust, M. Dresher, T. E. Harris, O. Helmer, J. C. C. McKinsey, L. S. Shapley, and R. N. Snow. See Bohnenblust.

——, and Herman Rubin, "A Bayes approach to a quality control model," *Annals of Mathematical Statistics*, **23**, 114–125, 1952.

Glicksberg, I., "Minimax theorem for upper and lower semicontinuous payoffs," *Research Memorandum RM*-478, The RAND Corporation, Santa Monica, 1950.

————, and O. Gross, "Notes on games over the square," in Kuhn and Tucker, pp. 173–184 [1953].

Goldman, A. J. and A. W. Tucker, "Theory of linear programming," in Kuhn and Tucker, pp. 53–97 [1956 *b*].

Good, I. J., *Probability and the Weighing of Evidence*, Charles Griffin and Company, London, and Hafner Publishing Company, New York, 1950.

Goodman, L. A., "On methods of amalgamation," in Thrall, Coombs, and Davis, pp. 39–48 [1954].

————, and Harry Markowitz, *Social Welfare Functions Based on Rankings*, Cowles Commission Discussion Paper, Economics, No. 2017, 1951.

————, and Harry Markowitz, "Social welfare functions based on individual rankings," *American Journal of Sociology*, **58**, 257–262, 1952.

Gross, O., and I. Glicksberg. See Glicksberg.

Guilbaud, G. T., "The theory of games," *Economie appliquée*, 1949, translated by A. L. Minkes, *International Economic Papers*, No. 1., pp. 37–65, The Macmillan Co., New York, 1951.

Hannan, J. F., *The Dynamic Theory of Decision and Games*, unpublished, 1957.

————, and H. E. Robbins, "Asymptotic solutions of the compound decision problem for two completely specified distributions," *Annals of Mathematical Statistics*, **26**, 37–51, 1955.

Harris, T. E., H. Bohnenblust, M. Dresher, M. A. Girshick, O. Helmer, J. C. C. McKinsey, L. S. Shapley, and R. N. Snow. See Bohnenblust.

Harsanyi, J. C., "Approaches to the bargaining problem before and after the theory of games: a critical discussion of Zeuthen's, Hick's, and Nash's theories," *Econometrica*, **24**, 144–157, 1956.

Hausner, Melvin, "Games of survival," *Research Memorandum RM-776*, The RAND Corporation, Santa Monica, 1952 (*a*).

————, "Optimal strategies in games of survival," *Research Memorandum RM-777*, The RAND Corporation, Santa Monica, 1952 (*b*).

————, "Multidimensional utilities," in Thrall, Coombs, and Davis, pp. 167–180 [1954].

Haywood, O. G., Jr., "Military decision and the mathematical theory of games," *Air University Quarterly Review*, **4**, 17–30, 1950.

————, "Military decision and game theory," *Journal of the Operations Research Society of America*, **2**, 365–385, 1954.

Helmer, Olaf, "Open problems in game theory" (report of a symposium on the theory of games, decision problems, and related topics), *Econometrica*, **20**, 90, 1952.

————, H. Bohnenblust, M. Dresher, M. A. Girshick, T. E. Harris, J. C. C. McKinsey, L. S. Shapley, and R. N. Snow. See Bohenblust.

Henderson, A., A. Charnes, and W. W. Cooper. See Charnes.

Herstein, I. N., and J. W. Milnor, "An axiomatic approach to measurable utility," *Econometrica*, **21**, 291–297, 1953.

Hildreth, Clifford, "Alternative conditions for social orderings," *Econometrica*, **21**, 81–94, 1953.

Hodges, J. L., Jr., and E. L. Lehmann, "The uses of previous experience in reaching statistical decisions," *Annals of Mathematical Statistics*, **23**, 396–407, 1952.

Hoffman, P. J., Leon Festinger, and D. H. Lawrence, "Tendencies toward group comparability in competitive bargaining," in Thrall, Coombs, and Davis, pp. 231–253 [1954].

Hurewicz, Witold, and Henry Wallman, *Dimension Theory*, Princeton University Press, Princeton, 1948.

Hurwicz, Leonid, *Optimality Criteria for Decision Making Under Ignorance*, Cowles Commission Discussion Paper, Statistics, No. 370, 1951 (mimeographed)   (*a*).

——, "Some specification problems and applications to econometric models," *Econometrica*, **19**, 343–344, 1951 (abstract)   (*b*).

——, "What has happened to the theory of games?" *American Economic Association*, **65**, 398–405, 1953.

——, A. G. Papandreou, O. H. Sauerlender, O. H. Brownlee, and William Franklin. See Papandreou.

Inada, Ken-ichi, "Elementary proofs of some theorems about the social welfare function," *Annals of the Institute of Statistical Mathematics*, **6**, 115–122, 1954.

——, "Alternative incompatible conditions for a social welfare function," *Econometrica*, **23**, 396–399, 1955.

Isbell, J. R., "A class of game solutions," *Proceedings of the American Mathematical Society*, **6**, 346–348, 1955.

Jeffreys, Harold, *Theory of probability*, second edition, Oxford University Press, London, 1948.

Johns, M. V., Jr., *Non-Parametric Empirical Bayes Procedures*, Ph. D. thesis, Department of Mathematical Statistics, Columbia University, 1956.

Kakutani, Shizuo, "A generalization of Brouwer's fixed point theorem," *Duke Mathematical Journal*, **8**, 457–458, 1941.

Kalisch, G. K., J. W. Milnor, J. F. Nash, and E. D. Nering, "Some experimental *n*-person games," *Research Memorandum RM*-948, The RAND Corporation, Santa Monica, 1952.

——, J. W. Milnor, J. F. Nash, and E. D. Nering, "Some experimental *n*-person games," in Thrall, Coombs, and Davis, pp. 301–327 [1954].

Kaplansky, Irving, "A contribution to von Neumann's theory of games," *Annals of Mathematics*, **46**, 474–479, 1945.

Karlin, Samuel, "Operator treatment of minmax principle," in Kuhn and Tucker, pp. 133–154 [1950].

——, "Continuous games," *Proceedings of the National Academy of Science, U. S. A.*, **37**, 220–223, 1951.

——, "Reduction of certain classes of games to integral equations," in Kuhn and Tucker, pp. 125–158 [1953]   (*a*).

——, "On a class of games," in Kuhn and Tucker, pp. 159–172 [1953] (*b*).

——, and H. F. Bohnenblust. See Bohnenblust.

——, H. F. Bohnenblust, and L. S. Shapley. See Bohnenblust.

——, and Melvin Dresher. See Dresher.

——, Melvin Dresher, and L. S. Shapley. See Dresher.

——, and L. S. Shapley, "Geometry of reduced moment spaces," *Proceedings of the National Academy of Sciences, U. S. A.*, **35**, 673–679, 1949.

Kaysen, Carl, "The minimax rule of the theory of games, and the choices of strategies under conditions of uncertainty," *Metroeconomica*, **4**, 5–14, 1952.

Keeping, E. S., "Statistical decisions," *American Mathematical Monthly*, **63**, 147–159, 1956.

Kemeny, J. G., Karel DeLeeuw, J. L. Snell, and G. L. Thompson, *Project Report* 1, Dartmouth Mathematics Project, Dartmouth College, Hanover, 1955.

Koopmans, T. C., editor, *Activity Analysis of Production and Allocation*, (proceedings of a conference), Cowles Commission Monograph 13, John Wiley & Sons, New York, 1951.

————, and G. W. Brown.   See Brown.

Krentel, W. D., J. C. C. McKinsey, and W. V. Quine, "A simplification of games in extensive form," *Duke Mathematical Journal*, **18**, 885–900, 1951.

Kuhn, H. W., "Extensive games," *Proceedings of the National Academy of Sciences, U. S. A.*, **36**, 570–576, 1950   (*a*).

————, "A simplified two-person poker," in Kuhn and Tucker, pp. 97–103 [1950]   (*b*).

————, *Lectures on the Theory of Games*, issued as a report of the Logistics Research Project, Office of Naval Research, Princeton University, Princeton, 1952.

————, editor, *Report of an Informal Conference on the Theory of n-Person Games*, Logistics Research Project, Department of Mathematics, Princeton University, 1953 (dittoed)   (*a*).

————, "Extensive games and the problem of information," in Kuhn and Tucker, pp. 193–216 [1953]   (*b*).

————, "On certain convex polyhedra," *Bulletin of the American Mathematical Society*, **61**, 557, 1955 (abstract 799).

————, David Gale, and A. W. Tucker.   See Gale.

————, and A. W. Tucker, editors, *Contributions to the Theory of Games*, **I**, Annals of Mathematics Studies, 24, Princeton University Press, Princeton, 1950.

————, and A. W. Tucker, editors, *Contributions to the Theory of Games*, **II**, Annals of Mathematics Studies, 28, Princeton University Press, Princeton, 1953.

————, and A. W. Tucker, "Theory of games," *The Encyclopaedia Britannica*, **10**, 5–10, 1956   (*a*).

————, and A. W. Tucker, *Linear Inequalities and Related Systems*, Annals of Mathematics Studies 38, Princeton University Press, Princeton, 1956   (*b*).

Laderman, J., "On the asymptotic behavior of decision procedures," *Annals of Mathematical Statistics*, **26**, 551–575, 1955.

Lawrence, D. H., P. F. Hoffman, and Leon Festinger.   See Hoffman.

Lehmann, E. L., "On the existence of least favorable distributions," *Annals of Mathematical Statistics*, **23**, 408–416, 1952.

————, and J. L. Hodges, Jr.   See Hodges.

Lemke, C. E., "The dual method of solving the linear programming problem," *Naval Research Logistics Quarterly*, **1**, 36–47, 1954.

Levitan, R. E., and J. G. March.   See March.

Loomis, L. H., "On a theorem of von Neumann," *Proceedings of the National Academy of Sciences, U. S. A.*, **32**, 213–215, 1946.

Luce, R. D., "A definition of stability for *n*-person games," *Annals of Mathematics*, **59**, 357–366, 1954.

————, "*ψ*-stability: a new equilibrium concept for *n*-person game theory," *Mathematical Models of Human Behavior*, proceedings of a symposium, pp. 32–44, Dunlap and Associates, Stamford, 1955   (*a*).

————, "*k*-stability of symmetric and of quota games," *Annals of Mathematics*, **62**, 517–527, 1955   (*b*).

————, *A Note on the Paper "Some Experimental n-Person Games,"* 1955 (dittoed)   (*c*).

————, "Semiorders and a theory of utility discrimination," *Econometrica*, **24**, 178–191, 1956   (*a*).

————, "A probabilistic theory of utility," *Technical Report* 14, Behavioral Models Project, Columbia University, New York, 1956   (*b*).

————, and E. W. Adams, "The determination of subjective characteristic functions in games with misperceived payoff functions," *Econometrica*, **24**, 158–171, 1956.

————, and A. A. Rogow, "A game theoretic analysis of congressional power distributions for a stable two-party system," *Behavioral Science*, **1**, 83–95, 1956.

——, and A. W. Tucker, *Contributions to the Theory of Games*, **IV**, Annals of Mathematics Studies, in preparation; expected date of publication, 1958.

Majumdar, Tapas, "Choice and revealed preference," *Econometrica*, **24**, 71–73, 1956.

March, J. G., and R. E. Levitan, *On the Normative Theory of Political Representation*, Graduate School of Industrial Administration, Carnegie Institute of Technology, 1955 (mimeographed).

Markowitz, Harry, "The utility of wealth," *Journal of Political Economy*, **60**, 151–158, 1952.

——, and L. A. Goodman. See Goodman.

Marschak, Jacob, "Neumann's and Morgenstern's new approach to static economics," *Journal of Political Economy*, **54**, 97–115, 1946.

——, "Rational behavior, uncertain prospects, and measurable utility," *Econometrica*, **18**, 111–141, 1950.

——, "Towards an economic theory of organization and information," in Thrall, Coombs, and Davis, pp. 187–220 [1954].

——, "Norms and habits of decision making under certainty," *Mathematical Models of Human Behavior*, Dunlap and Associates, pp. 45–54, 1955.

——, and Donald Davidson. See Davidson.

——, and Roy Radner. See Radner.

May, K. O., "A set of independent necessary and sufficient conditions for simple majority decision," *Econometrica*, **20**, 680–684, 1952.

——, "A note on the complete independence of the conditions for simple majority decision," *Econometrica*, **21**, 172–173, 1953.

——, "Intransitivity, utility, and the aggregation of preference patterns," *Econometrica*, **22**, 1–13, 1954.

Mays, W. J., "The valuation of risks," *The American Mathematical Monthly*, **52**, 138–148, 1945.

Mayberry, J. P., D. B. Gillies, and John von Neumann. See Gillies.

——, J. F. Nash, and Martin Shubik, "A comparison of treatments of a duopoly situation," *Econometrica*, **21**, 141–154, 1953.

McDonald, John, "Poker: an American game," *Fortune*, **37**, 128–131, 181–187, 1948.

——, "The theory of strategy," *Fortune*, **38**, 100–110, 1949.

——, *Strategy in Poker, Business, and War*, W. W. Norton and Co., New York, 1950.

——, "Strategy of the seller—or what businessmen won't tell," *Fortune*, **46**, 124–127, 197–198, 1952.

McKinsey, J. C. C., "Notes on games in extensive form," *Research Memorandum RM-157*, The RAND Corporation, Santa Monica, 1950   (a).

——, "Isomorphism of games, and strategic equivalence," in Kuhn and Tucker, pp. 117–130 [1950]   (b).

——, *Introduction to the Theory of Games*, McGraw-Hill Book Co., New York, 1952   (a).

——, "Some notions and problems in game theory," *Bulletin of the American Mathematical Society*, **58**, 591–611, 1952   (b).

——, H. Bohnenblust, M. Dresher, M. A. Girshick, T. E. Harris, O. Helmer, L. S. Shapley, and R. N. Snow. See Bohnenblust.

——, W. D. Krentel, and W. V. Quine. See Krentel.

Mills, W. H., "The four person game—edge of the cube," *Annals of Mathematics*, **59**, 367–378, 1954.

Milnor, J. W., "Games against nature," *Research Memorandum RM-679*, The RAND Corporation, Santa Monica, 1951.

——, "Reasonable outcomes for *n*-person games," *Research Memorandum RM-916*, The RAND Corporation, Santa Monica, 1952.

———, "Sums of positional games," in Kuhn and Tucker, pp. 291–302 [1953].

———, "Games against nature," in Thrall, Coombs, and Davis, pp. 49–60 [1954].

———, and I. N. Herstein. See Herstein.

———, G. K. Kalisch, J. F. Nash, and E. D. Nering. See Kalisch.

———, and L. S. Shapley, "On games of survival," *Research Memorandum* P-622, The RAND Corporation, Santa Monica, 1955.

Morgenstern, Oskar, "Oligopoly, monopolistic competition, and the theory of games," *Proceedings, American Economic Review*, **38**, 10–18, 1948.

———, "Economics and the theory of games," *Kyklos*, **3**, 294–308, 1949 (a).

———, "The theory of games," *Scientific American*, **180**, 22–25, 1949 (b).

———, "Theorie des Spiels," *Die Amerikanische Rundschau*, **5**, 76–87, 1949 (c).

———, "Die Theorie der Spiele und des Wirtschaftlichen Verhaltens, Part I," *Jahrbuch für Sozialwissenschaft*, **1**, 113–139, 1950.

———, and John von Neumann. See von Neumann.

Mosteller, Fredrick, and Philip Nogee, "An experimental measurement of utility," *Journal of Political Economy*, **59**, 371–404, 1951.

Motzkin, T. S., Howard Raiffa, G. L. Thompson, and R. M. Thrall, "The double description method," in Kuhn and Tucker, pp. 51–73 [1953].

Nagel, Ernest, "Principles of the theory of probability," *International Encyclopedia of Unified Science*, Vol. I, No. 6, University of Chicago Press, Chicago, 1939.

Nash, J. F., "Equilibrium points in n-person games," *Proceedings of the National Academy of Sciences, U. S. A.*, **36**, 48–49, 1950 (a).

———, "The bargaining problem," *Econometrica*, **18**, 155–162, 1950 (b).

———, "Non-cooperative games," *Annals of Mathematics*, **54**, 286–295, 1951.

———, "Two-person cooperative games," *Econometrica*, **21**, 128–140, 1953.

———, G. K. Kalisch, J. W. Milnor, and E. D. Nering. See Kalisch.

———, J. P. Mayberry, and Martin Shubik. See Mayberry.

———, and L. S. Shapley, "A simple three person poker game," in Kuhn and Tucker, pp. 105–116 [1950].

Nering, E. D., G. K. Kalisch, J. W. Milnor, and J. F. Nash. See Kalisch.

Neyman, J., and E. S. Pearson, "Contributions to the theory of testing statistical hypotheses," *Statistical Research Memoirs*, Parts I and II, 1936 and 1938.

Nogee, Philip, and Fredrick Mosteller. See Mosteller.

Norman, R. Z., "On the convex polyhedra of the symmetric traveling salesman problem," *Bulletin of the American Mathematical Society*, **61**, 559, 1955 (abstract 804).

Otter, Richard and J. J. Dunne, "Games with equilibrium points," *Proceedings of the National Academy of Sciences, U. S. A.*, **39**, 310–314, 1953.

Papandreou, A. G., "An experimental test of an axiom in the theory of choice," *Econometrica*, **21**, 447, 1953 (abstract).

———, O. H. Sauerlender, O. H. Brownlee, Leonid Hurwicz, and William Franklin, *A Test of a Proposition in the Theory of Choice*, University of Minnesota, 1954 (unpublished).

Pareto, Vilfredo, *Manuel d'économic politique*, first edition, second edition, Giard, Paris, 1909, 1927.

Paxson, E. W., "Recent developments in the mathematical theory of games," *Econometrica*, **17**, 72–73, 1949.

Pearson, E. S., and J. Neyman. See Neyman.

Peisakoff, M. P., "More on games of survival," *Research Memorandum* RM-884, The RAND Corporation, Santa Monica, 1952.

Quandt, R. E., "A probabilistic theory of consumer behavior," *The Quarterly Journal of Economics*, **70**, 507–536, 1956.

Quine, W. V., W. D. Krentel, and J. C. C. McKinsey. See Krentel.

Radner, Roy and Jacob Marschak, "Note on some proposed decision criteria," in Thrall, Coombs, and Davis, pp. 61–68 [1954].

Raiffa, Howard, "Arbitration schemes for generalized two-person games," Report M720-1, R30, Engineering Research Institute, University of Michigan, Ann Arbor, 1951.

———, "Arbitration schemes for generalized two-person games," in Kuhn and Tucker, pp. 361–387 [1953].

———, T. S. Motzkin, G. L. Thompson, and R. M. Thrall. See Motzkin.

Ramsey, F. P., *The Foundations of Mathematics and Other Logical Essays*, Chapter VII, Harcourt, Brace & Co., New York, 1931.

Richardson, Moses, "On weakly ordered systems," *Bulletin of the American Mathematical Society*, **52**, 113–116, 1946.

———, "Extension theorems for solutions of irreflexive relations," *Proceedings of the National Academy of Sciences, U. S. A.*, **39**, 649–655, 1953 (a).

———, "Solutions of irreflexive relations," *Annals of Mathematics*, **58**, 573–590, 1953 (b).

———, "Relativization and extension of solutions of irreflexive relations," *Pacific Journal of Mathematics*, **5**, 551–584, 1955.

———, "On finite projective games," *Proceedings of the American Mathematical Society*, **7**, 458–465, 1956.

Robbins, H. E., "Competitive estimation," *Annals of Mathematical Statistics*, **21**, 311–312, 1950 (abstract).

———, "Asymptotically subminimax solutions of compound statistical decision problems," *Proceedings of the Second Berkeley Symposium on Mathematical Statistics and Probability*, University of California Press, 1951.

———, and J. F. Hannan. See Hannan.

Robinson, Julia, "An iterative method of solving a game," *Annals of Mathematics*, **54**, 296–301, 1951.

Rogow, A. A., and R. D. Luce. See Luce.

Rubin, Herman, *The Existence of Measurable Utility and Psychological Probability*," Cowles Commission Discussion Paper, Statistics, No. 331, 1949.

———, and M. A. Girshick. See Girshick.

Samuelson, P. A., "Probability, utility, and the independence axiom," *Econometrica*, **20**, 670–678, 1952.

Sauerlender, O. H., A. G. Papandreou, O. H. Brownlee, Leonid Hurwicz, and William Franklin. See Papandreou.

Savage, L. J., "The theory of statistical decision, *Journal of the American Statistical Association*, **46**, 55–67, 1951.

———, *The Foundations of Statistics*, John Wiley & Sons, New York, and Chapman & Hall, London, 1954.

———, and Milton Friedman. See Friedman.

Scarf, Herbert, and L. S. Shapley, "Games with information lag," *Research Memorandum RM*-1320, The RAND Corporation, Santa Monica, 1954.

Seligman, B. B., "Games theory and collective bargaining," *Labor and Nation*, **8**, 50–52, 1952.

Shackle, G. L. S., *Expectation in Economics*, Cambridge University Press, Cambridge, 1949.

Shapley, L. S., "Information and the formal solution of many-moved games," *Proceedings of the International Congress of Mathematicians*, **I**, 574–575, American Mathematical Society, Providence, 1952 (a).

————, "Notes on the n-person game, III: some variants of the von Neumann-Morgenstern definition of solution," *Research Memorandum RM*-817, The RAND Corporation, Santa Monica, 1952   (*b*).

————, "n-person games, V: stable-set solutions including an arbitrary closed component," *Research Memorandum, RM*-1005, The RAND Corporation, Santa Monica, 1952   (*c*).

————, "Quota solutions of n-person games," in Kuhn and Tucker, pp. 343–359 [1953]   (*a*).

————, "A value for n-person games," in Kuhn and Tucker, pp. 307–317 [1953]   (*b*).

————, *Additive and Non-Additive Set Functions*, Ph.D. thesis, Department of Mathematics, Princeton University, 1953   (*c*).

————, "Stochastic games," *Proceedings of the National Academy of Sciences, U. S. A.*, **39**, 1095–1100, 1953   (*d*).

————, "A symmetric market game," *Research Memorandum RM*-1533, The RAND Corporation, Santa Monica, 1955.

————, H. F. Bohnenblust, and Samuel Karlin.   See Bohnenblust.

————, H. Bohnenblust, M. Dresher, M. A. Girshick, T. E. Harris, O. Helmer, J. C. C. McKinsey, and R. N. Snow.   See Bohnenblust.

————, Melvin Dresher, and Samuel Karlin.   See Dresher.

————, and Samuel Karlin.   See Karlin.

————, and J. W. Milnor.   See Milnor.

————, and J. F. Nash.   See Nash.

————, and Herbert Scarf.   See Scarf.

————, and Martin Shubik, "Solution of n-person games with ordinal utilities," *Econometrica*, **21**, 348, 1953 (abstract).

————, and Martin Shubik, "A method for evaluating the distribution of power in a committee system," *The American Political Science Review*, **48**, 787–792, 1954.

————, and R. N. Snow, "Basic solutions of discrete games," in Kuhn and Tucker, pp. 27–35 [1950].

Sherman, Seymour, "Games and sub-games," *Proceedings of the American Mathematical Society*, **2**, 186–187, 1951.

————, and David Gale.   See Gale.

Shiffman, Max, "Games of timing," in Kuhn and Tucker, pp. 97–123 [1953].

Shubik, Martin, "Information, theories of competition, and the theory of games," *Journal of Political Economy*, **60**, 145–150, 1952 (*a*).

————, "A business cycle model with organized labor considered," *Econometrica*, **20**, 284–294, 1952 (*b*).

————, "The role of game theory in economics," *Kyklos*, **6**, 21–34, 1953.

————, *Readings in Game Theory and Political Behavior*, Doubleday, Garden City, 1954.

————, "The uses of game theory in management," *Management Science*, **2**, 40–54, 1955.

————, *Competition, Oligopoly, and the Theory of Games*, in preparation, 1957.

————, J. P. Mayberry, and J. F. Nash.   See Mayberry.

————, and L. S. Shapley.   See Shapley.

Siegel, Sidney, Donald Davidson, and Patrick Suppes.   See Davidson.

Simmel, Georg, *The Sociology of Georg Simmel*, translated by K. H. Wolff, The Free Press, Glencoe, 1950.

Snell, J. L., J. G. Kemeny, Karel DeLeeuw, and G. L. Thompson.   See Kemeny.

Snow, R. N., H. Bohnenblust, M. Dresher, M. A. Girshick, T. E. Harris, O. Helmer, J. C. C. McKinsey, and L. S. Shapley.   See Bohnenblust.

————, and L. S. Shapley.   See Shapley.

Steinhaus, Hugo, "The problem of fair division," *Econometrica*, **16**, 101–104, 1948.

498   Bibliography

————, "Sur la division pragmatique," *Econometrica*, **17** (supplement), 315–319, 1949.

————, "Quality control by sampling (a plea for Bayes' rule)," *Colloquium Mathematicum*, **2**, 98–108, 1951.

Stewart, F. M., and David Gale. See Gale.

Stigler, G. J., "The development of utility theory," *Journal of Political Economy*, Part I, **58**, 307–327, 1950; Part II, **58**, 373–396, 1950.

Stone, R., "The theory of games," *Economic Journal*, **58**, 185–201, 1948.

Suppes, Patrick, Donald Davidson, and Sidney Siegel. See Davidson.

————, and Muriel Winet, "An axiomatization of utility based on the notion of utility differences," *Management Science*, **1**, 259–270, 1955.

Thomas, C. J., and T. E. Caywood. See Caywood.

Thompson, F. B., "Equivalence of games in extensive form," *Research Memorandum RM*-759, The RAND Corporation, Santa Monica, 1952.

Thompson, G. L., "Signaling strategies in *n*-person games," in Kuhn and Tucker, pp. 267–278 [1953] (*a*).

————, "Bridge and signaling," in Kuhn and Tucker, pp. 279–290 [1953] (*b*).

————, J. G. Kemeny, Karel DeLeeuw, and J. L. Snell. See Kemeny.

————, T. S. Motzkin, Howard Raiffa, and R. M. Thrall. See Motzkin.

Thrall, R. M., C. H. Coombs, and R. L. Davis, editors, *Decision Processes*, John Wiley & Sons, New York, 1954.

————, T. S. Motzkin, Howard Raiffa, and G. L. Thompson. See Motzkin.

Tucker, A. W., *Game theory and Programming*, Department of Mathematics, The Oklahoma Agricultural and Mechanical College, Stillwater, 1955 (mimeographed).

————, Melvin Dresher, and Philip Wolfe. See Dresher.

————, David Gale, and H. W. Kuhn. See Gale.

————, and A. J. Goldman. See Goldman.

————, and H. W. Kuhn. See Kuhn.

————, and R. D. Luce. See Luce.

Tukey, J. W., "A problem in strategy," *Econometrica*, **17**, 73, 1949 (abstract).

Vajda, S., *Theory of Games and Linear Programming*, John Wiley & Sons, New York, 1956.

Vickrey, William, *Strong and Weak Solutions in the Theory of Games*, Department of Economics, Columbia University, 1953 (dittoed).

Ville, Jean, "Noté sur la théorie générale des jeux où intervient l'habilité des jouers," in "Applications aux jeux de hasard," by Emile Borel and Jean Ville, Tome IV, Fascicule II, in Borel [1938].

Von Neumann, John, "Zur Theorie der Gesellschaftsspiele," *Mathematische Annalen*, **100**, 295–320, 1928.

————, "Uber ein ökonomisches Gleichungssystem und eine Verallgemeinerung des Brouwerschen Fixpunktsatzes," *Ergebnisse eines Mathematik Kolloquiums*, **8**, 73–83, 1937.

————, "A certain zero-sum two-person game equivalent to the optimum assignment problem," in Kuhn and Tucker, pp. 5–12 [1953].

————, "A numerical method to determine optimum strategy," *Naval Research Logistics Quarterly*, **1**, 109–115, 1954.

————, and G. W. Brown. See Brown.

————, and Maurice Fréchet. See Fréchet.

————, D. B. Gillies, and J. P. Mayberry. See Gillies.

————, and Oskar Morgenstern, *Theory of Games and Economic Behavior*, first edition, second edition, Princeton University Press, Princeton, 1944, 1947.

Wald, Abraham, "Contributions to the theory of statistical estimation and testing hypotheses," *Annals of Mathematical Statistics*, **10**, 299–326, 1939.

————, "On the principles of statistical inference," *Notre Dame Mathematics Lectures,* No. 1, University of Notre Dame, Indiana, 1942.

————, "Generalization of a theorem by von Neumann concerning zero-sum two-person games," *Annals of Mathematics,* **46,** 281–286, 1945 (*a*).

————, "Statistical decision functions which minimize the maximum risk," *Annals of Mathematics,* **46,** 265–280, 1945 (*b*).

————, "Theory of games and economic behavior by John von Neumann and Oskar Morgenstern," *The Review of Economic Statistics,* **39,** 47–52, 1947 (*a*).

————, "Foundation of a general theory of sequential decision functions," *Econometrica,* **15,** 279–313, 1947 (*b*).

————, *Statistical Decision Functions,* John Wiley & Sons, New York, 1950 (*a*).

————, "Note on zero-sum two-person games," *Annals of Mathematics,* **52,** 739–742, 1950 (*b*).

————, "Basic ideas of a general theory of statistical decision rules," *Proceedings of the International Congress of Mathematicians,* **I,** 231–243, American Mathematical Society, Providence, 1952.

————, Aryeh Dvoretsky, and Jacob Wolfowitz. See Dvoretsky.

————, and Jacob Wolfowitz, "Bayes solution of sequential decision problems," *Annals of Mathematical Statistics,* **21,** 82–99, 1950.

————, and Jacob Wolfowitz, "Two methods of randomization in statistics and theory of games," *Annals of Mathematics,* **53,** 581–586, 1951.

Wallman, Henry, and Witold Hurewicz. See Hurewicz.

Weldon, J. C., "On the problem of social welfare functions," *Canadian Journal of Economics and Political Science,* **18,** 452–463, 1952.

Weyl, Herman, "Elementary proof of a minimax theorem due to von Neumann," in Kuhn and Tucker, pp. 19–25 [1950] (*a*).

————, "The elementary theory of convex polyhedra," in Kuhn and Tucker, pp. 3–18 [1950] (*b*).

Williams, J. D., *The Compleat Strategyst, Being a Primer on the Theory of Games of Strategy,* McGraw-Hill Book Co., New York, 1954.

Winet, Muriel, and Patrick Suppes. See Suppes.

Wold, H., "Ordinal preferences or cardinal utility?" (with additional notes by G. L. S. Shackle, L. J. Savage, and H. Wold), *Econometrica,* **20,** 661–664, 1952.

Wolfe, Philip, editor, *Report of an Informal Conference on Recent Developments in the Theory of Games,* Logistics Research Project, Department of Mathematics, Princeton University, 1955 (mimeographed).

————, editor, *Report of the Third Conference on Games,* Logistics Research Project, Department of Mathematics, Princeton University, 1957 (mimeographed).

————, Melvin Dresher, and A. W. Tucker. See Dresher.

Wolfowitz, Jacob, "Minimax estimates of the mean of a normal distribution with known variance," *Annals of Mathematical Statistics,* **21,** 218–230, 1950.

————, Aryeh Dvoretsky, and Abraham Wald. See Dvoretsky.

————, and Abraham Wald. See Wald.

Zeuthen, Frederik, *Problems of Monopoly and Economic Warfare,* G. Routledge & Sons, London, 1930.

INDEX